Advances in

# VIRUS RESEARCH

VOLUME 64

# ADVISORY BOARD

# Advances in
# VIRUS RESEARCH

Edited by


**KARL MARAMOROSCH**
Department of Entomology
Rutgers University
New Brunswick, New Jersey

**AARON J. SHATKIN**
Center for Advanced Biotechnology
and Medicine
Piscataway, New Jersey


## VOLUME 64
## Virus Structure and Assembly

Edited by


**POLLY ROY**

Department of Infectious and Tropical Diseases
London School of Hygiene and Tropical Medicine
London, United Kingdom

ELSEVIER
ACADEMIC
PRESS

AMSTERDAM • BOSTON • HEIDELBERG • LONDON
NEW YORK • OXFORD • PARIS • SAN DIEGO
SAN FRANCISCO • SINGAPORE • SYDNEY • TOKYO

Elsevier Academic Press
525 B Street, Suite 1900, San Diego, California 92101-4495, USA
84 Theobald's Road, London WC1X 8RR, UK

This book is printed on acid-free paper. 

For all information on all Academic Press publications
visit our Web site at www.academicpress.com

ISBN: 0-12-039863-X

PRINTED IN THE UNITED STATES OF AMERICA
05 06 07 08 09 9 8 7 6 5 4 3 2 1

# CONTENTS

## Bacteriophage HK97: Assembly of the Capsid and Evolutionary Connections

ROGER W. HENDRIX



## Assembly of Double-Stranded RNA Bacteriophages

MINNA M. PORANEN, ROMAN TUMA, AND DENNIS H. BAMFORD



## Structure-Derived Insights into Virus Assembly

VIJAY S. REDDY AND JOHN E. JOHNSON

## Bluetongue Virus Proteins and Particles and Their Role in Virus Entry, Assembly, and Release

POLLY ROY



## Structure, Assembly, and Antigenicity of Hepatitis B Virus Capsid Proteins

ALASDAIR C. STEVEN, JAMES F. CONWAY, NAIQIAN CHENG, NORMAN R. WATTS, DAVID M. BELNAP, AUDRAY HARRIS, STEPHEN J. STAHL, AND PAUL T. WINGFIELD



## Molecular Interactions in the Assembly of Coronaviruses

CORNELIS A. M. DE HAAN AND PETER J. M. ROTTIER

## Mechanism of Membrane Fusion by Viral Envelope Proteins

Stephen C. Harrison



## Structure and Assembly of Icosahedral Enveloped RNA Viruses

Richard J. Kuhn and Michael G. Rossmann



## Kinetic and Mass Spectrometry-Based Investigation of Human Immunodeficiency Virus Type 1 Assembly and Maturation

Jason Lanman and Peter E. Prevelige, Jr.

## Role of Lipid Rafts in Virus Replication

Akira Ono and Eric O. Freed



## Polymorphism of Filovirus Glycoproteins

Viktor E. Volchkov, Valentina A. Volchkova, Olga Dolnik, Heinz Feldmann, and Hans-Dieter Klenk



## Influenza Virus Assembly and Budding at the Viral Budozone

Anthony P. Schmitt and Robert A. Lamb

# ACKNOWLEDGMENT

It is a pleasure for me to acknowledge the substantial contribution of Dr. Aaron Shatkin, who convinced me to edit this book. More importantly, Dr. Shatkin helped in the selection of each topic and provided considerable support in the daunting job of discussing and reviewing each chapter. He has been the most enthusiastic committee member of the Virus Structure and Assembly symposiums, which have been organized so successfully over the last 10 years and which, in part, have stimulated the production of this volume. I am also indebted to other members of the committee who have helped me organize these timely and exciting meetings at a number of unusual sites. It is only with all of their support that this book has become such an authoritative and comprehensive review.

# PREFACE

It is 70 years since Wendell Stanley succeeded in obtaining a virus, tobacco mosaic virus (TMV), in sufficient yield and purity to allow its crystallization—a tribute to both his perseverance and virus productivity. Now, with the availability of molecular clones for virus capsid proteins and a range of expression technologies, it is possible to obtain, even for viruses that do not grow so profusely, the yield and purity of particles, subparticles, or individual proteins that allow high-resolution analysis. The outcome of these developments is a wealth of data on precisely how viruses are put together and how the mechanisms of assembly and disassembly work. Moreover, this is now available for viruses with far more complex structures than TMV, including those that have lipid membranes in the final virion. The topic of virus assembly led to a small symposium (Virus Structure and Assembly, Marrakech, Morocco, 1995) that engaged biological and physical scientists in the problem. It was so successful that a biannual series followed and, from them, the suggestion of a symposium volume dedicated to recent progress. The result is witness to the remarkable advancement in the field of virus structure and assembly and allows a unique opportunity to compare and contrast the mechanisms adopted by viruses with a wide diversity of genome and host.

Chapters 1 and 2 deal with the assembly of two structurally different bacteriophages. It has been generally assumed that what is learned about one tailed phage will likely apply to all others, but as the vastness of the global phage population has been realized only recently, the security of such an assumption remains untested. Nevertheless, the handful of phages that have been studied to date has revealed much about phage assembly mechanisms. Bacteriophage HK97, the subject of Roger Hendrix's article, exhibits a number of assembly mechanisms, among them mechanisms that appear to be shared widely with other tailed phages and others that, if not unique to HK97, are at least unusual. A double-stranded RNA (dsRNA) bacteriophage, $\phi$6, is one of the best-characterized dsRNA viruses. Elegant studies combining molecular and biophysical techniques have revealed how the virion is constructed step by step. Bamford *et al.* summarize the fascinating story of this phage in Chapter 2. In

Chapter 3, Reddy and Johnson discuss the VIPER database, a repository for all high-resolution structures of simple nonenveloped, icosahedral viruses. The database is based on one standard convention that facilitates the subsequent rapid analysis of these structures. In Chapter 4, Polly Roy summarizes the current understanding of the three-dimensional structure of bluetongue virus (BTV) particles, subviral particles, and proteins and their role in the various stages of the virus life cycle as representative members of the *Reoviridae* family. They are grouped together by the characteristic segmented, double-stranded RNA genomes enclosed in complex, multilayered capsids. The *Reoviridae* family is one of the largest families of viruses and includes major human pathogens as well as other vertebrate, plant, and insect pathogens. BTV has the added dimension of being transmitted by *Culicoides* species and, because of its economic significance, BTV now represents one of the best-characterized viruses.

Steven *et al.*, in Chapter 5, review current information concerning the structure and assembly of hepatitis B virus (HBV), one of the major causes of liver disease in humans. In addition, they present insights into HBV antigenicity through a comprehensive coverage of historical and recent data and go on to describe the potential of HBV capsids for the cellular delivery of foreign epitopes. The next few chapters are dedicated to viruses with a positive-stranded RNA genome. In Chapter 6, de Haan and Rottier review the assembly of coronaviruses, with an updated overview of various aspects of coronavirus research in the assembly field, emphasizing particularly the interactions between the different structural components of the virus and the processes involved in formation of infectious virus particles. The subsequent two chapters also relate to enveloped RNA viruses, and both focus on virus envelope structure–function relationships. Harrison, in Chapter 7, reviews the structural biology that underpins the mechanism of membrane fusion by viral envelope proteins during virus entry. The focus is on two distinct classes of fusion protein typified by influenza virus and flavivirus, each of which catalyzes the fusion between the viral lipid bilayer and cellular membranes, diverse proteins with surprisingly similar mechanisms. The following chapter (Chapter 8), by Kuhn and Rossmann, provides an update on recent progress in determining the high-resolution structure and assembly of icosahedral enveloped RNA viruses, in particular, alphaviruses and flaviviruses. This has been possible only as a result of advances made in cryoelectron microscopy. Kuhn and

Rossmann describe how the fitting of atomic resolution structures from one group of viruses into the cryoelectron microscopy density maps of another has generated "pseudoatomic" structures for selected enveloped viruses. These have allowed an understanding of the dynamic process of virus assembly, the intermediates involved, and the entry pathway described at the molecular level. Lanman and Prevelige (Chapter 9) describe the application of mass spectrometry to the analysis of structures present in biological specimens, such as human immunodeficiency virus type 1 (HIV-1), which appear to be pleomorphic. They show how mass spectrometry could contribute to the understanding of complex macromolecular machines such as HIV-1 and serve as a complement to the traditional approaches of X-ray crystallography and electron microscopy.

In Chapter 10, Ono and Freed summarize the current understanding of the characteristics of lipid rafts, specific microdomains within biological lipid membranes. Lipid rafts have been the focus of intense interest as rafts appear to play a central role in a wide variety of cellular functions, including signal transduction and intracellular trafficking. It has become increasingly obvious that lipid rafts also play an important role in the replication of a number of viruses. The authors describe the connection between rafts and virus replication and the possible mechanisms by which these microdomains could promote virus entry and egress.

Klenk and colleagues in Chapter 11 discuss the polymorphism of two filoviral glycoproteins, those of Marburg and Ebola viruses. They compare and contrast the structural and functional polymorphism of these glycoproteins and discuss how each virus uses a distinct type of processing and posttranslational modification that, in turn, modulates the immune response of the hosts to virus infection. The closing chapter of the book concerns influenza virus, an enveloped virus with a segmented, single-stranded RNA genome that, because of the high incidence of influenza virus outbreaks with high morbidity, has been studied extensively. Here, Schmitt and Lamb review the progress that has been made toward understanding the various steps in the generation of infectious virus particles. Using influenza A as their model, they detail each step of the virus life cycle, including cell entry, organization and concentration of viral proteins at selected sites on the cell plasma membrane, recruitment of a full complement of eight RNA segments to the assembly sites, budding and release of particles by membrane fission, and spread of the virus from cell to cell.

Together, these chapters, all written by experts in their field, summarize our current state of knowledge in the field of virus structure and assembly. We are living through a time when many viral diseases are controlled by public health policies and vaccination. At the same time we face pernicious agents such as HIV and the challenge of emerging viruses. The molecular basis of virus assembly provides the basis for the rational design of intervention strategies that can only help to overcome these threats.

Polly Roy,
*London, United Kingdom, 2005*

ADVANCES IN VIRUS RESEARCH, VOL 64

# BACTERIOPHAGE HK97: ASSEMBLY OF THE CAPSID AND EVOLUTIONARY CONNECTIONS

Roger W. Hendrix

Pittsburgh Bacteriophage Institute and Department of Biological Sciences
University of Pittsburgh, Pittsburgh, Pennsylvania 15260



## I. Tailed Phages: Overview

Tailed bacteriophages are possibly the most abundant organisms on the planet (Wommack and Colwell, 2000). If material from an environmental source such as soil or water is plated on a given laboratory strain of bacteria the number of plaques obtained is generally quite small, but direct examination of such material for morphologically identifiable tailed phages by electron microscopy tells quite a different story. Coastal seawater from a variety of locations around the globe is seen to have around $10^7$ particles per milliliter, particle counts in some freshwater sources can be orders of magnitude higher than that, and tailed phages in surface soil have been measured at around $10^8$ particles per gram. In all these cases, the number of phage particles is fivefold to tenfold higher than the number of prokaryotic cells. The total number of tailed phages on Earth is estimated to be about $10^{31}$ individual particles, an almost unimaginably large number that implies that if all the phages were laid end to end the line of phages would stretch about 200 million light years.

Phages have been studied in the laboratory since their discovery nearly a century ago, but the number of distinct phage types submitted to detailed examination by genetic and biochemical techniques is quite small, perhaps as few as 20, and the number for which detailed structure and mechanism of assembly of the virion has been scrutinized is even smaller. It has generally been assumed that what is learned




0065-3527/05 $35.00
DOI: 10.1016/S0065-3527(05)64001-8

about one tailed phage will likely apply to other phages as well. This has been a useful approach, but it has only been in more recent years, with the advent of both environmental and genome sequencing approaches in phage biology, that we have on the one hand appreciated the extreme sparseness with which we have sampled the global phage population and on the other hand begun to have the means to estimate how well our work to date has illuminated more general questions about how virions are constructed and what assembly strategies are used to achieve those structures. Bacteriophage HK97, the subject of this article, illustrates a number of assembly mechanisms, among them mechanisms that appear to be shared widely with other tailed phages under study and others that, if not unique to HK97, are at least unusual. Comparisons of capsid protein sequences argue, moreover, that HK97 is a member of an evolutionarily related family of phages that may encompass the entirety of the tailed phages.

## II. Initial Stages of Capsid Assembly

HK97 is a tailed, double-stranded DNA (dsDNA) phage of *Escherichia coli*, with an isometric head 60 nm in diameter and a 180-nm-long noncontractile tail (Fig. 1). The head (capsid) has a triangulation number of 7 (Conway *et al.*, 1995) and a corresponding 420 copies of its single major capsid subunit (415 copies when the portal structure and attached tail take the place of one pentamer of the capsid subunit). As with most such viral capsids, the structure can be parsed into 12 rotationally symmetric pentamers, occupying the corners of the

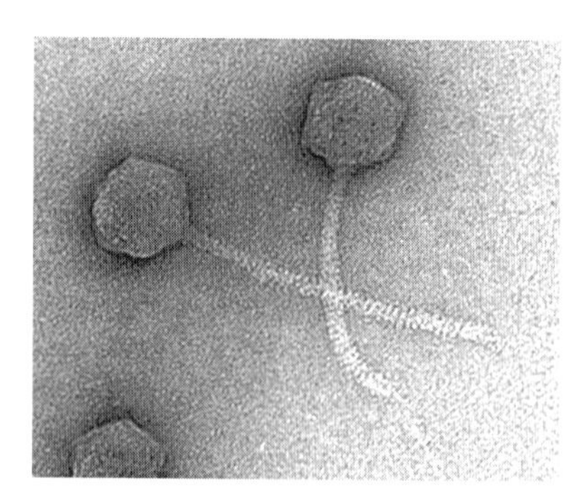

FIG 1. Electron micrograph of HK97 and HK022, negatively stained with uranyl formate. These two phages have identical capsid subunits but different tail proteins. HK97 is the phage with the longer tail. Scale: the HK97 tail is ~180 nm long.

icosahedral shell, and 60 hexamers, regularly arrayed between, with 3 hexamers on each of the 20 triangular faces of the icosahedron. HK97 has the additional unusual feature that covalent bonds form between adjacent subunits in the shell, linking all the subunits into a unitary nondissociable structure. This article reviews what we know about how the HK97 capsid subunits become assembled into a shell and transformed into a mature head. In addition, it considers to what extent we can expect the lessons learned from HK97 head assembly to be applicable more generally to other viruses.

The capsid subunit of HK97 is a 384-amino acid (42-kDa) protein (Duda *et al.*, 1995b). The protein folds with the help of the GroEL–GroES chaperone system, and although the interaction with GroEL–GroES involves only one HK97 subunit per GroEL–GroES complex, the first detectable form of the folded protein is oligomeric, a mixture of soluble pentamers and hexamers (Ding *et al.*, 1995; Xie and Hendrix, 1995). The two types of oligomers, collectively known as capsomers, correspond to the pentameric and hexameric groupings in the assembled capsid. The free capsomers appear to be produced in the cell at a ratio of pentamers to hexamers corresponding to this ratio in the assembled $T = 7$ capsids. Pentamers and hexamers can be interconverted *in vitro* in either direction by adjusting the properties of the solvent (Xie and Hendrix, 1995). It may be that pentamers and hexamers are in equilibrium *in vivo*.

Pentamers and hexamers will assemble accurately into $T = 7$ icosahedral shells *in vitro* if they are mixed and concentrated (Xie and Hendrix, 1995). An experiment in which *in vitro* assembly was carried out with different pentamer-to-hexamer ratios but with constant total protein concentration showed that both pentamers and hexamers are required for shell assembly (Xie and Hendrix, 1995). This experiment also ruled out assembly mechanisms in which the capsomers disassemble into smaller oligomers or monomers before they form the capsid. These observations led to assembly models, still not tested experimentally, in which the interactions between pentamers and hexamers form intermediate structures in which the proteins of the complex are marked by a conformational change (Conway *et al.*, 1995). This conformational change (according to this speculation) constrains the possibilities for subsequent assembly steps and thereby directs the proteins toward forming "correct" $T = 7$ shells and away from other possible assemblies.

Assembly from pentamers and hexamers is in striking contrast to the extensively characterized *in vitro* assembly of the capsids of phage P22, which also makes a $T = 7$ shell. The P22 capsid protein is found as a soluble monomer, and it seems clear that it assembles *in vitro* directly into shells without forming pentamer or hexamer intermediates (Prevelige and King, 1993). Another apparent difference between HK97 and P22 with regard to capsid assembly is that P22 assembly requires a second abundant protein, the scaffolding protein, which coassembles with the capsid subunit (King and Casjens, 1974). However, this difference may be more apparent than real in that the scaffolding protein function in HK97 may be carried out by the amino-terminal 102 amino acids of the capsid subunit. This amino-terminal part, known as the $\delta$ region, is removed by proteolysis and exits the capsid during capsid maturation, as does the scaffolding protein of P22 and many other phages.

## III. Role of Protease and Portal

The HK97 capsid protein will assemble efficiently and accurately into shells in the absence of any other phage-encoded proteins, as for example when the capsid protein is expressed at a high level from a plasmid clone of its gene, or when shells are assembled *in vitro* from purified pentamers and hexamers (Duda *et al.*, 1995b; Xie and Hendrix, 1995). Thus the capsid protein by itself carries all the information needed to accurately specify $T = 7$ capsids. However, in normal *in vivo* assembly there are two other phage-encoded proteins that also become part of the structure. These are the protease, encoded by the next gene upstream from the capsid gene, and the portal protein, encoded by the next gene upstream from the protease gene (Fig. 2).

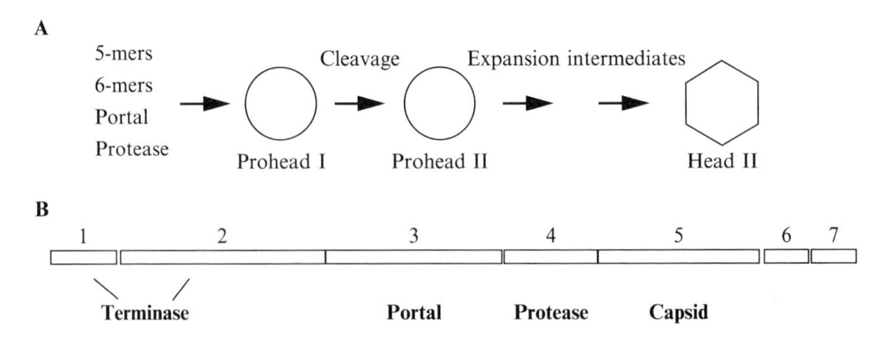


Fig 2. (A) Schematic HK97 head assembly pathway. (B) Head genes of HK97.

The portal protein is found in the head as a 12-subunit grommet-like structure sitting on one of the corners (fivefold symmetry axes) of the icosahedral shell and occupying the position that is occupied by a capsid protein pentamer at each of the other 11 corners. Although the function of the portal has not been studied extensively in HK97, it is clear in other systems that it has a central role in packaging the DNA into the preformed capsid, possibly rotating in its socket as DNA is pumped into the capsid under the motive force of ATP hydrolysis. The portal also serves as the site of tail joining, typically with one or more adaptor proteins located between the portal and the tail. This picture likely also applies to HK97, although the only relevant evidence in the case of HK97 is that the portal can be visualized by electron microscopy, attached to the top of the tail in virions that have lost their DNA. A long-standing mystery for all the tailed phages is how they end up with exactly 1 portal in the capsid, despite the presence of 12 fivefold corners on the capsid, each of which in principle could accommodate a portal. The simplest model that has been proposed is that the portal serves as a nucleator of capsid assembly, and that once nucleation is accomplished the rest of the shell assembles from capsid protein. Although plausible, this model has not been clearly confirmed for any phage. For HK97, if the portal acts as a nucleator, we need to explain how efficient assembly can occur both *in vivo* and *in vitro* in the absence of portal protein. A current speculation is that the need for a portal nucleator can be bypassed if the capsid protein concentration is sufficiently high.

The second protein required for the assembly of complete procapsids, beyond the capsid subunit itself, is the phage-coded protease. The protease is a 25-kDa protein that cleaves off the first 102 amino acids of each of the capsid proteins, reducing their mass from 42 to 31 kDa. The protease also degrades itself, and the resulting fragments of the protease as well as the fragments of the amino-terminal portion of the capsid protein are lost from the capsid, presumably by diffusion through small holes in the shell that are seen in cryoelectron microscopy (cryo-EM) images (Conway *et al.*, 1995). If proheads are assembled with a mutant form of the protease that is catalytically inactive but competent for assembly, there are roughly 50 molecules of protease per prohead, and this is presumed to be the same as the number of wild-type proteases per prohead. This number could plausibly indicate that there is 1 protease for each of the 72 capsomers in the prohead, but the actual locations of the proteases are not known. By a mechanism that is not yet clear, the protease is inactive until it is assembled into the prohead.

The proteolysis that removes 102 amino acids (the $\delta$ region) from the amino terminus of the capsid protein and leads as well to autoproteolysis of the protease is the first step in an elaborate series of rearrangements of the capsid subunits. This conformational and covalent transition takes the structure from a loose, easily dissociable grouping of hexamers and pentamers through a series of steps to a tough, unitary shell that is held firmly and irreversibly together by a combination of intricate intertwining of protein subunits and covalent cross-links that bind the entire shell into a single structure that cannot be dissociated without breaking covalent bonds (Duda *et al.*, 1995a). Before proteolysis, the capsid structure is known as Prohead I. Prohead I is ordinarily a transient structure, but it can be produced in a reasonably stable form if protease activity is absent. This situation can be achieved either by expressing the capsid protein in the absence of the protease gene or by using a mutant protease that lacks catalytic activity. Prohead I can be dissociated into a mixture of capsid protein pentamers and hexamers and those capsomers can subsequently be reassembled into Prohead I through rather moderate manipulations of solvent conditions and protein concentration. In contrast, the structure found after proteolysis of the $\delta$ region, known as Prohead II, is considerably more stable than Prohead I and in fact has been seen to disassemble only under strongly denaturing conditions. Despite the considerable difference in biochemical properties between Prohead I and Prohead II, the two structures have similar morphologies as viewed in moderate-resolution cryo-EM reconstructions (Conway *et al.*, 1995). However, Prohead II is poised to carry out the next steps in the capsid rearrangements.

## IV. DNA Packaging and Expansion

Prohead II is almost certainly the structure that packages phage DNA. DNA packaging has not been studied directly in HK97, but the strong conservation of the DNA-packaging mechanism across all the tailed phages studied, together with the conserved complement of head genes, leads to the strong expectation that DNA packaging in HK97 will be similar to that in phage $\lambda$, which is well studied in this regard. Assuming that to be the case, we would predict the following sequence of events in HK97 DNA packaging: the terminase, composed of proteins gp1 and gp2, binds to the replicated, concatemeric DNA at the *cos* packaging initiation site and cuts to form the 10-base 3′ extensions of the mature DNA ends; the terminase remains bound to the newly created end corresponding to the "left" end of the mature DNA, and that complex docks on the prohead, at the vertex containing the portal (gp3). During some of this part of the process, protein gp6 is associated with the terminase–DNA complex and probably serves as an assembly factor to assist in the docking of the terminase–DNA complex on the portal. Once the complex is docked, the terminase and portal together form the DNA-packaging motor, which hydrolyzes ATP and pumps the DNA into the head. When the packaging machine has finished pumping the full-length chromosome into the head and has reached the next *cos* site, the terminase cuts the DNA and leaves, remaining complexed to the end of the concatemer at the outboard side of the new cut. The final step in head assembly is thought, again by analogy with phages such as $\lambda$, to be the addition of a hexamer of protein gp7 to the portal, where it serves as a specific attachment site for the separately assembled tail.

In contrast to DNA packaging, a maturation process termed "expansion" that accompanies DNA packaging has been extensively studied in HK97 (Gan *et al.*, 2004; Hendrix and Duda, 1998). Expansion, as the name suggests, entails a change in the dimensions of the shell. In the case of HK97 this results in an approximately twofold increase in the internal volume of the shell, without any change in the number of protein subunits. In addition to this, there is a marked change in the overall shape and surface texture of the shell, from a nearly spherical structure with a lumpy surface to an angular icosahedron with a smooth surface. Other tailed phages that have been studied in this regard also undergo an analogous expansion reaction. In these cases the change in internal volume of the shell can be quite different from the twofold increase seen in HK97, with the internal volume of phage $\phi$29, for example, changing essentially not at all; what does appear to be common to all the expansion reactions is that the expanded structure is substantially more resistant to disruption than the unexpanded prohead. Because of these facts, it is assumed that the function of the expansion reaction is to increase the stability of the shell rather than to increase its internal volume. The detailed motions of the capsid subunits inferred for HK97 (see later) lend support to this view.

With HK97 it has been possible to carry out the prohead expansion reaction with purified proheads. Expansion is triggered by a variety of changes in the solvent, the most easily controlled being the lowering of pH to around 4 (Lata *et al.*, 2000). The rate of triggering is strongly dependent on pH in the vicinity of pH 4, suggesting that an important component of the triggering may be titration of a carboxylate group in the protein. Under conditions of acid expansion the capsids pause at four distinguishable intermediate states, and it has been possible to obtain structures for each of these by cryo-EM. This has made it possible in turn to infer a dynamic view of the expansion reaction at moderate cryo-EM resolution, in the form of a movie produced by interpolating between these structures (Lata *et al.*, 2000). A high-resolution X-ray structure is known for the mature Head II form of the capsid (Helgstrand *et al.*, 2003; Wikoff *et al.*, 2000), and this has been fit into the lower resolution movie to infer a "pseudo-atomic resolution" version of the movie (Conway *et al.*, 2001). The main feature of the expansion reaction, so visualized, is that the subunits move largely *en bloc*, with little or no refolding of the central domain that makes up the bulk of the subunit. This is remarkable in that individual subunits move as much as 58 Å and turn by as much as 39° relative to their starting positions, implying substantial changes in the intersubunit interactions. Two regions of the subunit, in contrast, engage in significant refolding during expansion. The amino-terminal ~30 amino acids are not visible and are presumably unfolded in the prohead but folded against and intertwined with neighboring subunits in the mature head. The second mobile region is a long $\beta$ hairpin, consisting of 34 amino acids and named the "E loop," which moves by a hinging motion at its base. The E loop is located on the outer surface of the capsid and in the prohead is free of contact with the rest of the subunit and pointing away from the surface. During expansion it descends and lands on the surface of the capsid with the tip of the E loop extending across the subunit boundary to lie on a part of the neighboring subunit.

## V. Cross-Linking

The capsid subunits in the prohead are joined exclusively by non-covalent interactions, but by the time the capsid is transformed into the mature head each of the 420 subunits is joined by covalent bonds to two of its neighbors. The bond is an amide bond—sometimes referred to in this context as an isopeptide bond—between the side chains of Lys-169 on one subunit and Asn-356 on the neighboring subunit (Duda *et al.*, 1995a). It was shown first in biochemical experiments and subsequently with the X-ray structure of the head that the individual subunits are joined covalently into closed rings of either five or six subunits and that these rings are linked topologically to each other into a unitary fabric called "viral chain mail" (Duda, 1998; Wikoff *et al.*, 2000). A surprising and elegant feature of the geometry of the chain-mail structure is that it is not the six (or five) subunits of an individual capsomer that join each other covalently to make a six (or five)-membered covalent circle. Rather, it is the six (or five) subunits immediately surrounding the capsomer, each derived from a different one of the neighboring capsomers, that form the covalent circle. A necessary consequence of this scheme of cross-linking is that the covalent protein circles are topologically linked at the 140 sites of threefold local and global icosahedral symmetry on the capsid.

Formation of the covalent cross-links is autocatalytic (Duda *et al.*, 1995a). That is, there is no separate enzyme to provide the necessary catalytic and positioning functions, which are instead supplied by the capsid proteins themselves. It is also clear that cross-link formation is triggered by the subunit rearrangements of the expansion reaction (Duda *et al.*, 1995a). The structural information available for the expansion reaction makes it possible to understand at least in broad terms how this works. The cross-linking lysine residue is located near the tip of the E loop and is therefore well away from the capsid surface in the prohead. During the course of the expansion the E loop descends onto the surface of the capsid with the cross-linking lysine juxtaposed with the asparagine on the neighboring subunit with which it will react. In addition, during the course of expansion a third subunit moves in such a way as to position the side chain of Glu-363 within about 2 Å of the cross-link bond (Wikoff *et al.*, 2000). This positioning suggests that the glutamate may have a catalytic role in the cross-linking, a supposition that gains support from a mutant in which the glutamate is changed to alanine. The mutation does not prevent the expansion reaction, but it is completely defective in cross-link formation (Wikoff *et al.*, 2000). Taken together, these observations argue that a function of expansion is to assemble the catalytic site for cross-linking, bringing together the three crucial residues (K169, N356, and E363), which are widely separated in the prohead, into a proximity that allows reaction.

This view of the role of expansion in allowing cross-linking suggests that cross-linking can commence once a catalytic site is assembled. Furthermore, it appears from the movie of expansion that not all seven of the quasi-equivalent classes of cross-linking sites achieve their final conformation at the same time. In particular, it appears that the E loops on the subunits of the pentamer capsomers are the last to descend and might therefore be expected to be the last to cross-link. Biochemical and structural experiments confirm this expectation (Gan *et al.*, 2004), providing corroboration for important features of the dynamic model of expansion embodied in the movie.

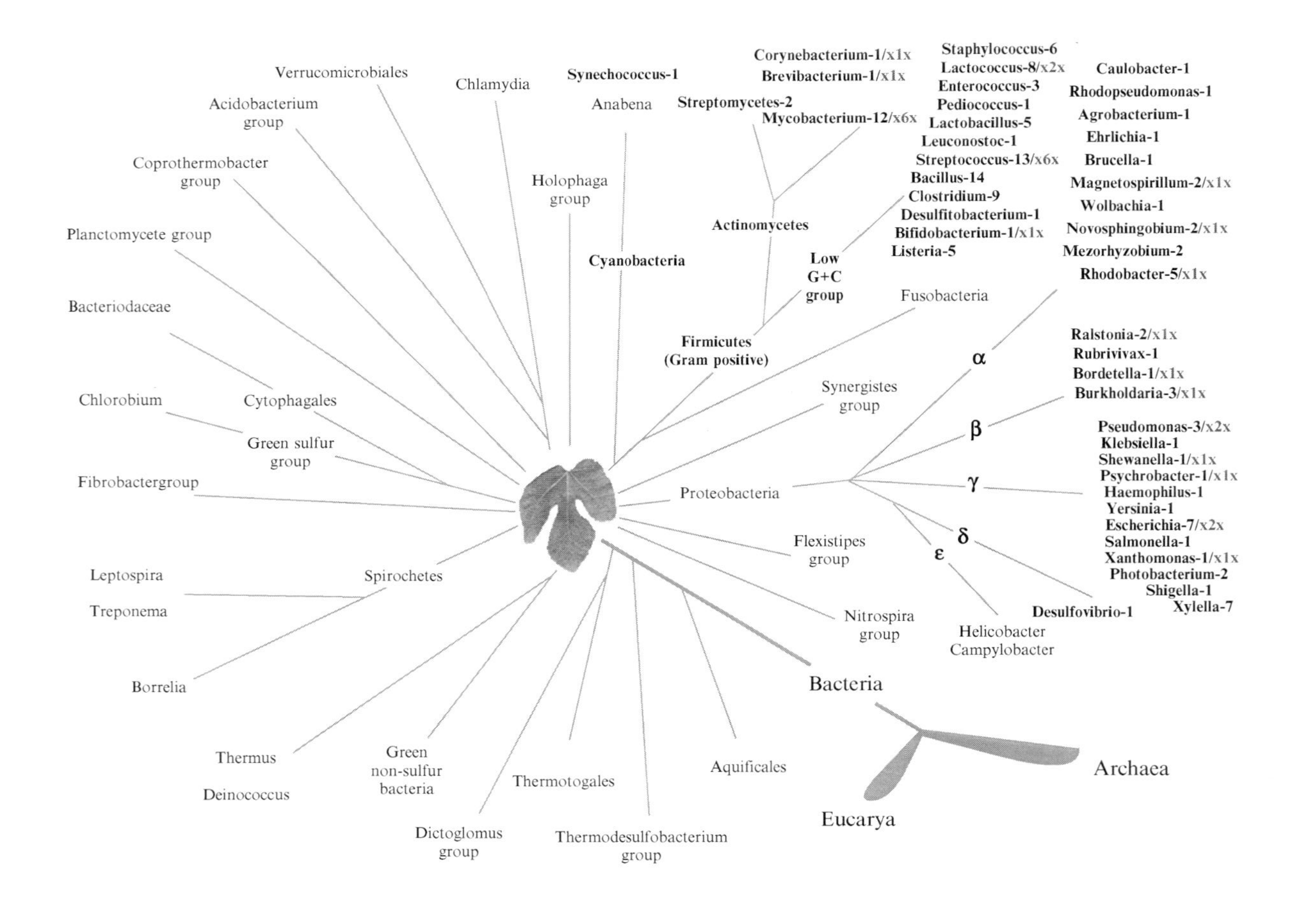

Verrucomicrobiales
Chlamydia
Synechococcus-1
Anabena
Corynebacterium-1/x1x
Brevibacterium-1/x1x
Streptomycetes-2
Mycobacterium-12/x6x
Staphylococcus-6
Lactococcus-8/x2x
Enterococcus-3
Pediococcus-1
Lactobacillus-5
Leuconostoc-1
Streptococcus-13/x6x
Bacillus-14
Clostridium-9
Desulfitobacterium-1
Bifidobacterium-1/x1x
Listeria-5
Caulobacter-1
Rhodopseudomonas-1
Agrobacterium-1
Ehrlichia-1
Brucella-1
Magnetospirillum-2/x1x
Wolbachia-1
Novosphingobium-2/x1x
Mezorhyzobium-2
Rhodobacter-5/x1x
Acidobacterium group
Coprothermobacter group
Holophaga group
Actinomycetes
Planctomycete group
Cyanobacteria
Low G+C group
Fusobacteria
Bacteriodaceae
Firmicutes (Gram positive)
α
Ralstonia-2/x1x
Rubrivivax-1
Bordetella-1/x1x
Burkholdaria-3/x1x
Synergistes group
Chlorobium
Cytophagales
β
Green sulfur group
Pseudomonas-3/x2x
Klebsiella-1
Shewanella-1/x1x
Psychrobacter-1/x1x
Haemophilus-1
Yersinia-1
Escherichia-7/x2x
Salmonella-1
Xanthomonas-1/x1x
Photobacterium-2
Shigella-1
Xylella-7
Fibrobactergroup
γ
Proteobacteria
Flexistipes group
δ
ε
Leptospira
Spirochetes
Treponema
Nitrospira group
Desulfovibrio-1
Helicobacter
Campylobacter
Borrelia
Bacteria
Thermus
Deinococcus
Green non-sulfur bacteria
Thermotogales
Aquificales
Archaea
Eucarya
Dictoglomus group
Thermodesulfobacterium group

## VI. Evolutionary Relationships

When the sequence of the HK97 capsid protein first became available, a sequence search against the public databases gave no matches. Since then, however, the number of phage capsid protein sequences in the databases has increased dramatically, and now a search with the HK97 sequence, using the PSI-BLAST algorithm, returns almost 200 hits. These sequences cover a wide range of similarity to HK97, with the most distant sequences showing as little as 12% amino acid sequence identity. Nevertheless, the sequences can be aligned plausibly, and it seems likely that they can be regarded as a single sequence family with common ancestry and, most probably, a common polypeptide fold and structure. This group of sequences includes a significant fraction of all known capsid sequences of tailed phages, but certainly not all. For example, the capsids of the well-studied *E. coli* phages $\lambda$ and T4 and *Salmonella* phage P22 are not detectably related to the HK97 sequence by these methods. We can ask whether the failure to find sequence similarity to HK97 in these and other cases means that the nonmatching capsid proteins represent different, independent lineages, or whether all the capsid proteins of tailed phages share common ancestry but have diverged so much in sequence that we can no longer recognize them as belonging to a single sequence family. At this point we cannot decide definitively between these alternatives. However, the many shared features of capsid assembly mechanisms and structure, even among phages whose capsid proteins have no detectable sequence difference, would seem to favor the view of a single highly diverged lineage. This view is also supported by a cryo-EM structure of phage P22 in which elements of protein secondary structure are visible. The structure strongly suggests that the P22 capsid protein has the same fold as the HK97 protein, despite the lack of detectable sequence similarity (Jiang *et al.*, 2003).

---

Fig 3. Phylogenetic distribution of HK97-like capsid protein sequences. The HK97 major capsid protein sequence, with the $\alpha$-helical $\delta$ domain trimmed off, was used as a probe in a PSI-BLAST search of GenBank sequences. The search yielded 136 sequences of both phage and prophage origin with $E$ value scores better than $10^{-20}$, and these are indicated in this 16S rRNA-based tree of bacterial taxa. The bacterial genera associated with these sequences are indicated in boldface type, with the following numbers indicating how many individual sequences are associated with that genus. The numbers flanked by $x$ indicate how many of the sequences in each group have the sequence features (see text), indicating that they engage in chain-mail cross-linking.

Among the HK97-like sequences that can be aligned, there are some—30 at this writing—in which the three residues known to be crucial for cross-linking in HK97 are conserved at the corresponding positions in the alignment. These 30 phages include all 10 in the alignment for which it is known experimentally that they cross-link their capsid proteins in a way that resembles chain-mail cross-linking in HK97. This observation suggests that it may be possible to predict from the sequence whether a capsid protein cross-links or not, and it argues that the phages that do cross-link do so by a conserved biochemical mechanism.

We can also ask how widely HK97-like capsid proteins are distributed across different bacterial taxa. For this purpose the author carried out a PSI-BLAST search using the HK97 capsid protein (cleaved form) as the search probe. The search yielded 136 capsid protein sequences with PSI-BLAST $E$ values better than $10^{-20}$, including both phage-encoded proteins and proteins in bacterial genomes known or presumed to be prophage genes. Figure 3 shows how the host associations of these genes are distributed on a tree representing all bacteria. The most striking result from this presentation is that the HK97-like sequences are spread across a wide swath of the bacterial tree, including the phyla *Proteobacteria, Firmicutes*, and *Cyanobacteria* (the latter including one example). The sequences among this group that are thought to engage in chain-mail cross-linking, on the basis of the conservation of the three critical residues in the alignment, are also distributed across the same range of bacterial genera rather than clustered in a small number of closely related bacteria. It is also apparent that there is a wide range of bacterial groups, on the left side of Fig. 3, for which we find no recognizable HK97-like capsid proteins. It is probably premature to make a conclusion about the significance of this negative result, because the numbers of both phage and bacterial genomes that have been sequenced in this part of the tree are substantially less than in the parts that do have HK97 matches.

## VII. Conclusions

HK97 provides a rich source of head assembly mechanisms, spanning from the overall organization of the pathway to atomic level details of the motions of the capsid subunits during capsid maturation. Many of the features of the pathway resemble those of other tailed phages for which capsid assembly has been studied in detail, despite the lack of detectable sequence similarity in the respective proteins. Some specific details of the HK97 pathway, such as the chain-mail cross-linking and the use of an attached "scaffolding protein," are unique to HK97 among the well-studied systems.

We would like to know how the various capsid assembly mechanisms are distributed in the global population of phages. Are the mechanisms and principles we are learning about in studying assembly of HK97, P22, T4, $\lambda$, and a small number of other systems broadly representative of the assembly strategies of tailed phages, or are we studying tiny enclaves of phage space with relevance primarily to themselves? The broad similarities of assembly strategies already noted suggest that the former view is more likely correct, and the sequence searches with HK97 reported here lend strength to this view. The bacterial hosts associated with the HK97 capsid protein sequences span a wide phylogenetic range, and the strength of sequence matches vary from identity to barely detectable. It is plausible (although not yet proved) that the common ancestor of the HK97-like sequences may be the common ancestor of all tailed phage capsid genes, some of which have drifted too far in sequence to be recognizably related. The "unique" feature of HK97 assembly, the chain-mail cross-linking, is seen by this analysis to be broadly distributed as well, apparently occurring in about one-quarter of all the HK97-like sequences. This lends credence to the view that all the tailed phages share the same basic capsid assembly mechanism but that there are numerous "variations on the theme" broadly distributed across the population. Consequently the author believes that the current strategy of combining detailed experimental studies of a few representative phages, with a broader survey of the population as a whole, using bioinformatics methods, is likely a good strategy to obtain a picture of the assembly mechanisms of tailed phages that is both detailed and representative of the entire population.

## Acknowledgments

I thank my colleague Robert Duda who, together with a number of graduate students, is responsible for much of the biochemical, genetic, and intellectual contributions to understanding HK97 head assembly. The structural work is the result of long-term fruitful collaborations with Alasdair Steven (cryo-EM), Jack Johnson (X-ray crystallography), and numerous past and present members of their laboratories. I thank Sherwood Casjens for the tree shown in Fig. 3 and Tom Godfrey for the fig leaf. Work in the author's laboratory is supported by NIH grant GM47795.

## References

Conway, J. F., Duda, R. L., Cheng, N., Hendrix, R. W., and Steven, A. C. (1995). Proteolytic and conformational control of virus capsid maturation: The bacteriophage HK97 system. *J. Mol. Biol.* **253:**86–99.

Conway, J. F., Wikoff, W. R., Cheng, N., Duda, R. L., Hendrix, R. W., Johnson, J. E., and Steven, A. C. (2001). Virus maturation involving large subunit rotations and local refolding. *Science* **292:**744–748.

Ding, Y., Duda, R. L., Hendrix, R. W., and Rosenberg, J. M. (1995). Complexes between chaperonin GroEL and the capsid protein of bacteriophage HK97. *Biochemistry* **34:**14918–14931.

Duda, R. L. (1998). Protein chainmail: Catenated protein in viral capsids. *Cell* **94:**55–60.

Duda, R. L., Hempel, J., Michel, H., Shabanowitz, J., Hunt, D., and Hendrix, R. W. (1995a). Structural transitions during bacteriophage HK97 head assembly. *J. Mol. Biol.* **247:**618–635.

Duda, R. L., Martincic, K., and Hendrix, R. W. (1995b). Genetic basis of bacteriophage HK97 prohead assembly. *J. Mol. Biol.* **247:**636–647.

Gan, L., Conway, J. F., Firek, B. A., Cheng, N., Hendrix, R. W., Steven, A. C., Johnson, J. E., and Duda, R. L. (2004). Control of crosslinking by quaternary structure changes during bacteriophage HK97 maturation. *Mol. Cell* **14:**559–569.

Helgstrand, C., Wikoff, W. R., Duda, R. L., Hendrix, R. W., Johnson, J. E., and Liljas, L. (2003). The refined structure of a protein catenane: The HK97 bacteriophage capsid at 3.44 Å resolution. *J. Mol. Biol.* **334:**885–899.

Hendrix, R. W., and Duda, R. L. (1998). Bacteriophage HK97 head assembly: A protein ballet. *Adv. Virus Res.* **50:**235–288.

Jiang, W., Li, Z., Zhang, Z., Baker, M. L., Prevelige, P. E., Jr., and Chiu, W. (2003). Coat protein fold and maturation transition of bacteriophage P22 seen at subnanometer resolutions. *Nat. Struct. Biol.* **10:**131–135.

King, J., and Casjens, S. (1974). Catalytic head assembling protein in virus morphogenesis. *Nature* **251:**112–119.

Lata, R., Conway, J. F., Cheng, N., Duda, R. L., Hendrix, R. W., Wikoff, W. R., Johnson, J. E., Tsuruta, H., and Steven, A. C. (2000). Maturation dynamics of a viral capsid: Visualization of transitional intermediate states. *Cell* **100:**253–263.

Prevelige, P. E., Jr., and King, J. (1993). Assembly of bacteriophage P22: A model for ds-DNA virus assembly. *Prog. Med. Virol.* **40:**206–221.

Wikoff, W. R., Liljas, L., Duda, R. L., Tsuruta, H., Hendrix, R. W., and Johnson, J. E. (2000). Topologically linked protein rings in the bacteriophage HK97 capsid. *Science* **289:**2129–2133.

Wommack, K. E., and Colwell, R. R. (2000). Virioplankton: Viruses in aquatic ecosystems. *Microbiol. Mol. Biol. Rev.* **64:**69–114.

Xie, Z., and Hendrix, R. W. (1995). Assembly *in vitro* of bacteriophage HK97 proheads. *J. Mol. Biol.* **253:**74–85.

ADVANCES IN VIRUS RESEARCH, VOL 64

# ASSEMBLY OF DOUBLE-STRANDED RNA BACTERIOPHAGES

Minna M. Poranen, Roman Tuma, and Dennis H. Bamford

Department of Biological and Environmental Sciences and Institute of Biotechnology
University of Helsinki, 00014 Helsinki, Finland



## I. Introduction

Bacteriophage $\phi 6$ is the type organism of the family *Cystoviridae* (http://www.ncbi.nlm.nih.gov/ICTV/). For a long time $\phi 6$ represented the only member of this family and was the only known double-stranded RNA (dsRNA) virus infecting bacteria. Eight similar viruses have been isolated (Mindich *et al.*, 1999). Some of the new isolates ($\phi 7$, $\phi 9$, $\phi 10$, and $\phi 11$) have clear sequence similarity with $\phi 6$, whereas others ($\phi 8$, $\phi 12$, and $\phi 13$) are more distantly related. However, the virion architecture and the genomic organization are conserved.

The similarities among dsRNA viruses are not restricted to viruses infecting bacterial hosts; instead, common features have been observed between cystoviruses and eukaryotic dsRNA viruses of the *Reoviridae* family (Bamford *et al.*, 2002). The similarity is evident in the innermost virion layer, which carries out genome packaging, replication, and transcription. On the other hand, the outer virion layers, which




0065-3527/05 $35.00
DOI: 10.1016/S0065-3527(05)64002-X

mediate host interactions, have diverged to assist recognition and entry into a variety of host cells.

In this review we focus on bacteriophage $\phi 6$ because most of the current knowledge about the structure, assembly, and molecular biology of dsRNA bacteriophages is derived from this model system. The other dsRNA bacteriophages, especially $\phi 8$, are also discussed when related information is available. The $\phi 6$ system has been of interest because its internal polymerase complex is a cytoplasmic molecular machine, translocating, replicating, and transcribing RNA in a highly specific manner. This multifunctional complex can now be reconstituted from its component proteins and the assembled complexes are fully functional both *in vitro* and *in vivo* (Poranen *et al.*, 2001). Such an *in vitro* assembly system has opened up the possibility to extract novel information about the sequence of molecular interactions operating during $\phi 6$ virion assembly.

### *A. Structural Organization of the Virion*

The virion of $\phi 6$ is composed of three concentric layers (Fig. 1). The innermost layer, the viral core (or in more general terms, the polymerase complex) is composed of four protein species, P1, P2, P4, and P7 (Bamford and Mindich, 1980; Mindich and Davidoff-Abelson, 1980). P1 is the major component of the polymerase complex forming the polyhedral skeleton of the particle (Ktistakis and Lang, 1987; Olkkonen and Bamford, 1987). There are 120 copies of protein P1 arranged as 60 dimers

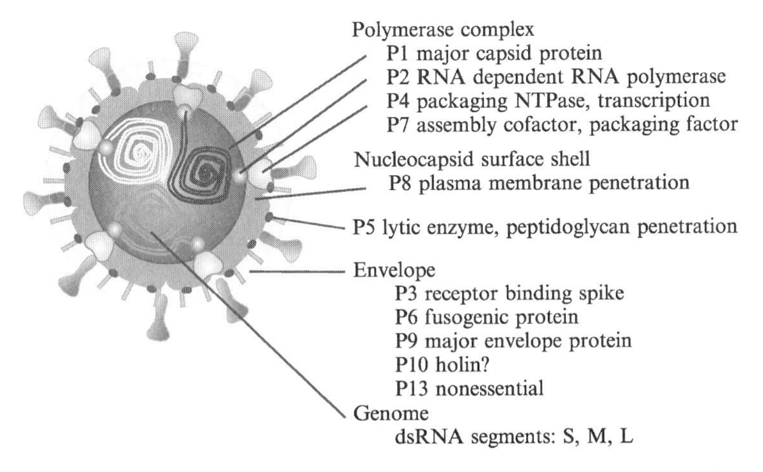


FIG 1. Architecture of $\phi 6$ virion and proposed functions of the proteins. (See Color Insert.)

on a $T = 1$ icosahedral lattice (Butcher *et al.*, 1997). Such a capsid organization exhibiting two chemically identical subunits in different conformations within the icosahedral asymmetric unit has been observed only in the cores of dsRNA viruses and is sometimes referred to as a "*T*2" structure (Grimes *et al.*, 1998). A hexameric protein, P4, forms turret-like protrusions on the P1 lattice, creating a symmetry mismatch at the fivefold vertices (Butcher *et al.*, 1997; de Haas *et al.*, 1999). Location at the fivefold and twofold symmetry axes of the polymerase complex has been proposed for P2 and P7, respectively (see Section II. B.1) (Ikonen *et al.*, 2003; Juuti and Bamford, 1997; Poranen *et al.*, 2001). In addition to the previously described symmetry mismatch, 1 of the 12 P4 hexamers is physically and functionally different (see Section III. B.2) (Pirttimaa *et al.*, 2002).

The core is surrounded by a second proteinaceous layer, called the nucleocapsid surface shell (Fig. 1), composed mainly of a single viral protein, P8 (Bamford and Mindich, 1980; van Etten *et al.*, 1976). P8 is arranged on a $T = 13$ lattice with the exception of the regions close to the fivefold symmetry axes, which are occupied by the hexameric P4 turrets extending from the polymerase complex (Butcher *et al.*, 1997). This double-shell assembly intermediate is designated as a nucleocapsid (NC). Protein P5 (a lytic enzyme) is loosely associated with the NC surface (Hantula and Bamford, 1988).

The outermost layer of the virion is a lipid bilayer (Fig. 1) (van Etten *et al.*, 1976). The envelope of $\phi 6$ contains phospholipids originating from the host plasma membrane (Laurinavicius *et al.*, 2004), and four virally encoded integral membrane proteins, P6, P9, P10, P13 (Gottlieb *et al.*, 1988a; Sinclair *et al.*, 1975; van Etten *et al.*, 1976), of which protein P6 anchors the receptor-binding spike protein P3 (Stitt and Mindich, 1983b). The envelope does not seem to follow icosahedral symmetry (F. de Haas, A. Paatero, S. Butcher, D. Bamford, and S. D. Fuller, personal communication).

The other members of the *Cystoviridae* family have an overall virion architecture similar to $\phi 6$ (Mindich *et al.*, 1999; Qiao *et al.*, 2000). However, $\phi 8$ lacks the layer corresponding to the $\phi 6$ NC surface shell (P8) (Hoogstraten *et al.*, 2000). Interestingly, all cystoviruses possess a lipid envelope, a unique feature among bacteriophages.

### *B. Genome*

The $\phi 6$ genome is tripartite, that is, composed of three distinct dsRNA segments (Semancik *et al.*, 1973; Van Etten *et al.*, 1973), and one copy of each is contained in every virion (Day and Mindich, 1980).

The genome segments are designated according to their size: small (2948 bp, GenBank accession number M12921), medium (4063 bp, M17462), and large (6374 bp, M17461). Abbreviations S, M, and L are used for the dsRNA genomic segments, whereas lower case letters s, m, and l indicate the single-stranded, plus-sense, RNA precursors. The dsRNA genome is always enclosed and processed within the polymerase complex, and only the single-stranded RNA (ssRNA) molecules are released into the host cell cytoplasm.

Each genomic segment encodes multiple viral polypeptides and the genes are distributed according to their function among the three segments (Gottlieb *et al.*, 1988a; McGraw *et al.*, 1986; Mindich *et al.*, 1988). The viral proteins are numbered according to their size (starting from the largest, P1) and their specific functions are summarized in Fig. 1. The L segment encodes proteins that assemble into the polymerase complex (P1, P2, P4, and P7, early expressed proteins). The M segment encodes membrane-associated proteins (P3, P6, P10, and P13, late expressed proteins) whereas the S segment contains genes encoding the NC shell protein (P8), the major membrane protein (P9, late protein), and proteins needed for host cell lysis (P5 and P11, late proteins). In addition, the genome encodes two nonstructural proteins, P12 (encoded by the S segment) and P14 (encoded by the L segment) (Casini and Revel, 1994), which are both expressed in the $\phi$6-infected cell.

The coding regions are flanked by distinct noncoding regions at both 5′ and 3′ termini of the segments (Gottlieb *et al.*, 1988a; McGraw *et al.*, 1986; Mindich *et al.*, 1988). These contain important signals for genome packaging and replication (see Section III).

The new *Cystoviridae* isolates all have tripartite genomes (Mindich *et al.*, 1999), and those for which the genomes have been sequenced share genome organization similar to that of $\phi$6 (Gottlieb *et al.*, 2002a,b; Hoogstraten *et al.*, 2000; Qiao *et al.*, 2000). Analogous genes have been recognized and the nomenclature follows that of $\phi$6. However, two rather than one polypeptide may carry out some of the functions. For example, the host attachment assembly of $\phi$8 and $\phi$13 consists of two peptides (P3a and P3b) (Hoogstraten *et al.*, 2000; Qiao *et al.*, 2000) instead of the one (P3) found in $\phi$6.

### *C. Life Cycle of* $\phi$*6*

The three structural layers of the $\phi$6 virion have distinct roles during the viral life cycle. Like other dsRNA viruses, $\phi$6 needs to deliver the virion core (containing RNA polymerase activity) into the host cytosol to initiate infection. During $\phi$6 entry the structural layers of the virion are sequentially exposed and dissociated as the virus passes through the triple-layered cell wall of its gram-negative host bacterium.

The receptor-binding spike (P3) mediates the initial host interaction with the bacterial pilus (Bamford *et al.*, 1976) and retraction of the pilus brings the virion into contact with the cell surface (Romantschuk and Bamford, 1985). The spike is removed, leading to activation of the fusogenic protein (P6), and fusion between the bacterial outer membrane and the virion envelope commences (Bamford *et al.*, 1987). The removal of the envelope releases the lytic enzyme P5 and NC penetrates the peptidoglycan barrier (Mindich and Lehman, 1979). Finally, the P8 shell of NC drives an endocytic-like plasma membrane penetration, which leads to the delivery of the virion core into the host cytosol (Poranen *et al.*, 1999; Romantschuk *et al.*, 1988).

Once in the cell cytosol the polymerase complex (core) is activated to synthesize full-length, single-stranded copies of the three genomic segments (transcription) (Kakitani *et al.*, 1980). These serve as mRNA templates for protein synthesis and are also packaged into the newly synthesized virions. The newly produced proteins that are encoded by the L segment (P1, P2, P4, and P7) assemble into empty dodecahedral particles (Bamford and Mindich, 1980) called procapsids (PCs) (Mindich and Davidoff-Abelson, 1980) or empty polymerase complexes. The PC has the same protein composition as the core but is devoid of RNA. PCs package the single-stranded copies of the viral genome segments and subsequently replicate the packaged ssRNA segments into the mature double-stranded form within the particle (Ewen and Revel, 1988, 1990). These dsRNA-filled particles start a new round of transcription that leads to the synthesis of late proteins (Rimon and Haselkorn, 1978). Protein P8 assembles onto the dsRNA-filled particles (virion cores) (Bamford and Mindich, 1980; Olkkonen *et al.*, 1991) and later the double-shelled NCs acquire the envelope and the host attachment spikes to form mature virions (Bamford *et al.*, 1976; Mindich *et al.*, 1979). The mature virions are released by host cell lysis (Mindich and Lehman, 1979; Vidaver *et al.*, 1973).

## II. Assembly of Empty Precursor Capsid

Early studies of $\phi$6-infected cells revealed that the assembly pathway involves formation of the PC (Fig. 2) (Bamford and Mindich, 1980). However, no assembly intermediates for the PC were detected in these

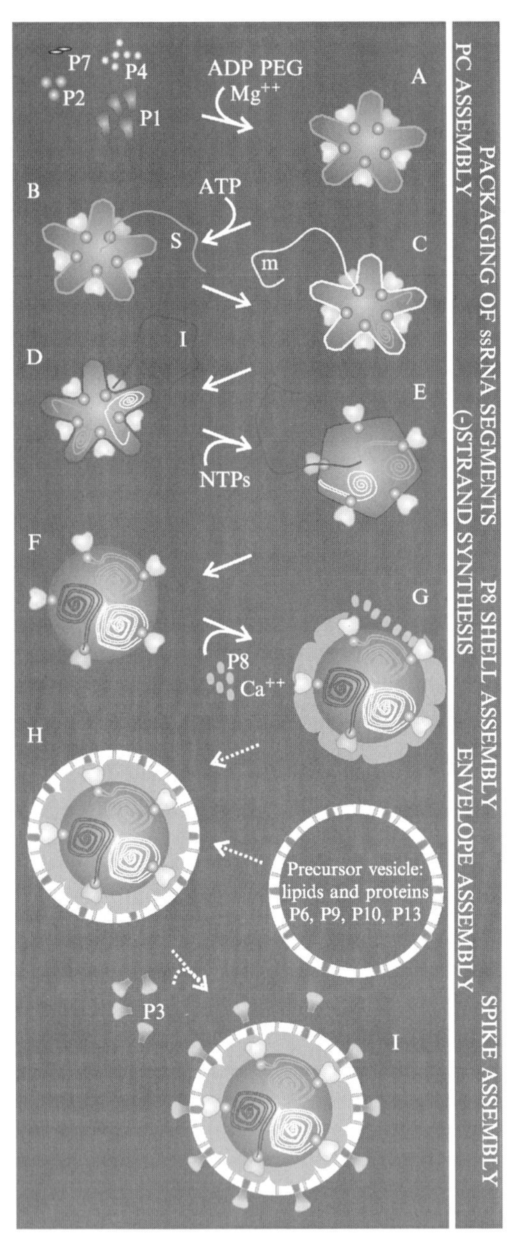


FIG 2. Overview of the $\phi6$ assembly pathway. The empty procapsid of $\phi6$ is composed of four proteins (A). P1 forms the structural skeleton of the particle, which is stabilized by P7. The enzymatic components, the packaging NTPase P4 hexamer and the polymerase

studies. Since then substantial information about the structural and functional organization of the PC has been obtained from analyses of assemblies formed during expression of different combinations of viral proteins (Gottlieb *et al.*, 1988b, 1990; Paatero *et al.*, 1998). Additional details on $\phi 6$ and $\phi 8$ polymerase complex morphogenesis are based on *in vitro* assembly assays (Kainov *et al.*, 2003b; Poranen *et al.*, 2001), which have facilitated elucidation of the sequence of protein–protein interactions and structural characterization of assembly intermediates.

### A. *Expression Studies*

When expressed within a host cell or in a heterologous organism, the polymerase complex proteins assemble into PCs similar to those found early during $\phi 6$ infection (Gottlieb *et al.*, 1988b). This indicates that PC assembly is not dependent on additional factors encoded by the host cell or by the M and S segments. PC formation is also independent of viral RNA.

In addition to the complete polymerase complexes, it is possible to produce various incomplete PCs, using similar heterologous expression

P2 monomer, are located at or near the fivefold axes. One of the P4 hexamers differs from the others in its physical properties so that P4 is more tightly bound to the procapsid (special vertex, green) than are the others (yellow). The special vertex is active during the packaging of the single-stranded genomic segments (B–E), while transcription is dependent on the P4 hexamers at the other vertices (F). Three structures have been described for the polymerase complex: the empty compressed particle (A), the expanded dodecahedral particle (E), and the mature particle with the rounded appearance (F). In addition, it is suggested that the particle undergoes conformational changes that switch the specificity of the RNA-binding site on the outer surface of the particle (B–D; indicated in green, yellow, and purple outlines). The packaging of the segments is initiated from the 5′ end, which contains a segment-specific *pac* site. The empty particle is preferentially in a conformation that has high-affinity binding site for the s segment (B). The binding or the packaging of the s segment induces a conformational change in the ssRNA-binding site so that the particle preferentially binds the m segment (C), which is subsequently packaged. The m segment then induces a switch in conformation that has high affinity for the l segment (D). The 5′ end of the l segment carries a signal that is needed for the initiation of minus-strand synthesis, and we suggest that the 5′ end of the l segment induces an expansion of the particle (E). After completion of minus-strand synthesis on the l segment the particle can initiate transcription. At this stage the dsRNA density within the expanded dodecahedral particle reaches its upper limits and we propose that the particle expands to the rounded conformation (F). The nucleocapsid shell (P8) is subsequently assembled around the polymerase complex (G) and the nucleocapsid is enveloped (H). Virions attain infectivity by acquisition of receptor-binding spikes (P3) (I). (See Color Insert.)

systems. Soluble particles missing protein P2, P4, or P7, or missing both P2 and P7, have been obtained. Thus, proteins P2, P4, and P7 can associate with the P1 skeleton independently of each other (Benevides *et al.*, 2002; de Haas *et al.*, 1999; Gottlieb *et al.*, 1988b; Juuti and Bamford, 1995; Paatero *et al.*, 1998). When P1 from $\phi 6$ is expressed alone, it forms spherical particles, which are aggregative and partially unfolded (Benevides *et al.*, 2002; Gottlieb *et al.*, 1988b; Paatero *et al.*, 1998).

### *B.* In Vitro *Assembly of Empty Precursor Particles*

The *in vitro* assembly of polymerase complexes of dsRNA bacteriophages $\phi 6$ and $\phi 8$ represented an important step in understanding the assembly of complex viruses and provided the first glimpse of early intermediates in the assembly pathway of a dsRNA virus. The $\phi 6$ and $\phi 8$ assembly systems are efficient; up to 50% of the precursor proteins are assembled into PCs when incubated under *in vitro* assembly conditions (Kainov *et al.*, 2003b; Poranen *et al.*, 2001). On the basis of their protein composition, sedimentation, morphology, and enzymatic activities (ssRNA packaging, replication, and transcription) the *in vitro*-assembled structures are indistinguishable from the recombinant particles purified from bacteria (Kainov *et al.*, 2003b; Poranen *et al.*, 2001). Furthermore, the *in vitro*-assembled polymerase complexes of $\phi 6$ are able to enter the host cell and initiate a new round of productive infection (Poranen *et al.*, 2001).

#### *1. Characterization of Building Blocks*

To set up an *in vitro* assembly system the component proteins must be isolated in an active and soluble form. Table I compares physicochemical properties of $\phi 6$ and $\phi 8$ PC building blocks. Despite negligible amino acid sequence homology each pair of PC proteins is similar in monomeric size (Hoogstraten *et al.*, 2000).

Protein P1 forms the dodecahedral framework of the viral core (Butcher *et al.*, 1997; Olkkonen and Bamford, 1987). In addition to its structural role, P1 is also involved in the specific binding and recognition of the viral genome segments (Qiao *et al.*, 2003a,b). The P1 framework is rather malleable and undergoes substantial and cooperative conformational changes (e.g., allostery) during packaging and replication (see Section III.D) (Butcher *et al.*, 1997). The allosteric properties most likely control the exact sequence of genome segment packaging and replication (Poranen *et al.*, 2005). We have purified protein P1 from two of the dsRNA bacteriophages, $\phi 6$ and $\phi 8$. P1 of

TABLE I
PROPERTIES OF PROCAPSID BUILDING BLOCKS

| Protein | Virus | Mass (kDa)[a] | Copies/ virion[b] | Multimeric status | Hydrodynamic radius (nm) | Function(s) |
|---|---|---|---|---|---|---|
| P1 | $\phi6$ | 85.0 | 120 | Monomer[c] | 3.8[c] | Major structural protein[d] |
| | $\phi8$ | 86.8 | 120 | Tetramer[e] | 5.5[e] | Major structural protein[f] |
| P2 | $\phi6$ | 74.8 | 12 | Monomer[g] | 3.3[c] | RNA-dependent RNA polymerase[g] |
| | $\phi8$ | 71.5 | 12 | Monomer[e] | 2.6[e] | RNA-dependent RNA polymerase[h] |
| P4 | $\phi6$ | 35.0 | 72 | Hexamer[i] | 5.9[i] | NTPase,[j] ssRNA packaging, transcription[k] |
| | $\phi8$ | 34.1 | 72 | Hexamer[e] | 5.1[e] | NTPase, helicase[l] |
| P7 | $\phi6$ | 17.2 | 60 | Dimer[m] | 3.7[c] | Assembly cofactor,[c] packaging factor[m] |
| | $\phi8$ | 19.0 | 60 | Dimer[e] | 2.8[e] | Not known |

[a] Calculated from the cDNA sequence (excluding N-terminal methionine that is removed).
[b] Estimated copy numbers, based on Day and Mindich (1980) and the symmetry of the virion.
[c] Poranen *et al.* (2001).
[d] Olkkonen and Bamford (1987).
[e] Kainov *et al.* (2003b).
[f] Hoogstraten *et al.* (2000).
[g] Makeyev and Bamford (2000).
[h] Yang *et al.* (2001).
[i] Juuti *et al.* (1998).
[j] Paatero *et al.* (1995).
[k] Pirttimaa *et al.* (2002).
[l] Kainov *et al.* (2003a).
[m] Juuti and Bamford (1997).

$\phi6$ is an elongated monomer exhibiting limited solubility (Benevides *et al.*, 2002; Poranen *et al.*, 2001), whereas P1 from $\phi8$ is a soluble tetramer (Kainov *et al.*, 2003b) (Table I).

Protein P2 contains the consensus sequence of known RNA-dependent RNA polymerases (Koonin *et al.*, 1989). Particles missing P2 (P1P4P7 particles) can package ssRNA but are unable to carry out genome replication and transcription (Juuti and Bamford, 1995). P2 has been purified as a monomer and is an active RNA-dependent RNA polymerase that can initiate complementary strand synthesis on the ssRNA template by a primer-independent mechanism

(Makeyev and Bamford, 2000). The structure of $\phi$6 P2 shares common features with the polymerases of hepatitis C virus and human immunodeficiency virus (Butcher *et al.*, 2001). The polymerase P2 is located under the fivefold vertices in the particle interior (Ikonen *et al.*, 2003).

Protein P4 of $\phi$6, as well as that of $\phi$8, contains sequence motifs of nucleoside triphosphate (NTP)-binding proteins (Walker motifs) (Gottlieb *et al.*, 1992a; Kainov *et al.*, 2003a). Purified P4 self-assembles into ring-like hexamers, which are stable in the case of $\phi$8 (Kainov *et al.*, 2003b) and $\phi$12 (Mancini *et al.*, 2004) but require ADP and divalent cations ($Mg^{2+}$ and $Ca^{2+}$) for stability in the case of $\phi$6 (Juuti *et al.*, 1998). $\phi$6 P4 is active only in the hexameric form and can hydrolyze all nucleotide triphosphates (rNTP, dNTP, and ddNTP). The NTPase activity of $\phi$6 P4 is weakly stimulated by ssRNA (Juuti *et al.*, 1998; Paatero *et al.*, 1995). On the other hand, the activity of $\phi$8 P4 is tightly coupled to ssRNA binding (Lisal *et al.*, 2004) and the hexameric protein also possesses RNA helicase activity (Kainov *et al.*, 2003a). Such activity has not been detected for purified P4 of $\phi$6, although $\phi$6 PC can displace short stretches of dsRNA during packaging (Kainov *et al.*, 2003a). $\phi$6 P4 is the viral packaging NTPase and, interestingly, it is also needed for transcription (Pirttimaa *et al.*, 2002) (see Section III.B.2).

Protein P7 is a minor component of the particle. Purified P7 is an elongated dimer (Juuti and Bamford, 1997) (Table I). The N-terminal part of $\phi$6 P7 forms the core of the dimer, whereas the C termini constitute flexible, protease-sensitive arms (S. J. Butcher, personal communication). The flexible part of the protein is not crucial for the stable incorporation of P7 into the polymerase complex, but it is essential for the viability of the virus (M. M. Poranen, unpublished data).

P7-null particles have poor packaging and transcription efficiencies (Juuti and Bamford, 1997). In addition, P7 is an assembly cofactor (see Section II.B.3). These diverse roles of P7 suggest that it might be either a particle-stabilizing (glue) protein or constitute a hinge that allows structural transitions within the P1 shell. The estimated copy number for P7 is 60 (Juuti and Bamford, 1997), and it has been suggested to be located near the twofold axes of the PC, so that the elongated P7 dimer could span the interface between two P1 dimers (Poranen *et al.*, 2001).

*2. Assembly Conditions*

The $\phi$6 and $\phi$8 PCs can be assembled by incubating the component proteins in the presence of 3–6% (w/v) polyethylene glycol 4000 or 6000 (Kainov *et al.*, 2003b; Poranen *et al.*, 2001). When high enough concentrations of $\phi$8 PC proteins (>1 mg/ml) were used, PC-like particles were also assembled in the absence of polyethylene glycol. The assembly of $\phi$6 PC is also dependent on low concentrations of ADP and divalent cations that stabilize the P4 hexamers.

In addition to the complete PCs, various incomplete particles can be produced *in vitro* by omitting one or more proteins from the assembly mixture (Kainov *et al.*, 2003b; Poranen *et al.*, 2001). At high concentrations P1 of $\phi$8 alone could assemble into PC-like dodecahedral cages (Kainov *et al.*, 2003b). However, the high critical concentration (4.6 g/liter for $\phi$8) and low stability of these cages indicated that other proteins were likely to control PC assembly via stabilization of intermediates and by lowering the critical concentration for assembly nucleation.

### 3. *Assembly Pathway for Empty Particles*

An *in vitro* system allows one to define the minimal set of components required for particle assembly and also to identify rate-limiting nucleation steps. Such systems are also amenable to time-resolved studies, which can provide information about the structure of metastable intermediates that are often populated early in the assembly pathway.

The formation of $\phi$6 PC-like structures is dependent on protein P4, suggesting that the P4 hexamer plays a crucial role in the nucleation step (Poranen *et al.*, 2001). Below critical concentration of $\phi$8 P1, the P2 polymerase was identified as necessary and sufficient for the formation of particle-like structures, although aberrant assemblies also appeared (Kainov *et al.*, 2003b). The addition of P4 increased the assembly rate and prevented aberrant structure formation. Thus, P2 and P4 are essential for the initiation of assembly. This ensures incorporation of the two key enzymes into the PC (Fig. 3).

The kinetics of $\phi$6 and $\phi$8 PC formation exhibit a distinct lag phase typical of nucleation-limited assembly (Prevelige *et al.*, 1993). Analyses of $\phi$6 kinetics identified formation of a P1 tetramer–P4 hexamer complex as the rate-limiting, nucleation step (Fig. 3) (Poranen *et al.*, 2001). In the $\phi$8 system P1 forms stable preassembled tetrameric building blocks and the nucleation complex resulted from interaction of several P1 tetramers with a P2 monomer (Fig. 3) (Kainov *et al.*, 2003b). When P7 was included in the $\phi$6 assembly system the rate of assembly was significantly accelerated. This suggests that P7 acts as an assembly cofactor, albeit a nonessential one. The presence of P7 deferred the rate-limiting step of $\phi$6 PC assembly and an intermediate composed of a tetrameric P1 connecting two P4 hexamers was

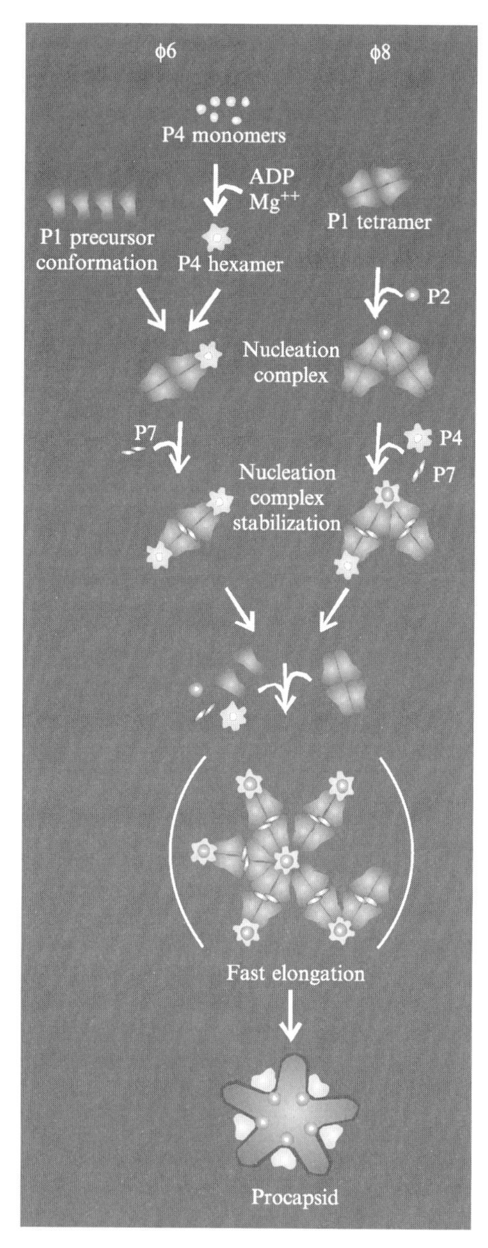


FIG 3. Summary and comparison of the $\phi6$ and $\phi8$ *in vitro* assembly pathways. Both pathways share a P1 tetramer as an obligatory intermediate and proceed via nucleation-limited polymerization. P1 and P4 association into a nucleation complex initiates $\phi6$

identified. Thus, P7 can stabilize the P1 tetramers and accelerate the assembly process (Poranen *et al.*, 2001). In $\phi$8 assembly, P7 had only a small effect on the assembly rate but P4 hexamers significantly increased the assembly rate, presumably playing a similar stabilizing role as $\phi$6 P7 (Kainov *et al.*, 2003b).

Using time-resolved dynamic light scattering, distinct intermediates with hydrodynamic radii between 10–12 nm (Poranen *et al.*, 2001) and 8–10 nm (Kainov *et al.*, 2003b) were observed for $\phi$6 and $\phi$8, respectively. These intermediates correspond to the stabilized nucleation complexes detected by kinetic methods (Fig. 3), that is, one P1 tetramer and one P4 hexamer in the case of $\phi$6 and two or three P1 tetramers together with P2 monomer and P4 hexamer in the case of $\phi$8.

A tetrameric P1 intermediate appears to be obligatory for both $\phi$6 and $\phi$8. The tetramer forms a bridge between two neighboring fivefold vertices and thus facilitates propagation of the structure (Fig. 3). In the $\phi$6 system it was possible to identify a nucleation complex containing two P4 hexamers, which suggests that the assembly proceeds via a twofold-centered assembly intermediate.

*4. Protein Folding During Procapsid Assembly*

Assembly-induced conformational rearrangements have been demonstrated for many viral systems either by high-resolution crystallographic studies or by spectroscopic methods. Such changes are often required to accommodate nonequivalent interactions in viral capsids (Dokland, 2000; Tuma *et al.*, 2001). The availability of incomplete $\phi$6 and $\phi$8 PCs has allowed the study of conformational changes associated with the interaction of particular subunits within the PCs. Large ordering of $\alpha$-helices has been shown to be characteristic of P1–P4 interactions (Benevides *et al.*, 2002; D. E. Kainov, R. Tuma, and G. J. Thomas, unpublished data). The C-terminal tail of P4, which is otherwise disordered in solution, most likely adopts an $\alpha$-helical structure. The same region has been shown essential for P4 incorporation into the PC (Paatero *et al.*, 1998). Such conformational switching may be necessary for accommodation of the symmetry mismatch between the fivefold vertex of the P1 and the P4 hexamer (de Haas *et al.*, 1999).

In addition to P4, P7 undergoes ordering on binding to PC. On the other hand, interactions of P2 do not lead to any significant changes

procapsid assembly, whereas the nucleation complex of $\phi$8 contains P1 and P2. In both cases the nucleation complex is then stabilized by P4 and P7 and the procapsid shell is rapidly completed by addition of individual building blocks. (See Color Insert.)

in secondary and tertiary structures (Benevides *et al.*, 2002). This is consistent with the fully ordered and compact three-dimensional structure of the polymerase monomer (Butcher *et al.*, 2001).

## III. Genome Encapsidation

The genome encapsidation process of $\phi$6 has been under considerable investigation (for a review see Mindich, 1999; Poranen *et al.*, 2005). It became apparent from early studies on infected cells that the genome was packaged into the empty precursor capsid as single-stranded, plus-sense RNA molecules, which are then replicated within the polymerase complex into the dsRNA genome (Ewen and Revel, 1988, 1990). Thus, genome replication in dsRNA bacteriophages is not performed in parallel with capsid assembly but is physically linked with virion morphogenesis.

The encapsidation mechanism of a segmented viral genome is an intriguing problem, which still poses many unanswered questions. The mature $\phi$6 virion contains only one copy of each segment and the specific infectivity is close to one (Day and Mindich, 1980), indicating that the packaging process is specific and tightly controlled. A breakthrough for $\phi$6 genome-packaging analyses was the isolation and purification of empty recombinant polymerase complexes (PCs) (Gottlieb *et al.*, 1988b), which could carry out the entire viral RNA replication cycle *in vitro* (Gottlieb *et al.*, 1990), that is, plus-sense ssRNA packaging, complementary minus-strand synthesis (replication), and transcription. To date the *in vitro* system of $\phi$6 has been the only one available to study the packaging of multisegmented viral genomes under defined conditions. The genome encapsidation process can be divided into three steps, namely ssRNA binding to the PC, ssRNA translocation into the PC, and synthesis of complementary minus stands within the polymerase complex particle. The biochemical requirements of each step are distinct and modifying the concentration of NTPs allows separation of the reactions *in vitro* (Fig. 2) (Frilander and Bamford, 1995; Juuti and Bamford, 1995; van Dijk *et al.*, 1995).

### *A. ssRNA Binding*

ssRNA binding to the PC occurs spontaneously *in vitro* (in the absence of NTPs) (Juuti and Bamford, 1995). A certain preference for $\phi$6-specific ssRNA has been detected. The binding affinities of the three segments vary: the s segment has the highest, followed by m segment and then the l segment (Juuti and Bamford, 1995). Although a nonspecific ssRNA, which does not contain the viral packaging signal, can bind to PCs, such binding does not lead to packaging, suggesting that only specific binding can initiate productive encapsidation. The bound ssRNA is sensitive to RNase, indicating that the segment-specific binding sites are located on the outer surface of the PC. Thus, selection of the ssRNA molecules could be accomplished before the energy-consuming steps of encapsidation (see Section III.B) (Poranen *et al.*, 2005). Protein P1 is likely to be involved in the initial binding and recognition of the genomic segments (Qiao *et al.*, 2003a,b).

### *B. ssRNA Packaging*

The ssRNA packaging into the PC is dependent on NTP hydrolysis by the viral NTPase P4 (Frilander and Bamford, 1995; Gottlieb *et al.*, 1991; Pirttimaa *et al.*, 2002). P4, in turn, can utilize all four NTPs, dNTPs, as well as ddNTPs to drive packaging (Frilander and Bamford, 1995). PCs that are void of protein P4 are completely inactive (Pirttimaa *et al.*, 2002). P7 is also needed for stable ssRNA packaging (Juuti and Bamford, 1997).

The location of the ring-shaped P4 hexamer at the fivefold vertices of the PC creates a symmetry mismatch within the assembly (de Haas *et al.*, 1999). Interestingly, similar symmetry mismatches are common for the packaging vertices of dsDNA bacteriophages (Bazinet and King, 1985; Casjens, 1985; Dube *et al.*, 1993; Lake and Leonard, 1974; Tsuprun *et al.*, 1994) and herpesviruses (Newcomb *et al.*, 2001). The recently solved atomic structure of $\phi$12 protein P4 provides a detailed mechanism for ATP-driven RNA translocation (Mancini *et al.*, 2004).

The PC of $\phi$6 can package all the plus-strand RNA molecules (s, m, and l) independently of each other (Frilander and Bamford, 1995; Gottlieb *et al.*, 1992b). However, the three ssRNA molecules have different packaging efficiencies. The packaging efficiencies follow the same order as observed for binding affinities ($s > m > l$) (Frilander and Bamford, 1995). It has also been shown that the packaging of s segment induces the packaging of m segment and that the packaging of m segment in turn enhances the packaging of l segment (Frilander and Bamford, 1995; Qiao *et al.*, 1995a). These observations suggest that the segments are preferentially packaged in an orderly fashion: s segment being packaged first, followed by m segment and l segment (Fig. 2). This is consistent with the appearance of particles containing only the s segment, both s and m segments, or all three segments but no other combination of segments in $\phi$6-infected cells (Ewen and Revel, 1990).

The binding affinities and packaging efficiencies indicate specific recognition of the three viral segments. Indeed, specific packaging sites (*pac* sites) have been located within the 5′ noncoding regions extending 273, 280, and 209 nucleotides from the 5′ end of the s, m, and l segment, respectively (Frilander *et al.*, 1992; Gottlieb *et al.*, 1994; Pirttimaa and Bamford, 2000; Qiao *et al.*, 1995b, 1997). Both the sequences and the predicted secondary structures of these *pac* sites show little homology apart from one conserved loop that is common to all three *pac* sites (Pirttimaa and Bamford, 2000). In addition, a few other loop structures are predicted to be similar between segment pairs.

### *1. How to Achieve Sequential Packaging?*

The precise incorporation of the three genomic segments into the virion apparently follows the specific recognition of the three segments and the sequential mode of packaging. How sequential packaging may be achieved is not completely understood. Although the three segments are preferentially packaged in an orderly manner (s, m, and l), the packaging of a segment is not an absolute prerequisite for the packaging of the next segment; the segments can be packaged independently of each other, albeit with different efficiencies (Frilander and Bamford, 1995). Therefore, it is likely that the empty particle can spontaneously adopt conformations in which s-, m-, or l-binding sites are exposed (Poranen *et al.*, 2005). The highest packaging efficiency, for the s segment, could reflect the fact that the conformation that preferentially packages s segment is the most favorable and abundant *in vitro* (and probably also *in vivo*), whereas the conformation with high affinity for l segment is only marginally populated. These results suggest that such switching between conformations does not involve high activation barriers and is fully reversible, for example, the conformations coexist in dynamic equilibrium. Thus, the sequential modulation of affinity by segment packaging could be achieved in a manner similar to allosteric regulation of ligand binding in smaller protein assemblies (Horovitz *et al.*, 2001).

### 2. *Genome Packaging via a Special Vertex*

Biochemical and genetic analyses have suggested that 1 of the 12 vertices in the $\phi$6 PC is physically and functionally different (Pirttimaa *et al.*, 2002). The P4 hexamer is more tightly bound to this special vertex and it is possible to obtain PCs that contain on average one hexamer per particle. These P4-deficient particles have the same packaging and RNA replication efficiencies as the PCs with the full complement of P4 (Pirttimaa *et al.*, 2002). Because P4-null PCs are completely inactive in packaging the activity of P4-deficient particles suggests that $\phi$6 PCs may contain one special vertex that is necessary and sufficient for genome encapsidation.

What would be the role of the hexamers at the other vertices? Although packaging and replication proceeded as in wild-type PCs the transcription activity of the P4-deficient particles was completely abolished. Furthermore, addition of P4 hexamers to the P4-deficient PCs fully restored transcription activity (Pirttimaa *et al.*, 2002). Thus, P4 hexamers at the other vertices are essential for transcription, probably assisting in the extrusion of the newly synthesized mRNA from the capsid. In contrast to the dsDNA bacteriophages, in which genome packaging and ejection are accomplished via a single specialized vertex (portal vertex), in $\phi$6 these processes are performed by distinct classes of vertices.

## *C. dsRNA Replication*

The packaged ssRNA segments are replicated into dsRNA form within the particle in order to produce genome-containing viral cores. Replication is performed by the viral polymerase P2 (Juuti and Bamford, 1995). The replication can be studied *in vitro* either using purified P2 (Makeyev and Bamford, 2000) or by allowing the *in vitro*-packaged polymerase complexes to proceed to minus-strand synthesis (van Dijk *et al.*, 1995). Interestingly, the template requirements for minus-strand synthesis differ between free P2 and P2 associated with the polymerase complex. Whereas efficient genome replication by the polymerase complex requires all genomic segments to contain specific 3′-end sequences (minus-strand synthesis initiation signals) (Frilander *et al.*, 1992; Mindich *et al.*, 1992, 1994), purified monomeric P2 is relatively unspecific and can efficiently replicate heterologous ssRNA (Makeyev and Bamford, 2000). The untranslated 3′ ends of the three $\phi$6 genomic segments that carry the minus-strand synthesis initiation signals share sequence homology and are predicted to fold into similar tRNA-like secondary structures (Mindich *et al.*, 1992). It has been suggested that within the particle, the minus-strand synthesis initiation signals might be needed for correct localization of the 3′ ends of the genome segments within the particle (e.g., in proximity to P2 polymerase) (Makeyev and Bamford, 2000).

#### *1. Initiation of Minus-Strand Synthesis*

The initiation of minus-strand synthesis is an important checkpoint during the genome encapsidation process (Poranen *et al.*, 2005). Systematic analyses of replication of different combinations of truncated, deleted, and modified ssRNA segments have identified the 5′ end of the l segment as the activator of efficient minus-strand synthesis (Frilander *et al.*, 1995; Poranen and Bamford, 1999). As the l segment is the last to be packaged it is natural that it also serves as the RNA replication initiation signal (Fig. 2). Interestingly, the very 5′-terminal fragment carrying the l segment packaging signal is sufficient for the replication of s and m segments without the requirement for complete packaging of the l segment (Poranen *et al.*, 2005). This is consistent with the sequential appearance of s, m, and l minus strands during $\phi$6 infection (Pagratis and Revel, 1990) and also suggests concurrent packaging of the l segment and replication of the s and m segments.

### *D. Capsid Maturation During Genome Encapsidation*

The polymerase complex of $\phi$6 is not a rigid structure but undergoes substantial conformational changes during maturation (Butcher *et al.*, 1997). Three different conformations have been described. These are (1) compressed empty polymerase complex (PC), (2) expanded polymerase complex, and (3) mature dsRNA-filled polymerase complex (see Fig. 2) (Butcher *et al.*, 1997). The PC has a dodecahedral appearance with fivefold facets recessed to form inwardly oriented "cups." Expanded polymerase complex is a dodecahedron with shallow depressions at the fivefold vertices. The expanded morphologies are found in PC preparations as well as in preparations of virion-derived polymerase complexes (viral core particles) that have lost the genome. The mature, dsRNA-filled polymerase complex is roughly spherical with a diameter (50 nm) that is larger than that of the PC (46 nm). These different forms of the polymerase complex are likely to represent intermediates during the genome encapsidation and replication processes.

### *1. Timing of Capsid Expansion*

The internal volumes of the three different forms of the polymerase complex have been estimated and the density of the packaged RNA has been calculated. These RNA densities were used to suggest at which stages of encapsidation and replication the conformational changes in the polymerase complex might occur (Fig. 2) (Poranen *et al.*, 2005). It appeared that the empty compressed particle had to expand at the latest during the packaging of the l segment (assuming that s and m segments have been packaged before the l segment). Thus either the packaging of s and m segments induces expansion, or the 5′ end of the l segment carries a signal that induces the conformational change (before the packaging of the full l segment). Because the 5′ end of the l segment is also needed for the activation of RNA synthesis (see Section III.C.1) (Frilander *et al.*, 1995), it is conceivable that the polymerase activation could also result from the expansion. For example, the polymerase (P2) active site could be hidden in the PC (collapsed) particle and becomes accessible only in the expanded form. Alternatively, the 3′ ends of the RNA segments would not be available to the polymerase in the collapsed form.

The upper limits of packaging density (packaging density equivalent to dsDNA packing in crystals) is reached within the expanded particle after the entire genome is replicated into dsRNA form (Poranen *et al.*, 2005). This indicates that expansion of the particle into the mature rounded conformation should occur at the latest before the initiation of transcription, as the highly active process of mRNA synthesis likely requires additional space to accommodate RNA movement and extrusion.

## IV. Nucleocapsid Surface Shell Assembly

The NC surface protein P8 can be purified from virions in a trimeric, highly $\alpha$-helical form (Poranen *et al.*, 2001; Tuma *et al.*, 1999). The purified protein reassembles onto dsRNA-filled polymerase complex isolated from virions as well as onto *in vitro*-reconstituted complexes (Fig. 2) (Olkkonen *et al.*, 1990, 1991; Poranen *et al.*, 2001). The assembly is calcium dependent and no detectable conformational change is associated with assembly of the P8 shell (Tuma *et al.*, 1999), suggesting that P8 subunits assemble in a quasi-equivalent manner.

At higher protein concentrations P8 can self-assemble into aberrant, shell-like structures with sizes varying according to the assembly conditions (M. M. Poranen, and D. H. Bamford, unpublished data). These assemblies are not topologically closed and appear as open spirals in electron micrographs. This suggests that assembly of closed P8 shells is dependent on the core template.

P8 shell assembly onto *in vitro*-assembled and packaged polymerase complexes yields *in vitro*-assembled NCs. Because NCs can infect host cell spheroplasts (i.e., cells with the outer membrane and peptidoglycan layers partially removed), *in vitro*-assembled polymerase complexes can be introduced into the host cell (Ojala *et al.*, 1990; Poranen *et al.*, 2001). The infectivity of the assembly products indicates that the formation of bacteriophage $\phi$6 NCs does not require any phage or host-encoded assembly factors.

## V. Envelope Assembly

The final step in $\phi$6 maturation is the envelope assembly around the NC (Fig. 2). So far no *in vitro*-based system has been devised to study this process and our knowledge about envelope formation is based on analyses of $\phi$6-infected cells, mutant studies, and various expression systems.

Of the five membrane-associated proteins only one, P9, is needed for the formation of the envelope (Stitt and Mindich, 1983a). A nonstructural protein, P12, is also essential for viral envelope morphogenesis (Johnson and Mindich, 1994; Mindich and Lehman, 1983; Mindich *et al.*, 1976). Interestingly, mutants in P5 do not affect particle assembly (Mindich *et al.*, 1976) (albeit P5 is located between the P8 shell and the envelope), although P5 is required for infection.

Envelope assembly apparently takes place within the cell cytoplasm. Both NCs and enveloped virions are detected in the central regions of the cell, not in contact with the plasma membrane (Bamford *et al.*, 1976; Ellis and Schlegel, 1974; Gonzalez *et al.*, 1977). The envelopment of $\phi$6 does not involve budding out from the host plasma membrane, which is typical of many animal viruses; rather, preformed lipid vesicles envelope the NCs and the mature virions are later released by host cell lysis (Stitt and Mindich, 1983a).

The point at which the other integral membrane proteins are associated with the P9–lipid vesicles is not clear. Also, it is not known whether P5 is associated with the membrane vesicles or the NC before envelopment. The host recognition and attachment protein P3, which is anchored to the envelope via protein P6, is expressed as a soluble protein and is the last component to be assembled onto the virions (Fig. 2) (Mindich *et al.*, 1979; Stitt and Mindich, 1983b).

## VI. Comparison with *Reoviridae*

The *Reoviridae*, infecting eukaryotic cells, and the *Cystoviridae*, infecting bacteria (including $\phi$6 and $\phi$8), share a similar genome replication strategy in which single-stranded genomic precursor molecules are replicated to their double-stranded forms within the viral capsid (Patton and Spencer, 2000; Poranen *et al.*, 2005). In addition, they share similar core architecture and it has been suggested that these two families may belong to the same evolutionary lineage (Bamford *et al.*, 2002).

Arrangement of the 120 P1 molecules within the core of $\phi$6 is similar to that of the bluetongue virus (BTV) VP3 layer (Grimes *et al.*, 1998), the reovirus $\lambda$1 layer (Reinisch *et al.*, 2000), the rotavirus VP2 layer (Lawton *et al.*, 1997; Prasad *et al.*, 1996, 2001), and the Gag protein of the yeast L-A virus (Naitow *et al.*, 2002). Arrangement of the fivefold vertices of the polymerase complex differs. Whereas $\phi$6 has the P4 packaging hexamer at these positions, reovirus has a turret of the RNA-capping pentamer ($\lambda$2) in equivalent locations (Reinisch *et al.*, 2000). BTV and rotavirus do not contain similar turret-like structures (Grimes *et al.*, 1998; Prasad *et al.*, 1996). The rotavirus and BTV RNA-processing and replication machinery has been localized within the core (Gouet *et al.*, 1999; Prasad *et al.*, 1996). Similarly, the P2 enzyme has been positioned inside the $\phi$6 core under the fivefold vertex (Ikonen *et al.*, 2003). Although the atomic structures of several eukaryotic dsRNA virus polymerase complexes are known, in the absence of an *in vitro* system, assembly intermediates have only been postulated. A decamer of VP3 was proposed as a BTV core assembly intermediate (Grimes *et al.*, 1998). It may well be an intermediate, albeit a late one, which is preceded by several smaller species. Results from the $\phi$6–$\phi$8 system suggest that a tetramer of the major structural protein P1 in complex with one or more of the key enzymes (the polymerase and the packaging ATPase) is essential for nucleation (Kainov *et al.*, 2003b; Poranen *et al.*, 2001). Interestingly, it was suggested that the viral polymerase of rotavirus (VP3) may nucleate the core assembly (Patton and Spencer, 2000).

The NC shell is built from P8 trimers in a fashion similar to the BTV VP7 and rotavirus VP6 layer (Grimes *et al.*, 1995; Lawton *et al.*, 1997). P8 as well as BTV VP7 need the core template for correct assembly (Grimes *et al.*, 1997). The outermost capsid layers (a membrane envelope among the members of *Cystoviridae*) accommodate host specificity and thus are not structurally related among different dsRNA viruses.

In contrast to $\phi 6$, the encapsidation and replication processes of the members of *Reoviridae* are dependent on several virally encoded nonstructural proteins and occur in cytoplasmic inclusions (viroplasm) (Patton and Spencer, 2000). Nevertheless, several features of packaging and replication may be shared between the *Cystoviridae* and the *Reoviridae*, namely: the presence of packaging and replication signal sequences at the 5′ and 3′ ends of genomic RNA segments (Chen and Patton, 1998), the existence of core precursors (precores) containing the RNA polymerase (Patton, 2001), and replication assisted by a nonspecific oligomeric NTPase (NSP2 in rotavirus, NS2 in BTV, and sNS in reovirus) (Jayaram *et al.*, 2002; Schuck *et al.*, 2001; Taraporewala *et al.*, 1999). The rotavirus NTPase NSP2 is an octamer with helix-destabilizing activity (Taraporewala and Patton, 2001). Similar activity has been detected for the P4 packaging hexamer of $\phi 8$ and for P4 associated with $\phi 6$ PCs (Kainov *et al.*, 2003a). Such similarity suggests that rotavirus may package ssRNA into empty precursor cores, using a helicase-like protein (Taraporewala and Patton, 2001). On the other hand, unlike $\phi 6$, rotavirus replication is nuclease sensitive and there is no evidence of ssRNA packaging before replication (Patton, 2001). At present it is not clear whether the sequential packaging model developed for $\phi 6$ (Mindich, 1999; Poranen *et al.*, 2005) is also applicable to rotavirus.

An alternative model for selective packaging has been proposed for reoviruses, in which decamers of the core subunit associate with different RNA segments and such intermediates then assemble into the core (Patton and Spencer, 2000). Such a coassembled core is proposed to undergo a conformational change, which in turn activates the polymerase and replication (Patton and Spencer, 2000). Because of the lack of *in vitro* assembly systems for the *Reoviridae* these hypotheses remain to be experimentally verified.

## VII. Conclusions

The assembly of enveloped dsRNA bacteriophages can be dissected into four distinct steps: (1) assembly of empty precursor capsid, (2) genome encapsidation involving ssRNA packaging and synthesis of complementary strand, (3) NC surface shell assembly, and

(4) envelopment. The first three steps can be studied under defined *in vitro* conditions, and we are in the process of obtaining information about the molecular details of the assembly pathway. Formation of the complete, infectious, fully functional, NC *in vitro* is a demonstration of the self-assembly principle: all the information and functions needed for the formation of the final structure are encoded by and carried out by the structural constituents. No viral or host encoded accessory factors are needed.

The formation of NCs constitutes a nonbranched, linear assembly pathway with distinct obligatory intermediates. How the process is controlled is not known in detail. However, correct assembly apparently involves sequential interactions between virion constituents and conformational switching (allostery) within the intermediates during assembly. The sequential stimulation of segment packaging, which is mediated by allostery, ensures exact packaging of all three genome segments. The proposed gradual expansion of the polymerase complex likely plays a crucial regulatory role during maturation. Correct assembly of the P8 shell depends on the core template.

Envelope formation is a separate assembly process that merges with the NC assembly line to produce enveloped, mature virions.

## References


Bamford, D. H., Burnett, R. M., and Stuart, D. I. (2002). Evolution of viral structure. *Theor. Popul. Biol.* **61:**461–470.

Bamford, D. H., and Mindich, L. (1980). Electron microscopy of cells infected with nonsense mutants of bacteriophage $\phi$6. *Virology* **107:**222–228.

Bamford, D. H., Palva, E. T., and Lounatmaa, K. (1976). Ultrastructure and life cycle of the lipid-containing bacteriophage $\phi$6. *J. Gen. Virol.* **32:**249–259.

Bamford, D. H., Romantschuk, M., and Somerharju, P. J. (1987). Membrane fusion in prokaryotes: Bacteriophage $\phi$6 membrane fuses with the *Pseudomonas syringae* outer membrane. *EMBO J.* **6:**1467–1473.

Bazinet, C., and King, J. (1985). The DNA translocating vertex of dsDNA bacteriophage. *Annu. Rev. Microbiol.* **39:**109–129.

Benevides, J. M., Juuti, J. T., Tuma, R., Bamford, D. H., and Thomas, G. J., Jr. (2002). Raman difference spectroscopy of the $\phi$6 procapsid: Characterization of subunit-specific interactions in a dsRNA virus. *Biochemistry* **41:**11946–11953.

Butcher, S. J., Dokland, T., Ojala, P. M., Bamford, D. H., and Fuller, S. D. (1997). Intermediates in the assembly pathway of the double-stranded RNA virus $\phi$6. *EMBO J.* **16:**4477–4487.

Butcher, S. J., Grimes, J. M., Makeyev, E. V., Bamford, D. H., and Stuart, D. I. (2001). A mechanism for initiating RNA-dependent RNA polymerization. *Nature* **410:**235–240.

Casini, G., and Revel, H. R. (1994). A new small low-abundant nonstructural protein encoded by the L segment of the dsRNA bacteriophage $\phi$6. *Virology* **203:**221–228.

Casjens, S. (1985). Nucleic acid packaging by viruses. *In* "Virus Structure and Assembly" (S. Casjens, ed.), pp. 88–95. Jones and Bartlett, Boston.
Chen, D., and Patton, J. T. (1998). Rotavirus RNA replication requires a single-stranded 3′ end for efficient minus-strand synthesis. *J. Virol.* **72:**7387–7396.
Day, L. A., and Mindich, L. (1980). The molecular weight of bacteriophage $\phi$6 and its nucleocapsid. *Virology* **103:**376–385.
de Haas, F., Paatero, A. O., Mindich, L., Bamford, D. H., and Fuller, S. D. (1999). A symmetry mismatch at the site of RNA packaging in the polymerase complex of dsRNA bacteriophage $\phi$6. *J. Mol. Biol.* **294:**357–372.
Dokland, T. (2000). Freedom and restraint: Themes in virus capsid assembly. *Struct. Fold. Des.* **8:**R157–R162.
Dube, P., Tavares, P., Lurz, R., and van Heel, M. (1993). The portal protein of bacteriophage SPP1: A DNA pump with thirteenfold symmetry. *EMBO J.* **12:**1303–1309.
Ellis, L. F., and Schlegel, R. A. (1974). Electron microscopy of *Pseudomonas* $\phi$6 bacteriophage. *J. Virol.* **14:**1547–1551.
Ewen, M. E., and Revel, H. R. (1988). *In vitro* replication and transcription of the segmented double-stranded RNA bacteriophage $\phi$6. *Virology* **165:**489–498.
Ewen, M. E., and Revel, H. R. (1990). RNA–protein complexes responsible for replication and transcription of the double-stranded RNA bacteriophage $\phi$6. *Virology* **178:**509–519.
Frilander, M., and Bamford, D. H. (1995). *In vitro* packaging of the single-stranded RNA genomic precursors of the segmented double-stranded RNA bacteriophage $\phi$6: The three segments modulate each other's packaging efficiency. *J. Mol. Biol.* **246:**418–428.
Frilander, M., Gottlieb, P., Strassman, J., Bamford, D. H., and Mindich, L. (1992). Dependence of minus-strand synthesis on complete genomic packaging in the double-stranded RNA bacteriophage $\phi$6. *J. Virol.* **66:**5013–5017.
Frilander, M., Poranen, M., and Bamford, D. H. (1995). The large genome segment of dsRNA bacteriophage $\phi$6 is the key regulator in the *in vitro* minus and plus strand synthesis. *RNA.* **1:**510–518.
Gonzalez, C. F., Langenberg, W. G., Van Etten, J. L., and Vidaver, A. K. (1977). Ultrastructure of bacteriophage $\phi$6: Arrangement of the double-stranded RNA and envelope. *J. Gen. Virol.* **35:**353–359.
Gottlieb, P., Metzger, S., Romantschuk, M., Carton, J., Strassman, J., Bamford, D. H., Kalkkinen, N., and Mindich, L. (1988a). Nucleotide sequence of the middle dsRNA segment of bacteriophage $\phi$6: Placement of the genes of membrane-associated proteins. *Virology* **163:**183–190.
Gottlieb, P., Potgieter, C., Wei, H., and Toporovsky, I. (2002a). Characterization of $\phi$12, a bacteriophage related to $\phi$6: Nucleotide sequence of the large double-stranded RNA. *Virology* **295:**266–271.
Gottlieb, P., Qiao, X., Strassman, J., Frilander, M., and Mindich, L. (1994). Identification of the packaging regions within the genomic RNA segments of bacteriophage $\phi$6. *Virology* **200:**42–47.
Gottlieb, P., Strassman, J., Bamford, D. H., and Mindich, L. (1988b). Production of a polyhedral particle in *Escherichia coli* from a cDNA copy of the large genomic segment of bacteriophage $\phi$6. *J. Virol.* **62:**181–187.
Gottlieb, P., Strassman, J., Frucht, A., Qiao, X. Y., and Mindich, L. (1991). *In vitro* packaging of the bacteriophage $\phi$6 ssRNA genomic precursors. *Virology* **181:**589–594.
Gottlieb, P., Strassman, J., and Mindich, L. (1992a). Protein P4 of the bacteriophage $\phi$6 procapsid has a nucleoside triphosphate-binding site with associated nucleoside triphosphate phosphohydrolase activity. *J. Virol.* **66:**6220–6222.

Gottlieb, P., Strassman, J., Qiao, X., Frilander, M., Frucht, A., and Mindich, L. (1992b). *In vitro* packaging and replication of individual genomic segments of bacteriophage $\phi 6$ RNA. *J. Virol.* **66:**2611–2616.

Gottlieb, P., Strassman, J., Qiao, X. Y., Frucht, A., and Mindich, L. (1990). *In vitro* replication, packaging, and transcription of the segmented double-stranded RNA genome of bacteriophage $\phi 6$: Studies with procapsids assembled from plasmid-encoded proteins. *J. Bacteriol.* **172:**5774–5782.

Gottlieb, P., Wei, H., Potgieter, C., and Toporovsky, I. (2002b). Characterization of $\phi 12$, a bacteriophage related to $\phi 6$: Nucleotide sequence of the small and middle double-stranded RNA. *Virology* **293:**118–124.

Gouet, P., Diprose, J. M., Grimes, J. M., Malby, R., Burroughs, J. N., Zientara, S., Stuart, D. I., and Mertens, P. P. (1999). The highly ordered double-stranded RNA genome of bluetongue virus revealed by crystallography. *Cell* **97:**481–490.

Grimes, J., Basak, A. K., Roy, P., and Stuart, D. (1995). The crystall structure of bluetongue virus VP7. *Nature* **373:**167–170.

Grimes, J. M., Burroughs, J. N., Gouet, P., Diprose, J. M., Malby, R., Zientara, S., Mertens, P. P., and Stuart, D. I. (1998). The atomic structure of the bluetongue virus core. *Nature* **395:**470–478.

Grimes, J. M., Jakana, J., Ghosh, M., Basak, A. K., Roy, P., Chiu, W., Stuart, D. I., and Prasad, B. V. (1997). An atomic model of the outer layer of the bluetongue virus core derived from X-ray crystallography and electron cryomicroscopy. *Structure* **5:**885–893.

Hantula, J., and Bamford, D. H. (1988). Chemical crosslinking of bacteriophage $\phi 6$ nucleocapsid proteins. *Virology* **165:**482–488.

Hoogstraten, D., Qiao, X., Sun, Y., Hu, A., Onodera, S., and Mindich, L. (2000). Characterization of $\phi 8$, a bacteriophage containing three double-stranded RNA genomic segments and distantly related to $\phi 6$. *Virology* **272:**218–224.

Horovitz, A., Fridmann, Y., Kafri, G., and Yifrach, O. (2001). Review: Allostery in chaperonins. *J. Struct. Biol.* **135:**104–114.

Ikonen, T., Kainov, D., Timmins, P., Serimaa, R., and Tuma, R. (2003). Locating the minor components of double-stranded RNA bacteriophage $\phi 6$ by neutron scattering. *J. Appl. Crystallogr.* **36:**525–529.

Jayaram, H., Taraporewala, Z., Patton, J. T., and Prasad, B. V. (2002). Rotavirus protein involved in genome replication and packaging exhibits a HIT-like fold. *Nature* **417:**311–315.

Johnson, M.D., 3rd, and Mindich, L. (1994). Isolation and characterization of nonsense mutations in gene 10 of bacteriophage $\phi 6$. *J. Virol.* **68:**2331–2338.

Juuti, J. T., and Bamford, D. H. (1995). RNA binding, packaging and polymerase activities of the different incomplete polymerase complex particles of dsRNA bacteriophage $\phi 6$. *J. Mol. Biol.* **249:**545–554.

Juuti, J. T., and Bamford, D. H. (1997). Protein P7 of phage $\phi 6$ RNA polymerase complex, acquiring of RNA packaging activity by *in vitro* assembly of the purified protein onto deficient particles. *J. Mol. Biol.* **266:**891–900.

Juuti, J. T., Bamford, D. H., Tuma, R., and Thomas, G. J., Jr. (1998). Structure and NTPase activity of the RNA-translocating protein (P4) of bacteriophage $\phi 6$. *J. Mol. Biol.* **279:**347–359.

Kainov, D. E., Pirttimaa, M., Tuma, R., Thomas, G. J., Jr., Butcher, S.J., Bamford, D. H., and Makeyev, E. V. (2003a). RNA packaging device of double-stranded RNA bacteriophages, possibly as simple as hexamer of P4 protein. *J. Biol. Chem.* **278:**48084–48091.

Kainov, D. E., Butcher, S. J., Bamford, D. H., and Tuma, R. (2003b). Conserved intermediates on the assembly pathway of dsRNA bacteriophages. *J. Mol. Biol.* **328:**791–804.

Kakitani, H., Iba, H., and Okada, Y. (1980). Penetration and partial uncoating of bacteriophage $\phi 6$ particle. *Virology* **101:**475–483.

Koonin, E. V., Gorbalenya, A. E., and Chumakov, K. M. (1989). Tentative identification of RNA-dependent RNA polymerases of dsRNA viruses and their relationship to positive strand RNA viral polymerases. *FEBS Lett.* **252:**42–46.

Ktistakis, N. T., and Lang, D. (1987). The dodecahedral framework of the bacteriophage $\phi 6$ nucleocapsid is composed of protein P1. *J. Virol.* **61:**2621–2623.

Lake, J. A., and Leonard, K. R. (1974). Bacteriophage structure: Determination of head–tail symmetry mismatch for *Caulobacter crescentus* phage $\phi$CbK. *Science* **183:**744–747.

Laurinavicius, S., Kakela, R., Bamford, D. H., and Somerharju, P. (2004). The origin of phospholipids of the enveloped bacteriophage $\phi 6$. *Virology* **326:**182–190.

Lawton, J. A., Zeng, C. Q., Mukherjee, S. K., Cohen, J., Estes, M. K., and Prasad, B. V. (1997). Three-dimensional structural analysis of recombinant rotavirus-like particles with intact and amino-terminal-deleted VP2: Implications for the architecture of the VP2 capsid layer. *J. Virol.* **71:**7353–7360.

Lisal, J., Kainov, D. E., Bamford, D. H., Thomas, G. J., Jr., and Tuma, R. (2004). Enzymatic mechanism of RNA translocation in dsRNA bacteriophages. *J. Biol. Chem.* **279:**1343–1350.

Makeyev, E. V., and Bamford, D. H. (2000). Replicase activity of purified recombinant protein P2 of double-stranded RNA bacteriophage $\phi 6$. *EMBO J.* **19:**124–133.

Mancini, E. J., Kainov, D. E., Grimes, J. M., Tuma, R., Bamford, D. H., and Stuart, D. I. (2004). Atomic snapshots of an RNA packaging motor reveal comformational changes linking ATP hydrolysis to RNA translocation. *Cell* **118:**743–755.

McGraw, T., Mindich, L., and Frangione, B. (1986). Nucleotide sequence of the small double-stranded RNA segment of bacteriophage $\phi 6$: Novel mechanism of natural translational control. *J. Virol.* **58:**142–151.

Mindich, L. (1999). Precise packaging of the three genomic segments of the double-stranded RNA bacteriophage $\phi 6$. *Microbiol. Mol. Biol. Rev.* **63:**149–160.

Mindich, L., and Davidoff-Abelson, R. (1980). The characterization of a 120 S particle formed during $\phi 6$ infection. *Virology* **103:**386–391.

Mindich, L., and Lehman, J. (1979). Cell wall lysin as a component of the bacteriophage $\phi 6$ virion. *J. Virol.* **30:**489–496.

Mindich, L., and Lehman, J. (1983). Characterization of $\phi 6$ mutants that are temperature sensitive in the morphogenetic protein P12. *Virology* **127:**438–445.

Mindich, L., Lehman, J., and Huang, R. (1979). Temperature-dependent compositional changes in the envelope of $\phi 6$. *Virology* **97:**171–176.

Mindich, L., Nemhauser, I., Gottlieb, P., Romantschuk, M., Carton, J., Frucht, S., Strassman, J., Bamford, D. H., and Kalkkinen, N. (1988). Nucleotide sequence of the large double-stranded RNA segment of bacteriophage $\phi 6$: Genes specifying the viral replicase and transcriptase. *J. Virol.* **62:**1180–1185.

Mindich, L., Qiao, X., Onodera, S., Gottlieb, P., and Frilander, M. (1994). RNA structural requirements for stability and minus-strand synthesis in the dsRNA bacteriophage $\phi 6$. *Virology* **202:**258–263.

Mindich, L., Qiao, X., Onodera, S., Gottlieb, P., and Strassman, J. (1992). Heterologous recombination in the double-stranded RNA bacteriophage $\phi 6$. *J. Virol.* **66:**2605–2610.

Mindich, L., Qiao, X., Qiao, J., Onodera, S., Romantschuk, M., and Hoogstraten, D. (1999). Isolation of additional bacteriophages with genomes of segmented double-stranded RNA. *J. Bacteriol.* **181:**4505–4508.

Mindich, L., Sinclair, J. F., and Cohen, J. (1976). The morphogenesis of bacteriophage $\phi$6: Particles formed by nonsense mutants. *Virology* **75:**224–231.

Naitow, H., Tang, J., Canady, M., Wickner, R. B., and Johnson, J. E. (2002). L-A virus at 3.4 Å resolution reveals particle architecture and mRNA decapping mechanism. *Nat. Struct. Biol.* **9:**725–728.

Newcomb, W. W., Juhas, R. M., Thomsen, D. R., Homa, F. L., Burch, A. D., Weller, S. K., and Brown, J.C. (2001). The UL6 gene product forms the portal for entry of DNA into the herpes simplex virus capsid. *J. Virol.* **75:**10923–10932.

Ojala, P. M., Romantschuk, M., and Bamford, D. H. (1990). Purified $\phi$6 nucleocapsids are capable of productive infection of host cells with partially disrupted outer membranes. *Virology* **178:**364–372.

Olkkonen, V. M., and Bamford, D. H. (1987). The nucleocapsid of the lipid-containing double-stranded RNA bacteriophage $\phi$6 contains a protein skeleton consisting of a single polypeptide species. *J. Virol.* **61:**2362–2367.

Olkkonen, V. M., Gottlieb, P., Strassman, J., Qiao, X. Y., Bamford, D. H., and Mindich, L. (1990). *In vitro* assembly of infectious nucleocapsids of bacteriophage $\phi$6: Formation of a recombinant double-stranded RNA virus. *Proc. Natl. Acad. Sci. USA* **87:**9173–9177.

Olkkonen, V. M., Ojala, P. M., and Bamford, D. H. (1991). Generation of infectious nucleocapsids by *in vitro* assembly of the shell protein on to the polymerase complex of the dsRNA bacteriophage $\phi$6. *J. Mol. Biol.* **218:**569–581.

Paatero, A. O., Mindich, L., and Bamford, D. H. (1998). Mutational analysis of the role of nucleoside triphosphatase P4 in the assembly of the RNA polymerase complex of bacteriophage $\phi$6. *J. Virol.* **72:**10058–10065.

Paatero, A. O., Syväoja, J. E., and Bamford, D. H. (1995). Double-stranded RNA bacteriophage $\phi$6 protein P4 is an unspecific nucleoside triphosphatase activated by calcium ions. *J. Virol.* **69:**6729–6734.

Pagratis, N., and Revel, H. R. (1990). Detection of bacteriophage $\phi$6 minus-strand RNA and novel mRNA isoconformers synthesized *in vivo* and *in vitro*, by strand-separating agarose gels. *Virology* **177:**273–280.

Patton, J. T. (2001). Rotavirus RNA replication and gene expression. *Novartis Found. Symp.* **238:**64–77.

Patton, J. T., and Spencer, E. (2000). Genome replication and packaging of segmented double-stranded RNA viruses. *Virology* **277:**217–225.

Pirttimaa, M., Paatero, A. O., Frilander, M., and Bamford, D. H. (2002). Nonspecific NTPase P4 of dsRNA bacteriophage $\phi$6 polymerase complex is required for ssRNA packaging and transcription. *J. Virol.* **76:**10122–10127.

Pirttimaa, M. J., and Bamford, D. H. (2000). RNA secondary structure of the bacteriophage $\phi$6 packaging regions. *RNA* **6:**880–889.

Poranen, M. M., and Bamford, D. H. (1999). Packaging and replication regulation revealed by chimeric genome segments of double-stranded RNA bacteriophage $\phi$6. *RNA* **5:**446–454.

Poranen, M. M., Daugelavicius, R., Ojala, P. M., Hess, M. W., and Bamford, D. H. (1999). A novel virus–host cell membrane interaction: Membrane voltage-dependent endocytic-like entry of bacteriophage $\phi$6 nucleocapsid. *J. Cell Biol.* **147:**671–682.

Poranen, M. M., Paatero, A. O., Tuma, R., and Bamford, D. H. (2001). Self-assembly of a viral molecular machine from purified protein and RNA constituents. *Mol. Cell* **7:**845–854.

Poranen, M. M., Pirttimaa, M. J., and Bamford, D. H. (2005). Bacteriophage $\phi$6. *In* "Viral Genome Packaging" (C. Catalano, ed.). Landes Bioscience, Georgetown, TX.

Prasad, B. V., Crawford, S., Lawton, J. A., Pesavento, J., Hardy, M., and Estes, M. K. (2001). Structural studies on gastroenteritis viruses. *Novartis Found. Symp.* **238:**26–37.

Prasad, B. V., Rothnagel, R., Zeng, C. Q., Jakana, J., Lawton, J. A., Chiu, W., and Estes, M. K. (1996). Visualization of ordered genomic RNA and localization of transcriptional complexes in rotavirus. *Nature* **382:**471–473.

Prevelige, P. E., Jr., Thomas, D., and King, J. (1993). Nucleation and growth phases in the polymerization of coat protein and scaffolding subunits into icosahedral procapsid shells. *Biophys. J.* **64:**824–835.

Qiao, J., Qiao, X., Sun, Y., and Mindich, L. (2003a). Isolation and analysis of mutants of double-stranded-RNA bacteriophage $\phi$6 with altered packaging specificity. *J. Bacteriol.* **185:**4572–4577.

Qiao, X., Casini, G., Qiao, J., and Mindich, L. (1995a). *In vitro* packaging of individual genomic segments of bacteriophage $\phi$6 RNA: Serial dependence relationships. *J. Virol.* **69:**2926–2931.

Qiao, X., Qiao, J., and Mindich, L. (1995b). Interference with bacteriophage $\phi$6 genomic RNA packaging by hairpin structures. *J. Virol.* **69:**5502–5505.

Qiao, X., Qiao, J., and Mindich, L. (1997). Stoichiometric packaging of the three genomic segments of double-stranded RNA bacteriophage $\phi$6. *Proc. Natl. Acad. Sci. USA* **94:**4074–4079.

Qiao, X., Qiao, J., and Mindich, L. (2003b). Analysis of specific binding involved in genome packaging of the double-stranded-RNA bacteriophage $\phi$6. *J. Bacteriol.* **185:**6409–6414.

Qiao, X., Qiao, J., Onodera, S., and Mindich, L. (2000). Characterization of $\phi$13, a bacteriophage related to $\phi$6 and containing three dsRNA genomic segments. *Virology* **275:**218–224.

Reinisch, K. M., Nibert, M. L., and Harrison, S. C. (2000). Structure of the reovirus core at 3.6 Å resolution. *Nature* **404:**960–967.

Rimon, A., and Haselkorn, R. (1978). Transcription and replication of bacteriophage $\phi$6 RNA. *Virology* **89:**206–217.

Romantschuk, M., and Bamford, D. H. (1985). Function of pili in bacteriophage $\phi$6 penetration. *J. Gen. Virol.* **66:**2461–2469.

Romantschuk, M., Olkkonen, V. M., and Bamford, D. H. (1988). The nucleocapsid of bacteriophage $\phi$6 penetrates the host cytoplasmic membrane. *EMBO J.* **7:**1821–1829.

Schuck, P., Taraporewala, Z., McPhie, P., and Patton, J. T. (2001). Rotavirus nonstructural protein NSP2 self-assembles into octamers that undergo ligand-induced conformational changes. *J. Biol. Chem.* **276:**9679–9687.

Semancik, J. S., Vidaver, A. K., and Van Etten, J. L. (1973). Characterization of segmented double-helical RNA from bacteriophage $\phi$6. *J. Mol. Biol.* **78:**617–625.

Sinclair, J. F., Tzagoloff, A., Levine, D., and Mindich, L. (1975). Proteins of bacteriophage $\phi$6. *J. Virol.* **16:**685–695.

Stitt, B. L., and Mindich, L. (1983a). Morphogenesis of bacteriophage $\phi$6: A presumptive viral membrane precursor. *Virology* **127:**446–458.

Stitt, B. L., and Mindich, L. (1983b). The structure of bacteriophage $\phi$6: Protease digestion of $\phi$6 virions. *Virology* **127:**459–462.

Taraporewala, Z., Chen, D., and Patton, J. T. (1999). Multimers formed by the rotavirus nonstructural protein NSP2 bind to RNA and have nucleoside triphosphatase activity. *J. Virol.* **73:**9934–9943.

Taraporewala, Z. F., and Patton, J. T. (2001). Identification and characterization of the helix-destabilizing activity of rotavirus nonstructural protein NSP2. *J. Virol.* **75:**4519–4527.

Tsuprun, V., Anderson, D., and Egelman, E. H. (1994). The bacteriophage $\phi$29 head–tail connector shows thirteenfold symmetry in both hexagonally packed arrays and as single particles. *Biophys. J.* **66:**2139–2150.

Tuma, R., Bamford, J. K., Bamford, D. H., and Thomas, G. J., Jr. (1999). Assembly dynamics of the nucleocapsid shell subunit (P8) of bacteriophage $\phi$6. *Biochemistry* **38:**15025–15033.

Tuma, R., Tsuruta, H., Benevides, J. M., Prevelige, P. E., Jr., and Thomas, G. J., Jr. (2001). Characterization of subunit structural changes accompanying assembly of the bacteriophage P22 procapsid. *Biochemistry* **40:**665–674.

van Dijk, A. A., Frilander, M., and Bamford, D. H. (1995). Differentiation between minus- and plus-strand synthesis: Polymerase activity of dsRNA bacteriophage $\phi$6 in an *in vitro* packaging and replication system. *Virology* **211:**320–323.

Van Etten, J. V., Lane, L., Gonzalez, C., Partridge, J., and Vidaver, A. (1976). Comparative properties of bacteriophage $\phi$6 and $\phi$6 nucleocapsid. *J. Virol.* **18:**652–658.

Van Etten, J. L., Vidaver, A. K., Koski, R. K., and Semancik, J. S. (1973). RNA polymerase activity associated with bacteriophage $\phi$6. *J. Virol.* **12:**464–471.

Vidaver, A. K., Koski, R. K., and Van Etten, J. L. (1973). Bacteriophage $\phi$6: A lipid-containing virus of *Pseudomonas phaseolicola*. *J. Virol.* **11:**799–805.

Yang, H., Makeyev, E. V., and Bamford, D. H. (2001). Comparison of polymerase subunits from double-stranded RNA bacteriophages. *J. Virol.* **75:**11088–11095.

ADVANCES IN VIRUS RESEARCH, VOL 64

# STRUCTURE-DERIVED INSIGHTS INTO VIRUS ASSEMBLY

Vijay S. Reddy and John E. Johnson

Department of Molecular Biology, Scripps Research Institute, La Jolla, California 92037



## I. Introduction

Viruses are self-assembling macromolecular complexes made up of nucleoprotein components; in the case of enveloped viruses, the nucleoprotein complexes are surrounded by lipid bilayers. The genetic efficiency seen in viruses is manifested in their self-assembling nature, as multiple copies of a few proteins, making repeated equivalent interactions, assemble into symmetric closed shells encapsidating their own genome (Crick and Watson, 1956). The structural work that followed the observations by Crick and Watson revealed that spherical capsids indeed display icosahedral symmetry (Finch and Klug, 1959; Klug and Finch, 1960). Because the majority of viral capsids contain more than 60 subunits, Caspar and Klug, in their seminal work, proposed a method to organize a larger number of coat protein subunits into closed shells and characterized these capsids by the triangulation ($T$) number (Caspar and Klug, 1962). The assembly of simple nonenveloped viruses occurs rapidly and spontaneously with a high degree of fidelity. The latter implies that the "instructions" to form closed shells (capsids) are built into their components, mainly the coat protein subunits (Caspar, 1965). If this is true, then it should be possible to gain insights into virus assembly on the basis of the structure and organization of the protein subunits in those types of capsids, which do not require any auxiliary or scaffolding proteins for the assembly.

Many nonenveloped isometric viruses form a homogeneous population of virions, and hence are amenable to crystallization and




0065-3527/05 $35.00
DOI: 10.1016/S0065-3527(05)64003-1

subsequent high-resolution structural studies by X-ray crystallography. Since the first report of the virus structure of tomato bushy stunt virus at high resolution (Harrison *et al.*, 1978), a significant number (74) of distinct virus/capsid structures representing 21 different families have been determined (Table I). A Web site and a database [Virus Particle Explorer (VIPER), http://mmtsb.scripps.edu/viper/] were created as a repository for all high-resolution capsid structures and their structural analyses (Reddy *et al.*, 2001). The major driving force for such a database is to have all the virus/capsid structures oriented in one standard icosahedral convention that facilitates the subsequent rapid analysis of these structures. These structures have been analyzed in terms of protein–protein interactions, quasi-equivalence, and self-assembly pathways (Damodaran *et al.*, 2002; Reddy *et al.*, 1998). In this article we focus on the architecture and self-assembly of simple nonenveloped viruses whose structures have been determined at near atomic resolution.

## II. Architecture of Viral Capsids

The capsid morphology of nonenveloped viruses can be broadly classified into one of four types: spherical (icosahedral), rod shaped (helical), filamentous (flexuous), and bacilliform.

### A. *Icosahedral Viruses*

The simplest viruses that display icosahedral symmetry have 60 copies of a single coat protein subunit (single gene product) arranged in equivalent (identical) environments related by fivefold, threefold, and twofold axes of symmetry (e.g., satellite tobacco necrosis virus). This type of capsid is commonly referred to as a $T = 1$ capsid. The triangulation ($T$) number usually refers to the number of unique/distinct structural environments or simply the number of coat protein subunits present in an icosahedral asymmetric unit (1/60th of an icosahedron). According to the quasi-equivalence theory, $T = h^2 + hk + k^2$, where $h$ and $k$ are any integers (Caspar and Klug, 1962; Johnson and Speir, 1997). Caspar and Klug's theory describes how to arrange multiples of 60 subunits (e.g., 180, 240, etc.) on an icosahedral lattice with minor distortions in subunit conformation and intersubunit bonding. According to this theory, based on geometric principles, only capsids with certain $T$ numbers are possible to build, for example, $T = 1, 3, 4, 7, 9....$ A useful resource illustrating possible $T$ numbers

TABLE I

LIST OF VIRUSES AND THEIR SUBUNIT FOLDS DETERMINED AT HIGH RESOLUTION

| Virus family/name | *T* number | Subunit fold | Resolution (Å) | PDB ID |
|---|---|---|---|---|
| *Bromoviridae* | | $\beta$ Sandwich | | |
| Alfalfa mosaic virus | 1 | | 4.0 | N/A |
| Brome mosaic virus | 3 | | 3.4 | 1JS9 |
| Cowpea chlorotic mottle virus | 3 | | 3.2 | 1CWP |
| Cucumber mosaic virus | 3 | | 3.2 | 1FL5 |
| Tomato aspermy virus | 3 | | 3.4 | 1LAJ |
| *Caliciviridae* | | | | |
| Norwalk virus capsid | 3 | | 3.4 | 1IHM |
| *Comoviridae* | | $\beta$ Sandwich | | |
| Bean pod mottle virus | P3 | | 3.0 | 1BMV |
| Cowpea mosaic virus | P3 | | 2.8 | 1NY7 |
| Red clover mottle virus | P3 | | 2.4 | N/A |
| Tobacco ringspot virus | P3 | | 3.5 | 1A6C |
| *Dicistroviridae* | | $\beta$ Sandwich | | |
| Cricket paralysis virus | P3 | | 2.4 | 1B35 |
| *Hepadnaviridae* | | Helical | | |
| Hepatitis B virus | 4 | | 3.6 | 1QGT |
| *Leviviridae* | | $\alpha + \beta$ | | |
| Bacteriophage FR | 3 | | 3.5 | 1FRS |
| Bacteriophage GA | 3 | | 3.4 | 1GAV |
| Bacteriophage MS2 | 3 | | 2.8 | 2MS2 |
| Bacteriophage PP7 | 3 | | 3.5 | 1DWN |
| Bacteriophage Q$\beta$ | 3 | | 3.5 | 1QBE |
| *Microviridae* | | $\beta$ Sandwich | | |
| Bacteriophage G4 | 1 | | 3.0 | 1GFF |
| Bacteriophage $\phi \times 174$ | 1 | | 3.5 | 2BPA |
| Bacteriophage $\phi \times 174$ + scaffold | 1 | | 3.5 | 1AL0 |
| *Necrovirus* | | | | |
| Tobacco necrosis virus | 3 | $\beta$ Sandwich | 2.25 | 1C8N |
| *Nodaviridae* | | $\beta$ Sandwich + $\alpha$ domain | | |
| Black beetle virus | 3 | | 2.8 | 2BBV |
| Flock house virus | 3 | | 3.0 | N/A |
| Nodamura virus | 3 | | 3.3 | 1NOV |
| Pariacoto virus | 3 | | 3.0 | 1F8V |

*(continues)*

TABLE I (*continued*)

| Virus family/name | *T* number | Subunit fold | Resolution (Å) | PDB ID |
|---|---|---|---|---|
| *Papillomaviridae* | | β Sandwich | | |
| Human papillomavirus 16 L1 capsid | 1 | | 3.5 | 1DZL |
| *Parvoviridae* | | β Sandwich | | |
| Adeno-associated virus | 1 | | 3.0 | 1LP3 |
| Canine parvovirus (CPV) empty | 1 | | 3.3 | 2CAS |
| Densovirus | 1 | | 3.6 | 1DNV |
| Feline panleukopenia virus | 1 | | 3.0 | 1C8E |
| Murine minute virus | 1 | | 3.5 | 1MVM |
| Porcine parvovirus capsid | 1 | | 3.5 | 1K3V |
| *Picornaviridae* | | β Sandwich | | |
| Bovine enterovirus | P3 | | 3.0 | 1BEV |
| Coxsackievirus B3 | P3 | | 3.0 | 1COV |
| Cricket paralysis virus 1 | P3 | | 2.4 | 1B35 |
| Echovirus 1 | P3 | | 3.55 | 1EV1 |
| Echovirus 11 | P3 | | 2.9 | 1H8T |
| Foot-and-mouth disease virus | P3 | | 3.0 | 1BBT |
| HRV16 | P3 | | 2.15 | 1AYM |
| HRV1A | P3 | | 3.0 | 1R1A |
| HRV serotype 2 | P3 | | 2.6 | 1FPN |
| HRV serotype 3 | P3 | | 3.0 | 1RHI |
| HRV serotype 14 | P3 | | 3.0 | 4RHV |
| HRV14 chimera | P3 | | 2.7 | 4RHV |
| Mengo encephalomyocarditis virus | P3 | | 3.0 | 2MEV |
| Poliovirus type 1, Mahoney strain, crystal structure | P3 | | 2.2 | 1HXS |
| Poliovirus type 1, Mahoney strain | P3 | | 2.9 | 2PLV |
| Poliovirus type 1, empty capsid | P3 | | 2.88 | 1POV |
| Poliovirus type 1 at −170°C | P3 | | 2.9 | 1ASJ |
| Poliovirus type 2 Lansing | P3 | | 2.9 | 1EAH |
| Poliovirus type 3 | P3 | | 2.4 | 1PVC |

(*continues*)

TABLE I (*continued*)

| Virus family/name | *T* number | Subunit fold | Resolution (Å) | PDB ID |
|---|---|---|---|---|
| Swine vesicular disease virus | P3 | | 3.0 | 1OOP |
| Theiler MEV DA | P3 | | 2.8 | 1TME |
| Theiler MEV BeAn | P3 | | 3.5 | 1TMF |
| *Polyomaviridae* | | $\beta$ Sandwich | | |
| Murine polyomavirus | 7d | | 3.7 | 1SID |
| Simian virus 40 (SV40) | 7d | | 3.1 | 1SVA |
| *Reoviridae* | | $\beta$ Sandwich and $\alpha$ domain | | |
| Bluetongue virus | 13 | | 3.5 | 2BTV |
| Reovirus core | 13 | | 3.6 | 1EJ6 |
| Rice dwarf virus | 13 | | 3.5 | 1UF2 |
| *Satellites* | | $\beta$ Sandwich | | |
| Satellite panicum mosaic virus | 1 | | 1.9 | 1STM |
| Satellite tobacco mosaic virus | 1 | | 1.8 | 1A34 |
| Satellite tobacco necrosis virus | 1 | | 2.5 | 2STV |
| *Siphoviridae* | | $\alpha + \beta$ | | |
| Bacteriophage HK97 | 7l | | 3.6 | 1FH6 |
| *Sobemovirus* | | $\beta$ Sandwich | | |
| Cocksfoot mottle virus | 3 | | 2.7 | 1NG0 |
| Rice yellow mottle virus | 3 | | 3.0 | 1F2N |
| Sesbania mosaic virus | 3 | | 2.9 | 1SMV |
| Southern bean mosaic virus | 3 | | 2.8 | 4SBV |
| *Tetraviridae* | | $\beta$ Sandwich | | |
| Nudaurelia capensis $\beta$ virus | 4 | | 2.8 | 1OHF |
| *Tombusviridae* | | $\beta$ Sandwich | | |
| Carnation mottle virus | 3 | | 3.2 | 1OPO |
| Tomato bushy stunt virus | 3 | | 2.9 | 2TBV |
| *Totiviridae* | | $\alpha + \beta$ | | |
| L-A virus | 1 | | 3.6 | 1M1C |
| *Tymoviridae* | | $\beta$ Sandwich | | |
| Desmodium yellow mottle virus | 3 | | 2.7 | 1DDL |
| Physalis mottle virus | 3 | | 3.8 | 1QJZ |
| Turnip yellow mosaic virus | 3 | | 3.2 | 1AUY |

*Abbreviation*: N/A, not available.

and the associated icosahedral lattices can be found at the icosahedral server (http://mmtsb.scripps.edu/viper/icos_server.php) developed by C. Qu. There are, however, exceptions to the allowed $T$ numbers, as envisioned by Caspar and Klug (1962). The inner capsids of bluetongue, reovirus, and rice dwarf virus cores (Grimes *et al.*, 1998; Nakagawa *et al.*, 2003; Reinisch *et al.*, 2000), and the protein coat of L-A virus (Naitow *et al.*, 2002) have been shown to contain only 120 copies of a single subunit, resulting in "$T = 2$" capsids. Polyoma viral capsids, on the other hand, contain 360 subunits arranged on a $T = 7d$ lattice, instead of 420 subunits required for the formation of a conventional $T = 7$ capsid as envisioned by Caspar and Klug. In the polyoma-like viruses, the 60 hexavalent positions that in principle should be occupied by the subunit hexamers are occupied by the pentamers of subunits, resulting in a pentamers-only capsid (Liddington *et al.*, 1991; Stehle and Harrison, 1996; Stehle *et al.*, 1996). Figure 1A shows schematic diagrams of $T = 1$, 3, and 4 capsids. Whereas all the subunits in $T = 1$ capsids assume a single environment (A), they occupy three (A, B, and C) and four (A, B, C, and D) distinct environments in $T = 3$ and $T = 4$ capsids, respectively. Similarly, the subunits in $T = 7$ and $T = 13$ capsids occupy 7 and 13 distinct environments with respect to icosahedral symmetry axes. Table I shows the list of viruses, whose structures have been determined at high resolution by X-ray crystallography, compiled from the VIPER Web site (Reddy *et al.*, 2001). There are currently 74 unique capsid structures from 21 different virus families available (Table I). Different icosahedral capsids that have been characterized structurally to date at high resolution display capsid architectures with the $T$ numbers 1, 3, 4, 7, and 13. However, even within a category of capsids identified by the same $T$ number (e.g., $T = 3$), there are significant variations in subunit associations in forming the respective capsids (Damodaran *et al.*, 2002). The capsid protein subunits of the majority of capsids from animal, bacterial, and plant viruses have the canonical "jelly roll" $\beta$-sandwich fold (Fig. 1B) (Rossmann and Johnson, 1989), whereas the remaining subunits display a combination of $\beta$-sheet and/or helical motifs (Table I). Figure 2 shows the non-$\beta$-barrel subunit folds found in a number of viral capsids.

### *B. Rod-Shaped Viruses*

The most thoroughly studied rod-shaped virus is tobacco mosaic virus (TMV) a plant virus that belongs to the tobamovirus family. The virion of TMV is a rigid rod, consisting of ~2130 identical coat

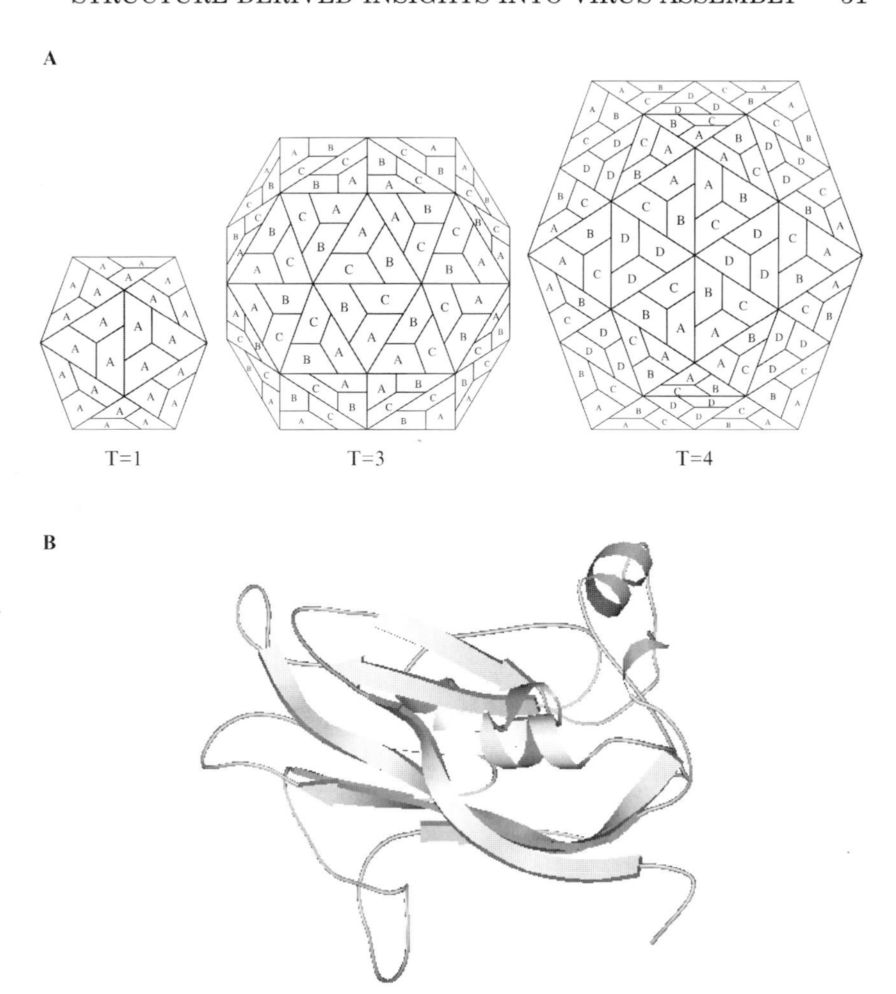


FIG 1. (A) Schematic representation of $T = 1$, $T = 3$, and $T = 4$ icosahedral lattices. Each trapezoid corresponds to a subunit. The letters A, B, C, and D identify the distinct subunit environments. Whereas all the subunits occupy an equivalent environment, namely "A" in $T = 1$ capsids, they occupy three (letters A, B, and C) and four (letters A, B, C, and D) distinct environments in $T = 3$ and $T = 4$ capsids, respectively. (B) A ribbon diagram showing the tertiary structure of the capsid protein subunit of Southern bean mosaic virus (SBMV) as an illustration of the canonical $\beta$-barrel motif.

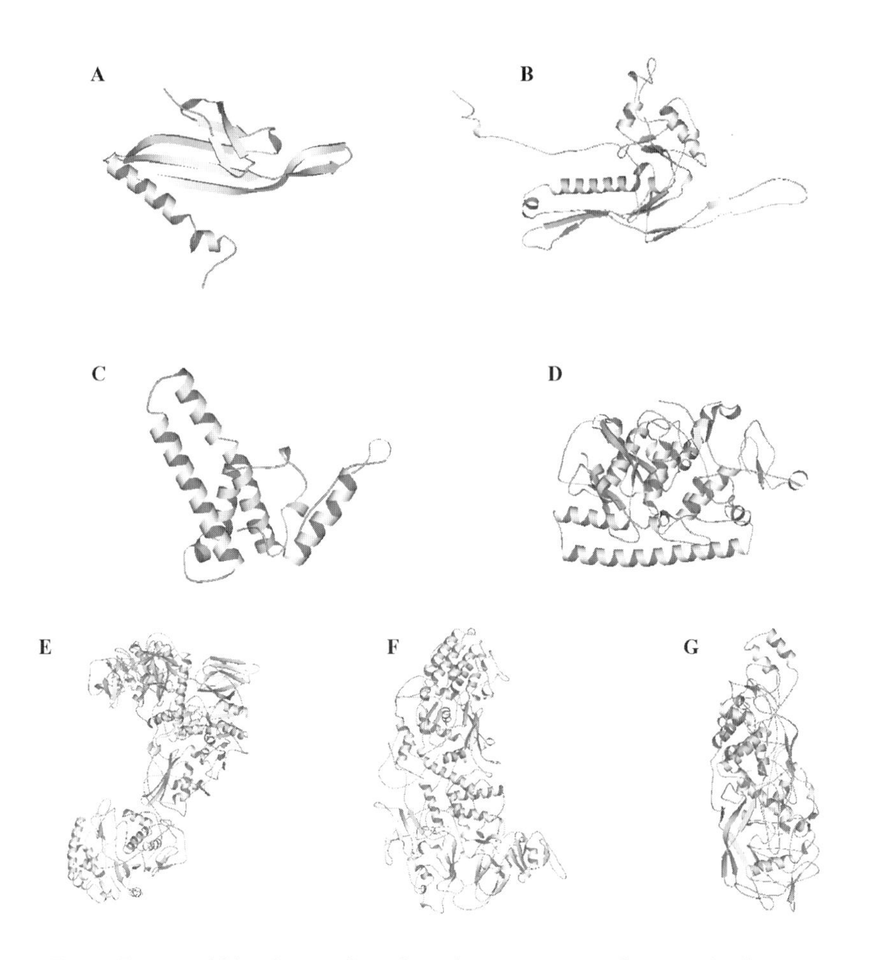


FIG 2. Tertiary folds of non-$\beta$-barrel viral coat protein subunits. (A) Coat protein subunit of MS2; (B) processed subunit of HK97 head II; (C) hepatitis B subunit; (D–F) $\sigma 3$, $\lambda 2$, and $\lambda 1$ subunits of reovirus cores, respectively; (G) subunit of L-A virus.

protein subunits of molecular mass 17.5 kDa assembled in a right-handed helix around a single strand of RNA (Butler, 1976, 1999; Pattanayek and Stubbs, 1992; Stubbs, 1999). The pitch (raise per turn) of the helix is 2.3 nm with 16.3 to 17.3 subunits per turn. The sizes of TMV virions range from 70 to 300 nm in length and 18 nm in diameter. The structure of the complete virus was determined at 2.9-Å resolution by X-ray fiber diffraction methods (Namba *et al.*, 1989; Namba and Stubbs, 1986). An illustration of the TMV virion is shown in Fig. 3.

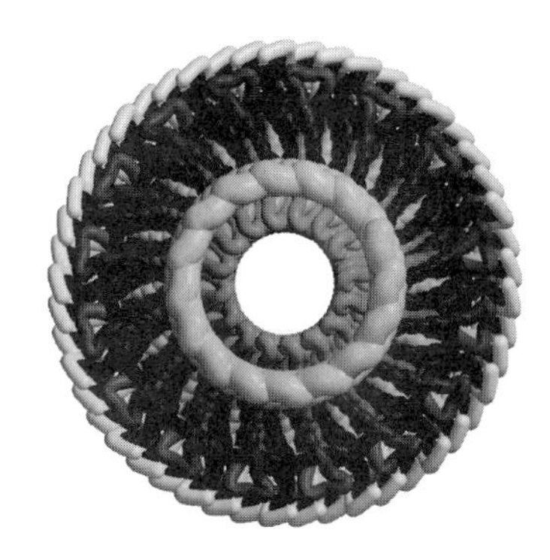

FIG 3. Illustration of TMV based on the structure, determined by Pattanayek and Stubbs (1992). *Left*: A view perpendicular to the helix axis. *Right*: A view down the helix axis. The TMV subunits are shown in a tube representation with a color gradient going from black to light gray. The spiral in the middle, shown as a thick gray tube, corresponds to ssRNA.

### *C. Bacilliform Viruses*

Alfalfa mosaic virus (AMV) is structurally well characterized among the viruses that form bacilliform particles (Cremers *et al.*, 1981; Hull *et al.*, 1969; Kumar *et al.*, 1997; Mellema, 1975). On the basis of electron microscopic analysis, a model for the organization of the AMV subunits into bacilliform particles has been proposed (Mellema, 1975). According to this model, the tubular part of the virion displays cylindrical *P*6 symmetry with its ends capped with two half-icosahedral shells with their threefold axis coincident with the cylindrical axis of the tube. Figure 4 shows such a model generated on the basis of the AMV coat protein subunit structure (Kumar, 1997).

## III. STRUCTURE-DERIVED ASSEMBLY PATHWAYS

### *A. Protein–Protein Interactions*

Viral capsids provide an excellent resource for studying protein–protein interactions in homo-oligomers that form symmetric closed shells. Association and self-assembly of the capsid protein subunits are common to all viruses. Analysis of protein–protein interactions in viral capsids may reveal the principles of self-assembly and the nature of interactions that lead to capsid formation. Interactions between the coat protein subunits determine the size and the robustness of the

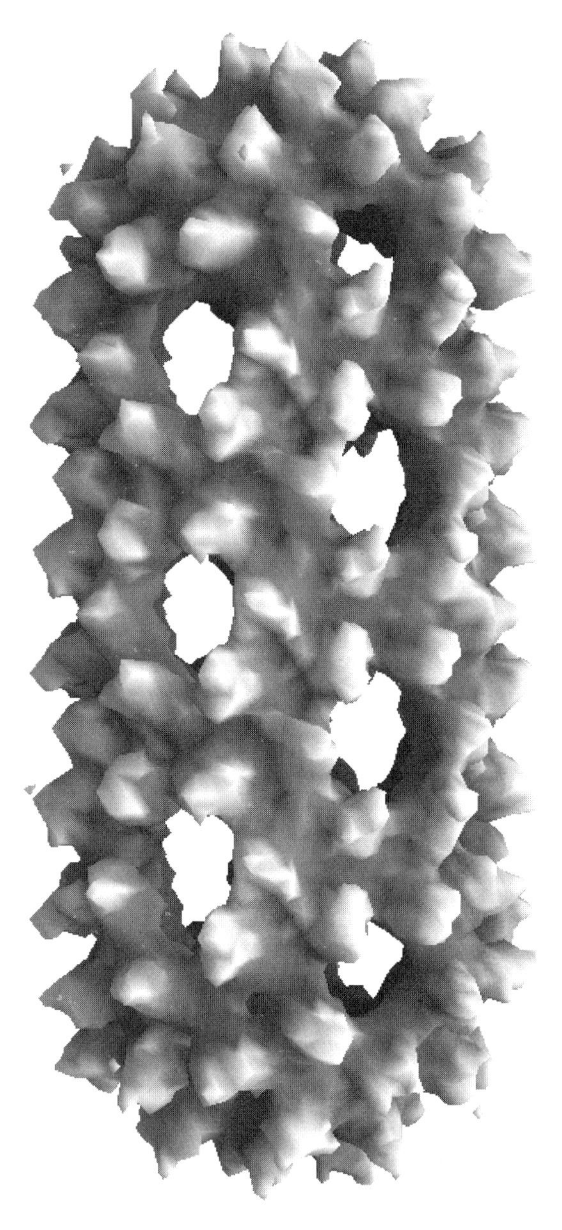

FIG 4. Surface representation of the bacilliform particle of AMV generated on the basis of the structure of the AMV coat protein subunit (Kumar *et al.*, 1997). The shaft of the bacilliform particle obeys cylindrical $P6$ symmetry, where the ends are capped with half-icosahedral $T = 1$ particles. The pitch of the tube is 10 nm with 3.45 hexameric units per turn.

capsid. Other elements important in capsid formation are protein–nucleic acid interactions, interactions between the coat protein and auxiliary/scaffolding proteins, and the particle maturation or postprocessing of the capsid protein subunits. Table II lists a representative group of viruses and the total accessible surface area buried on capsid formation as a measure of the extent of protein–protein interactions. Figure 5 shows the mostly linear increase in buried surface area (protein–protein interactions) as a function of capsid size. The picornaviral capsids display stronger protein–protein interactions whereas the "$T = 2$" capsids have weaker interactions (Table II). The extent of these protein–protein interactions at the subunit interfaces within a capsid can be used to delineate the most likely pathway of virus assembly (Horton and Lewis, 1992; Reddy *et al.*, 1998). Stronger protein–protein interactions between the subunits may also point to the coupling of protein folding and assembly (Umashankar *et al.*, 2003).

### *B. Assembly Pathways*

Coat protein subunits carry instructions to form a particular type of capsid with high fidelity (Caspar, 1965; Prevelige, 1998). The structure and the organization of the subunits obtained from the complete capsid structures can be used to decipher these "instructions" (Reddy *et al.*, 1998). This approach identifies the repeating units (building blocks) of the assembly and any potential intermediates that may be formed during the course of capsid formation. A substructure of $n$ subunits is considered a likely intermediate or a preferred substructure during the course of assembly, when the association/binding energy of formation of the substructure is favored significantly over any other combinations of $n$ subunit associations (Reddy *et al.*, 1998). In monitoring the association of a full complement of subunits (e.g., 60, 180, etc.) that form a closed shell, identifying the preferred substructures along the way may suggest the most likely pathway of capsid formation. The stronger the protein–protein interactions, the greater the chance of identifying the assembly intermediates with certainty.

Table IIIA shows the unique subunit–subunit interfaces in human rhinovirus 16 (PDB ID: 1AYM) and the associated buried surface areas and association energies taken from the VIPER Web site (Reddy *et al.*, 2001). The association energies were calculated by a simple approach (Eisenberg and McLachlan, 1986) that uses the buried surface area between the reference and symmetry-related subunits (Table IIIA) multiplied by the atomic solvation parameters. Table IIIB shows the association energies between the protomers, which consist of Vp1, Vp0

TABLE II

EXTENT OF BURIED SURFACE AREAS AND SIZES OF A REPRESENTATIVE LIST OF VIRUSES

| S. No. | Virus name | Family | Genus | Average Diameter (Å) | Total BSA[a] (Å) | PDB-ID |
|---|---|---|---|---|---|---|
| 1 | Satellite tobacco mosaic virus | Satellites | Satellite virus | 168 | $4.82 \times 10^5$ | 1A34 |
| 2 | L-A virus | *Totiviridae* | *Totivirus* | 416 | $6.26 \times 10^5$ | 1M1C |
| 3 | Densovirus | *Parvoviridae* | *Densovirus* | 252 | $6.66 \times 10^5$ | 1DNV |
| 4 | Cucumber mosaic virus | *Bromoviridae* | *Cucumovirus* | 294 | $6.86 \times 10^5$ | 1F15 |
| 5 | Hepatitis B virus | *Hepadnaviridae* | *Orthohepadnavirus* | 334 | $8.76 \times 10^5$ | 1QGT |
| 6 | Cowpea chlorotic mottle virus | *Bromoviridae* | *Bromovirus* | 278 | $8.80 \times 10^5$ | 1CWP |
| 7 | Turnip yellow mosaic virus | *Tymoviridae* | *Tymovirus* | 292 | $8.97 \times 10^5$ | 1AUY |
| 8 | Sesbania mosaic virus | | *Sobemovirus* | 292 | $9.06 \times 10^5$ | 1SMV |
| 9 | Bacteriophage MS2 | *Leviviridae* | *Levivirus* | 268 | $9.99 \times 10^5$ | 2MS2 |
| 10 | Adeno-associated virus | *Parvoviridae* | *Dependovirus* | 258 | $1.02 \times 10^6$ | 1LP3 |
| 11 | Feline panleukopenia virus | *Parvoviridae* | *Parvovirus* | 270 | $1.02 \times 10^6$ | 1C8E |
| 12 | Cowpea mosaic virus | *Comoviridae* | *Comovirus* | 278 | $1.09 \times 10^6$ | 1NY7 |
| 13 | Tomato bushy stunt virus | *Tombusviridae* | *Tombusvirus* | 336 | $1.17 \times 10^6$ | 2TBV |
| 14 | Flock house virus | *Nodaviridae* | *Alphanodavirus* | 326 | $1.25 \times 10^6$ | 1FHV |
| 15 | Norwalk virus capsid | *Caliciviridae* | *Norovirus* | 396 | $1.47 \times 10^6$ | 1IHM |
| 16 | Bacteriophage $\phi \times 174$ | *Microviridae* | *Microvirus* | 320 | $1.47 \times 10^6$ | 2BPA |
| 17 | Pariacoto virus | *Nodaviridae* | *Alphanodavirus* | 320 | $1.81 \times 10^6$ | 1F8V |
| 18 | Foot-and-mouth disease virus | *Picornaviridae* | *Aphthovirus* | 296 | $2.08 \times 10^6$ | 1BBT |
| 19 | Human rhinovirus 16 | *Picornaviridae* | *Rhinovirus* | 306 | $2.22 \times 10^6$ | 1AYM |
| 20 | Poliovirus type 1 | *Picornaviridae* | *Enterovirus* | 304 | $2.64 \times 10^6$ | 1HXS |
| 21 | Nudaurelia capensis $\beta$ virus | *Tetraviridae* | *Omegatetravirus* | 416 | $3.14 \times 10^6$ | 1NWV |
| 22 | Bacteriophage HK97 | *Siphoviridae* | $\lambda$-like | 598 | $3.87 \times 10^6$ | 1FH6 |
| 23 | Bluetongue virus, $T = 13$ core | *Reoviridae* | *Orbivirus* | 692 | $4.23 \times 10^6$ | 2BTV |
| 24 | Simian virus 40 (SV40) | *Polyomaviridae* | *Polyomavirus* | 488 | $5.08 \times 10^6$ | 1SVA |

*Abbreviations:* BSA, buried surface area.

[a] Sorted by increasing order of BSA.

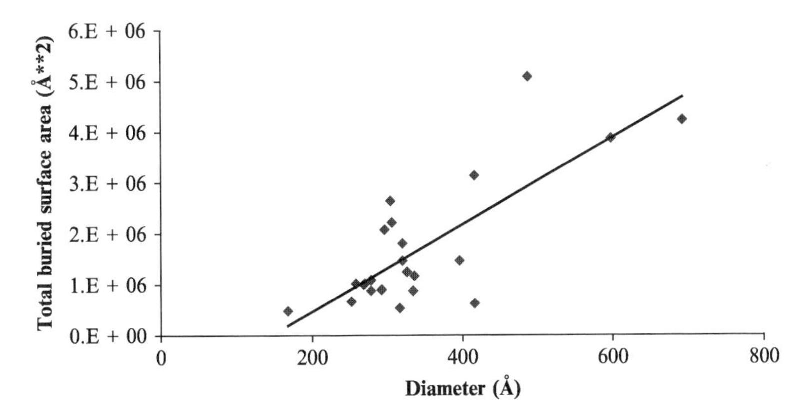


FIG 5. Plot showing the linear increase in total buried surface area of the constituting subunits in the respective capsids shown in Table II as a function of the average diameter of the capsid.

(Vp2 + Vp4), and Vp3 subunits together as a single structural unit, related by the icosahedral symmetry. These energies were used as the basis set in the combinatorial assembly method for associating subunits in the context of known quaternary associations that display the local and icosahedral symmetry (Reddy *et al.*, 1998).

Briefly, for each subunit or building block (a group of subunits) in the capsid, a second subunit/building block is added in all possible ways allowed in the capsid, that is, the known quaternary association of subunits found within the capsid. For each of the two-subunit (dimeric) associations enumerated, the corresponding association energies were obtained from the basis set as a table look-up procedure. The association energies of higher (>2) oligomeric structures were calculated as the combinatorial addition of the association energies for all the subunit pairs that occur in the substructure. Once all the possible structures were enumerated, they were sorted according to increasing energy (decreasing stability). After this step, the most stable structures are at the top and the least likely (unstable) structures are at the bottom. If the absolute difference in the energy ($\Delta\Delta G_{12}$) between the best (first) structure and second best structure is significant, then the first structure is considered the preferred structure among the $n$ subunit assemblies and a potential intermediate. At each stage up to 9000 configurations of $n$ subunits were enumerated and selected as the seeds for enumerating assemblies containing $n + 1$ subunits, of which thermodynamically preferred structure(s) were identified. This procedure was repeated until a full complement of subunits in

TABLE III

Human Rhinovirus 16 (1AYM) Subunit Interfaces and Contacting Protomers, Association Energies, and Buried Surface Areas

| Subunit interface[a] | Association energy (kcal/mol) | Buried surface area (Å$^2$) |
|---|---|---|
| 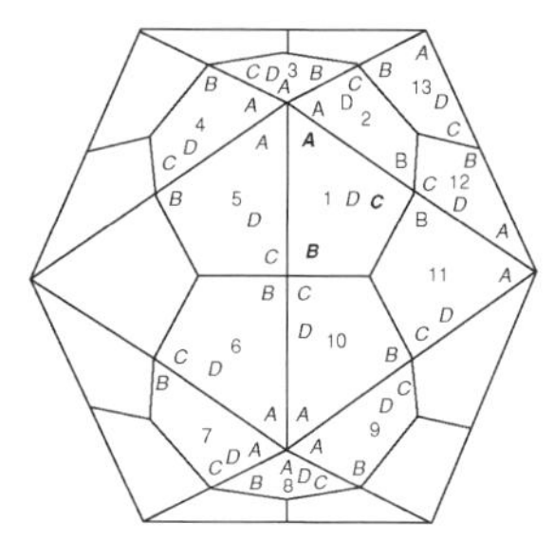 | | |
| **A. Unique Protein–Protein Interfaces and Corresponding Binding Energies and Buried Surface Areas** | | |
| A1_A2 | −36.0 | 2217 |
| A1_B1 | −180.0 | 8975 |
| A1_B2 | −54.0 | 2960 |
| A2_B1 | −12.0 | 896 |
| A1_C1 | −110.0 | 5714 |
| A2_C1 | − 25.0 | 1457 |
| A1_C10 | −20.0 | 1319 |
| B1_B2 | −35.0 | 1975 |
| B1_B3 | −3.0 | 432 |
| B1_C1 | −81.0 | 4310 |
| B2_C1 | −29.0 | 1702 |
| B1_B10 | −50.0 | 2949 |
| C1_C2 | −5.0 | 582 |
| C1_C12 | −24.0 | 1540 |
| **B. Association Energies and Buried Surface Areas of the Various Contacting Protomers (Units)** | | |
| Unit1_Unit2 | −195.0 | 9976 |
| Unit1_Unit3 | −12.0 | 868 |
| Unit1_Unit10 | −72.0 | 4085 |
| Unit1_Unit12 | −24.0 | 1540 |

[a] *Key:* A, Vp1; B, Vp3; C, Vp2; D, Vp4. Numbers 1–13 identify different icosahedral asymmetric units.

the respective capsids (e.g., 60, 180, etc.) was assembled. Plotting the $|\Delta\Delta G_{12}|$ as a function of number of associating subunits highlights the most preferred structures during capsid assembly (Figs. 6 and 7).

*1. Picornaviridae*

The picornaviruses form $T = 1$ capsids made up of 60 copies of a protomer containing VP1, VP2, VP3, and VP4 protein chains. Picornavirus particles have an average diameter of 300 Å. The three-dimensional structures of human rhinovirus 16 and foot-and-mouth disease virus were determined at high resolution (Hadfield *et al.*, 1997; Fry *et al.*, 1993). Figure 6A shows the calculated capsid assembly profiles for human rhinovirus 16 (HRV16; PDB ID: 1AYM), and foot-and-mouth disease virus (FMDV; PDB ID: 1BBT), which belong to the genera *Rhinovirus* and *Aphthovirus*, respectively. In each case, the subunit protomer was used as the building block in the assembly calculations. The pentamer of protomers stands out as the first significant intermediate ( $|\Delta\Delta G_{12}|^{HRV} = 270$ kcal; $|\Delta\Delta G_{12}|^{FMDV} = 212$ kcal). The similarity between the two assembly profiles is quite high, as the calculated statistical similarity indicators, that is, the correlation coefficient (0.999) and residual (0.007) of energies for the preferred structures at each stage, imply that the pathways of assembly are nearly identical for these capsids. The previously described results strongly suggest that the capsid assembly of picornaviruses, in general, proceeds by the addition of pentamers of protomers where the pentameric blocks form the assembly intermediates in agreement with experimental results (Rueckert, 1985).

*2. Bromoviridae*

Structures of several viruses from the *Bromovirus* and *Cucumovirus* genera of the *Bromoviridae* family have been determined at near atomic resolution (Lucas *et al.*, 2002a,b; Speir *et al.*, 1995; Smith *et al.*, 2000). These viruses form $T = 3$ capsids that consist of 180 subunits with the canonical $\beta$-barrel fold. The calculated protein–protein interactions between the protein subunits of bromoviruses (e. g., cowpea chlorotic mottle virus [CCMV] and brome mosaic virus [BMV]) suggest that the twofold-related coat protein dimers ($C–C_2$; $A–B_5$) form stable oligomers in agreement with previous observations (Zhao *et al.*, 1995). The derived assembly pathways for bromoviruses suggest that the first preferred structure is a trimer of dimers (6-mer), followed by a pentamer of dimers (10-mer) and a decamer of dimers (20-mer) surrounding the fivefold axes of symmetry (Fig. 6B). Although, the results are in agreement with experimental data (Zlotnick

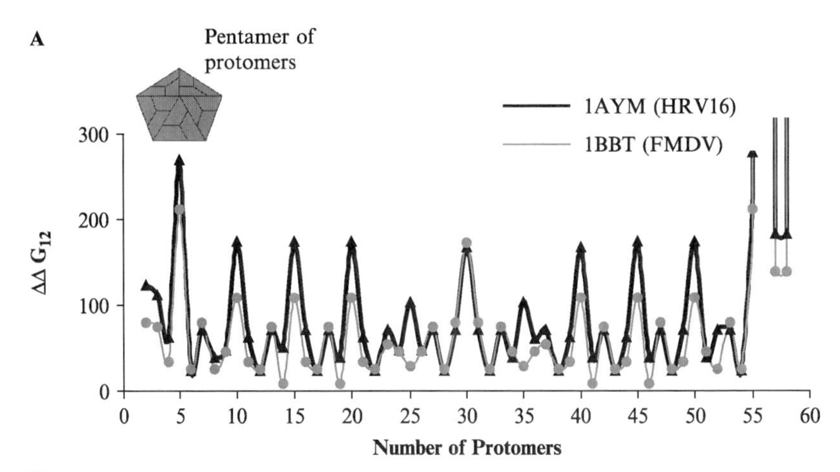


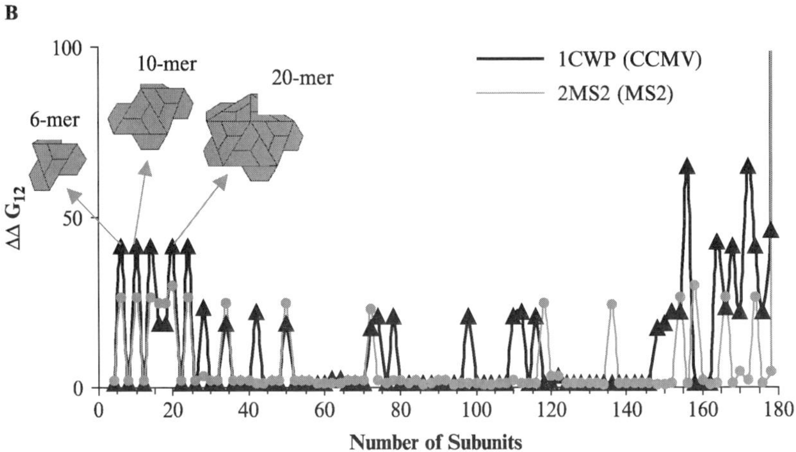


FIG 6. Assembly profiles: absolute difference in $\Delta\Delta G$ of the top two preferred configurations as a function of number of associating chains. Peaks correspond to stable structural oligomers or intermediates in capsid assembly. (A) Assembly profiles of human rhinovirus 16 (PDB ID: 1AYM), shown as a thick black line, and of foot-and-mouth disease virus (PDB ID: 1BBT), shown as a thin gray line. The preferred intermediate, pentamer of protomers is shown as the schematic of the pentameric vertex. (B) Assembly profiles of CCMV (PDB ID: 1CWP) and bacteriophage MS2 (PDB ID: 2MS2) are shown as thick black and thin gray lines, respectively. In both cases the assembly profiles were calculated by the addition of subunit dimers as the building blocks.

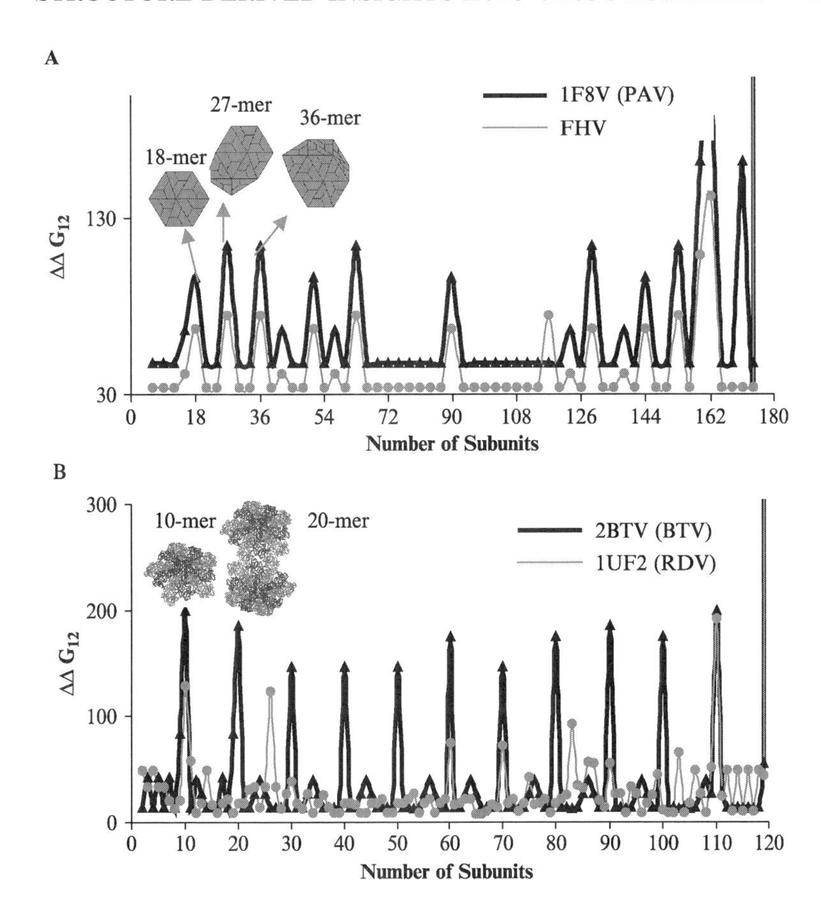


FIG 7. Assembly profiles of nodaviruses and inner cores of reoviruses. (A) Assembly profiles of pariacoto virus and flock house virus are shown as thick black and thin gray lines, respectively. The hexamer of subunit trimers (18-mer) appears to be the initial intermediate. (B) Assembly profiles of bluetongue and rice dwarf virus inner cores shown as thick black and thin gray lines, respectively. The decamer of subunits is the major intermediate and association of these decamers leads to complete capsid assembly.

*et al.*, 2000) that the 10-mer is a potential intermediate, the preferred decamer in our analysis is different from that reported by Zlotnick and colleagues. The reported 10-mer structure with the five A–$B_5$ dimers surrounding the fivefold axes of symmetry is at least 40 kcal less stable than the decamer structure made up of three A–$B_5$ dimers tightly cojoined with two C–$C_2$ dimers identified by our analysis (Fig. 6B). Furthermore, the first preferred structure, a 6-mer, made up of two A–$B_5$ dimers and one C–$C_2$ dimer, is the repeating unit that forms the hexagonal tube of the bacilliform capsids formed by alfalfa mosaic virus, another member of *Bromoviridae* family, which also preferentially forms subunit dimers in solution (Kumar, 1997). Thus our previously described results suggest that the assembly of bromoviruses occurs by forming the 6-mer, a trimer of dimers, as the nucleating structure followed by addition of dimers to complete a pentameric vertex representing a decamer of dimers (20-mer) (Fig. 6B). Further addition of dimers to the latter structure leads to formation of the complete capsid.

Remarkably, cucumber mosaic virus (CMV) of the genus *Cucumovirus* appears to follow an entirely different assembly pathway. Cucumoviruses preferentially form hexamers and pentamers of the coat protein subunits instead of dimers as seen in bromoviruses (data not shown). Particle assembly seems to proceed by the addition of hexamers and pentamers, as in the geometric description of building capsids put forward by Caspar and Klug (1962). The structure-derived assembly pathway suggests that the first preferred oligomeric structure in cucumovirus assembly is a hexamer of subunits located at the quasi-sixfold axes (icosahedral threefold axes). Although the pentamer of subunits is not the most stable structure among five-subunit associations, it could be in equilibrium with the hexamers or precursors of hexamers. Cucumovirus assembly occurs as the addition of hexamers and pentamers instead of as dimers as found in the case of bromoviruses.

### 3. *Hepatitis B Virus*

Hepatitis B virus (HBV) forms a mixture of $T = 3$ and $T = 4$ capsids (Kenney *et al.*, 1995; Stannard and Hodgkiss, 1979). The structure of hepatitis B virus $T = 4$ capsid was determined at high resolution (Wynne *et al.*, 1999). The subunit structure of HBV is unique, being entirely made up of helices, unlike other viral capsid proteins (Fig. 2C). The capsid protein subunits preferentially form dimers, much like bromoviruses. Interestingly, the structure-derived assembly pathway is also similar to that of the bromoviruses, initially forming the preferred structures surrounding the fivefold axes of symmetry. However, in forming the $T = 4$ capsids, there are two 6-mer (trimer of dimer) structures that have comparable free energies ($\Delta\Delta G_{12} = 9$ kcal). One is formed with two A–$B_5$ dimers and a C–D dimer, located at the quasi-threefold axes of symmetry, whereas the other is formed with three CD dimers positioned at the icosahedral threefold axes. These two 6-mer structures are probably in equilibrium and/or indistinguishable. However, further addition of dimers to these precursor structures significantly favors the stable 20-mer structure around the fivefold axes as in bromoviruses.

As expected, in the $T = 3$ capsids of HBV, there is only one type of 6-mer (trimer of dimer), located at the axes of quasi-threefold symmetry. This quasi-threefold-related 6-mer stands out as the preferred structure or as an intermediate, in agreement with the results derived by Zlotnick and colleagues (1999). Addition of dimers to these precursor structures leads to a stable 20-mer structure around the fivefold axis and further leads to capsid assembly, quite analogous to that of bromoviruses. Interestingly, in both $T = 3$ and $T = 4$ capsid assembly, the assembly seems to occur preferentially through a pentameric vertex. The difference may arise because of an incorporation of 6-mers between these 20-mer structures, either through chance or weak preference, thus forming a mixture of both types of capsid during the assembly.

#### 4. *Leviviridae*

Leviviruses are small RNA phages that form $T = 3$ capsids with a diameter of 270 Å (Golmohammadi *et al.*, 1993, 1996). The subunit structures of *Leviviridae* also have a non-$\beta$-barrel fold (Fig. 2A). Because the coat protein subunits form dimers, the assembly pathway of these phages resembles that of the bromoviruses and HBV (Fig. 6B). Indeed, the capsids that preferentially form coat protein dimers appear to follow similar assembly paths. Capsids that form stable dimers and stabilize predominantly by protein–protein interactions are likely to follow assembly pathways similar to that of leviviruses and bromoviruses.

#### 5. *Nodaviridae*

The $T = 3$ capsids of alphanodaviruses are composed of 180 subunits displaying a rhombic triacontahedral quaternary architecture (Fig. 1). Earlier assembly analysis performed on black beetle virus (BBV) suggested that the assembly occurs as the addition of subunit trimers with the hexamer of trimers (18-mer) as a significant intermediate (Reddy *et al.*, 1998). Although there are no experimental data available on the assembly pathway of nodaviruses, the structure-derived assembly pathway is consistent with the observation that nucleic acid is required for particle formation as the potential intermediate (18-mer), which contains both the protein and nucleic acid found in the capsid structure. Similar analyses performed on the other known nodaviruses, that is, flock house virus (FHV) and pariacoto virus (PAV) (PDB ID: 1F8V), suggest that the same assembly pathway is preferred for all the alphanodaviruses. Analyses of FHV and PAV also suggest that structures containing 27 and 36 subunits that are derived from the 18-mer structures also form stable structures (Fig. 7A).

### 6. *Parvoviridae*

Parvoviruses form $T = 1$ capsids with 60 copies of either 60- or 80-kDa protein subunits. The 60-kDa protein is a proteolytically processed version of the 80-kDa protein. The subunit structure of parvoviruses has an extended $\beta$-barrel fold with large loops inserted between strands (Simpson *et al.*, 2000). The calculated assembly profile for parvoviruses suggests that they assemble from multiples of trimers, where the three subunits are related by the icosahedral threefold axes of symmetry (data not shown). The preferred trimer of subunits forms a face of the icosahedron. There is also a significant preference for the structure containing 45 subunits (three-fourths of a capsid) that is missing a pentameric vertex (15-mer), suggesting that this could be a potential intermediate during parvovirus assembly.

### 7. *"$T = 2$" Capsid Assembly*

The inner capsids of the reoviruses contain 120 subunits, with two copies of the same gene product occupying the icosahedral asymmetric unit (Grimes *et al.*, 1998; Nakagawa *et al.*, 2003; Reinisch *et al.*, 2000). Although the number 2 is not permitted as a triangulation ($T$) number according to quasi-equivalence theory, all the inner cores of three of the reovirus structures determined so far have 120 subunit capsids. The structure of L-A virus is the first stand-alone capsid comprising 120 subunits (Naitow *et al.*, 2002). These reports show that "$T = 2$" or 120-subunit capsids certainly occur in nature. The calculations performed on the inner capsid of bluetongue virus (BTV) suggest that the 10-mer containing 5 sets of A and B subunits surrounding the fivefold axes is a potential assembly intermediate, and 12 such 10-mers come together to form the entire capsid (Fig. 7B). This is in agreement with the suggested pathway for rice dwarf virus (RDV; Nakagawa *et al.*, 2003). Although calculations suggest that the 10-mer of subunits is the likely assembly intermediate in the cases of bluetongue virus and RDV, this is not quite apparent for the reovirus and L-A virus structures. This may be due to the significant number of residues that are disordered in the latter two X-ray structures. About 240 residues at the N terminus are disordered in the A-type subunits in reovirus cores, whereas 30 residues at the C terminus are disordered in the A and B subunits in L-A virus. This also brings out an important limitation of the method: the subunit structures must be considered in full detail for the success of these assembly calculations.

## IV. Conclusion

Viral capsid structures provide information about the tertiary structure of capsid protein subunits as well as their quaternary organization. This information can be used to derive preferred pathways for assembling subunits into closed capsids. By calculating the energies of protein–protein interactions that occur in the viral capsids and employing the combinatorial assembly method, we derived the thermodynamically preferred assembly pathways for a representative group of capsids. In most of these cases, the elucidated assembly pathways are in agreement with experimental data. For example, excellent agreement can be seen in the case of picornaviruses, among which the pentamer of protomers unambiguously stands out as the potential intermediate. In the case of bromoviruses, our results suggest that a 6-mer of subunits is the initially preferred structure followed by a 10-mer, formed of three $A–B_5$ dimers and two $C–C_2$ dimers. This 10-mer is at least 40 kcal more favorable than the 10-mer structure representing the pentamer of $A–B_5$ dimers. Combining this computational approach with the experimental data provides a powerful method to identify the preferred substructures or intermediates in the assembly pathway. In the case of alphanodaviruses, the results point to a trimeric unit made up of A, B, and C subunits as the repeating unit of assembly. These trimeric units associate to form hexamers of trimers in the presence of duplex RNA. This structure is an important intermediate in nodavirus assembly. Lack of formation of this structure might prevent assembly. This agrees with the requirements for nucleic acid and full-length N and C termini, which interact with the nucleic acid for proper capsid assembly in alphanodaviruses (Krishna *et al.*, 2003; Marshall and Schneemann, 2001; Schneemann and Marshall, 1998).

It is interesting to note that bromovirus and levivirus proteins, which form coat protein dimers preferentially in solution, appear to follow a similar pathway of virus assembly. In the case of hepadnaviruses, the formation of $T = 3$ and $T = 4$ capsids could be traced to incorporation of 6-mers (trimer of dimers) at one or two locations, respectively. This structure-based approach is likely to be most useful for capsids, where strong protein–protein interactions determine particle formation. Furthermore, this method, applied to multiple capsids from the same family, can be used to derive a consensus pathway of assembly for that family of capsids. This approach is currently being used to investigate the role of variations in protein–protein interactions and the importance of structural elements on virus assembly.

## Acknowledgments

The authors thank Professor Anette Schneemann for valuable suggestions and for critical reading of the manuscript. This work was supported by the NIH research resource: Multiscale Modeling Tools for Structural Biology (MMTSB, RR12255).

## References

Butler, P. J. (1976). Assembly of tobacco mosaic virus. *Philos. Trans. R. Soc. Lond. B Biol. Sci.* **276:**151–163.

Butler, P. J. (1999). Self-assembly of tobacco mosaic virus: The role of an intermediate aggregate in generating both specificity and speed. *Philos. Trans. R. Soc. Lond. B Biol. Sci.* **354:**537–550.

Caspar, D. L. D. (1965). Design principles in virus particle construction. *In* "Viral and Rickettsial Infections of Man" (F. L. J. Horsfall and I. Tamm, eds.), 4th Ed., pp. 51–93. Lippincott, Philadelphia.

Caspar, D. L. D., and Klug, A. (1962). Physical principles in the construction of regular viruses. *Cold Spring Harb. Symp. Quant. Biol.* **27:**1.

Cremers, A. F., Oostergetel, G. T., Schilstra, M. J., and Mellema, J. E. (1981). An electron microscopic investigation of the structure of alfalfa mosaic virus. *J. Mol. Biol.* **145:**545–561.

Crick, F. H. C., and Watson, J. D. (1956). The structure of small viruses. *Nature* **177:**473–475.

Damodaran, K. V., Reddy, V. S., Johnson, J. E., and Brooks, C. L., III (2002). A general method to quantify quasi-equivalence in icosahedral viruses. *J. Mol. Biol.* **324:**723–737.

Eisenberg, D., and McLachlan, A. D. (1986). Solvation energy in protein folding and binding. *Nature* **319:**199–203.

Finch, J. T., and Klug, A. (1959). Structure of poliomyelitis virus. *Nature* **183:**1709–1714.

Fry, E., Acharya, R., and Stuart, D. (1993). Methods used in the structure determination of foot-and-mouth disease virus. *Acta Crystallogr. A* **49:**45–55.

Golmohammadi, R., Fridborg, K., Bundule, M., Valegård, K., and Liljas, L. (1996). The crystal structure of bacteriophage Q $\beta$ at 3.5 Å resolution. *Structure* **4:**543–554.

Golmohammadi, R., Valegård, K., Fridborg, K., and Liljas, L. (1993). The refined structure of bacteriophage MS2 at 2.8 Å resolution. *J. Mol. Biol.* **234:**620–639.

Grimes, J. M., Burroughs, J. N., Gouet, P., Diprose, J. M., Malby, R., Zientara, S., Mertens, P. P., and Stuart, D. I. (1998). The atomic structure of the bluetongue virus core. *Nature* **395:**470.

Hadfield, A. T., Lee, W., Zhao, R., Oliveira, M. A., Minor, I., Rueckert, R. R., and Rossmann, M. G. (1997). The refined structure of human rhinovirus 16 at 2.15 Å resolution: Implications for the viral life cycle. *Structure* **5:**427–441.
Harrison, S. C., Olson, A. J., Schutt, C. E., and Winkler, F. K. (1978). Tomato bushy stunt virus at 2.9 Å resolution. *Nature* **276:**368–373.
Horton, N., and Lewis, M. (1992). Calculation of the free energy of association for protein complexes. *Protein Sci.* **1:**169–181.
Hull, R., Hills, G. J., and Markham, R. (1969). Studies on alfalfa mosaic virus. II. The structure of the virus components. *Virology* **37:**416–428.
Johnson, J. E., and Speir, J. A. (1997). Quasi-equivalent viruses: A paradigm for protein assemblies. *J. Mol. Biol.* **269:**665–675.
Kenney, J. M., von Bonsdorff, C. H., Nassal, M., and Fuller, S. D. (1995). Evolutionary conservation in the hepatitis B virus core structure: Comparison of human and duck cores. *Structure* **3:**1009–1019.
Klug, A., and Finch, J. T. (1960). The symmetries of the protein and nucleic acid in turnip yellow mosaic virus: X-ray diffraction studies. *J. Mol. Biol.* **2:**201–215.
Krishna, N. K., Marshall, D., and Schneemann, A. (2003). Analysis of RNA packaging in wild-type and mosaic protein capsids of flock house virus using recombinant baculovirus vectors. *Virology* **305:**10–24.
Kumar, A. (1997). Structural studies on alfalfa mosaic virus coat protein. Ph.D. thesis. Purdue University, West Lafayette, IN.
Kumar, A., Reddy, V. S., Yusibov, V., Chipman, P. R., Hata, Y., Fita, I., Fukuyama, K., Rossmann, M. G., Loesch-Fries, L. S., Baker, T. S., and Johnson, J. E. (1997). The structure of alfalfa mosaic virus capsid protein assembled as a $T = 1$ icosahedral particle at 4.0-Å resolution. *J. Virol.* **71:**7911–7916.
Liddington, R. C., Yan, Y., Moula, J., Sahli, R., Benjamin, T. L., and Harrison, S. C. (1991). Structure of simian virus 40 at 3.8-Å resolution. *Nature* **354:**278–284.
Lucas, R. W., Larson, S. B., Canady, M. A., and McPherson, A. (2002a). The structure of tomato aspermy virus by X-ray crystallography. *J. Struct. Biol.* **139:**90–102.
Lucas, R. W., Larson, S. B., and McPherson, A. (2002b). The crystallographic structure of brome mosaic virus. *J. Mol. Biol.* **317:**95–108.
Marshall, D., and Schneemann, A. (2001). Specific packaging of nodaviral RNA2 requires the N-terminus of the capsid protein. *Virology* **285:**165–175.
Mellema, J.E. (1975). Model for the capsid structure of alfalfa mosaic virus. *J. Mol. Biol.* **94:**643–648.
Naitow, H., Tang, J., Canady, M., Wickner, R. B., and Johnson, J. E. (2002). L-A virus at 3.4 Å resolution reveals particle architecture and mRNA decapping mechanism. *Nat. Struct. Biol.* **9:**725–728.
Nakagawa, A., Miyazaki, N., Taka, J., Naitow, H., Ogawa, A., Fujimoto, Z., Mizuno, H., Higashi, T., Watanabe, Y., Omura, T., Cheng, R. H., and Tsukihara, T. (2003). The atomic structure of rice dwarf virus reveals the self-assembly mechanism of component proteins. *Structure (Camb)* **11:**1227–1238.
Namba, K., Pattanayek, R., and Stubbs, G. (1989). Visualization of protein–nucleic acid interactions in a virus: Refined structure of intact tobacco mosaic virus at 2.9 Å resolution by X-ray fiber diffraction. *J. Mol. Biol.* **208:**307–325.
Namba, K., and Stubbs, G. (1986). Structure of tobacco mosaic virus at 3.6 Å resolution: Implications for assembly. *Science* **231:**1401–1406.
Pattanayek, R., and Stubbs, G. (1992). Structure of the U2 strain of tobacco mosaic virus refined at 3.5 Å resolution using X-ray fiber diffraction. *J. Mol. Biol.* **228:**516–528.

Prevelige, P. E., Jr. (1998). Inhibiting virus-capsid assembly by altering the polymerisation pathway. *Trends Biotechnol.* **16:**61–65.

Reddy, V. S., Giesing, H. A., Morton, R. T., Kumar, A., Post, C. B., Brooks, C. L., III, and Johnson, J. E. (1998). Energetics of quasiequivalence: Computational analysis of protein–protein interactions in icosahedral viruses. *Biophys. J.* **74:**546–558.

Reddy, V. S., Natarajan, P., Okerberg, B., Li, K., Damodaran, K.V., Morton, R. T., Brooks, C. L., III, and Johnson, J. E. (2001). Virus Particle Explorer (VIPER), a Website for virus capsid structures and their computational analyses. *J. Virol.* **75:**11943–11947.

Reinisch, K. M., Nibert, M. L., and Harrison, S. C. (2000). Structure of the reovirus core at 3.6 Å resolution. *Nature* **404:**960–967.

Rossmann, M. G., and Johnson, J. E. (1989). Icosahedral RNA virus structure. *Annu. Rev. Biochem.* **58:**533–573.

Rueckert, R. (1985). Picornaviruses and the replication. *In* "Fields Virology" (B. N. Fields, D. M. Knipe, and P. M. Howley, eds.), pp. 705–738. Raven Press, New York.

Schneemann, A., and Marshall, D. (1998). Specific encapsidation of nodavirus RNAs is mediated through the C terminus of capsid precursor protein $\alpha$. *J. Virol.* **72:**8738–8746.

Simpson, A. A., Chandrasekar, V., Hebert, B., Sullivan, G. M., Rossmann, M. G., and Parrish, C.R. (2000). Host range and variability of calcium binding by surface loops in the capsids of canine and feline parvoviruses. *J. Mol. Biol.* **300:**597–610.

Smith, T. J., Chase, E., Schmidt, T., and Perry, K. L. (2000). The structure of cucumber mosaic virus and comparison to cowpea chlorotic mottle virus. *J. Virol.* **74:**7578–7586.

Speir, J. A., Munshi, S., Wang, G., Baker, T. S., and Johnson, J. E. (1995). Structures of the native and swollen forms of cowpea chlorotic mottle virus determined by X-ray crystallography and cryo-electron microscopy. *Structure* **3:**63–78.

Stannard, L. M., and Hodgkiss, M. (1979). Morphological irregularities in Dane particle cores. *J. Gen. Virol.* **45:**509–514.

Stehle, T., Gamblin, S. J., Yan, Y., and Harrison, S. C. (1996). The structure of simian virus 40 refined at 3.1 Å resolution. *Structure* **4:**165–182.

Stehle, T., and Harrison, S. C. (1996). Crystal structures of murine polyomavirus in complex with straight-chain and branched-chain sialyloligosaccharide receptor fragments. *Structure* **4:**183–194.

Stubbs, G. (1999). Tobacco mosaic virus particle structure and the initiation of disassembly. *Philos. Trans. R. Soc. Lond. B Biol. Sci.* **354:**551–557.

Umashankar, M., Murthy, M. R., and Savithri, H. S. (2003). Mutation of interfacial residues disrupts subunit folding and particle assembly of *Physalis* mottle tymovirus. *J. Biol. Chem.* **278:**6145–6152.

Wynne, S. A., Crowther, R. A., and Leslie, A. G. W. (1999). The crystal structure of the human hepatitis B virus capsid. *Mol. Cell* **3:**771–780.

Zhao, X., Fox, J. M., Olson, N. H., Baker, T. S., and Young, M. J. (1995). *In vitro* assembly of cowpea chlorotic mottle virus from coat protein expressed in *Escherichia coli* and *in vitro*-transcribed viral cDNA. *Virology* **207:**486–494.

Zlotnick, A., Aldrich, R., Johnson, J. M., Ceres, P., and Young, M. J. (2000). Mechanism of capsid assembly for an icosahedral plant virus. *Virology* **277:**450–456.

Zlotnick, A., Johnson, J. M., Wingfield, P. W., Stahl, S. J., and Endres, D. (1999). A theoretical model successfully identifies features of hepatitis B virus capsid assembly. *Biochemistry* **38:**14644–14652.

ADVANCES IN VIRUS RESEARCH, VOL 64

# BLUETONGUE VIRUS PROTEINS AND PARTICLES AND THEIR ROLE IN VIRUS ENTRY, ASSEMBLY, AND RELEASE

Polly Roy

London School of Hygiene and Tropical Medicine
London WC1E 7HT, United Kingdom



## I. Introduction

Bluetongue disease in sheep, goats, cattle, and other domestic animals, as well as wild ruminants (e.g., blesbuck, white-tailed deer, elk, and pronghorn antelope), was first described in the late eighteenth century. In sheep, the disease is acute and mortality is accordingly high. For many decades, the disease was believed to be confined to Africa and by early 1900 the disease was reported to be caused by a virus. To date, bluetongue virus (BTV) has been isolated from many tropical, subtropical, and temperate zones and some 24 different serotypes have been identified from different parts of the world. BTV




0065-3527/05 $35.00
DOI: 10.1016/S0065-3527(05)64004-3

and other related viruses with similar morphological (doughnut-shaped capsomers) and physiochemical properties were grouped together into a distinct genus, the orbiviruses. Orbiviruses occur within the family *Reoviridae* because of the characteristic double-stranded and segmented features of their RNA genomes (Verwoerd, 1969; Verwoerd *et al.*, 1970, 1972). The *Reoviridae* family is one of the largest families of viruses and includes major human pathogens (e.g., rotavirus) as well as other vertebrate, plant, and insect pathogens. However, unlike the mammalian reoviruses, orbiviruses, comprising 14 serogroups, are vectored to (Verwoerd, 1969) a variety of vertebrates by arthropod species (e.g., gnats, mosquitoes, and ticks) and replicate in both hosts. BTV is transmitted by *Culicoides* species and because of its economic significance BTV has been the subject of extensive molecular, genetic, and structural study. As a consequence it now represents one of the best characterized viruses.

Unlike the reovirus and rotavirus particles, the mature BTV particle is relatively fragile and the infectivity of BTV is lost easily under mildly acidic conditions. BTV virions (550S) are architecturally complex structures composed of seven discrete proteins organized into two concentric shells, the inner and outer capsids (Hewat *et al.*, 1992b; Mertens *et al.*, 1987; Verwoerd, 1969; Verwoerd *et al.*, 1972) (Fig. 1). The virion proteins encapsidate a genome of 10 double-stranded (ds) RNA segments. The outer capsid, which is composed of two major structural proteins (VP2 and VP5), is involved in cell attachment and virus penetration during the initial stages of infection (Eaton and Crameri, 1989; Hassan and Roy, 1999; Hassan *et al.*, 2001). Shortly after infection, BTV is uncoated (VP2 and VP5 are removed) to yield a transcriptionally active 470S core particle which is composed of two major proteins (VP7 and VP3) and three minor proteins (VP1, VP4, and VP6) in addition to the dsRNA genome (Huismans *et al.*, 1987b). There is no evidence that any trace of the outer capsid remains associated with these cores, as has been described for reovirus (Silverstein *et al.*, 1972). The cores may be further uncoated to form 390S subcore particles that lack VP7, also in contrast to reovirus. Subviral particles are probably akin to cores derived *in vitro* from virions by physical or proteolytic treatments that remove the outer capsid and cause activation of the BTV transcriptase (Van Dijk and Huismans, 1980). In addition to the seven structural proteins, three nonstructural (NS) proteins, NS1, NS2, and NS3 (and a related protein, NS3A), are synthesized in BTV-infected cells (Huismans and Els, 1979; Roy, 1996, 2001). Of these, NS3/NS3A is involved in the egress of the progeny virus. The two remaining nonstructural proteins, NS1

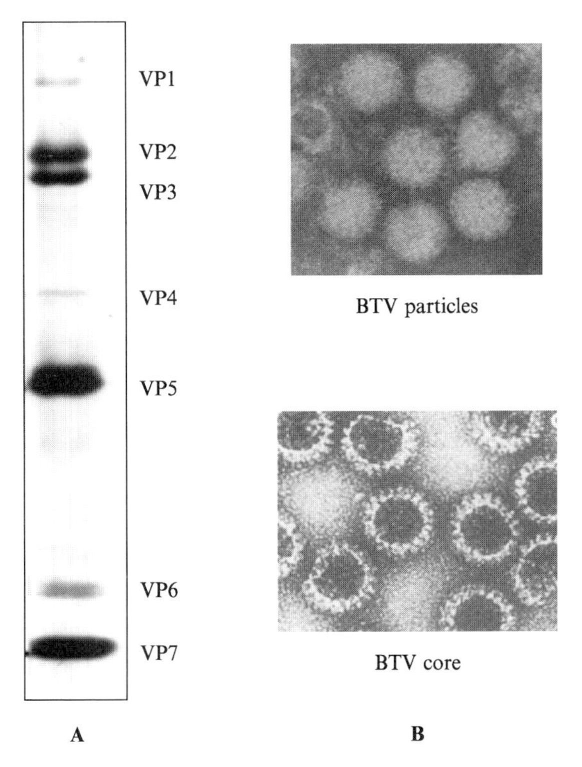


FIG 1. Bluetongue virus particles and proteins: (A) Purified proteins of BTV-10 showing the migration patterns of seven structural proteins (VP1–VP7) on an SDS–10% polyacrylamide gel. (B) Electron micrographs of negatively stained complete BTV particles and cores.

and NS2, are produced at high levels in the cytoplasm and are believed to be involved in virus replication, assembly, and morphogenesis. This article summarizes the current understanding of the three-dimensional structure of BTV particles, subviral particles, and proteins and their role in the various stages of the virus life cycle.

## II. Structural Arrangement of BTV Outer Capsid Proteins and Their Involvement in Virus Entry

BTV and other members of the *Reoviridae* family are nonenveloped particles and thus lack the virus-encoded integral membrane proteins of enveloped viruses that are responsible for membrane fusion.

Therefore, the proteins that constitute the outer surface of the members of the family must adopt a structural organization such that they can perform essentially the same function as the lipid-embedded proteins of enveloped viruses (e.g., the transport of virus through the lipid bilayer from the extracellular medium and delivery of the virus capsid or genomes into the host cell).

For the majority of animal viruses the initial stage of the entry process is the binding of a viral attachment protein to a generalized receptor such as heparin sulfate followed by interaction with a particular host cell receptor. Subsequently, fusion between the membranes of the enveloped viruses and the cell occurs before the virus core formally penetrates. Envelope glycoproteins are typically synthesized as "inactive" precursors that undergo proteolytic cleavage in order to become fully active. After receptor interaction, a conformational change, sometimes pH triggered, is necessary to expose a hydrophobic "fusion peptide" that is able to interact with the cell membrane and mediate membrane fusion (Cohen and Melikyan, 2001; Jahn *et al.*, 2003; Peisajovich and Shai, 2002; Tamm *et al.*, 2002). Fusion appears to be driven, in many cases, by a coiled-coil structure intimately involved with the conformational changes that accompany the process (Bentz, 2000; Matthews *et al.*, 2000; Weissenhorn *et al.*, 1999). Because the opportunity for membrane–membrane fusion does not arise for nonenveloped viruses, the penetration proteins of many of these viruses either possess a myristoyl group at their amino terminus and/or undergo autolytic cleavage as well as structural and conformational changes to trigger cell permeabilization (Chandran and Nibert, 2003; Greber, 2002; Greber *et al.*, 1993; Hogle, 2002; Jane-Valbuena *et al.*, 2002; Liemann *et al.*, 2002; Nemerow and Stewart, 1999; Rossman *et al.*, 2000). Studies using three-dimensional structural determination and molecular/biological analysis of the BTV outer capsid proteins have given some insight into how the outer capsid could function in an analogous manner to classic fusion proteins in this process.

Cryoelectron microscopy (cryo-EM) and image reconstruction of BTV particles have been used to deduce the general relationships of each BTV protein within the virus particle. Initially a low-resolution (45-Å) three-dimensional reconstruction of BTV particles was obtained that revealed a well-ordered morphology of the complete virion with icosahedral symmetry (Hewat *et al.*, 1992b, 1994). A higher resolution (22-Å) reconstruction has been obtained that not only confirmed the initial virion structure but also provided details of the essential organization of the two outer capsid proteins (Nason *et al.*, 2004). This is in contrast to the morphology deduced by earlier negative-staining EM

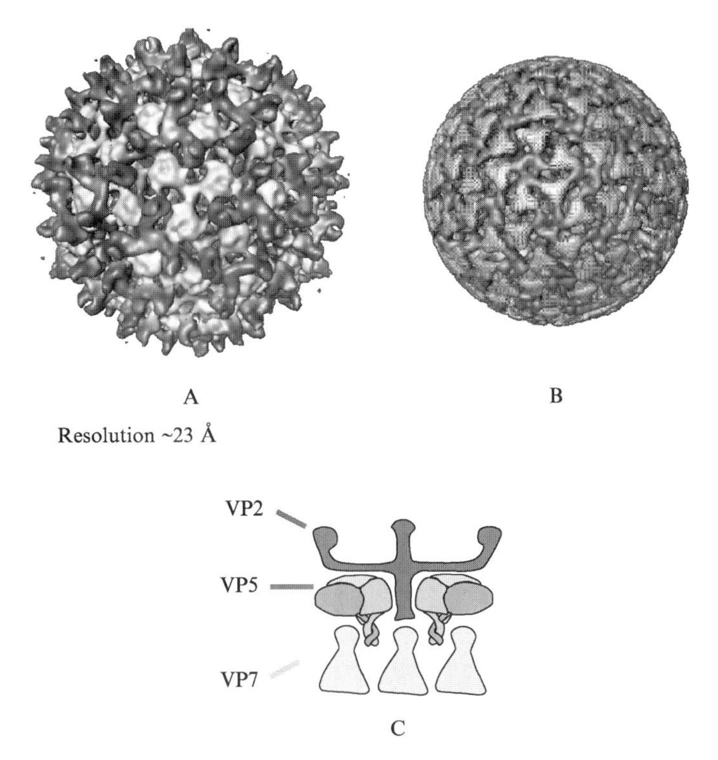


FIG 2. Surface representations of the three-dimensional cryo-EM structures (22-Å resolution) of BTV. (A) Whole particle showing sail-shaped triskelion propellers (VP2 trimers) in red and globular domains (VP5 trimers) in yellow. (B) A particle with cut-off VP2 density, showing arrangement of globular domains (magenta) close to the surface VP7 layer of the core (purple). (C) Schematic showing a side view of VP2 and VP5 organization in relation to each other and the core surface, depicted from image reconstruction of cryo-EM analysis. (See Color Insert.)

methods that gave a characteristically fuzzy appearance to the outer capsid, suggesting it is highly sensitive to preparation conditions (Els and Verwoerd, 1969). Whereas the initial low-resolution structural study revealed the general shape and configuration of the two outer capsid proteins VP2 and VP5, the high-resolution data have provided an accurate position of the two proteins in relation to each other and to the underlying VP7 layer of the core (Fig. 2). Their juxtaposition and structural features give insight into their role in virus attachment to the host cell receptor and membrane penetration, entirely consistent with the available biochemical data.

The outer capsid of BTV has an icosahedral configuration, and exhibits a unique organization made up of two distinct shapes: triskelion and globular. The outer layer occupies the region between radii of 380 and 440 Å and consists of a total of 60 triskelion densities and 120 globular densities (Fig. 2). Thus, unlike rotaviruses and reoviruses, and other notable members of the *Reoviridae* family, the outer capsid of this virus does not conform to a $T = 13$ icosahedral symmetry. The most external part of the outer capsid is the propeller-shaped triskelion motif. Each blade of the propeller is ~75 Å in length and 28 Å wide. At the tip of each blade, the molecule broadens out to be 60 Å wide and, at this point, bends upward perpendicular to the plane of the virus. These bent tips give the entire virion a diameter of ~880 Å and extend from the main body of the particle by 30 Å. Interspersed between the triskelions and lying more internally between radii of 380 and 420 Å are globular densities with a diameter of ~60 Å. These are also entirely exposed in the virion. Both proteins make extensive contacts with the core surface of VP7 trimers underneath (discussed later), indicating that the two proteins assemble independently on the assembled core surface. The currently accepted model of how the core particles of BTV gain access to the cytoplasm, where they become transcriptionally active, involves receptor-mediated endocytosis and pH-triggered conformational change inside endosomal vesicles, which activate a membrane-permeabilizing protein. The two outer capsid proteins, particularly VP2, are easily removed and degraded (Huismans *et al.*, 1987a; Verwoerd *et al.*, 1972). It is clear from studies using soluble recombinant protein that the 110-kDa VP2 is the cellular receptor-binding protein and mediates attachment of the virion to the cell surface and its subsequent internalization (Eaton and Crameri, 1989; Hassan and Roy, 1999). It leads to receptor-mediated endocytosis of virions and is most likely proteolytically degraded once the virions reach endosomal compartments (Eaton *et al.*, 1990; Hassan and Roy, 1999). Indeed, it is the most variable of the BTV proteins, eliciting virus-neutralizing antibody, and is responsible both for serotype specificity and virus hemagglutination activity (French and Roy, 1990; Fukusho *et al.*, 1987; Ghiasi *et al.*, 1987; Gould, 1988; Huismans and Erasmus, 1981; Inumaru and Roy, 1987; Kahlon *et al.*, 1983; Roy, 1992; Roy *et al.*, 1990). The functional features of VP2 correlate well with the structural features of the triskelion motifs on the outer capsid as the sailed-shaped VP2 is the most accessible protein on the virion surface. VP2 oligomerizes into a trimer when expressed as a recombinant protein (Hassan and Roy, 1999) and the volume calculation of the structure closely matches that of each observed triskelion; thus, each is composed of three VP2 molecules. The assignment of VP2 to the triskelions implies that the globular motif is made solely of the second outer capsid protein VP5 (Nason *et al.*, 2004). The volume calculations from the cryo-EM map also indicated that each globular motif is composed of three molecules of VP5, again confirming the biological data obtained with recombinant VP5 in solution (Hassan *et al.*, 2001). After internalization, the clathrin coat of the vesicles is rapidly lost resulting in the formation of large endocytic vesicles within which either VP2 or VP2 and VP5 are degraded (Hyatt and Eaton, 1990). Removal of the VP2 layer exposes VP5, and the partly denuded BTV virion causes destabilization of the vesicle membrane to allow the penetration of the now uncoated core particle into the cytoplasm. The dependency on acidic pH is clear, as the addition of compounds that raise the lysosomal or endosomal pH prevent endocytosed virus particles from entering the cytoplasm (Forzan *et al.*, 2004; Hyatt *et al.*, 1989).

The structural features of VP5 and it location support its involvement in membrane-permeabilizing activity. Indeed studies on the biological activity of VP5 confirm this role unequivocally (Forzan *et al.*, 2004; Hassan *et al.*, 2001). At the sequence level VP5 is essentially divided into an amino-terminal coiled-coil domain and a carboxy-terminal globular domain, with a flexible hinge region in between (Fig. 3A). VP5 does not have an N-terminal myristoyl group nor is it cleaved autocatalytically like the equivalent proteins of reoviruses and rotaviruses. However, the amino terminus of VP5 could also potentially form two amphipathic helices (helix 1 and helix 2), each with approximately equal separation of hydrophobic and hydrophilic faces followed by a putative fusion peptide (Fig. 3A). Fusion proteins of enveloped viruses can be grouped into several classes based on their three-dimensional structure. Despite the difference in overall structure, all fusion proteins have certain features in common, most notably the presence of heptad repeat regions that mediate oligomerization and the presence of a hydrophobic region that can insert into lipid bilayers (Peisajovich and Shai, 2003). VP5 shares both of these features, having an N-terminal membrane-inserting hydrophobic region followed by a long heptad repeat region, emphasizing the concept that VP5 could plausibly induce membrane destabilization and permeabilization (Purdy *et al.*, 1986). Direct evidence has also demonstrated that VP5 has intrinsic membrane "pore formation" ability in keeping with its supposed biological role in penetration (Hassan *et al.*, 2001). The cytotoxicity of VP5 was mapped to the N-terminal sequences that potentially form the amphipathic helices.

Direct evidence of the role of VP5 in membrane penetration has been confirmed by one study (Forzan *et al.*, 2004). Using an experimental system for cell surface expression of VP2 and VP5, it was shown that

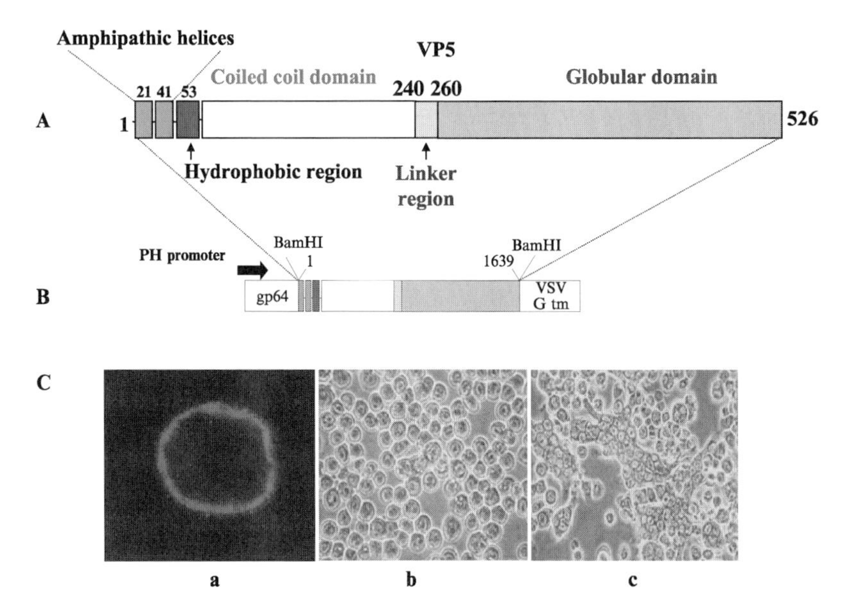


FIG 3. Arrangement of VP5 domains and fusion activity. (A) Schematic organization of VP5 protein, showing the predicted positions of two amphipathic helices, the coiled-coil domain, and the globular domain that are linked by a hinged region. (B) Schematic of a recombinant baculovirus transfer vector, showing the positions of a signal peptide of the baculovirus gp64 and the C-terminal part of vesicular stomatitis virus glycoprotein (VSV G) flanking the BTV-10 VP5 gene. (C) Confocal microscopy showing the cell surface expression of chimeric VP. (a) Fusion activity of VP5 in *Sf*9 cells after a low-pH shift; (b) note the formation of syncytia (seen by confocal microscopy); (c) absence of fusion activity before the pH shift. (See Color Insert.)

VP5 has the ability to interact with host cell membranes and to induce cell–cell fusion in a manner similar to that seen for the fusion proteins of enveloped viruses (Fig. 3B and C). VP5 exhibits its membrane-interacting properties only after it has undergone a low pH-triggered activation step, which presumably alters the conformation to fusion competence. Deletion of the first 40 residues (the two amphipathic helices) abolished fusion activity and, notwithstanding the precise mechanism, it is clear that VP5 can functionally substitute for a typical viral fusion protein. Although VP5 shows the same arrangement of membrane-inserting peptide and heptad repeat region as class I fusion proteins (Colman and Lawrence, 2003), it does not require a proteolytic activation step to render it fully functional. The fusion peptide analog is already located at the N terminus of the native protein. Maturation of the functional homolog of VP5 in the related reovirus, however, includes the proteolytic activation of a precursor protein, which results in exposure of an internal fusion peptide at the N terminus of the C-terminal cleavage product, similar to the situation observed in class I fusion proteins (see Harrison, 2005).

In light of the striking similarities between class I viral fusion proteins and membrane-permeabilizing proteins of members of the *Reoviridae* family it is tempting to speculate that these two classes of proteins have evolved from a common ancestor. Although it is not yet clear whether fusion induced by VP5 follows the same mechanisms proposed for enveloped viruses (Peisajovich and Shai, 2002), it seems probable that the amphipathic helices insert into the target membrane. This early interaction and the formation of the initial pore might be mechanistically different for VP5 when compared with other fusion proteins, as the amphipathic helices of VP5 more closely resemble pore-forming antimicrobial peptides than typical fusion peptides (Merrifield *et al.*, 1982; Putsep *et al.*, 1999).

The available data are consistent with the protein coat hypothesis (Kozlov and Chernomordik, 2002), in which the driving force for several consecutive stages of the fusion process is generated by a dense interconnected layer of proteins including those outside the initial contact zone. The dense layer of fusion proteins has an intrinsic tendency to bend away from the viral particle and to adopt a curvature that effectively results in the formation of an outside-in proteinaceous vesicular structure. The protein coat hypothesis is consistent with the highly localized expression of VP5 on the cell surface. In the BTV particle the VP5 protein makes contact with the underlying VP7 layer and also with the spike protein VP2, which protrudes above VP5 (see Fig. 2). Combining the structural and biochemical data, a plausible hypothesis would be that the low pH inside endosomes induces a rearrangement and conformational change in VP5, loosening the interactions of VP5 and VP7 and simultaneously allowing VP5 to form a protein layer with intrinsic outside-in curvature. Proteolytic removal, or pH-induced conformational change of VP2, then allows the VP5 amphipathic helices to freely interact with membranes and initiate the permeabilization process.

In summary, the accumulating current data point toward a model whereby VP2 makes initial contact with the host cell and VP5 mediates the penetration of the host cell membrane by destabilizing the endosomal membrane. This allows core particles that lack both VP2 and VP5 proteins to be released into the cytoplasm from the endosomal vesicles.

## III. Structure of the Core and Its Components

Unlike the whole virion, the BTV core is a highly stable and robust particle reflecting its biological role in virus multiplication. In addition to the 10 segments of genomic dsRNA it contains all the proteins necessary for genome replication within a two-layered protective coat. The first visualization of the three-dimensional core structure and its protein organization came from a relatively low-resolution (40-Å) cryo-EM analysis (Prasad *et al.*, 1992). Nevertheless, it was possible to delineate the organization of the surface layer, which revealed that the core (estimated at 470S), with a diameter of 70 nm, conforms to an icosahedral symmetry with a triangulation number of 13 ($T = 13$) in a left-handed configuration. The core consists of two concentric layers of density surrounding a central density. More detail of arrangement of each layer was obtained by a subsequent high-resolution (22-Å) cryo-EM analysis followed by an atomic structure of the core particle (solved at 3.6 Å) as discussed later (Grimes *et al.*, 1997, 1998).

The outer layer of core is made up of clusters of VP7 trimers, which appear as prominent triangle-shaped protrusions in the image reconstruction, and occupies between 260 and 345 Å from the center of the core. These trimers are arranged around 132 distinctive channels as six-member rings, with five-member rings at the vertices of the icosahedrons (Fig. 4). There are a total of 780 VP7 molecules per particle,

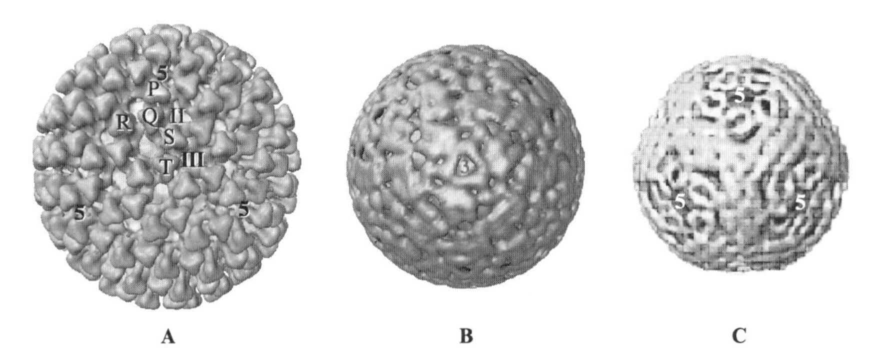


Fig 4. Surface representations of the three-dimensional cryo-EM structures of BTV-10 core. (A) The whole core (700 Å in diameter) viewed along the icosahedral threefold axis, showing the trimers of VP7 (in blue). The five quasi-equivalent trimers (P, Q, R, S, and T) and the locations of channels II and III are marked. (B) The middle layer of VP3 molecules (210–290 Å) cut away from the core structure. (C) The innermost density of three minor proteins in association with the genome. Ordered strands of genomic RNA are visible at 22-Å resolution. (See Color Insert.)

forming 260 trimers that are easily visible as protrusions by negative staining of the particles under EM, giving a bristly appearance to cores described by Els and Verwoerd as capsomers arranged in ring-like structures (Els and Verwoerd, 1969). These ring structures were alluded to in the genus name *Orbis* (ring or circle). The organization of the five quasi-equivalent trimers that form a protomeric unit (P, Q, R, S, and T) could be seen clearly. The aqueous channels that are surrounded by these trimers are about 8 nm wide at the outer surface and 8 nm deep. On the basis of their locations with respect to the icosahedral symmetry axes these channels can be grouped into three types; type I channels run along the icosahedral fivefold axes, type II channels surround the fivefold axes, and type III channels are located around the icosahedral threefold axes. Some of these channels extend inward, penetrating through to the inner layer.

By cryo-EM analysis the inner layer beneath the VP7 layer occurs between 230 and 260 Å and appears as a relatively featureless, almost spherical shape but exhibits an angular appearance due to variations in thickness of the protein shell (Fig. 4). There are some small holes that do not align with the channels in the above VP7 layer. It was assigned as the VP3 layer as VP3 is the second major protein of the core. The volume is consistent with 120 molecules of the 103-kDa VP3. Further confirmation of the positions and arrangement of the VP7 and VP3 layers was obtained from a cryo-EM analysis of empty core-like particles that were generated by assembling only VP7 and VP3 (Hewat *et al.*, 1992a). In the absence of other components of the core it was possible to visualize the overall flower-shaped VP3 molecules underneath the VP7 layer, but it was not possible to identify the precise positions of the 120 molecules. The density in the central portion of the native core structure occupies a radius of less than 21 nm and contains the remaining three minor proteins, VP1, VP4, and VP6, and the genomic RNA.

### A. *VP7 Structure*

Three different crystallographic structures of BTV-10 VP7 have been solved by X-ray diffraction, two of which were complete VP7 trimers, one solved at high resolution (2.6 Å) and the other only at low resolution (5.6 Å) (Basak *et al.*, 1992, 1996, 1997; Grimes *et al.*, 1995). A third structure was derived from a cleavage product consisting of only the top domain. The three structures together have allowed complete understanding of the arrangement of each monomer within the VP7 trimer and also indicated the structural constraints and flexibility of

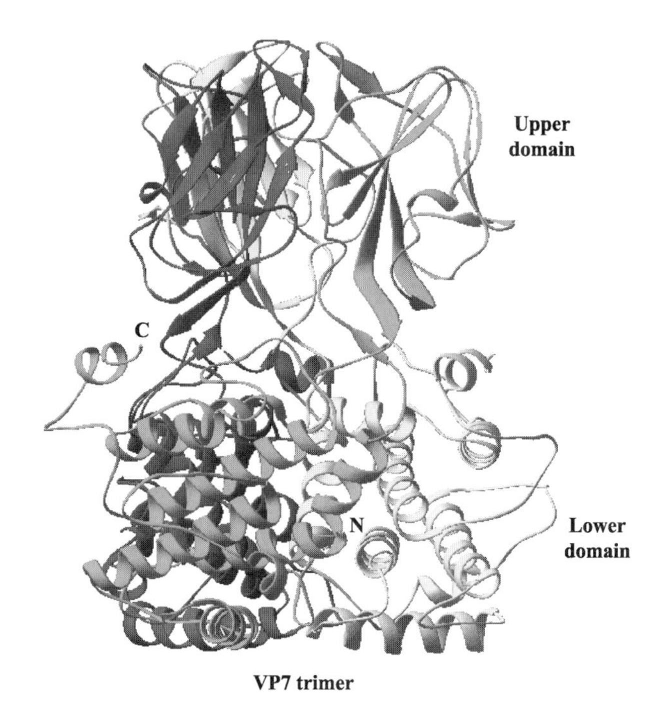


FIG 5. Trimer image of the VP7 atomic structure solved at 2.8-Å resolution. Two domains of the molecule are indicated. The carboxyl and amino termini are indicated. The view is shown from the side. Note that the flat base of the trimer lies in a horizontal plane in this view. (See Color Insert.)

the trimers that are required during virus assembly and morphogenesis. The VP7 trimer revealed a unique molecular architecture. Each monomer consists of two distinct domains, the "upper" and "lower," which are twisted such that the top domain of one monomer rests on the lower domain of a threefold-related subunit in a clockwise direction (Fig. 5). The interactions between subunits are extensive both between the upper and lower domains, involving both hydrophobic as well as specific hydrogen bond interactions. The smaller upper domain of VP7 forms the "head region" of the trimer, which appears as a knobbly projection on the outer surface of the core by cryo-EM and contains the central one-third of the polypeptide chain of the molecule (amino acids 121–249). The upper domain of the VP7 molecule is folded into an antiparallel $\beta$ sandwich. In the trimer, the three upper domains are intimately associated both with each other and with all possible threefold-related subunits via two extended loops from each subunit.

The three monomers of this domain form a stable trimer. Indeed, two other X-ray structures of upper VP7 domains, one from BTV serotype 10 (BTV-10, amino acids 123–253) and one from African horsesickness virus serotype 4 (AHSV-4), which were solved subsequently, also revealed a similar structural arrangement, indicating that it is likely that the VP7 proteins of all orbiviruses have similar structures (Basak *et al.*, 1996).

The lower, broader domain of the VP7 monomer is composed of nine $\alpha$-helices and long extended loops and is formed by both the N terminus (amino acids 1–120) and the C terminus (amino acids 250–349) of the molecule (Fig. 5). The interjunction region between the two domains is exposed, and is easily accessible to proteolysis. A single lysine residue of BTV-10 VP7 is situated at this junction between the two domains. Mutation at this site has a deleterious effect on core assembly (Le Blois and Roy, 1993) and indicates that the interactions between VP7 monomers are rigid and can easily be destabilized by a single point mutation at key positions (Le Blois and Roy, 1993; Limn *et al.*, 2000). Despite this, the low-resolution VP7 structure (5.4 Å) indicated a different orientation of the helical lower domain from that of the high-resolution structure, suggesting an "unwound" or "open" form compared with the tightly packed helices found in the high-resolution form (Basak *et al.*, 1997). Two radically different structures indicate that VP7 has the potential for some internal flexibility in the lower domain that attaches to the VP3 scaffold and suggests that substantial conformational change in VP7 occurs at some stage in the viral life cycle.

An indication of which form of the VP7 trimer is the final form present in the core initially came from fitting the X-ray structure of the individual VP7 trimer into the 22-Å cryo-EM core structure. Thus, an atomic model of the VP7 layer within the core was constructed, which allowed visualization of the structure and organization of VP7 and VP3 within the core particle (Grimes *et al.*, 1997). It was clear that the "closed" trimers but not the "open, unwound" trimers are the final form in the core particles. The data demonstrated that of the two domains of VP7, the $\beta$-barrel domains are external to the core and the $\alpha$-helical domains of VP7 interact with the underlying VP3 layer. Adjacent VP7 trimers interact principally through well-defined regions in the broader lower domains and conform well to the expectations of quasi-equivalence for the assembly of complex systems.

X-ray crystallographic structures at 3.5-Å resolution for core particles of two different BTV serotypes (BTV-1 and BTV-10) confirmed the precise organization of the 260 VP7 trimers (Gouet *et al.*, 1999; Grimes *et al.*, 1995). The threefold axis of the trimer (85 Å long) is

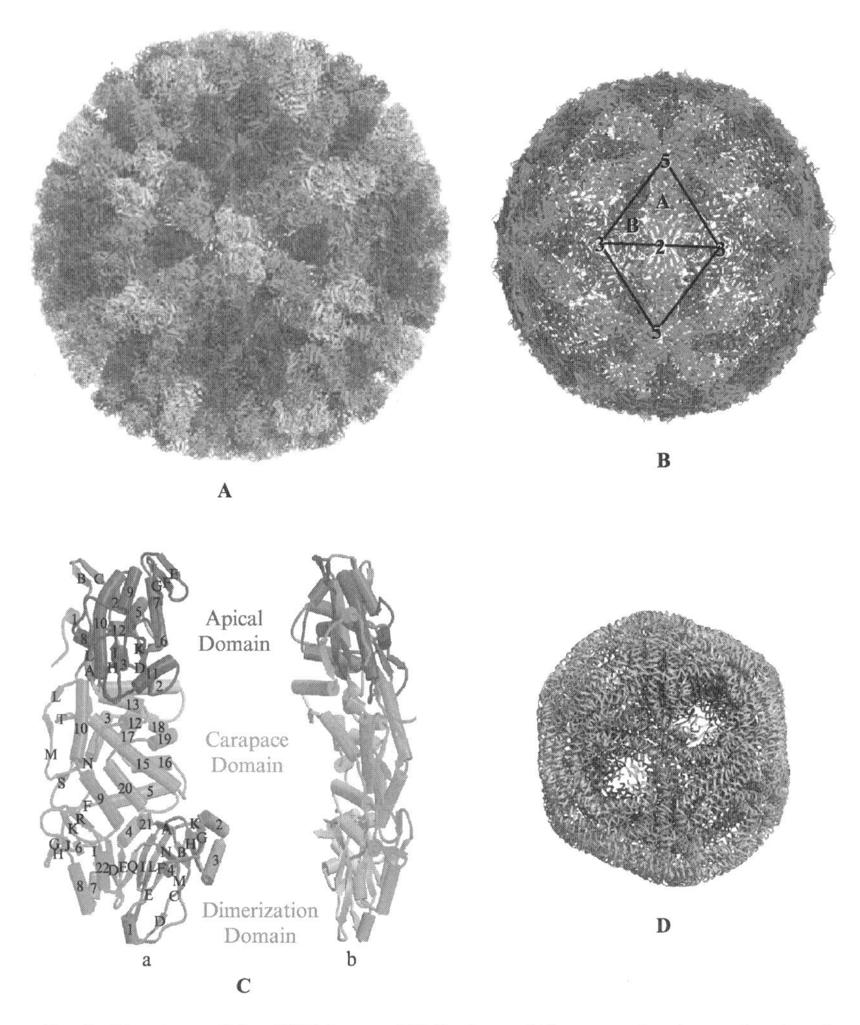


FIG 6. Structure of the BTV-1 core. (A) Surface of the core structure, showing the arrangement of 260 trimers. Structurally equivalent trimers have been given the same color. (B) Structure of the VP3 layer, showing the arrangement of 120 copies of two conformationally distinct types of VP3 molecules, "A" shown in green and "B" shown in red. Sixty copies of each of the two molecules together form 12 decamers that are linked together, forming a thin protein shell. The icosahedral fivefold, threefold, and symmetry axes are indicated. (C) Structures of two types of VP3 molecule, a (*left*) and b (*right*), as observed in the BTV-1 core. The molecules are similar in overall shape of thin triangular plates, with slight differences in conformation. Note some "curling up" of the b molecule, which facilitates it packing around the icosahedral threefold axes. The details of the fold within the two molecules differ at several places. Note the differences at the top of the molecules (closest to the icosahedral fivefold axes), where a compact structure is seen in

perpendicular to the core surface, and the broader base of the trimer (65 Å) contacts the internal network of the VP3 of the subcore (Fig. 6A). The channels or "pores" are small (7 Å in diameter) at the icosahedral threefold axes and slightly larger (9 Å in diameter) at the fivefold axes; they act as portals to release the newly synthesized transcripts. Whereas the "T" trimers of VP7 sit tightly on the pores at the threefold axes, the "P" trimers are located loosely above the pores at fivefold axes.

The interactions between the different trimers are facilitated by a hydrophobic patch, which leads to some small deviations between the 13 copies in the contact patches. The interface between the VP7 and VP3 trimer layers is principally formed by the side-to-side packing of short hydrophobic side chains from helix 4 and its twofold or quasi-twofold partner in the neighboring trimers. The interaction between various VP7 trimers and the VP3 layer underneath is extensive and relatively tight. The lower domains of VP7 pack down against 6 corners of each VP3 molecule such that 12 monomers of VP7 (4 trimers) are directly in association with the VP3 molecules. However, the thirteenth monomer of VP7 (organized in the icosahedron as $T = 13$) is not directly in contact with the VP3 molecule, but is raised slightly and in contact with monomers of adjacent trimers.

### *B. VP3 Structure*

VP3 molecules showed a disk-shaped appearance and the dimers of VP3 serve as the unit building block to form the scaffold for deposition of the VP7 layer. The 120 molecules of the VP3 shell are organized as an icosahedral lattice with $T = 2$ symmetry (Fig. 6B) and are arranged as two sets (A and B) of 60 subunits (Grimes *et al.*, 1998). Five of the "A" group of molecules are arranged as a five-pointed star around each fivefold axis. Five additional "B" group molecules are positioned, one between each of the points of the star, but at a further distance from the fivefold axis. Thus, VP3 molecules form a convex disk shape in which 12 decamers, positioned 1 at each of the fivefold axes, are arranged end to end, so as to form the complete VP3 shell. The VP3 shell has a maximum diameter of 59 nm, forming a central space with an internal diameter of 38 nm that contains both the genome and the minor core proteins.

molecule a and a more extended structure is seen in molecule b. (D) The ordered genomic dsRNA in the core of BTV. The diagram shows the model that has been built into the four layers of electron density, accounting for approximately 80% of the genome. The empty volume close to the icosahedral fivefold axes is the putative location for transcription complex. (See Color Insert.)

Each VP3 molecule (both A and B) forms a flat, elongated triangular wedge-shaped structure (13–35 Å thick × 75 Å wide × 130 Å long) with three distinct domains and contains many $\alpha$-helices as well as $\beta$-strands (Fig. 6C). The three domains have quite different folds from each other. The carapace domain (amino acids 7–297, 588–698, and 855–901) forms a rigid plate that represents the majority of the surface of the subcore shell, whereas the apical domain (amino acids 298–587) is situated closest to the fivefold axes in the intact particle and the dimerization domain (amino acids 699–854) is situated furthest away from the fivefold axes. This domain forms quasi-twofold interactions between individual A group molecules, and appears also to be responsible for the interactions of A and B subunits between different decamers. The inner surface of the VP3 shell has relatively few charged residues and has a series of shallow grooves.

### *C. Structure of the RNA Genome*

The X-ray structure of the core exhibited electron density distributed throughout the internal space of the core within the VP3 shell, as had been visualized by cryo-EM analyses of BTV and other dsRNA viruses (Dryden *et al.*, 1998; Gouet *et al.*, 1999; Grimes *et al.*, 1998; Prasad *et al.*, 1992, 1996; Shaw *et al.*, 1996). The density map showed features consistent with the layers of RNA being organized in a highly ordered fashion (Gouet *et al.*, 1999). Approximately 80% of the entire genome of 19,219 base pairs (Fukusho *et al.*, 1989; Roy, 1989) can be modeled as four distinct concentric layers that have center-to-center spacing between RNA strands of 26–30 Å (see Fig. 6D). As a consequence of the spacing of the A and B forms of VP3, the grooves in the inner surface of the subcore shell form a spiral around the fivefold axis. The outer dsRNA layer appears to interact with these grooves. It is uncertain how the RNA molecules pass inward to the next and subsequent concentric shells. The density detected in the inner layers of RNA progressively weakens, suggesting that its organization and icosahedral symmetry are progressively less rigid, although each layer maintains an overall spiral organization (Gouet *et al.*, 1999).

### *D. Visualization of the Transcription Complex Within the Core*

The high-resolution crystal structure of the BTV core did not define the structure of the enzymatic minor proteins that are closely associated with the viral genome (Gouet *et al.*, 1999; Grimes *et al.*, 1995).

However, in both the virion and the core reconstructions there is a significant amount of density internal to the VP3 layer. The density that is due to the internal proteins has been determined by examining the structures of core-like particles (CLPs) of BTV with and without the internal proteins, in particular VP1 (the polymerase; see later) and VP4 (the mRNA-capping enzyme; see later) generated by recombinant baculoviruses (see Fig. 7). In all CLP reconstructions, the VP7 and VP3 layers exhibit identical features (Nason *et al.*, 2004). These features are similar to those found in the native core structure, except that in CLPs there are only 200 VP7 trimers instead of the expected 260 trimers. Compared with the virion and core reconstructions virtually no internal density was seen in the CLPs with only VP3 and VP7. However, in the particles obtained by the coexpression of VP3

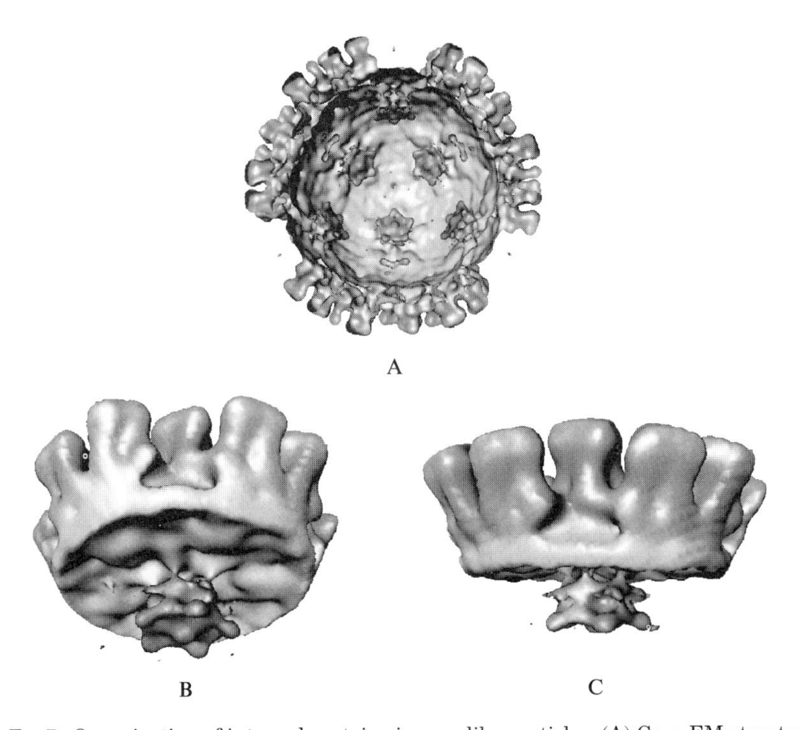


FIG 7. Organization of internal proteins in core-like particles. (A) Cryo-EM structure of a recombinant core-like particle, showing the inside view of the CLP reconstruction with VP3, VP7, VP1, and VP4. A flower-shaped density feature (red) attached to the inside surface of VP3 (green) at all the fivefold axes is clearly seen. (B and C) Conical cutaway from the reconstruction, providing close-up views of the flower-shaped structure and its interaction with the VP3 layer. (See Color Insert.)

and VP7 along with VP1 and VP4, a flower-shaped density directly beneath the icosahedral fivefold axes and attached to the underside of the VP3 layer was clearly observed (Fig. 7A–C). A similar feature had been observed in recombinant rotavirus particles containing homologous proteins (Prasad *et al.*, 1996). The cryo-EM structural data were strongly supported by coimmunoprecipitation studies, which indicated that VP1 and VP4 not only interact strongly with one another but also with VP3. Such close interactions between the polymerase and the capping enzyme may be required to cap the nascent mRNAs as they emerge from the polymerase. In rotavirus particles it has been shown that transcripts as short as seven nucleotides are capped, indicating that the capping site is only seven nucleotides away from the enzymatic site in the polymerase (Lawton *et al.*, 2001).

The individual contributions of VP1 and VP4 to the flower-shaped density were difficult to delineate from the reconstruction of CLPs. However, on the basis of results from reovirus (Zhang *et al.*, 2003) it is likely that the central mass is predominantly due to the VP1 polymerase and that some of the sideward projecting density is due to VP4. From the volume of this fivefold averaged flower-shaped mass, the most likely interpretation is that each flower-shaped complex contains one molecule each of VP1 and VP4. However, biochemical analysis suggested that VP4 is a dimer in solution (Ramadevi *et al.*, 1998b). Perhaps only examination of a polymerase complex that also includes the VP6 protein could reveal the exact stoichiometry of each protein in the complex. The location of the polymerase complex is consistent with the model proposed for genome organization in BTV suggested from X-ray crystallographic studies (Gouet *et al.*, 1999). It is noteworthy that in the presence of magnesium ions and nucleoside triphosphate (NTP) substrates the BTV core-associated enzymes are activated and conformational changes can be seen around the fivefold axis of the BTV core. This results in outward movement of VP3 and VP7, allowing opening of a pore in the VP3 layer at the fivefold axes (Gouet *et al.*, 1999). The mRNAs synthesized by the core are extruded through this pore and subsequently pass out through the central space in the pentameric rings of VP7 trimers (P trimers). There are also pores in the VP3 layer at the threefold axes, but in the intact core particle these are completely blocked by the T trimer of VP7 as discussed previously.

Although it is only in BTV and rotavirus particles that the location of the polymerase complex has been unequivocally demonstrated, using recombinant particles, similar density is seen in the reconstructions of several other dsRNA viruses including viruses from outside the

*Reoviridae* family (Nakagawa *et al.*, 2003; Pesavento *et al.*, 2001; Reinisch *et al.*, 2000; Zhou *et al.*, 2001). Thus in all these viruses, the overall structural mechanism of endogenous transcription is likely to be similar. The unique icosahedral organization of the innermost layer, found in the majority of dsRNA viruses including those of bacterial (Butcher *et al.*, 1997) and fungal origin (Naitow *et al.*, 2002), may be a critical requirement for the appropriate positioning of the transcription enzyme complex and genome organization. In viruses with multiple dsRNA segments a consensus model consistent with the available structural and biochemical data is that each segment is associated with a transcription complex at the fivefold axis to allow for its independent and simultaneous transcription.

## IV. Enzymatic Activities of the Transcription Complex

It is presumed that the overall strategy of BTV and related orbivirus mRNA transcription is similar to that of reoviruses and rotaviruses. The mRNA species are capped and methylated during the transcription process. They are extruded from subviral particles into the cytoplasm and program the synthesis of virus proteins. *In vitro*, BTV mRNA synthesis can be activated by proteolytic digestion of the outer capsid of the virus (Van Dijk and Huismans, 1980, 1988). This facilitates access of NTPs to the genome within the core and the extrusion of newly synthesized mRNAs to the cytoplasm through the channels discussed previously. The structural integrity of the core particle appears to be essential for maintaining efficient transcriptional activity. Studies using individual recombinant proteins *in vitro* have allowed the delineation of the specific roles of the proteins of the transcription complex within the core and are summarized later.

### A. *Helicase Activity*

Transcription of the dsRNA genome by members of the *Reoviridae* family occurs by a fully conservative process, that is, dsRNA is separated, nascent RNA synthesis is initiated, and the newly synthesized RNA strand is separated from the template, allowing the parent strands to reanneal (Banerjee and Shatkin, 1970). It is logical that this process would involve a helicase protein either to unwind the dsRNA ahead of the transcriptase protein or to separate the parental and newly synthesized RNAs after transcription. BTV is the only *Reoviridae* member where helicase activity has been located to a minor structural component (VP6, 35.7 kDa) of the core (Stauber *et al.*, 1997). For reovirus this activity is associated with the major structural protein λ1 (Bisaillon *et al.*, 1997) and for rotavirus no similar activity has been described. It was found that incubation of purified BTV VP6 with dsRNA in the presence of ATP leads to the unwinding of the dsRNA template. Unusually, this activity was detected not only for templates that were complete duplexes but also for those that had short 5′ or 3′ overhangs (Stauber *et al.*, 1997).

Most models explaining helicase activity require two features: ATP-driven conformational rearrangements of the protein that, in turn, generate kinetic energy, and possession of at least two nucleic acid-binding sites in order to maintain contact with nucleic acid (Lohman *et al.*, 1998). This requirement seems indispensable for the unwinding reaction, because it allows the protein to bind both single-strand and duplex regions of a polynucleotide simultaneously or two complementary single strands in an unwinding fork. The majority of known helicase proteins function as a hexamer. By various biochemical and biophysical analyses it has now been confirmed that VP6 exists as a hexamer in solution (Kar and Roy, 2003). Furthermore, incubation of purified VP6 with RNA resulted in the formation of protein–RNA complexes that are visible as a ring-like structure by electron microscopy (Kar and Roy, 2003). By sequence analysis it was possible to identify regions of VP6 that share homology with the nucleic acid-binding and ATPase domains of known helicase proteins (Kar and Roy, 2003). These comparisons provided a basis for mutagenesis and the data obtained demonstrated that mutation of the putative RNA-binding motif at the C terminus of VP6 (R205Q VP6) was sufficient to severely reduce RNA binding. Similarly, mutation of the conserved lysine in the putative ATPase domain (AxxGxGK110V) was found to reduce the ability of VP6 to bind ATP. Unsurprisingly, the same mutant was unable to convert ATP to ADP. Helicase activity was altered in both mutants.

### *B. RNA-Dependent RNA Polymerase Activity*

The replication of viral dsRNA occurs in two distinct steps. First, plus-strand RNA (mRNA) is transcribed, using the dsRNA genome segments as templates, and extruded from the core particle (Banerjee and Shatkin, 1970; Cohen, 1977; Van Dijk and Huismans, 1980). Second, the plus-strand RNAs serve as templates for the synthesis of new minus strands at an undefined stage in the assembly of a new virus core particle. *Cis*-acting sequences required for the viral

RNA-dependent RNA polymerase (RdRP) to replicate the plus-strand RNAs are believed to be located at their 3′ ends (Chen and Patton, 1998; Chen *et al.*, 1994; Patton *et al.*, 1996). The polymerase activity present in the BTV core particle presents the constituent proteins (VP1, VP3, VP4, VP6, and VP7) as candidates, either individually or in a complex, for the viral transcriptase and the viral replicase.

The largest viral protein, VP1, $M_r$ 149,588 Da, is a highly basic protein and is present at a low molar ratio in the virion (Roy *et al.*, 1988). On the basis of its size, location, molar ratio in the core, and predicted amino acid sequence, this protein is the prime candidate for the virion RdRP. The primary sequence of VP1 contains a GDD motif (positions 287–289), characteristic of other RNA polymerases. Indeed, extracts of cells that were infected with a recombinant baculovirus expressing VP1 protein showed poly(A) polymerase activity in an *in vitro* assay (Urakawa *et al.*, 1989).

Studies of replicase activity in other members of the *Reoviridae* family have utilized a particulate replicase, in which the catalytic subunit is proposed to possess replicase activity only in the context of a subviral particle (Chen and Patton, 1998; Chen *et al.*, 1994, 2001; Patton *et al.*, 1997; Tortorici *et al.*, 2003). Purified rotavirus VP1 has been shown to bind to the 3′ end of *in vitro*-synthesized rotavirus plus-strand RNA, but replicase activity was detected only when the remaining core proteins were present or in the context of subviral particles (Patton, 1996; Patton *et al.*, 1997). Purified recombinant reovirus λ3 has been shown to be a poly(C)-dependent poly(G) polymerase, but does not possess replicase activity (Starnes and Joklik, 1993). The processive replicase activity of reovirus λ3 has been reported only for λ3 crystals containing short oligonucleotide templates (Tao *et al.*, 2002). It has been suggested that the generation of a functional viral replicase occurs only within the shell of the assembling virus core in order to avoid the triggering of an antiviral response, induced via dsRNA surveillance pathways.

However, processive replicase activity associated with soluble BTV VP1 in the absence of other viral proteins (unlike the rotavirus VP1 polymerase) has been reported (Boyce *et al.*, 2004). In addition to elongating the nascent minus strand, the synthesis of a unit-length labeled strand shows that purified BTV VP1 initiates minus-strand synthesis *de novo*, rather than extending a fortuitous secondary structure present at the 3′ end of the plus-strand template. It is unlikely that there is a fundamental difference in the way that core assembly and genome replication occur in different genera of the family and this apparent discrepancy may be attributable to the varied experimental systems used to assay replicase activity. Indeed, the replicase activity associated with recombinant BTV VP1 is low, implying that the activity of VP1 may be modulated by other viral proteins present in the assembling core particle.

BTV VP1 in the absence of other core components exhibited little sequence specificity for BTV "plus"-strand template, suggesting that specificity for viral template comes from location within the viral core. The lack of specificity for the template 3′ end of VP1 is consistent with a lack of specificity reported using crystals of reovirus $\lambda 3$ complexed with a template oligonucleotide (Tao *et al.*, 2002). Synthesis of a dsRNA oligonucleotide was possible when the template oligonucleotide did not recapitulate the 3′ end of a reovirus "plus" or "minus" strand. The specificity of the replicase activity for virion "plus" strands may be determined by other viral core proteins or nonstructural proteins, or the specificity may occur during packaging of the plus strands, without further discrimination occurring during replication. The latter possibility seems unlikely, given the template specificity found for rotavirus plus-strand 3′ end sequences in a replicase system, which used rotavirus cores and exogenously supplied template. It seems more likely that the replicase activity of VP1 is modulated *in vivo*, as has been shown *in vitro* with hepatitis C polymerase and the viral proteins NS3, NS4A, and NS5A (Piccininni *et al.*, 2002; Shirota *et al.*, 2002). Such modulation may take the form of inhibition of the replicase activity of soluble VP1 in the cytosol, in order to prevent the formation of free dsRNA. Template specificity may be added to the replicase activity of VP1 by viral proteins present in the assembling virus core, as shown in a reconstituted influenza virus replicative transcription complex (Lee *et al.*, 2002).

After dsRNA synthesis is complete, viral core proteins may trigger the switch from replicase activity to further transcriptase activity.

### *C. BTV mRNA-Capping Activity*

The transcripts produced by BTV cores have "cap" structures at their 5′ termini. By using recombinant purified VP4 protein (76.4 kDa), it has been demonstrated that VP4, in the absence of any other viral proteins, is sufficient for the generation of a cap structure at the 5′ end of BTV transcripts. Both purified recombinant VP4 and encapsidated VP4 in recombinant CLPs react with GTP and covalently bind G via a phosphoamide linkage, a characteristic feature of guanylyltransferase enzyme. VP4 also catalyzes a GTP–$PP_i$ exchange reaction, indicating that this protein is the guanylyltransferase of BTV. In addition, VP4 possesses RNA 5′-triphosphatase activity, which catalyzes the first step in the RNA-capping sequence. The protein also has the ability to catalyze the conversion of unmethylated GpppG or *in vitro*-produced, uncapped BTV RNA transcripts to m7GpppGm in the presence of *S*-adenosyl-l-methionine, indicating that VP4 is also a methyltransferase enzyme. Analysis of the methylated products identified both methyltransferase type 1 and type 2 activities associated with VP4, demonstrating that the complete BTV cap reaction is associated with this single core protein (Martinez-Costas *et al.*, 1998; Ramadevi and Roy, 1998; Ramadevi *et al.*, 1998b). These data are particularly significant as the capping activities associated with vaccinia virus had shown that an enzyme complex of three proteins was required to accomplish all the reactions (Venkatesan *et al.*, 1980). In addition, an X-ray structure of a fragment of Dengue virus NS5 clearly demonstrates that this capping polymerase enzyme has (nucleoside-2′-*O*-)-methyltransferase activity. Other structural studies have focused specifically on guanylyltransferase activities (Hakansson and Wigley, 1998; Hakansson *et al.*, 1997). VP4 therefore represents a model enzyme, which has both methyltransferase and guanylyltransferase in addition to RNA triphosphatase activities in a single protein that is distinct from the viral RNA-dependent RNA polymerase. Furthermore, through gel filtration, mutagenesis, and sucrose gradient analyses it was shown that purified VP4 is dimeric in solution and that dimerization is dependent on a leucine zipper motif located between amino acids 523 and 551 of the protein. Mutations that abolish the ability of VP4 to dimerize in solution also abolish the ability of the protein to be integrated into the structure of CLP when coexpressed with VP1, VP3, and VP7 in baculovirus-infected insect cells (Ramadevi *et al.*, 1998a).

## V. Virus Assembly Site: Inclusion Bodies and Nonstructural Protein NS2

For a number of animal and plant viruses, replication complexes, transcription complexes, replication and assembly intermediates, as well as nucleocapsids and virions accumulate in specific locations within the host cell in structures described as "virus assembly factories" or "virus inclusion bodies." Virus inclusion bodies (VIBs) are also seen in the cytoplasm of BTV-infected as well as reovirus- and rotavirus-infected cells (Altenburg *et al.*, 1980; Rhim *et al.*, 1962). Early in infection the VIBs appear as granular material scattered throughout the cell, but they later coalesce to form a prominent inclusion with a perinuclear location (Brookes *et al.*, 1993; Eaton and Hyatt, 1989; Hyatt and Eaton, 1988).

The final product of disassembly for all members of the family is the transcriptionally active core structure that initiates the transcription of genomic RNAs. The newly synthesized single-stranded RNA species are subsequently released into the cytosol and in turn serve both as templates for viral dsRNA genome synthesis and also act as messengers for the synthesis of viral proteins within the cytoplasm.

During infection, a fibrillar network develops around the infecting core particle in the cytoplasm. This resembles the VIBs that are so pronounced later during the infection course. VIBs are rich in mRNA and phosphorylated NS2 protein (Bowne, 1967; Thomas *et al.*, 1990). In addition, viral capsid proteins, such as core proteins VP3 and VP7, are also present in the VIB matrix, indicating that this may be the assembly site of core particles either within or at the periphery of the VIBs (Hyatt and Eaton, 1988). By contrast, outer capsid proteins VP2 and VP5 are either present at the periphery of the VIBs or at other locations (Hyatt and Eaton, 1988). The 41-kDa NS2 protein, which is synthesized at high levels early in BTV-infected cells, is associated predominantly with VIBs (Brookes *et al.*, 1994; Eaton and Hyatt, 1989; Hyatt and Eaton, 1988). When NS2 is expressed by baculovirus vectors, VIBs are formed readily in insect cells. The morphology is similar to that found in BTV-infected cells and suggests that neither other viral proteins nor a full complement of viral RNA is necessary for VIB formation (Thomas *et al.*, 1990). In addition to electron microscopy data, our biochemical analysis of coexpression experiments using the recombinant NS2 and virion proteins in different combinations has confirmed the association of NS2 with the structural proteins and assembling particles (our unpublished observations). The association of the individual core proteins with expressed VIBs in the cytoplasm has also been confirmed by confocal microscopy (Fig. 8A). These data together with the observation of the presence of electron-dense virus-derived immature particles inside VIBs has led to the belief that core particles may be synthesized within VIBs. It is likely that the further stages of mature virion development occur either in the cytosol or after binding of the particles to the cytoskeleton. The data emphasize the role of NS2 VIBs in capsid assembly.

Viruses with a segmented RNA genome face a challenging task in the recruitment and assortment of specific virally encoded RNA species in the infected cells. How are the 10–12 RNA segments specifically recruited and transported to the virus replication and assembly sites within the cytoplasm and are any specific sequences involved?

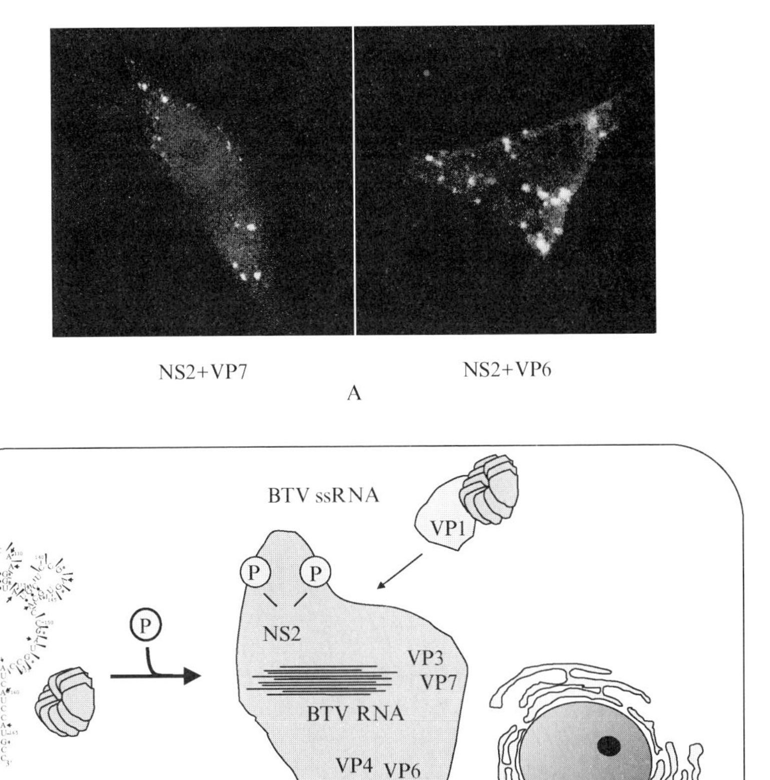


FIG 8. Interaction of NS2 and viral components. (A) Colocalization of core proteins and NS2 in BTV-infected cells. (B) Schematic showing the interaction of NS2/VIB with BTV proteins and the secondary structure of BTV ssRNA. "P" represents phosphorylation of NS2, which is responsible for the formation of VIBs. (See Color Insert.)

It is commonly believed that each of the viral segments must possess some specific sequence or structures, which are recognized by one or more virally encoded proteins to facilitate these processes. Generally, viral RNA-binding proteins are distinguished by being in one of two categories. The first includes proteins that are associated with the nucleocapsid (nucleoproteins) (Portela and Digard, 2002) and are involved in the replication process (transcription and packaging) of the viral genome (Kouznetzoff *et al.*, 1998; Shubsda *et al.*, 2002). The second includes proteins that play an essential role in recruiting, transporting (Najera *et al.*, 1999), modifying (Chiu *et al.*, 2001), and translating (Deo *et al.*, 2002) viral RNA. These proteins can also interact with cellular RNA to suppress the expression of regulatory genes (Brown *et al.*, 1997), thereby protecting the viral RNA from cellular recognition mechanisms in order to use the cellular machinery for virus propagation (Najera *et al.*, 1999).

NS2 has a strong affinity for ssRNA but not for dsRNA, suggesting it may have a role in the recruitment of RNA for replication (Thomas *et al.*, 1990). In this respect NS2 is similar to rotavirus NSP3 protein and reovirus σNS protein, which also form oligomeric structures, bind ssRNA, and exist in complexes with viral RNA in infected cells (Gillian *et al.*, 2000). Whereas the amino terminus, but not the carboxy terminus, of NS2 is essential for ssRNA binding, neither the amino terminus (up to 92 residues) nor the carboxy terminus (the last 130 amino acids) of the protein is required for oligomerization (Zhao *et al.*, 1994). NS2 is the only virus-specific protein that is phosphorylated in BTV-infected cells, although the significance of the modification, which occurs on specific serine residues, is not known. Generally, phosphorylation has been shown to play a role in the multimerization and protein–protein interactions of a number of viral proteins (Gao and Lenard, 1995; Montenarh and Muller, 1987; Spadafora *et al.*, 1996). Evidence has been obtained that the phosphorylation of NS2 is important for VIB formation but not for interaction with other viral proteins (J. Modrof and P. Roy, unpublished observation). Consistent with this interpretation, mutagenesis of the key serine residues has shown phosphorylation not to be essential for RNA binding (Thomas *et al.*, 1990).

The activities of NS2 (formation of VIBs, phosphorylation, and RNA binding) suggest that it might be involved in recruiting and packaging of BTV ssRNA before encapsidation. In a study using recombinant purified NS2 in combination with BTV RNA species and a range of specific and nonspecific ssRNA competitors, it was possible to demonstrate that NS2 has high affinity for specific BTV RNA structures that are unique in each RNA segment (Lymperopoulos *et al.*, 2003). The data suggest that it may be involved in the identification and possibly packaging of the 10 RNA segments into assembling virions within VIBs. If not directly involved in assortment, NS2 could alternatively serve as a site for RNA condensation, allowing RNA–RNA interactions to form. There may be particular site(s) within NS2 solely responsible for binding BTV RNA, consistent with the data that indicated that NS2 may have multiple RNA-binding domains (Fillmore *et al.*, 2002; Zhao *et al.*, 1994). In other segmented RNA genome viruses such as influenza virus, it has been shown that replication and packaging are mediated by *cis*-acting signals, which are located at the 3′ and 5′ ends of the viral segments (Luytjes *et al.*, 1989; Tchatalbachev *et al.*, 2001). The 3′- and 5′-terminal noncoding regions of each RNA segment have short, highly conserved sequences that form a partial duplex structure (the panhandle stem–loop structure) and longer nonconserved sequences. However, by using a reverse genetic system it has been shown that sequences within the coding region of the neuraminidase viral RNA of influenza A possess a signal that drives incorporation of this segment into virions (Fujii *et al.*, 2003). This salient study, in contrast to previous reports, suggests that unique sequences within the coding regions of individual viral RNA segments are responsible for the selective mechanism of viral RNA recruitment into virions. Unlike influenza virus, in members of *Reoviridae* neither the RNA-packaging mechanism nor the role of specific signals is clearly understood. However, study has provided some understanding of the RNA-packaging and assembly mechanism for a much simpler virus, bacteriophage $\varphi$6, which has a genome of only three dsRNA segments (Mindich, 1999; Patton and Spencer, 2000; Pirttimaa and Bamford, 2000). The packaging signals of bacteriophage $\phi$6 are located within the 5′ noncoding regions, and although packaging signals of all segments share similar elements, each folds into a different secondary structure (Mindich, 1999).

Analogous to the RNA genome segments of influenza virus, all RNA segments of BTV and other orbiviruses have short conserved 5′ and 3′ termini that are shared by each of the RNA segments. In addition, each noncoding sequence also includes nonconserved sequences of variable lengths. However, neither the short conserved noncoding sequences and the potential panhandle structures (formed by the partial complementary sequences of the 5′ and 3′ termini) nor the variable noncoding sequences had any specificity for NS2 binding. In contrast, sequences within the coding region of each RNA segment were responsible for a specific interaction between NS2 and BTV RNA. Further, particular secondary structures rather than primary sequence are critical for the recognition of BTV RNA by NS2 (Lymperopoulos *et al.*, 2003). Chemical and enzymatic structure probing indeed confirmed that the NS2-binding region of the BTV RNA transcripts folds into unique hairpin–loop secondary structures. The

NS2–hairpin interaction was further confirmed by using hairpin variants. But it is not known whether the protection afforded by bound NS2 is a consequence of the binding of a single NS2 molecule, or multiple NS2 proteins.

A number of RNA-binding proteins specifically recognize the structure of RNA and include the Gag polyproteins of spleen necrosis virus (SNV) and murine leukemia virus (MLV). These proteins recognize packaging signals that are contained in hairpin structures with no apparent sequence homology (Beasley and Hu, 2002). Also, the RNA-dependent RNA polymerase of influenza A virus exhibits structural specificity for an internal RNA loop of the viral promoter (Bae *et al.*, 2001). However, RNA-binding proteins of none of the other members of the *Reoviridae* family, such as reovirus or rotavirus, have been shown to have RNA structural specificity, although rotavirus NSP3 appears to bind a linear sequence found at the 3′ end of all rotavirus RNA segments (Poncet *et al.*, 1993).

It is unclear how the packaging of the segmented RNA genome of members is achieved. However, the fact that temperature-sensitive (ts) mutants of the rotavirus NSP2 protein (ssRNA-binding protein and a major component of VIBs) produce virions that are "empty" of viral RNA at nonpermissive temperatures (Ramig and Petrie, 1984) suggests that nonstructural proteins may play a role in this process. It is tempting to speculate that NS2 may be responsible for the sequestration and assortment of viral ssRNA molecules that are subsequently packaged by a separate viral protein or protein complex, for example, the polymerase complex. This would account for the nonspecific substrate specificity shown by the polymerase protein of $\phi$6 (Makeyev and Bamford, 2000) and BTV polymerase protein VP1 (Boyce *et al.*, 2004), as template selectivity would have been achieved by selective encapsidation before polymerase action.

It can therefore be speculated that VIBs are formed by the multimerization of NS2 and that BTV ssRNA is specifically recruited to such assemblies to interact with VP1, VP4, and VP6, the transcriptase complex, and subsequently be encapsulated by a single shell of VP3 to form the initial core of the assembling virus (see schematic in Fig. 8B). Second-strand synthesis to form the mature genome would then occur within the cores. NS2 binding to ssRNA and multimerization into a VIB may make an RNA molecule unavailable for translation. It would therefore act as a control mechanism determining whether an RNA molecule is translated or packaged. Interestingly, purified NS2 has phosphohydrolase activity: it could bind ATP and GTP and hydrolyze both nucleotides to their corresponding

NMPs, with a higher efficiency for the hydrolysis of ATP in the presence of divalent cations (Horscroft and Roy, 1997). The NTPase activity described for NS2 may play a role in providing energy for the assortment, movement, packaging, or condensation of bound ssRNA, similar to that found for bacteriophage $\phi$6 P4 protein (Pirttimaa *et al.*, 2002; Poranen and Tuma, 2004).

## VI. Capsid Assembly Pathway

Viral capsid assembly is a precise process that is fundamental to the success of the viral infection. The assembly of the virion particles, particularly the inner capsid or core of various members of the *Reoviridae* family, has been a major focus at both the structural and molecular levels. The assembly of multilayered viral cores is especially intriguing as both protein–protein and protein–RNA interactions are required in order to complete the core particle.

Cryo-EM in conjunction with X-ray crystallography has provided structural details of several viruses belonging to the various genera of *Reoviridae*. Overall there are similar structural arrangements for the inner capsids of this diverse range of viruses, indicating the structural constraints on certain proteins (Grimes *et al.*, 1998; Hewat *et al.*, 1992b; Hill *et al.*, 1999; Lu *et al.*, 1998; Prasad *et al.*, 1992, 1996; Reinisch *et al.*, 2000; Shaw *et al.*, 1996). BTV has been used as a model system to understand the complex nature of the protein–protein interactions necessary for the assembly of these large, icosahedral, multilayered particles and, to a certain extent, to determine the order of protein assembly. Considerable progress toward understanding the role of specific residues necessary to preserve the stable capsid structure and of the intermediates of the assembly pathway that achieve the final spherical icosahedral structure has been made. This has been possible mainly because of the generation of virus-like particles (VLPs) and core-like particles (CLPs) that are produced by coexpression of BTV proteins, using the baculovirus expression system (French and Roy, 1990; French *et al.*, 1990). Assembly of BTV subviral particles by recombinant proteins in different combinations has given us the first indication of how the four major capsid proteins are assembled (Fig. 9). These assay systems, in combination with extensive site-specific and deletion mutagenesis of selected capsid proteins, have mapped the essential assembly pathway of the capsid proteins, particularly the assembly of VP3 and VP7 and their compatibility between related orbiviruses.

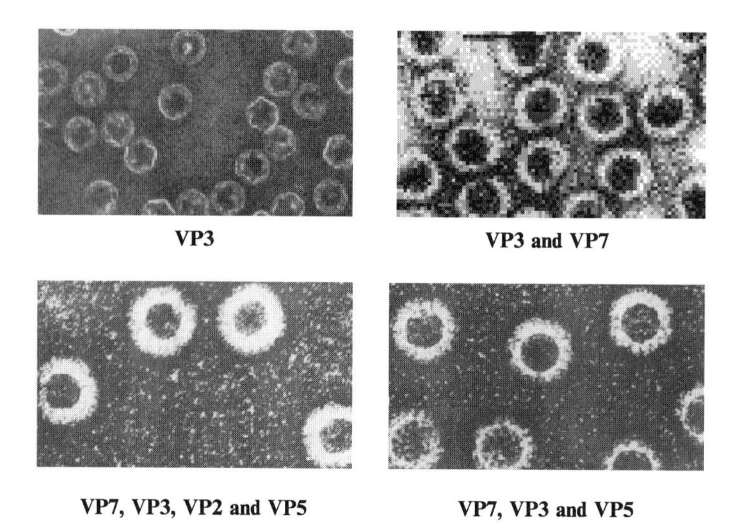


FIG 9. Baculovirus-expressed BTV-like particles. Negatively stained electron micrographs of VP3 subcore-like particles, VP3 plus VP7 core-like particles, and virus-like particles of three or four proteins. Note each type of particle is essentially empty.

### A. *Assembly of VP1, VP4, and VP6*

Unlike the extensive knowledge gained on the enzymatic activity attributed to each of the three internal proteins, we have limited understanding of the structure, organization, and order of assembly of these proteins. However, results obtained from earlier coexpression studies indicated that both VP1 and VP4 proteins are independently associated with the VP3 layer, although data suggest that VP4 protein has more direct contact with VP3 (Le Blois *et al.*, 1992; Loudon *et al.*, 1992; Nason *et al.*, 2004). In addition, the two proteins are tightly associated with each other, forming a stable complex that can be visualized by cryo-EM analysis. Further, this complex directly interacts with the VP3 decamer in solution, supporting the cryo-EM analyses.

The interaction of VP1 and VP4 in the absence of BTV genome or VP3 and the association of VP1 and VP4 with the VP3 decamer suggest that assembly of the BTV capsid probably initiates with the complex formed by these two enzymes, which simultaneously or rapidly associates with the VP3 decamers and hence the VP3 subcore. VP3 decamers as assembly intermediates are most likely involved in the recruitment of the polymerase complex before completion of the assembly of the VP3 subcore.

Unlike VP1 and VP4, it has not been possible to confirm the location of VP6 in the core, although it is likely that VP6 is also located either in the vicinity or together with VP1 and VP4 at the fivefold axis of the VP3 layer directly beneath the decamer. Nevertheless, VP6 can be encapsidated within the CLPs when coexpressed together with VP3 and VP7, indicating that VP6 must possess a signal for encapsidation (Le Blois *et al.*, 1991). After removal of the VP7 layer by treatment with hypotonic solution, VP1, VP4, or VP6 remains within the subcore and thus confirm that all three proteins specifically associate with VP3, not VP7 (Le Blois *et al.*, 1991; Loudon and Roy, 1991). The C-terminal residues of VP6 were shown to be essential for this encapsidation (Yi *et al.*, 1996). By NMR analysis purified VP6 protein shows an open arm configuration (P. Roy, unpublished observation). However, VP6 forms a defined hexamer in the presence of BTV transcripts and assembles into distinct ring-like structures that could be isolated by glycerol gradient centrifugation (Kar and Roy, 2003). Such characteristics are common with many other helicase proteins. However, it is not currently known whether such structures are present within the core and incorporated either together with VP1 and VP4 into the VP3 decamer intermediates or independently during the assembly of the 12 VP3 decamers.

### *B. VP3 Assembly*

The highly conservative nature of the VP3 (100 kDa) sequence across 24 BTV serotypes as well as considerable similarities with the sequence of other related orbiviruses (epizootic hemorrhagic disease virus [EHDV] and AHSV) highlight the structural constraints that may be essential for virus assembly (Iwata *et al.*, 1992; Roy, 1996). The functional importance of the VP3 layer is emphasized further by the conservative nature of the overall architecture of this shell, which is shared by all members of *Reoviridae* and other viruses with segmented dsRNA genomes (Grimes *et al.*, 1998; Hewat *et al.*, 1992b; Prasad *et al.*, 1992, 1996; Reinisch *et al.*, 2000). In each case 120 copies of the equivalent protein ($\lambda$1 for orthoreovirus, VP2 for rotavirus) per core of that virus form the icosahedral shell, which packages genome and transcription complex components. In the absence of any other core protein, VP3 alone can assemble as an icosahedral subcore-like structure (see Fig. 9), indicating that assembly of the VP3 layer initiates the formation of core. The equivalent protein of rotavirus (VP2) also shares this feature (Labbe *et al.*, 1991). By low-resolution cryo-EM analysis, the VP3 subcore exhibits organization of VP3 molecules

(Hewat *et al.*, 1992a) similar to that seen for the subsequent atomic structure of VP3 within the authentic core, indicating that VP3 is probably assembled first and then serves a primary role in formation of the virus capsid.

VP3 has a particularly high hydrophobic amino acid content (Purdy *et al.*, 1984), but several internal hydrophobic stretches can be deleted without much effect on CLP formation (i.e., when coexpressed in the presence of VP7) (Tanaka and Roy, 1994). Moreover, these regions can accommodate insertion of short foreign sequences, implying flexibility in VP3 folding. Similarly, deletion of 10 residues at the carboxy terminus did not affect the VP3–VP3 and VP3–VP7 interactions or formation of CLPs (Tanaka *et al.*, 1995). In contrast, specific single substitution mutations (e.g., the methionine at position 500, arginine at position 502) appear to be critical for core assembly, emphasizing their critical position in the molecule. Interestingly, these residues are within the apical domain of the molecule, situated nearer to the fivefold axes in the particle.

The arrangements of the VP3 molecules within the core as seen in the final structure showed that the amino terminus of the VP3 A molecule is disordered. This has been hypothesized to accommodate the minor rotation of the apical domain that abuts the main body of the protein, which in turns allows assembly via a proposed decameric intermediate. The crucial role of the amino terminus in BTV core assembly was verified by biochemical analysis of N-terminal deletion mutants (Kar *et al.*, 2004). The deletion of the first 10 residues (Δ10 N mutant) significantly affected the efficiency of VP3 assembly and its interaction with VP7 trimers, although up to 5 residues were dispensable both for VP3 assembly and its interaction with the internal proteins. This is in contrast to other members, in which N-terminal residues have also been studied in a number of homologous proteins (Kim *et al.*, 2002; Zeng *et al.*, 1998). Like the N-terminal 56 residues of BTV VP3 protein, the 230 N-terminal residues of orthoreovirus $\lambda1$ are not clearly resolved by X-ray crystallography (Reinisch *et al.*, 2000); similarly, the N-terminal 89 residues of rotavirus VP2 are also not clearly located by the cryo-EM (Lawton *et al.*, 1997). The 230 N-terminal residues of $\lambda1$ and 89 residues of rotavirus VP2 are dispensable (Dryden *et al.*, 1998; Zeng *et al.*, 1998) for viral core formation; but the intact VP2 N terminus is needed for the incorporation of VP1 and VP3 and as well as for RNA binding (Labbe *et al.*, 1994; Shaw *et al.*, 1996). Likewise, in orthoreovirus $\lambda1$, an intact N terminus is proposed to be involved in the binding and encapsidation of RNA (Kim *et al.*, 2002; Labbe *et al.*, 1994).

A key question in VP3 assembly is whether the decamer present in the final assembled particle is an identifiable intermediate in the assembly process or arises only on assembly. Deletion of the dimerization domain showed abolition of subcore formation, but did not perturb decamer or dimer formation. Decamers were still highly stable but, presumably because of their hydrophobic nature, a higher order aggregation of decamers was also evident from the cryo-EM and dynamic light-scattering experiments. These data are consistent with the hypothesis that interaction between the decamers via this region is an intermediate stage of assembly (Kar and Roy, 2003).

As discussed previously, VP3 decamers retain the ability to interact with the minor proteins as the VP3 subcore layer does. However, decamers did not bind BTV RNA under conditions in which the intact VP3 bound RNA efficiently. This is consistent with the model of Gouet *et al.* (1999), which suggested that the position in VP3, compatible with RNA–protein interaction, lies in a single loop, the $\beta$J/$\beta$K loop (residues 790–817), within the dimerization domain (Grimes *et al.*, 1998).

These data suggest that decamers are the first stable assembly intermediates of the virion and that VP3 incorporates VP1, VP4, and VP6 at this stage (see Fig. 10). The viral genome, by contrast, wraps around the VP1–VP4 and possibly VP6 complex while the subcores are assembling. An interesting extension of this model is that it suggests that infectious BTV, assembled entirely *in vitro*, may be a realistic possibility as further aspects of the assembly pathway become equally clear.

### C. *VP7 Assembly*

The core particle contains 780 molecules of the 38-kDa VP7 protein that encapsidate and interact with 120 molecules of the 100-kDa VP3 protein. The mismatch between the number of subunits in the VP3 and VP7 layers poses an interesting problem as to how these layers reconcile to form an intact icosahedral structure. In the absence of VP3, VP7 forms trimers, but these trimers do not assemble as icosahedral particles. The construction of the $T = 13$ icosahedral shell requires polymorphism in the association of the VP7 subunits, each of which has two domains that contributes to trimer formation. In solution, VP7 forms hexamers and pentamers of trimers. Although both domains contribute substantially to the formation of the trimer, it is the lower helical domain that controls exclusively both the formations of hexamers and pentamers and their interaction with

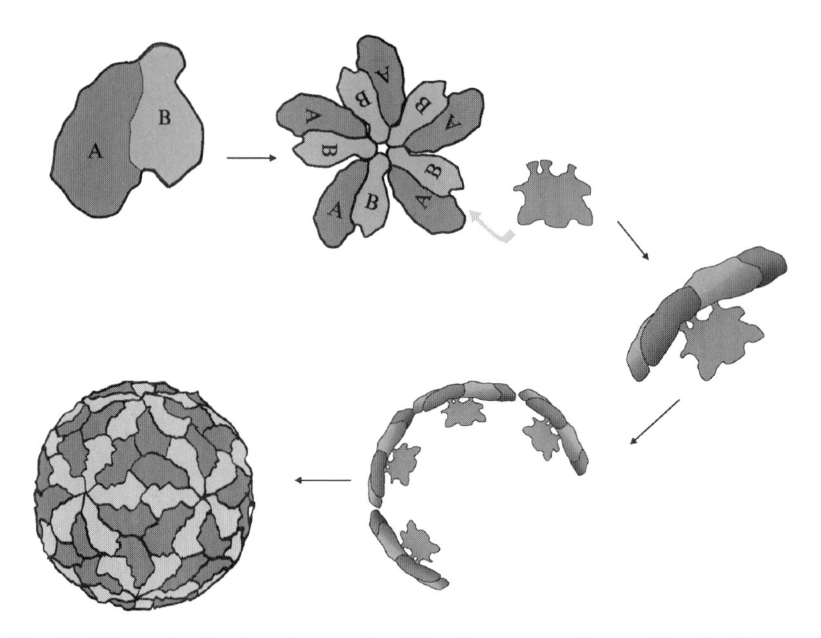


FIG 10. Schematic showing the growing VP3 decamers and assembly of polymerase complex with the VP3 decamer before formation of the VP3 protein shell. (See Color Insert.)

the scaffolding layer of VP3. The precise juxtaposition of the five quasi-equivalent trimers varies slightly in accordance with the requirements of quasi-equivalence. The structural and relative sequence information has guided an investigation of how such a complex structure is achieved during virus assembly and what residues are required to form a stable capsid (Limn and Roy, 2003; Limn *et al.*, 2000). Extensive site-directed mutagenesis in combination with various biological assay systems has given insight into the order of steps in the assembly pathway of VP7 and stable core formation that subsequently serve as a foundation for the deposition of the outer capsid.

*1. VP7 Monomer–Monomer Interaction*

Because the lower domains are directly in contact with the VP3 layer, and the interactions between the lower domains within the trimers are extensive, a series of VP7 mutations was generated among the residues of the lower domain, which appeared to be involved in intramolecular (within the VP7 subunit) and intermolecular (between the VP7 subunits) interactions (Limn *et al.*, 2000). These mutations were particularly targeted at the residues of three helices, namely, helices 5, 6, and 8, and associated loops on the basis of their position and conservative nature among the three orbivirus VP7 proteins. The effects of these mutations on VP7 solubility, ability to trimerize, and formation of CLPs in the presence of the VP3 scaffold were investigated. Some of these VP7 mutants (e.g., R111F, W119D, and T321R) severely affected the stability of CLPs, whereas other mutants (e.g., F268R and D318N) had lesser effects or had no apparent effect (e.g., E104W) on the formation of stable capsids. Some had a drastic effect, such as insolubility of the protein (particularly when the conserved Tyr-271 was substituted with arginine) implying an acute alteration in the folding of the molecule despite the prediction that such a change would be accommodated. However, all soluble mutants, including those that did not form stable CLPs, assembled into stable trimers, indicating that CLP assembly is not an automatic consequence of trimerization. It is clear that a single substitution at a key position could destabilize BTV core assembly without drastically disrupting trimer formation. The effects of these VP7 mutations on CLP formation also demonstrated that VP7 trimer formation could not accommodate most of the designed point mutations without any functional effect on capsid assembly (Limn *et al.*, 2000). The instability of the CLPs in these experiments was likely due to weak interactions between the various mutant trimers, indicating that subtle alterations (such as bonding angle or subunit conformation) within a trimer could have a profound effect on trimer–trimer interaction although formation of stable trimers was still achieved. Thus, the formation of VP7 lattice on the core surface requires exact fitting of 260 VP7 trimers. In addition, the mutations might also have influenced indirectly the interactions between the VP7 trimers and the VP3 subcore by slightly changing the flat surface of the bottom of the trimer.

The requirement for a correct trimeric structure (presumably a correct shape) of the lower domain as crucial for interaction with the underlying VP3 scaffold was also illustrated in a complementary study in which chimeric domain-switched VP7 proteins were synthesized, using BTV and AHSV (Monastyrskaya *et al.*, 1995). Although as expected these chimeric molecules could readily oligomerize as trimers in solution they did not interact with VP3 and assemble as CLPs. The studies also highlight some of the key residues that are crucial for BTV core assembly and illustrate how the structure of VP7 in isolation underrepresents the dynamic nature of the assembly process at the biological level.

### 2. *Trimer–Trimer Interaction*

Because VP7 molecules oligomerize into trimers when expressed alone, in the absence of VP3 protein, it is possible that attachment of preformed trimers onto the VP3 subcore and subsequent formation of the VP7 layer would be directed by side-to-side interaction of adjacent trimers. Interactions between neighboring trimers at the twofold axis appear to be through a thin band around the lower domains.

A series of single and multiple substitution mutants of VP7 was created to target these regions in order to examine their involvement in assembly (Limn and Roy, 2003). Of the various mutations at different sites, the importance of at least two particular residues, P338 and A346 of helix 9 (located at the carboxyl end of the molecule), in CLP assembly was shown after biochemical and morphological assay. Whereas single-site mutants P338Q and A346R could still form reasonably stable CLPs, the double mutations G337Q/P338Q and A342R/A346R resulted in poor CLP morphology, with only ~60 VP7 molecules retained by the VP3 surface of each CLP isolated in the presence of high salt. As helix 9 was not involved in direct interaction with the VP3 subcore (Grimes *et al.*, 1998), the effect of these mutations relates to the later packing of VP7 trimers on the VP3 surface. Although these changes confirmed the role of the helix in trimer–trimer interaction, trimer formation per se was not altered, suggesting that trimer assembly alone is not sufficient for completion of VP7 attachment to the subcore.

### 3. *VP7–VP3 Interactions in Core Assembly*

X-ray crystallization data have shown that the VP7 trimers contact the underlying VP3 lattice through a range of hydrophobic residues that occur on the relatively flat underside of the trimers (Grimes *et al.*, 1998). A series of single or multiple site-specific substitution mutations have been introduced into the regions of the flat undersurface of VP7 trimers that adhere closely to the VP3 surface, in particular, helix 2 and its associated loops. These mutations consisted of substitution, using either related AHSV sequences or nonrelated sequences, predominantly substituting hydrophobic for charged residues. Another group of mutations targeted within this region are certain sequences that are highly conserved within the different orbiviruses. These changes were designed not to disturb the folding pattern of the molecule but were expected to interfere with the protein–protein contacts.

When CLP assembly was examined with each of these VP7 mutants, all viruses with mutations that mimicked the AHSV sequence behaved much like BTV—with no detectable effect on overall folding or the molar ratio of VP7 to VP3 in CLPs, which emphasizes the conservative structural nature of the two molecules. In contrast, mutations that changed the character of the residue (e.g., hydrophobic to charged or vice versa) resulted in unstable, sometimes severely unstable, CLP assembly. CLPs were more stable when isolated in high salt, suggesting that solvation of the charged groups can prevent a substantial mismatch between the VP7 and VP3 proteins, while the assembly is driven by the hydrophobic interactions.

The effects of the second group of mutations on CLP assembly, in which the changes were introduced at conserved residues, were more profound and had a dramatic effect on the stable attachment of VP7 to the VP3 scaffold, retaining fewer VP7 proteins than in native CLPs. In particular, a single change, asparagine at residue 38 to aspartate, abolished CLP assembly, which may be explained by the long side chain of the aspartate hindering contact between the VP3 layer and VP7 trimers.

In summary, from a detailed analysis of an extensive range of targeted VP7 mutations it has been possible to precisely identify the critical residues responsible for core assembly and to reveal that core assembly depends not only on trimer formation but also on the precise "shape" of the trimers involved. In terms of the overall assembly pathway of the BTV core, a combination of cryo-EM and X-ray structures has revealed that, of the 13 unique contacts made between the VP3 and VP7 shells, the contact that aligns the VP7 trimer axis with the icosahedral threefold axis of the VP3 layer is the strongest. This suggests that it may nucleate the assembly of the VP7 lattice on the VP3 subcore once the first trimer is anchored. It has been postulated that preformed hexamers of VP7 may propagate around the initial VP7 trimer, forming a sheet that loosely covers the VP3 layer (Gouet *et al.*, 1999). The data obtained from mutagenesis studies did not support a gradient of trimer association from a single nucleation site, as mutations that destabilize the CLP particle still allow assembly of some VP7 lattice on the VP3 shell. An alternative model for assembly suggests that multiple sheets of VP7 may form at different nucleation sites (Limn and Roy, 2003). A likely pathway of core assembly is one involving multiple preferred initiation sites for VP7 assembly on VP3 (as illustrated in Fig. 11) and a second set of weaker interactions that then "fill the gaps." There is a clear sequential order of trimer attachment on the VP3 scaffold. The T trimers, which are at the threefold axis of the icosahedron, locate first, whereas P trimers, which are further from the threefold axis and near the fivefold axis, are the last to

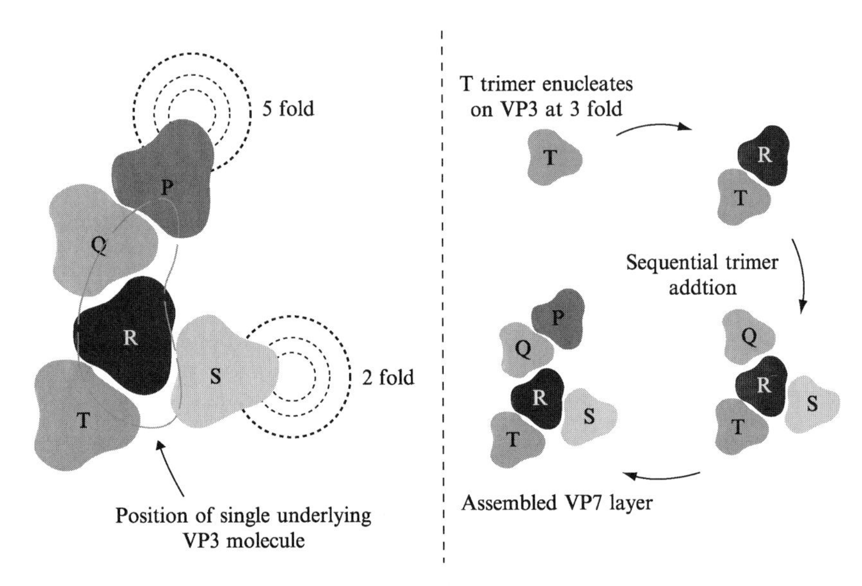


FIG 11. Schematic diagram showing the possible pathway of VP7 trimer assembly. *Left*: Position of VP7 protomeric unit and position of VP3 layer. *Right*: Sequential assembly of various VP7 trimers on VP3 scaffold.

attach. The distinction between those VP7 trimers that first occupy the subcore and those that follow is intriguingly subtle, as it must be sufficient to allow variation in packing yet not prevent the overall biological purpose of virus assembly.

### *D. Interactions Between the Outer Capsid Layer and the Surface VP7 Layer of the Core*

Once the stable core is assembled, the two outer capsid proteins VP2 and VP5 are acquired by interaction with the core surface, both proteins being attached to the VP7 layer of the core. The assembly of VP2 and/or VP5 with the core is easily mimicked either by baculovirus expression systems (see Fig. 9) or in an *in vitro* transcription–translation system (French *et al.*, 1990; Liu *et al.*, 1992; Loudon and Roy, 1991). Both proteins can be incorporated with the core independent of each other, implying that the two proteins are directly attached to the core and do not intimately interact with each other. Once VP2 and VP5 are attached, the particles are no longer transcriptionally active. Instead, they are available for release from the infected cell.

A

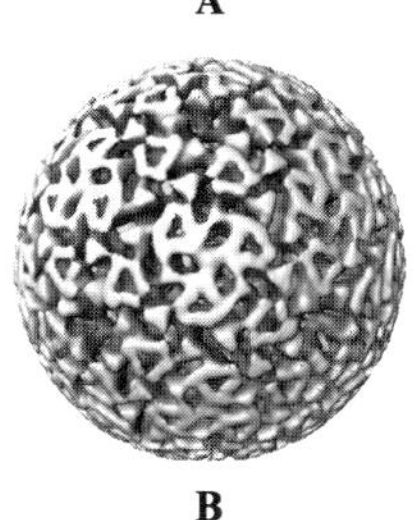

B

FIG 12. VP7 residues involved in the interactions with outer layer proteins. (A) Top head-on view (*left*) and a side view (*right*). The X-ray structure of the VP7 trimer is shown in ribbon representation. The trimeric subunits are colored red, blue, and green. The locations on VP7 where the triskelion motifs interact are shown in white and where the globular domains interact is shown in yellow. (B) Virion reconstruction, at approximately the same outer radius of the core structure, showing the triangular feature at the type II and type III channels. A set of three fivefold axes is also indicated. (See Color Insert.)

The structural organization of the outer layer represents a drastic mismatch with the underlying VP7 layer. The VP7 trimers with a triangular top portion serve as the platform for the deposition of VP2 and VP5. Each triskelion motif (VP2 trimers) essentially interacts with four VP7 trimers by their underside in four places. The base of the triskelion interacts with the Q-type VP7, whereas the propeller-like arms make three other connections with the P, R, and S VP7 trimers. By docking the X-ray structure of the VP7 trimer into the cryo-EM virion reconstruction it was possible to identify the VP7 residues that are in contact with VP2. The base of the VP2 trimer interacts exclusively with the upper flattish surface of the VP7 trimer, which includes amino acid residues 141–143, 164–166, 195–205, and 238–241 (Fig. 12A). All the VP7 molecules of the core are covered by the connections that are made with the tips of the propeller except for VP7 at the icosahedral threefold axis (T type). The top of the VP7 trimer at this position is thus clearly exposed to the exterior.

The globular densities (VP5 trimers), on the other hand, sit right above the type II and type III channels, between the VP7 trimers, making contacts with the sides of the VP7 trimers that face these channels. Thus, the channels are filled with triangle-shaped densities that are connected to the VP5 trimers (Fig. 12B). The inner triangle-shaped density interacts mainly with the lower portion of the $\beta$-barrel domain and includes residues 168–173, 210–215, and 226–234 of the distal $\beta$-barrel domain of VP7 (Fig. 12A). Thus VP5 has much stronger interaction with VP7, which is consistent with biochemical data demonstrating that VP2 can be dissociated from the virion easily without removal of VP5 (Huismans *et al.*, 1987a; Verwoerd *et al.*, 1972).

## VII. What Is Known About BTV Egress and the Involvement of Host Protein(s)

Unlike enveloped viruses, BTV and other orbiviruses lack lipid envelopes and are released predominantly by cell lysis of the infected vertebrate hosts. By contrast, BTV release from insect vector cells is nonlytic and the virus particles leave the infected cell by passage, either individually or in groups, through a locally disrupted plasma membrane (Hyatt *et al.*, 1989). Biologically, a nonlytic infection of the insect allows it to retain its ability to act as a vector, but how the virus controls this differential egress is one of the most interesting aspects of BTV replication.

The nonstructural protein NS3 is synthesized in two forms, NS3 (229 amino acids) and NS3A (216 amino acids), as a result of translation initiation at a second in-frame initiation codon in RNA segment 10 (French *et al.*, 1989; Lee and Roy, 1986). The NS3/NS3A proteins accumulate to only low levels in BTV-infected mammalian cells, but in invertebrate cells the expression of these proteins is high (French *et al.*, 1989; Guirakhoo *et al.*, 1995). The correlation between high NS3/NS3A expression and nonlytic virus release suggests a significant functional role for NS3 in virus egress from invertebrate cells. The newly synthesized NS3 proteins are transported to the Golgi apparatus and then to the cell membrane, and both proteins exist in glycosylated and nonglycosylated forms (Wu *et al.*, 1992). The proteins have two cytoplasmic domains, which are linked via two hydrophobic domains, each spanning the cell membrane, consistent with a role in membrane perturbation and a short extracellular domain. The NS3 proteins have also been found in association with smooth-surfaced, intracellular vesicles (Brookes *et al.*, 1993). When NS3/NS3A is expressed using recombinant baculoviruses it facilitates the release of VLPs (acting as surrogates for authentic virions) from host cells, and NS3 protein is localized at the site of the membrane from which VLPs are released (Hyatt *et al.*, 1993). This suggests that the integration of NS3/NS3A into the plasma membrane may facilitate mature virion release, a function that is similar to that described for NSP4, a non-structural glycoprotein of rotaviruses, which interacts with viral cores to facilitate their transport across the rough endoplasmic reticulum of infected cells (Meyer *et al.*, 1989). More importantly, data suggest that NS3 may take advantage of the normal cellular exocytosis pathway to facilitate the nonlytic release of mature virions from insect cells (Beaton *et al.*, 2002).

By using the two-hybrid system it has been shown that the cytoplasmic amino-terminal domain of NS3 specifically interacts with the calpactin light chain (p11) of the cellular Annexin II complex, a member of the S100 family (Beaton *et al.*, 2002). The protein p11 commonly forms a heterotetramer (the Annexin II complex) with Annexin II heavy chain (p36). The complex has been implicated in membrane-related events along the endocytic and regulated secretory pathways (for reviews see Creutz, 1992; Gerke, 2001; Raynal and Pollard, 1994), including the trafficking of vesicles. In particular, later steps in the $Ca^{2+}$-regulated pathway of secretion in chromaffin cells and the peripheral positioning of early endosomes in polarized epithelial cells appear to be affected by Annexin II (Puisieux *et al.*, 1996; Sarafian *et al.*, 1991). In addition, p11 light chain was shown to be the critical component involved in the plasma membrane translocation of NAV1.8, a subunit of the tetrodotoxin-resistant sodium channel found in sensory neurons (Okuse *et al.*, 2002). It is possible p11 represents a general effector molecule for plasma membrane localization and that BTV, and perhaps other viruses, make use of this cellular pathway for virus egress. NS3 contains a putative amphipathic helix in its first 13 amino acids that is responsible for binding p11, suggesting that the interaction seen is probably highly specific. The specificity of such interaction was further confirmed by *in vitro* pull-down assay and by *in vivo* colocalization studies (Fig. 13A). Additional studies of protein–protein interactions analysis combined with deletion and site-specific mutagenesis studies have further supported the specific nature of p11–NS3NH complex formation and demonstrated that NS3 interacts at

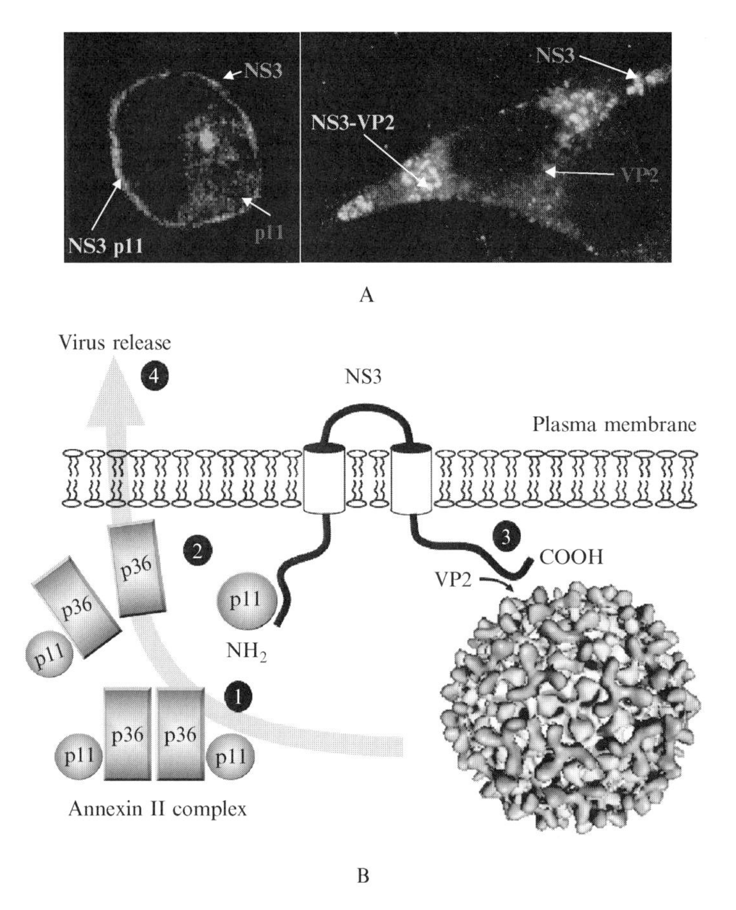


FIG 13. (A) Confocal microscopy showing (*left*) the merging of p11 and NS3 in BTV-infected HeLa cells and (*right*) the merging of NS3 and VP2 in BTV-infected BHK cells. (B) Schematic model of host protein and NS3 interaction during virus release from the infected cell. (See Color Insert.)

the same site of p11 as the p36 (Beaton *et al.*, 2002). Thus NS3 seems to mimic the interaction of p36 with p11 and the NS3–p11 complex could play a role in the budding of viral particles by linking the membranes at the neck of the nascent viral particle, thereby facilitating the eventual fusion of the opposing membranes. That the interaction of NS3 with p11 is a physiologically relevant interaction for virus release of progeny virions from infected insect vector cells was demonstrated by the inhibitory effects of an NS3 peptide (mimicking the sequences of the p11-binding domain) on virus release. These data suggest that interaction of p11 with NS3 may help direct NS3 to sites of active cellular exocytosis, or it could be part of an active extrusion process. Furthermore, there are some indications that cytoskeletal material was seen at sites of BTV egress (Hyatt *et al.*, 1989), which may be Annexin II being drawn through the membrane during the extrusion process, as it is still associated with NS3.

It is most intriguing that the N-terminal cytoplasmic domain of NS3 contains so-called late domain motifs that serve to recruit cellular proteins involved in the formation of multivesicular bodies (MVBs) and sorting of ubiquitinated cargo into these vesicles. MVBs are specialized endocytic compartments, which can either function as transport intermediates in the lysosomal degradation pathway or as storage intermediates for intracellular vesicles that can be released as exosomes. The significance of the involvement of the MVB machinery in virus budding derives from the fact that formation of MVBs shares the same topology as budding of virus particles, that is, away from the cytoplasm. Formation of MVBs requires a multitude of cellular proteins, which seem to be recruited in a sequential manner. One of the proteins that is recruited at an early stage of MVB formation is Tsg101, which has been shown to bind to the late domain motif PTAP of HIV (Ono and Freed, 2004). Our study has shown that Tsg101 also binds to NS3 (C. Wirblich and P. Roy, unpublished observation) via a PSAP (amino acids 41–44) motif in its N-terminal domain. This interaction is particularly interesting because the heavy chain of calpactin I (Annexin II) is clearly involved in the regulation of MVB biogenesis (Mayran *et al.*, 2003). In addition to p11 and Tsg101 NS3 also seems to interact with the dynein light chain 2A (A. R. Beaton and P. Roy, unpublished data). It is intriguing that dynein also plays an important role in endosomal/lysosomal membrane dynamics (Harrison *et al.*, 2003). Thus, it seems that NS3 acts as a platform that recruits cellular machinery. A condition of NS3 playing a role in virus release is that it must also associate with the maturing virions during morphogenesis. Using yeast two-hybrid screening of a viral protein library, it was demonstrated that NS3 is capable of interacting with outer capsid protein VP2 but not with VP5 via its cytoplasmic carboxy-terminal domain (Beaton *et al.*, 2002). The VP2–NS3 interaction in infected cell cytoplasm was further supported by visualizing the colocalization of NS3 and VP2 (Fig. 13A). It is likely that interaction between NS3 and assembling viruses is part of an active transportation process. Our combined results enable us to propose a model (Fig. 13B) for the function of NS3 based on its membrane residency, known topology, and interaction with p11 and VP2. In this model, NS3 forms a "bridge" between the maturing virions and p11, most likely as part of the Annexin II complex, directing virus to the cellular exocytic machinery. It has not been possible to elucidate the exact method of virus release but it is likely that the link between p11 and virus, via NS3, allows active transport of virus across the membrane. These data form a platform from which further understanding of the mechanisms of nonlytic versus lytic release of BTV may be approached and may also lead to the establishment of general mechanisms whereby arboviruses modulate their relative cytotoxicity.

## VIII. Concluding Remarks

Significant progress has been made in relation to the structure–function relationships among the BTV proteins. The virus is now well characterized both genetically (the sequence was completed in 1989) and structurally. These advances have been possible mainly because of the availability of each BTV protein in highly purified form and in biologically active native configuration. In addition, multiprotein complexes representing subviral particles have facilitated an understanding of the mechanism of protein–protein interaction and the virus assembly pathway. A combination of mutagenesis and reexpression of BTV proteins has allowed for the mapping of the precise function of many of them, in particular, the virion structural proteins. These data, together with the success of obtaining atomic structures for the complex BTV core and core proteins, have provided a sound foundation for a more detailed analysis of the role of each protein in virus replication. Latterly, studies have concentrated on the fundamental mechanisms that are used by the virus to invade, replicate in and escape from susceptible host cells. Progress has been made in understanding the structure and entry of intact virus particles and the critical interactions that occur between the viral nonstructural proteins and viral RNA, and host cell proteins. Moreover, they have discovered experimental evidence for the role of cellular proteins in nonenveloped virus egress.

Despite these advances, some critical questions remain unanswered for the BTV life cycle and a more complete understanding of the interactions between the virus and the host cell are required for these to be addressed. For example, it is clear that VP2 and VP5 have activities that are functionally equivalent to the fusion proteins of enveloped viruses, that VP2 in trimeric form mediates virus attachment to the cell and that structural changes are triggered in VP5 oligomers in response to the low-pH environment of the endosome. However, neither the details of the BTV entry process into host cells, nor the structural rearrangements in the VP5 protein after cell entry are currently known; combined molecular and structural studies of the BTV core have allowed an understanding of the structure and assembly of the most abundant proteins of the core and definition of enzymatic roles of its internal minor proteins. However, there are some significant gaps in our understanding of this complex molecular machine, because structures of the three minor proteins that are essential components of the viral polymerase complex are not yet available at the atomic level. However, it may not be too long before some structural information is available for these proteins, due to the ongoing research efforts in several laboratories. That information, when available, will allow dissection of the assembly and molecular mechanisms of those enzymes; progress has been made in the identification of signals for the recruitment of BTV RNA segments into the virion assembly site in the host cell cytoplasm, but it has not been possible yet to determine how exactly each genome segment is packaged into the progeny virus. It is also not apparent when and how these genome segments wrap around the polymerase complex once the RNA has been recruited; little is known of the intracellular trafficking of newly generated virions although there are some indications of involvement of the cytoskeleton, intermediate filaments, and vimentin during BTV morphogenesis. Host–virus interactions during virus trafficking will be one of the future areas needing intense attention. In the same way that there appear to be parallels in the entry processes of enveloped viruses and BTV virions, a similar common pathway for export of BTV may exist because there is clear indication that BTV uses resident proteins of the cellular exocytic pathway. However, the exact role of these proteins (Annexin II and p11, Tsg101 and MVB) and their contribution in the release of nonenveloped viruses, such as BTV, needs to be defined.

One of the major drawbacks of research with BTV and other members of *Reoviridae* has been the lack of availability of a suitable system (i.e., reverse genetics) that allows genetic manipulation of the virus. This has been a major obstacle to date in understanding the replication processes of these viruses. There is no doubt that it is one of the dominant research areas and that many laboratories are currently involved. Once in place, molecular and structural studies of individual BTV proteins can be placed in the context of the whole virus.

## References

Altenburg, B. C., Graham, D. Y., and Estes, M. K. (1980). Ultrastructural study of rotavirus replication in cultured cells. *J. Gen. Virol.* **46:**75–85.

Bae, S. H., Cheong, H. K., Lee, J. H., Cheong, C., Kainosho, M., and Choi, B. S. (2001). Structural features of an influenza virus promoter and their implications for viral RNA synthesis. *Proc. Natl. Acad. Sci. USA* **98:**10602–10607.

Banerjee, A. K., and Shatkin, A. J. (1970). Transcription *in vitro* by reovirus associated ribonucleic acid-dependent polymerase. *J. Virol.* **6:**1–11.

Basak, A. K., Gouet, P., Grimes, J., Roy, P., and Stuart, D. (1996). Crystal structure of the top domain of African horse sickness virus VP7: Comparisons with bluetongue virus VP7. *J. Virol.* **70:**3797–3806.

Basak, A. K., Grimes, J. M., Gouet, P., Roy, P., and Stuart, D. I. (1997). Structures of orbivirus VP7: Implications for the role of this protein in the viral life cycle. *Structure* **5:**871–883.

Basak, A. K., Stuart, D. I., and Roy, P. (1992). Preliminary crystallographic study of bluetongue virus capsid protein, VP7. *J. Mol. Biol.* **228:**687–689.

Beasley, B. E., and Hu, W. S. (2002). *Cis*-acting elements important for retroviral rna packaging specificity. *J. Virol.* **76:**4950–4960.

Beaton, A. R., Rodriguez, J., Reddy, Y. K., and Roy, P. (2002). The membrane trafficking protein calpactin forms a complex with bluetongue virus protein NS3 and mediates virus release. *Proc. Natl. Acad. Sci. USA* **99:**13154–13159.

Bentz, J. (2000). Membrane fusion mediated by coiled coils: A hypothesis. *Biophys. J.* **78:**886–900.

Bisaillon, M., Bergeron, J., and Lemay, G. (1997). Characterization of the nucleoside triphosphate phosphohydrolase and helicase activities of the reovirus $\lambda_1$ protein. *J. Biol. Chem.* **272:**18298–18303.

Boyce, M., Wehrfritz, J., Noad, R., and Roy, P. (2004). Purified recombinant bluetongue virus VP1 exhibits RNA replicase activity. *J. Virol.* **78:**3994–4002.

Brookes, S. M., Hyatt, A. D., and Eaton, B. T. (1993). Characterization of virus inclusion bodies in bluetongue virus infected cells. *J. Gen. Virol.* **74:**525–530.

Brookes, S. M., Hyatt, A. D., and Eaton, B. T. (1994). The use of immuno-gold silver staining in bluetongue virus adsorption and neutralisation studies. *J. Virol. Methods* **46:**117–132.

Brown, D., Brown, J., Kang, C., Gold, L., and Allen, P. M. (1997). Single-stranded RNA recognition by the bacteriophage T4 translational repressor, RegA. *J. Biol. Chem.* **272:**14969–14974.

Browne, J. G., and Jochim, M. M. (1967). Cytopathologic changes and development of inclusion bodies in cultured cells infected with bluetongue virus. *Am. J. Veter. Res.* **28:**1091–1105.

Butcher, S. J., Dokland, T., Ojala, P. M., Bamford, D. H., and Fuller, S. D. (1997). Intermediates in the assembly pathway of the double-stranded RNA virus $\varphi$6. *EMBO J.* **16:**4477–4487.

Chandran, K., and Nibert, M. L. (2003). Animal cell invasion by a large nonenveloped virus: Reovirus delivers the goods. *Trends Microbiol.* **11:**374–382.

Chen, D., and Patton, J. T. (1998). Rotavirus RNA replication requires a single-stranded 3′ end for efficient minus-strand synthesis. *J. Virol.* **72:**7387–7396.

Chen, D., Zeng, C. Q., Wentz, M. J., Gorziglia, M., Estes, M. K., and Ramig, R. F. (1994). Template-dependent, *in vitro* replication of rotavirus RNA. *J. Virol.* **68:**7030–7039.

Chen, D., Barros, M., Spencer, E., and Patton, J. T. (2001). Features of the 3′ consensus sequence of rotavirus mRNAs critical to minus strand synthesis. *Virology* **282:**221–229.

Chiu, Y.-L., Coronel, E., Ho, C. K., Shuman, S., and Rana, T. M. (2001). HIV-1 Tat protein interacts with mammalian capping enzyme and stimulates capping of TAR RNA. *J. Biol. Chem.* **276:**12959–12966.

Cohen, F. S., and Melikyan, G. B. (2001). Implications of a fusion peptide structure. *Nat. Struct. Biol.* **8:**653–655.

Cohen, J. (1977). Ribonucleic acid polymerase activity associated with purified calf rotavirus. *J. Gen. Virol.* **36:**395–402.

Colman, P. M., and Lawrence, M. C. (2003). The structural biology of type I viral membrane fusion. *Nat. Rev. Mol. Cell. Biol.* **4:**309–319.

Creutz, C. E. (1992). The Annexins and exocytosis. *Science* **258:**924–931.

Deo, R. C., Groft, C. M., Rajashankar, K. R., and Burley, S. K. (2002). Recognition of the rotavirus mRNA 3′ consensus by an asymmetric NSP3 homodimer. *Cell* **108:**71–81.

Dryden, K. A., Farsetta, D. L., Wang, G., Keegan, J. M., Fields, B. N., Baker, T. S., and Nibert, M. L. (1998). Internal/structures containing transcriptase-related proteins in top component particles of mammalian orthoreovirus. *Virology* **245:**33–46.

Eaton, B. T., and Crameri, G. S. (1989). The site of bluetongue virus attachment to glycophorins from a number of animal erythrocytes. *J. Gen. Virol.* **70:**3347–3353.

Eaton, B. T., and Hyatt, A. D. (1989). Association of bluetongue virus with the cytoskeleton. *Subcell. Biochem.* **15:**233–273.

Eaton, B. T., Hyatt, A. D., and Brookes, S. M. (1990). The replication of bluetongue virus. *Curr. Top. Microbiol. Immunol.* **162:**89–118.

Els, H. J., and Verwoerd, D. W. (1969). Morphology of bluetongue virus. *Virology* **38:**213–219.

Fillmore, G. C., Lin, H., and Li, J. K.-K. (2002). Localization of the single-stranded RNA-binding domains of bluetongue virus nonstructural protein NS2. *J. Virol.* **76:**499–506.

Forzan, M., Wirblich, C., and Roy, P. (2004). A capsid protein of nonenveloped bluetongue virus exhibits membrane fusion activity. *Proc. Natl. Acad. Sci. USA* **101:**2100–2105.

French, T. J., Inumaru, S., and Roy, P. (1989). Expression of two related nonstructural proteins of bluetongue virus (BTV) type 10 in insect cells by a recombinant baculovirus: Production of polyclonal ascitic fluid and characterization of the gene product in BTV-infected BHK cells. *J. Virol.* **63:**3270–3278.

French, T. J., Marshall, J. J., and Roy, P. (1990). Assembly of double-shelled, viruslike particles of bluetongue virus by the simultaneous expression of four structural proteins. *J. Virol.* **64:**5695–5700.

French, T. J., and Roy, P. (1990). Synthesis of bluetongue virus (BTV) corelike particles by a recombinant baculovirus expressing the two major structural core proteins of BTV. *J. Virol.* **64:**1530–1536.

Fujii, Y., Goto, H., Watanabe, T., Yoshida, T., and Kawaoka, Y. (2003). Selective incorporation of influenza virus RNA segments into virions. *Proc. Natl. Acad. Sci. USA* **100:**2002–2007.

Fukusho, A., Ritter, G. D., and Roy, P. (1987). Variation in the bluetongue virus neutralization protein VP2. *J. Gen. Virol.* **68:**2967–2973.

Fukusho, A., Yamaguchi, Y., and Roy, P. (1989). Completion of the sequence of bluetongue virus serotype 10 by the characterization of structural protein VP5 and a nonstructural protein, NS2. *J. Gen. Virol.* **70:**1677–1689.

Gao, Y., and Lenard, J. (1995). Cooperative binding of multimeric phosphoprotein (P) of vesicular stomatitis virus to polymerase (L) and template: Pathways of assembly. *J. Virol.* **69:**7718–7723.
Gerke, V. (2001). Annexins and membrane organisation in the endocytic pathway. *Cell Mol. Biol. Lett.* **6**(2)**:**204.
Ghiasi, H., Fukusho, A., Eshita, Y., and Roy, P. (1987). Identification and characterization of conserved and variable regions in the neutralization VP2 gene of bluetongue virus. *Virology* **160:**100–109.
Gillian, A. L., Schmaechel, S. C., Livny, J., Schiff, L. A., and Nibert, M. L. (2000). Reovirus protein $\sigma$ NS binds in multiple copies to single stranded RNA and shares properties with single stranded DNA binding proteins. *J. Virol.* **74:**5939–5948.
Gouet, P., Diprose, J. M., Grimes, J. M., Malby, R., Burroughs, J. N., Zientara, S., Stuart, D. I., and Mertens, P. P. (1999). The highly ordered double-stranded RNA genome of bluetongue virus revealed by crystallography. *Cell* **97:**481–490.
Gould, A. R. (1988). Conserved and non-conserved regions of the outer coat protein, VP2, of the Australian bluetongue serotype 1 virus, revealed by sequence comparison to the VP2 of North American BTV serotype 10. *Virus Res.* **9:**145–158.
Greber, U. F. (2002). Signalling in viral entry. *Cell. Mol. Life Sci.* **59:**608–626.
Greber, U. F., Willetts, M., Webster, P., and Helenius, A. (1993). Stepwise dismantling of adenovirus 2 during entry into cells. *Cell* **75:**477–486.
Grimes, J., Basak, A. K., Roy, P., and Stuart, D. (1995). The crystal structure of bluetongue virus VP7. *Nature* **373:**167–170.
Grimes, J. M., Burroughs, J. N., Gouet, P., Diprose, J. M., Malby, R., Zientara, S., Mertens, P. P. C., and Stuart, D. I. (1998). The atomic structure of the bluetongue virus core. *Nature* **395:**470–478.
Grimes, J. M., Jakana, J., Ghosh, M., Basak, A. K., Roy, P., Chiu, W., Stuart, D. I., and Prasad, B. V. V. (1997). An atomic model of the outer layer of the bluetongue virus core derived from X-ray crystallography and electron cryomicroscopy. *Structure* **5:**885–893.
Guirakhoo, F., Catalan, J. A., and Monath, T. P. (1995). Adaptation of bluetongue virus in mosquito cells results in overexpression of NS3 proteins and release of virus particles. *Arch. Virol.* **140:**967–974.
Hakansson, K., Doherty, A. J., Shuman, S., and Wigley, D. B. (1997). X-ray crystallography reveals a large conformational change during guanyl transfer by mRNA capping enzymes. *Cell* **89:**545–553.
Hakansson, K., and Wigley, D. B. (1998). Structure of a complex between a cap analogue and mRNA guanylyl transferase demonstrates the structural chemistry of RNA capping. *Proc. Natl. Acad. Sci. USA* **95:**1505–1510.
Harrison, R. E., Bucci, C., Vieira, O. V., Schroer, T. A., and Grinstein, S. (2003). Phagosomes fuse with late endosomes and/or lysosomes by extension of membrane protrusions along microtubules: Role of Rab7 and RILP. *Mol. Cell. Biol.* **23:**6494–6506.
Harrison, S. C. (2005). Mechanism of membrane fusion by viral envelope proteins. *Adv. Virus Res.* **64**.
Hassan, S. H., and Roy, P. (1999). Expression and functional characterization of bluetongue virus VP2 protein: Role in cell entry. *J. Virol.* **73:**9832–9842.
Hassan, S. H., Wirblich, C., Forzan, M., and Roy, P. (2001). Expression and functional characterization of bluetongue virus VP5 protein: Role in cellular permeabilization. *J. Virol.* **75:**8356–8367.
Hewat, E. A., Booth, T. F., Loudon, P. T., and Roy, P. (1992a). Three-dimensional reconstruction of baculovirus expressed bluetongue virus core-like particles by cryo-electron microscopy. *Virology* **189:**10–20.

Hewat, E. A., Booth, T. F., and Roy, P. (1992b). Structure of bluetongue virus particles by cryoelectron microscopy. *J. Struct. Biol.* **109:**61–69.
Hewat, E. A., Booth, T. F., and Roy, P. (1994). Structure of Bluetongue virus-like particles by cryo-electron microscopy. *J. Struct. Biol.* **109:**61–69.
Hill, C. L., Booth, T. F., Prasad, B. V. V., Grimes, J. M., Mertens, P. P., Sutton, G. C., and Stuart, D. I. (1999). The structure of a cypovirus and the functional organization of dsRNA viruses. *Nat. Struct. Biol.* **6:**565–568.
Hogle, J. M. (2002). Poliovirus cell entry: Common structural themes in viral cell entry pathways. *Annu. Rev. Microbiol.* **56:**677–702.
Horscroft, N. J., and Roy, P. (1997). Thermal denaturation of proteins for SDS–PAGE analysis by microwave irradiation. *Biotechniques* **22:**224–226.
Huismans, H., and Els, H. J. (1979). Characterization of the tubules associated with the replication of three different orbiviruses. *Virology* **92:**397–406.
Huismans, H., and Erasmus, B. J. (1981). Identification of the serotype-specific and group-specific antigens of bluetongue virus. *Onderstepoort J. Vet. Res.* **48:**51–58.
Huismans, H., van der Walt, N. T., Cloete, M., and Erasmus, B. J. (1987a). Isolation of a capsid protein of bluetongue virus that induces a protective immune response in sheep. *Virology* **157:**172–179.
Huismans, H., Van Dijk, A. A., and Els, H. J. (1987b). Uncoating of parental bluetongue virus to core and subcore particles in infected L cells. *Virology* **157:**180–188.
Hyatt, A. D., and Eaton, B. T. (1988). Ultrastructural distribution of the major capsid proteins within bluetongue virus and infected cells. *J. Gen. Virol.* **69:**805–815.
Hyatt, A. D., and Eaton, B. T. (1990). Virological applications of the grid-cell-culture technique. *Electron Microsc. Rev.* **3:**1–27.
Hyatt, A. D., Eaton, B. T., and Brookes, S. M. (1989). The release of bluetongue virus from infected cells and their superinfection by progeny virus. *Virology* **173:**21–34.
Hyatt, A. D., Zhao, Y., and Roy, P. (1993). Release of bluetongue virus-like particles from insect cells is mediated by BTV nonstructural protein NS3/NS3A. *Virology* **193:**592–603.
Inumaru, S., and Roy, P. (1987). Production and characterization of the neutralization antigen VP2 of bluetongue virus serotype 10 using a baculovirus expression vector. *Virology* **157:**472–479.
Iwata, H., Yamagawa, M., and Roy, P. (1992). Evolutionary relationships among the gnat-transmitted orbiviruses that cause African horse sickness, bluetongue, and epizootic hemorrhagic disease as evidenced by their capsid protein sequences. *Virology* **191:**251–261.
Jahn, R., Lang, T., and Sudhof, T. C. (2003). Membrane fusion. *Cell* **112:**519–533.
Jane-Valbuena, J., Breun, L. A., Schiff, L. A., and Nibert, M. L. (2002). Sites and determinants of early cleavages in the proteolytic processing pathway of reovirus surface protein sigma3. *J. Virol.* **76:**5184–5197.
Kahlon, J., Sugiyama, K., and Roy, P. (1983). Molecular basis of bluetongue virus neutralization. *J. Virol.* **48:**627–632.
Kar, A. K., Ghosh, M., and Roy, P. (2004). Mapping the assembly of bluetongue virus scaffolding protein VP3. *Virology* **324:**387–399.
Kar, A. K., and Roy, P. (2003). Defining the structure–function relationships of bluetongue virus helicase protein VP6. *J. Virol.* **77:**11347–11356.
Kim, J., Zhang, X., Centonze, V. E., Bowman, V. D., Noble, S., Baker, T. S., and Nibert, M. L. (2002). The hydrophilic amino-terminal arm of reovirus core shell protein lambda1 is dispensable for particle assembly. *J. Virol.* **76:**12211–12222.

Kouznetzoff, A., Buckle, M., and Tordo, N. (1998). Identification of a region of the rabies virus N protein involved in direct binding to the viral RNA. *J. Gen. Virol.* **79:**1005–1013.

Kozlov, M. M., and Chernomordik, L. V. (2002). The protein coat in membrane fusion: Lessons from fission. *Traffic* **3:**256–267.

Labbe, M., Baudoux, P., Charpilienne, A., Poncet, D., and Cohen, J. (1994). Identification of the nucleic acid binding domain of the rotavirus VP2 protein. *J. Gen. Virol.* **75:**3423–3430.

Labbe, M., Charpilienne, A., Crawford, S. E., Estes, M. K., and Cohen, J. (1991). Expression of rotavirus VP2 produces empty corelike particles. *J. Virol.* **65:**2946–2952.

Lawton, J. A., Estes, M. K., and Prasad, B. V. V. (2001). Identification and characterization of a transcription pause site in rotavirus. *J. Virol.* **75:**1632–1642.

Lawton, J. A., Zeng, C. Q., Mukherjee, S. K., Cohen, J., Estes, M. K., and Prasad, B. V. V. (1997). Three-dimensional structural analysis of recombinant rotavirus-like particles with intact and amino-terminal-deleted VP2: Implications for the architecture of the VP2 capsid layer. *J. Virol.* **71:**7353–7360.

Le Blois, H., Fayard, B., Urakawa, T., and Roy, P. (1991). Synthesis and characterization of chimaeric particles between epizootic haemorrhagic disease virus and bluetongue virus: Functional domains are conserved on the VP3 protein. *J. Virol.* **65:**4821–4831.

Le Blois, H., French, T., Mertens, P. P. C., Burroughs, J. N., and Roy, P. (1992). The expressed VP4 protein of bluetongue virus binds GTP and is the candidate guanylyltransferase of the virus. *Virology* **189:**757–761.

Le Blois, H., and Roy, P. (1993). A single point mutation in the VP7 major core protein of bluetongue virus prevents the formation of core-like particles. *J. Virol.* **67:**353–359.

Lee, J. W., and Roy, P. (1986). Nucleotide sequence of a cDNA clone of RNA segment 10 of bluetongue virus (serotype 10). *J. Gen. Virol.* **67:**2833–2837.

Lee, M. T., Bishop, K., Medcalf, L., Elton, D., Digard, P., and Tiley, L. (2002). Definition of the minimal viral components required for the initiation of unprimed RNA synthesis by influenza virus RNA polymerase. *Nucleic Acids Res.* **30:**429–438.

Liemann, S., Chandran, K., Baker, T. S., Nibert, M. L., and Harrison, S. C. (2002). Structure of the reovirus membrane-penetration protein, $\mu$1, in a complex with its protector protein, $\sigma$3. *Cell* **108:**283–295.

Limn, C.-H., Staeuber, N., Monastyrskaya, K., Gouet, P., and Roy, P. (2000). Functional dissection of the major structural protein of bluetongue virus: Identification of key residues within VP7 essential for capsid assembly. *J. Virol.* **74:**8658–8669.

Limn, C. K., and Roy, P. (2003). Intermolecular interactions in a two-layered viral capsid that requires a complex symmetry mismatch. *J. Virol.* **77:**11114–11124.

Liu, H. M., Booth, T. F., and Roy, P. (1992). Interactions between bluetongue virus core and capsid proteins translated *in vitro*. *J. Gen. Virol.* **73:**2577–2584.

Lohman, T. M., Thom, K., and Vale, R. D. (1998). Staying on track: Common features of DNA helicases and microtubule motors. *Cell* **93:**9–12.

Loudon, P. T., Liu, H. M., and Roy, P. (1992). Gene to complex structures of bluetongue viruses: Structure–function relationships of bluetongue virus proteins. *In* "Bluetongue, African Horse Sickness and Related Orbiviruses" (T. E. Walton and B. I. Osburn, eds.), pp. 383–389. CRC Press, Boca Raton, FL.

Loudon, P. T., and Roy, P. (1991). Assembly of five bluetongue virus proteins expressed by recombinant baculoviruses: Inclusion of the largest protein VP1 in the core and virus-like proteins. *Virology* **180:**798–802.

Lu, G., Zhou, Z. H., Baker, M. L., Jakana, J., Cai, D., Wei, X., Chen, S., Gu, X., and Chiu, W. (1998). Structure of double-shelled rice dwarf virus. *J. Virol.* **72:**8541–8549.

Luytjes, W., Krystal, M., Enami, M., Pavin, J., and Palese, P. (1989). Amplification, expression, and packaging of foreign gene by influenza virus. *Cell* **59:**1107–1113.

Lymperopoulos, K., Wirblich, C., Brierley, I., and Roy, P. (2003). Sequence specificity in the interaction of Bluetongue virus non-structural protein 2 (NS2) with viral RNA. *J. Biol. Chem.* **278:**31722–31730.

Makeyev, E. V., and Bamford, D. H. (2000). Replicase activity of purified recombinant protein P2 of double-stranded RNA bacteriophage $\varphi$6. *EMBO J.* **19:**124–133.

Martinez-Costas, J., Sutton, G., and Roy, P. (1998). Guanylyltranferase and RNA5-triphosphatase activities of the purified expressed VP4 protein of bluetongue virus. *J. Mol. Biol.* **280:**859–866.

Matthews, J. M., Young, T. F., Tucker, S. P., and Mackay, J. P. (2000). The core of the respiratory syncytial virus fusion protein is a trimeric coiled coil. *J. Virol.* **74:**5911–5920.

Mayran, N., Parton, R. G., and Gruenberg, J. (2003). Annexin II regulates multivesicular endosome biogenesis in the degradation pathway of animal cells. *EMBO J.* **22:**3242–3253.

Merrifield, R. B., Vizioli, L. D., and Boman, H.G. (1982). Synthesis of the antibacterial peptide cecropin A (1–33). *Biochemistry* **21:**5020–5031.

Mertens, P. P., Burroughs, J. N., and Anderson, J. (1987). Purification and properties of virus particles, infectious subviral particles, and cores of bluetongue virus serotypes 1 and 4. *Virology* **157:**375–386.

Meyer, J. C., Bergmann, C. C., and Bellamy, A. R. (1989). Interaction of rotavirus cores with the non-structural glycoprotein NS28. *Virology* **171:**98–107.

Mindich, L. (1999). Precise packaging of the three genomic segments of the double-stranded-RNA bacteriophage $\varphi$6. *Microbiol. Mol. Biol. Rev.* **63:**149–160.

Monastyrskaya, K., Gould, E. A., and Roy, P. (1995). Characterization and modification of the carboxy-terminal sequences of bluetongue virus type 10 NS1 protein in relation to tubule formation and location of an antigenic epitope in the vicinity of the carboxy terminus of the protein. *J. Virol.* **69:**2831–2841.

Montenarh, M., and Muller, D. (1987). The phosphorylation of Thr 124 of simian virus 40 large T antigen is crucial in cell entry. *J. Virol.* **221:**199–204.

Naitow, H., Tang, J., Canady, M., Wickner, R. B., and Johnson, J. E. (2002). L-A virus at 3.4 Å resolution reveals particle architecture and mRNA decapping mechanism. *Nat. Struct. Biol.* **9:**725–728.

Najera, I., Krieg, M., and Karn, J. (1999). Synergistic stimulation of HIV-1 rev-dependent export of unspliced mRNA to the cytoplasm by hnRNPA1. *J. Mol. Biol.* **285:**1951–1964.

Nakagawa, A., Miyazaki, N., Taka, J., Naitow, H., Ogawa, A., Fujimoto, Z., Mizuno, H., Higashi, T., Watanabe, Y., Omura, T., Cheng, H. R., and Tsukihara, T. (2003). The atomic structure of rice dwarf virus reveals the self-assemby mechanism of component proteins. *Structure* **11:**1227–1238.

Nason, E., Rothnagel, R., Muknerge, S. K., Kar, A. K., Forzan, M., Prasad, B. V. V., and Roy, P. (2004). Interactions between the inner and outer capsids of bluetongue virus. *J. Virol.* **78:**8059–8067.

Nemerow, G. R., and Stewart, P. L. (1999). Role of $\alpha$ integrins in adenovirus cell entry and gene delivery. *Microbiol. Mol. Biol. Rev.* **63:**725–734.

Okuse, K., Malik-Hall, M., Baker, M. D., Poon, W. Y., Kong, H., Chao, M. V., and Wood, J. N. (2002). Annexin II light chain regulates sensory neuron-specific sodium channel expression. *Nature* **417:**653–656.

Ono, A., and Freed, E. O. (2004). Cell-type-dependent targeting of human immunodeficiency virus type 1 assembly to the plasma membrane and the multivesicular body. *J. Virol.* **78:**1552–1563.

Patton, J., and Spencer, E. (2000). Genome replication and packaging of segmented double-stranded RNA viruses. *Virology* **277:**217–225.

Patton, J. T. (1996). Rotavirus VP1 alone specifically binds to the 3′ end of viral mRNA, but the interaction is not sufficient to initiate minus-strand synthesis. *J. Virol.* **70:**7940–7947.

Patton, J. T., Jones, M. T., Kalbach, A. N., He, Y. W., and Xiaobo, J. (1997). Rotavirus RNA polymerase requires the core shell protein to synthesize the double-stranded RNA genome. *J. Virol.* **71:**9618–9626.

Patton, J. T., Wentz, M., Xiaobo, J., and Ramig, R. F. (1996). *Cis*-acting signals that promote genome replication in rotavirus mRNA. *J. Virol.* **70:**3961–3971.

Peisajovich, S. G., and Shai, Y. (2002). New insights into the mechanism of virus-induced membrane fusion. *Trends Biochem. Sci.* **27:**183–190.

Peisajovich, S. G., and Shai, Y. (2003). Viral fusion proteins: Multiple regions contribute to membrane fusion. *Biochim. Biophys. Acta* **1614:**122–129.

Pesavento, J. B., Lawton, J. A., Estes, M. E., and Venkataram Prasad, B. V. (2001). The reversible condensation and expansion of the rotavirus genome. *Proc. Natl. Acad. Sci. USA* **98:**1381–1386.

Piccininni, S., Varaklioti, A., Nardelli, M., Dave, B., Raney, K. D., and McCarthy, J. E. (2002). Modulation of the hepatitis C virus RNA-dependent RNA polymerase activity by the non-structural (NS) 3 helicase and the NS4B membrane protein. *J. Biol. Chem.* **277:**45670–45679.

Pirttimaa, M., and Bamford, D. (2000). RNA secondary structures of the bacteriophage $\varphi$6 packaging regions. *RNA* **6:**880–889.

Pirttimaa, M., Paatero, A., Frilander, M., and Bamford, D. (2002). Nonspecific nucleoside triphosphatase P4 of double-stranded RNA bacteriophage $\varphi$6 is required for single-stranded RNA packaging and transcription. *J. Virol.* **76:**10122–10127.

Poncet, D., Aponte, C., and Cohen, J. (1993). Rotavirus protein NSP3 (NS34) is bound to the 3′ end consensus sequence of viral mRNAs in infected cells. *J. Virol.* **67:**3159–3165.

Poranen, M., and Tuma, R. (2004). Self-assembly of double-stranded RNA bacteriophages. *Virus Res.* **101:**93–100.

Portela, A., and Digard, P. (2002). The influenza virus nucleoprotein: A multifunctional RNA-binding protein pivotal to virus replication. *J. Gen. Virol.* **83:**723–734.

Prasad, B. V. V., Rothnagel, R., Zeng, C. Q., Jakana, J., Lawton, J. A., Chiu, W., and Estes, M. K. (1996). Visualization of ordered genomic RNA and localization of transcriptional complexes in rotavirus. *Nature* **382:**471–473.

Prasad, B. V. V., Yamaguchi, S., and Roy, P. (1992). Three-dimensional structure of single-shelled bluetongue virus. *J. Virol.* **66:**2135–2142.

Puisieux, A., Ji, J., and Ozturk, M. (1996). Annexin II up-regulates cellular levels of p11 protein by a post-translational mechanisms. *Biochem. J.* **313:**51–55.

Purdy, M., Petre, J., and Roy, P. (1984). Cloning of the bluetongue virus L3 gene. *J. Virol.* **51:**754–759.

Purdy, M. A., Ritter, G. D., and Roy, P. (1986). Nucleotide sequence of cDNA clones encoding the outer capsid protein, VP5, of bluetongue virus serotype 10. *J. Gen. Virol.* **67:**957–962.

Putsep, R. A., Branden, C. I., Boman, H. G., and Normark, S. (1999). Antibacterial peptide from *H. pylori*. *Nature* **398:**671–672.

Ramadevi, N., and Roy, P. (1998). Bluetongue virus core protein VP4 has nucleoside triphosphate phosphohydrolase activity. *J. Gen. Virol.* **79:**2475–2480.

Ramadevi, N., Burroughs, J. N., Mertens, P. P. C., Jones, I. M., and Roy, P. (1998a). Capping and methylation of mRNA by purified recombinant VP4 protein of bluetongue virus. *Proc. Natl. Acad. Sci. USA* **95:**13537–13542.

Ramadevi, N., Rodriguez, J., and Roy, P. (1998b). A leucine zipper-like domain is essential for dimerization and encapsidation of bluetongue virus nucleocapsid protein VP4. *J. Virol.* **72:**2983–2990.

Ramig, R. F., and Petrie, B. L. (1984). Characterisation of temperature-sensitive mutants of simian rotavirus SA11: Protein synthesis and morphogenesis. *J. Virol.* **49:**665–673.

Raynal, P., and Pollard, H. B. (1994). Annexins: The problem of assessing the biological role for a gene family of multifunctional calcium- and phospholipid-binding proteins. *Biochim. Biophys. Res. Acta* **1197:**63–93.

Reinisch, K. M., Nibert, M. L., and Harrison, S. C. (2000). Structure of the reovirus core at 3.6 Å resolution. *Nature* **404:**960–967.

Rhim, J. S., Jordan, L. E., and Mayor, H. D. (1962). Cytochemical, fluorescent-antibody and electron microscopic studies on the growth of reovirus (ECHO 10) in tissue culture. *Virology* **17:**342–355.

Rossman, M. G., Bella, J., Kolatkaer, P. R., He, Y., Wimmer, E., Kuhn, R. J., and Baker, T. S. (2000). Cell recognition and entry by rhino- and enteroviruses. *Virology* **269:**239–247.

Roy, P. (1989). Bluetongue virus genetics and genome structure [review article]. *Virus Res.* **13:**179–206.

Roy, P. (1992). Bluetongue virus proteins. *J. Gen. Virol.* **73:**3051–3064.

Roy, P. (1996). Orbiviruses and their replication. *In* "Field's Virology" (B. N. Fields, ed.), 3rd Ed., Vol. 1, pp. 1709–1734. Lippincott-Raven, Philadelphia, PA.

Roy, P. (2001). Orbiviruses. *In* "Field's Virology" (D. M. Knipe and P. M. Howley, eds.), 4th Ed., pp. 1835–1869. Lippincott Williams & Wilkins, Philadelphia, PA.

Roy, P., Fukusho, A., Ritter, G. D., and Lyon, D. (1988). Evidence for genetic relationship between RNA and DNA viruses from the sequence homology of a putative polymerase gene of bluetongue virus with that of vaccinia virus: Conservation of RNA polymerase genes from diverse species. *Nucleic Acids Res.* **16:**11759–11767.

Roy, P., Marshall, J. J., and French, T. J. (1990). Structure of the bluetongue virus genome and its encoded proteins. *Curr. Top. Microbiol. Immunol.* **162:**43–87.

Sarafian, T., Pradel, L. A., Henry, J. P., Aunis, D., and Bader, M. F. (1991). The participation of Annexin II (calpactin I) in calcium-evoked exocytosis requires protein kinase C. *J. Cell Biol.* **114:**1135–1148.

Shaw, A. L., Samal, S. K., Subramanian, K., and Prasad, B. V. V. (1996). The structure of aquareovirus shows how the different geometries of the two layers of the capsid are reconciled to provide symmetrical interactions and stabilization. *Structure* **4:**957–967.

Shirota, Y., Luo, H., Qin, W., Kaneko, S., Yamashita, T., Kobayashi, K., and Murakami, S. (2002). Hepatitis C virus (HCV) NS5A binds RNA-dependent RNA polymerase (RdRP) NS5B and modulates RNA-dependent RNA polymerase activity. *J. Biol. Chem.* **277:**11149–11155.

Shubsda, M. F., Paoletti, A. C., Hudson, B. S., and Borer, P. N. (2002). Affinities of packaging domain loops in HIV-1 RNA for the nucleocapsid protein. *Biochemistry* **41:**5276–5282.

Silverstein, S. C., Astell, C., Levin, D. H., Schonberg, M., and Acs, G. (1972). The mechanisms of reovirus uncoating and gene activation *in vivo*. *Virology* **47:**797–806.

Spadafora, D., Canter, D. M., Jackson, R. L., and Perrault, J. (1996). Constitutive phosphorylation of the vesicular stomatitis virus P protein modulates polymerase complex formation but is not essential for transcription or replication. *J. Virol.* **70:**4538–4548.

Starnes, M. C., and Joklik, W. K. (1993). Reovirus protein λ3 is a poly(C)-dependent poly (G)polymerase. *Virology* **193:**356–366.

Stauber, N., Martinez-Costas, J., Sutton, G., Monastyrskaya, K., and Roy, P. (1997). Bluetongue virus VP6 protein binds ATP and exhibits an RNA-dependent ATPase function and a helicase activity that catalyze the unwinding of double-stranded RNA substrates. *J. Virol.* **71:**7220–7226.

Tamm, L. K., Han, X., Li, Y., and Lai, A. L. (2002). Structure and function of membrane fusion peptides. *Biopolymers* **66:**249–260.

Tanaka, S., Mikhailov, M., and Roy, P. (1995). Synthesis of bluetongue virus chimeric VP3 molecules and their interactions with VP7 protein to assemble into virus core-like particles. *Virology* **214:**593–601.

Tanaka, S., and Roy, P. (1994). Identification of domains in bluetongue virus VP3 molecules essential for the assembly of virus cores. *J. Virol.* **68:**2795–2802.

Tao, Y., Farsetta, D. L., Nibert, M. L., and Harrison, S. C. (2002). RNA synthesis in a cage-structural studies of reovirus polymerase λ3. *Cell* **111:**733–745.

Tchatalbachev, S., Flick, R., and Hobom, G. (2001). The packaging signal of influenza viral RNA molecules. *RNA* **7:**979–989.

Thomas, C. P., Booth, T. F., and Roy, P. (1990). Synthesis of bluetongue virus-encoded phosphoprotein and formation of inclusion bodies by recombinant baculovirus in insect cells: It binds the single-stranded RNA species. *J. Gen. Virol.* **71:**2073–2083.

Tortorici, M. A., Broering, T. J., Nibert, M. L., and Patton, J. T. (2003). Template recognition and formation of initiation complexes by the replicase of a segmented double-stranded RNA virus. *J. Biol. Chem.* **3:**3.

Urakawa, T., Ritter, D. G., and Roy, P. (1989). Expression of largest RNA segment and synthesis of VP1 protein of bluetongue virus in insect cells by recombinant baculovirus: Association of VP1 protein with RNA polymerase activity. *Nucleic Acids Res.* **17:**7395–7401.

Van Dijk, A. A., and Huismans, H. (1980). The *in vitro* activation and further characterization of the bluetongue virus-associated transcriptase. *Virology* **104:**347–356.

Van Dijk, A. A., and Huismans, H. (1988). *In vitro* transcription and translation of bluetongue virus mRNA. *J. Gen. Virol.* **69:**573–581.

Venkatesan, S., Gershowitz, A., and Moss, B. (1980). Modification of the 5′ end of mRNA: Association of RNA triphosphatase with the RNA guanylyltransferase-RNA (guanine-7-)methyltransferase complex from vaccinia virus. *J. Biol. Chem.* **255:**903–908.

Verwoerd, D. W. (1969). Purification and characterization of bluetongue virus. *Virology* **38:**203–212.

Verwoerd, D. W., Els, H. J., De Villiers, E. M., and Huismans, H. (1972). Structure of the bluetongue virus capsid. *J. Virol.* **10:**783–794.

Verwoerd, D. W., Louw, H., and Oellermann, R. A. (1970). Characterization of bluetongue virus ribonucleic acid. *J. Virol.* **5:**1–7.

Weissenhorn, W., Dessen, A., Calder, L. J., Harrison, S. C., Skehel, J. J., and Wiley, D. C. (1999). Structural basis for membrane fusion by enveloped viruses. *Mol. Membr. Biol.* **16:**3–9.

Wu, X., Chen, S. Y., Iwata, H., Compans, R. W., and Roy, P. (1992). Multiple glycoproteins synthesized by the smallest RNA segment (S10) of bluetongue virus. *J. Virol.* **66:**7104–7112.

Yi, C. K., Bansal, O. B., Hong, M. L., Chatterjee, S., and Roy, P. (1996). Sequences within the VP6 molecule of bluetongue virus that determine cytoplasmic and nuclear targeting of the protein. *J. Virol.* **70:**4778–4782.

Zeng, C. Q., Estes, M. K., Charpilienne, A., and Cohen, J. (1998). The N terminus of rotavirus VP2 is necessary for encapsidation of VP1 and VP3. *J. Virol.* **72:**201–208.

Zhang, X., Walker, S. B., Chipman, P. R., Nibert, M. L., and Baker, T. S. (2003). Reovirus polymerase $\lambda3$ localized by cryo-electron microscopy of virions at a resolution of 7.6 Å. *Nat. Struct. Biol.* **10:**1011–1018.

Zhao, Y., Thomas, C., Bremer, C., and Roy, P. (1994). Deletion and mutational analyses of bluetongue virus NS2 protein indicate that the amino but not the carboxy terminus of the protein is critical for RNA–protein interactions. *J. Virol.* **68:**2179–2185.

Zhou, Z. H., Baker, M. L., Jiang, W., Dougherty, M., Jakana, J., Dong, G., Lu, G., and Chiu, W. (2001). Electron cryomicroscopy and bioinformatics suggest protein fold models for rice dwarf virus. *Nat. Struct. Biol.* **8:**868–873.

ADVANCES IN VIRUS RESEARCH, VOL 64

# STRUCTURE, ASSEMBLY, AND ANTIGENICITY OF HEPATITIS B VIRUS CAPSID PROTEINS

Alasdair C. Steven,* James F. Conway,‡,§ Naiqian Cheng,* Norman R. Watts,† David M. Belnap,*,|| Audray Harris,* Stephen J. Stahl,† and Paul T. Wingfield†

*Laboratory of Structural Biology and †Protein Expression Laboratory
National Institute of Arthritis, Musculoskeletal and Skin Diseases
National Institutes of Health, Bethesda, Maryland 20892
‡Laboratoire de Microscopie Electronique, Institut de Biologie Structurale J. P. Ebel
38027 Grenoble, France



§ Present Address: Structural Biology Program, University of Pittsburgh School of Medicine, Pittsburgh, Pennsylvania 15261

|| Present Address: Department of Chemistry and Biochemistry, Brigham Young University, Provo, Utah B4602




0065-3527/05 $35.00
DOI:10.1016/S0065-3527(05)64005-5

## I. Introduction

Hepatitis B virus (HBV) is one of five distinct families of viruses that cause liver disease in humans. Despite success with the development of vaccines—HBV was the first virus for which a vaccine was developed on the basis of recombinant expression of one of its structural proteins (Valenzuela *et al.*, 1982)—and antiviral drugs (Honkoop and De Man, 2003), it remains a public health problem on a vast scale, particularly in Asia (Blumberg, 1997). Accordingly, there remains a strong interest in research to further elucidate its replication cycle and natural history with a view to identifying alternative intervention points for antivirals (Deres *et al.*, 2003), and its antigenic properties with a view to producing more efficacious vaccines (Baumeister *et al.*, 2000) and immunodiagnostic reagents (Belnap *et al.*, 2003).

HBV belongs to the hepadnavirus family, which has a relatively narrow host range, confined to rodent and avian species and primates (Seeger and Mason, 2000). Although it has not yet been possible to identify an established cell line in which HBV may be propagated, viral genes have been expressed in a variety of recombinant systems (Acs *et al.*, 1987; Chang *et al.*, 1987; Sureau *et al.*, 1986). HBV has a small (3.2 kb) DNA genome, although this modest genetic endowment is amplified by a variety of strategies, including alternative expression products of the same gene (Ganem, 1996).

In the replication cycle of HBV, the genome is initially incorporated into the assembling virus particle as a single-stranded RNA molecule—the pregenome—that is subsequently retrotranscribed *in situ* by the viral reverse transcriptase (RT). The DNA-containing nucleocapsid subsequently becomes enveloped by a membrane containing the viral glycoprotein—surface antigen (sAg), of which there are three size variants called S, M, and L, respectively—to yield the completely assembled and infectious virion (also known as the Dane particle), Fig. 1.

In the serum of infected individuals, membranous aggregates of sAg without nucleocapsids are found in large abundance and sAg is an important diagnostic indicator. Two other viral antigens are routinely used for clinical diagnoses. Both are variants of the capsid protein. Core antigen (cAg) represents assembled capsids whereas e antigen (eAg) is an unassembled form of capsid protein. cAg and eAg have different termini, reflecting differences in initiation site and posttranslational processing (Fig. 2). A preponderance of eAg in patient serum is indicative of a state of high infectivity (Hollinger, 1996).

HBV capsid protein plays a key role in the replication cycle. Not only does the capsid provide a vehicle for transferring genetic material between

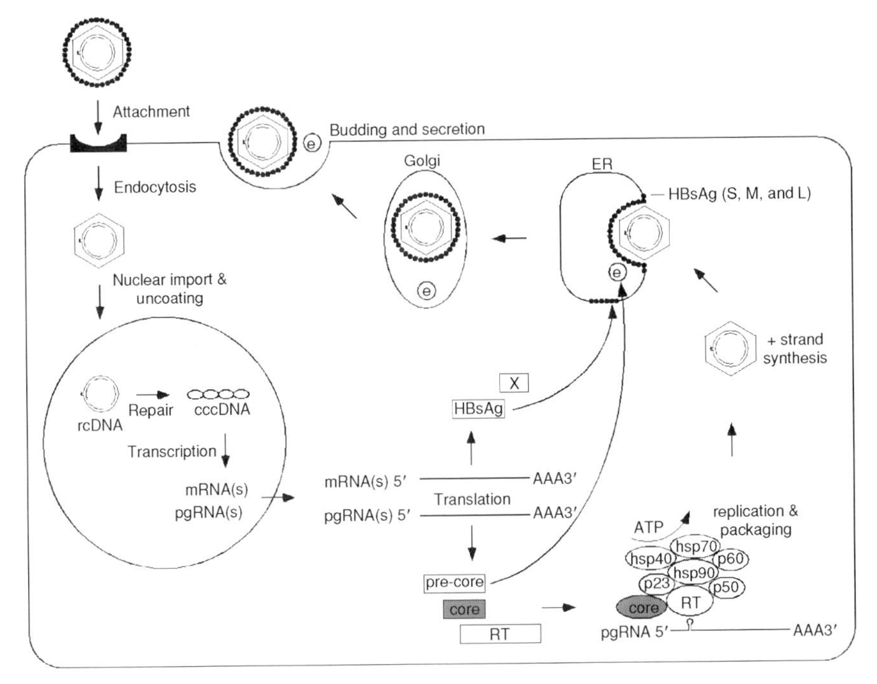


FIG 1. Biosynthetic pathways of HBV capsid proteins in the viral replication cycle. cccDNA, covalently closed circular DNA; e, e antigen; ER, endoplasmic reticulum; HBsAg, hepatitis B surface antigen; pgRNA, pregenomic RNA; rcDNA, relaxed circular DNA; RT, reverse transcriptase.

compartments within an infected cell and then infected to uninfected cells (Fig. 1), but it also constitutes a specialized compartment where retro-transcription takes place. In spite of the importance of HBV as a human pathogen, it is only more recently that the basic structure of its capsid has been worked out, in large part through studies by electron microscopy and image reconstruction, and latterly, X-ray crystallography.

The capsid protein turned out to have several unexpected properties. It was found to have a novel fold, rich in $\alpha$ helix, and quite distinct from the eight-stranded $\beta$ barrel that was common to the first dozen or so capsid proteins to be solved (from plant, animal, and bacterial viruses) and the other capsid protein folds that have been determined more recently (Chapman and Liljas, 2003). The building block is a dimer whose dimerization motif is a four-helix bundle—one of the most common protein folds. In this system, the bundle is formed by the pairing of $\alpha$-helical hairpins from the two interacting subunits. HBV capsids are of two sizes corresponding to $T$ numbers of 3 (90 dimers) and 4 (120 dimers),

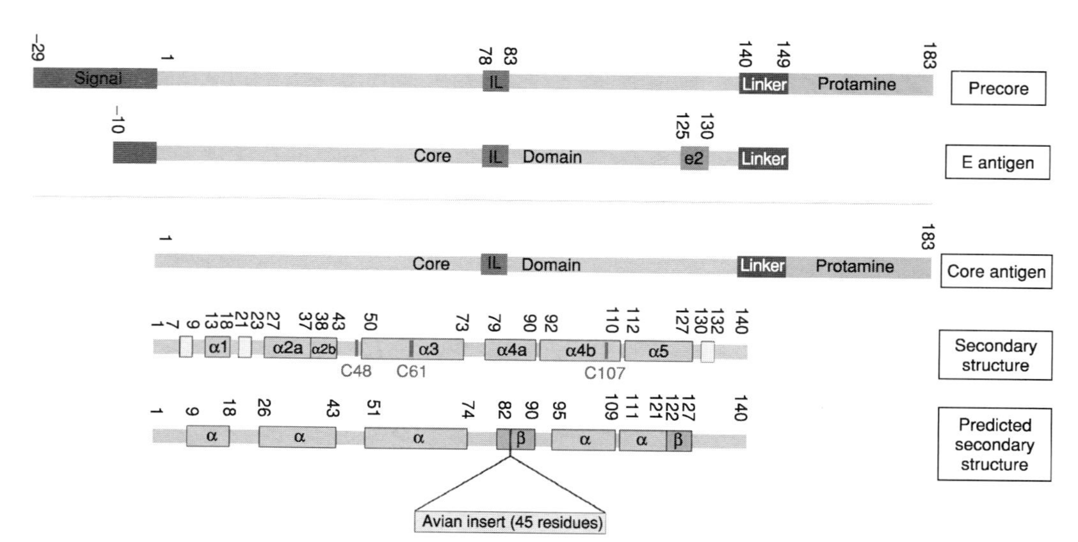


FIG 2. Primary and secondary structures of HBV capsid proteins with various physiologically relevant sites marked. The secondary structure prediction is from Bringas (1997). (See Color Insert.)

respectively. This property is unusual in that—with a few exceptions such as alfalfa mosaic virus, a plant virus that packages its segmented genome in particles of multiple sizes (Driedonks *et al.*, 1978)—the $T$ number tends to be a uniquely defined feature of a given virus. Quaternary structure appears to be a stable property, slow to mutate. In many instances, the $T$ numbers of distantly related viruses remain unchanged despite the divergence of their sequences during the course of evolution, for example, with herpesviruses, which are all $T = 16$ (Booy *et al.*, 1996); and the 120-subunit inner capsids of double-stranded RNA (dsRNA) viruses (e.g., Cheng *et al.*, 1994), which are $T = 1$ with two nonequivalent subunits as protomer.

The purpose of this article is to review current information pertaining to the structure and assembly properties of HBV capsid protein, as well as the insights into its antigenicity and other interactions, for example, during envelopment, that have emerged from the structural information. Comprehensive coverage of earlier work and related topics has been given in accounts of nucleocapsid assembly (Nassal, 1996; Nassal and Schaller, 1993), immunological properties (Seifer and Standring, 1995; Vanlandschoot *et al.*, 2003), and the biotechnology of adapting the HBV capsid for presentation of foreign epitopes (Pumpens and Grens, 2001; Pumpens *et al.*, 1995).

## II. The Two Forms of HBV Capsid Protein, cAg and eAg: Primary Structure

The primary structures of both proteins are diagrammed in Fig. 2 and the sites of posttranslational modifications are marked. cAg is a polypeptide of 183 residues, consisting of two domains—the "core" domain (residues 1–140) and a "protamine" domain (residues 150–183) connected by a linker peptide (residues 141–149). The core domain is capable of self-assembly into capsids, whereas further truncation to Cp138 abrogates assembly (Birnbaum and Nassal, 1990; Metzger and Bringas, 1998). Some erosion from the N terminus is also tolerated, for example, a construct lacking residues 1–5 remains assembly-competent (P. T. Wingfield, unpublished results). The protamine domain is highly basic—half of its residues are arginine—and it secures nucleic acid to the inner surface of the capsid. eAg lacks the protamine domain but retains, at its N terminus, a 10-residue fragment of the 29-residue propeptide. The C-terminal portion of core domain has a cluster of prolines (5 of 10 residues between amino acids 129 and 138).

The HBV capsid protein is highly $\alpha$-helical. Circular dichroism measurements estimated the $\alpha$-helical content of a core domain construct at 55–60% (i.e., 79–86 out of 144 residues) (Wingfield *et al.*, 1995). This conformation was also anticipated by secondary structure prediction methods (Bringas, 1997). When the crystal structure was determined (Wynne *et al.*, 1999), both observations were confirmed. The predicted and observed distributions of α-helical segments are compared in Fig. 2.

Two peptide segments have been implicated in epitopes by immunochemical mapping and competition experiments between antibodies. One, lying near the middle of the protein, ~residues 78–83, was predicted to be nonhelical and turned out to be so, and is called the "immunodominant loop" (Salfeld *et al.*, 1989; Sällberg *et al.*, 1991). This region has been associated with (distinct) conformational epitopes of cAg and eAg, and sizable insertions in this region do not abrogate the ability of the protein to assemble (Kratz *et al.*, 1999; Milich *et al.*, 2001; Schödel *et al.*, 1992). In the same vein, the capsid proteins of avian hepadnaviruses have insertions of about 45 residues at this site (Fig. 2) (Bringas, 1997; Kenney *et al.*, 1995). The segment covering residues 128–133, another nonhelical peptide, has been associated with a linear eAg epitope (e2) (Salfeld *et al.*, 1989). Other peptide segments that are involved in epitopes are considered further later (see Section VIII).

### A. *Nomenclature*

Traditional usage equates cAg (residues 1 to 183) with particulate material (capsids)—the term "core particles" is also used—and eAg (residues −10 to 149) with nonparticulate material. However, in light of the observations that eAg can also form particles under certain conditions (see later), and that cAg has three parts—core domain, linker, and protamine domain—of which the core domain alone is necessary and sufficient for assembly, we follow a more conventional usage, that is, capsid for an assembled shell, regardless of the protein from which it is built, and nucleocapsid for a capsid containing nucleic acid. "Core domain construct" refers to any fragment lacking the protamine domain, for example, Cp145 refers to the core domain plus the first five residues of the linker. In a traditional misnomer, "precore protein" (residues −29 to 149) refers to the precursor to eAg, not cAg, and "preC region" to the signal sequence, residues −29 to −1.

## III. Capsid Assembly and Disassembly *In Vivo*

Early events in the replication cycle remain obscure. The viral receptor on hepatocytes is still unknown although several candidates have been identified (Ryu *et al.*, 2000). The virus probably enters the host cell by pH-independent membrane fusion after which the genome-containing nucleocapsid is transported to the nucleus, where uncoating occurs and replication begins (Fig. 1): for reviews, see Ganem (1996) and Seeger and Mason (2000).

After uncoating, the 3.2-kb relaxed circular DNA (rcDNA) genome, which consists of the minus-strand DNA with the reverse transcriptase (RT) covalently attached at the 5′ end and a partially complete plus strand with a short RNA primer at its 5′ end, is freed of RT and RNA and converted to covalently closed circular DNA (cccDNA). cccDNA is the template for two classes of transcripts: a subgenomic transcript that encodes the surface proteins and protein X, and a pregenomic transcript (pgRNA) that also encodes the 25-kDa eAg precursor, 21-kDa cAg protein, and RT. Eventually this pgRNA is selected for packaging (Ganem, 1996; Seeger and Mason, 2000). The eAg precursor is targeted to the endoplasmic reticulum, where its first 19 residues and the protamine domain are removed, yielding eAg (17 kDa), which is secreted (Ou, 1997). cAg assembles into nucleocapsids in the cytoplasm.

HBV replication and genome packaging are intimately linked. The current model (Ganem, 1996; Hu *et al.*, 1997; Seeger and Mason, 2000) involves the RT and pgRNA, as well as components of the hsp90 chaperone complex (and possibly others, including members of the immunophilin isomerases), and ATP. RT has an N-terminal domain (terminal protein, TP) that is involved in packaging of pgRNA and protein priming of minus-strand DNA synthesis, and a C-terminal domain with RNase H activity that is involved in reverse transcription. pgRNA has a stem–loop structure (termed ε) at its 5′ end to which the RT-TP attaches. The corresponding structure at the 3′ end is not recognized, thus restricting packaging to pgRNA. In the model, initiation of nucleocapsid assembly and the priming of viral DNA synthesis are simultaneous, dynamic, energy-driven processes wherein the chaperone complex maintains the RT in several alternate conformations. hsp90 and its cofactor p23 act on the RT to induce a conformational change that allows binding to pgRNA, which can then recruit cAg protein to initiate nucleocapsid assembly and, subsequently, DNA synthesis. This process depends on the activity of the ATPase hsp70 and its partner hsp40, and is stimulated by p48. p50 mediates the interaction between RT and hsp90 while p60 mediates the interaction of hsp90 and hsp70. The eventual incorporation of the hsp90 complex into the nucleocapsid raises the possibility that these proteins may have functions additional to initiating genome replication and capsid assembly.

After completion of rcDNA synthesis, the nucleocapsid matures in a way that facilitates envelopment. The nature of this maturation signal is not known but it may involve phosphorylation. cAg protein may be phosphorylated at several C-terminal residues by an endogenous kinase, and its phosphorylation state appears to be important in both RNA packaging and DNA replication (Lan *et al.*, 1999). Because the C-terminal domain is located inside the nucleocapsid (Machida *et al.*, 1991; Watts *et al.*, 2002; Zlotnick *et al.*, 1997), it may be that the capsid shell undergoes a conformational change in response to these phosphorylation event(s) that facilitates interaction between its outer surface and sAg. As a result, nucleocapsids bud into the lumen of the endoplasmic reticulum (ER) and are subsequently secreted as virions (see Fig. 1).

In view of the great stability of HBV nucleocapsids assembled from full-length capsid protein (Wingfield *et al.*, 1995), their uncoating *in vivo* poses something of an enigma. The first step of uncoating should involve destabilization, perhaps by loss or perturbation of the encapsidated nucleic acid, which would render the C termini (protamine domain plus linker) susceptible to proteolysis. Degradation of the C termini would render the capsid to a state comparable to core domain particles, which are less stable. It is possible that stability is enhanced by interdimer cross-linking via disulfide bonds between Cys-183 residues, and this effect would be annulled by degradation of the C termini.

Phosphorylation may also play a role. C183 is phosphorylated at residues Ser-162 and Ser-170 by host cell kinases before nucleocapsid assembly (Daub *et al.*, 2002). It appears that kinases are also present in the virion (Gerlich *et al.*, 1982), and these enzymes could contribute to the uncoating process. For example, further phosphorylations might weaken protein–nucleic acid interactions sufficiently to allow transient exposure of C termini outside of the capsid shell, where they would be accessible to host proteases.

## IV. Expression and Assembly of Recombinant Capsid Proteins

Vectors encoding HBV capsid protein constructs have been expressed in a variety of cells (Kniskern *et al.*, 1986; Takehara *et al.*, 1988; Zhou and Standring, 1992; additional references cited in Pumpens *et al.*, 1995). In each case, capsids of the correct size and shape were produced, leaving little doubt that the morphogenic information resides in the capsid protein per se and, specifically, in its core domain. It follows that the extensive involvement of chaperones and other factors in nucleocapsid assembly *in vivo* must reflect regulatory requirements, for example, to ensure that needed internal components (RT and pgRNA) are incorporated, and perhaps to forestall premature assembly into empty capsids. Conversely, although eAg is nonparticulate *in vivo*, we have found that this protein is able to form capsids *in vitro* (Milich *et al.*, 1988; P. T. Wingfield, unpublished data), albeit less readily than other constructs. Its unassembled state *in vivo* may also reflect regulatory constraints, for example, failure to build up a high enough local concentration of protein.

Both cAg and eAg have been expressed successfully in *Escherichia coli*, contrasting in this respect with the other HBV proteins—for instance, sAg expression requires a eukaryotic system (Hitzeman *et al.*, 1983). The Cp183 expression product consists of soluble, fully assembled capsids (Cohen and Richmond, 1982) containing bacterial RNA in amounts equivalent to viral pgRNA (Wingfield *et al.*, 1995). This was the first example of a multisubunit protein produced in bacteria by recombinant DNA technology (Stahl *et al.*, 1982). The protein is expressed at high levels and its purification is facilitated by its high molecular mass. Electron micrographs of Cp149 and Cp183 imaged by three different methods are compared in Fig. 3.

Many structural studies have been carried out on Cp149 capsids, which also assemble in *E. coli* but, unlike Cp183, do not package RNA. Highly purified protein may be obtained by first isolating capsids by gel filtration and then dissociating them into dimers by low concentrations (~1.5–2.0 M) of urea at pH 9.5, and further purifying them by a second round of gel filtration (Wingfield *et al.*, 1995). Finally, dimers are reassembled into capsids by adjusting to pH 7.5 in the presence of 0.15–0.25 M NaCl. The two capsids, $T = 3$ and $T = 4$, may be partially resolved by gel filtration on Sepharose 4B or, more efficiently, by velocity sedimentation on sucrose gradients (Zlotnick *et al.*, 1996, 1999b). Alternatively, bacterially expressed Cp149 capsids may be isolated directly by gel filtration and sucrose density gradient centrifugation (Wynne *et al.*, 1999).

Secretory eAg retains 10 residues of the signal sequence, and we refer to the corresponding construct as Cp(−10)149. Its N-terminal appendage, SKLCLGWLWG, is hydrophobic and contains a cysteine. Cp(−10)149 was expressed in *E. coli* by Schödel *et al.* (1992). Although a detailed biophysical characterization was not presented, the purified material was shown to be a soluble, low molecular weight protein at pH 7.2 and 9.6. We have also expressed Cp(−10)149 at high levels and found that, unlike Cp149 and Cp183, the protein does not accumulate as assembled capsids and little of it is soluble. This material

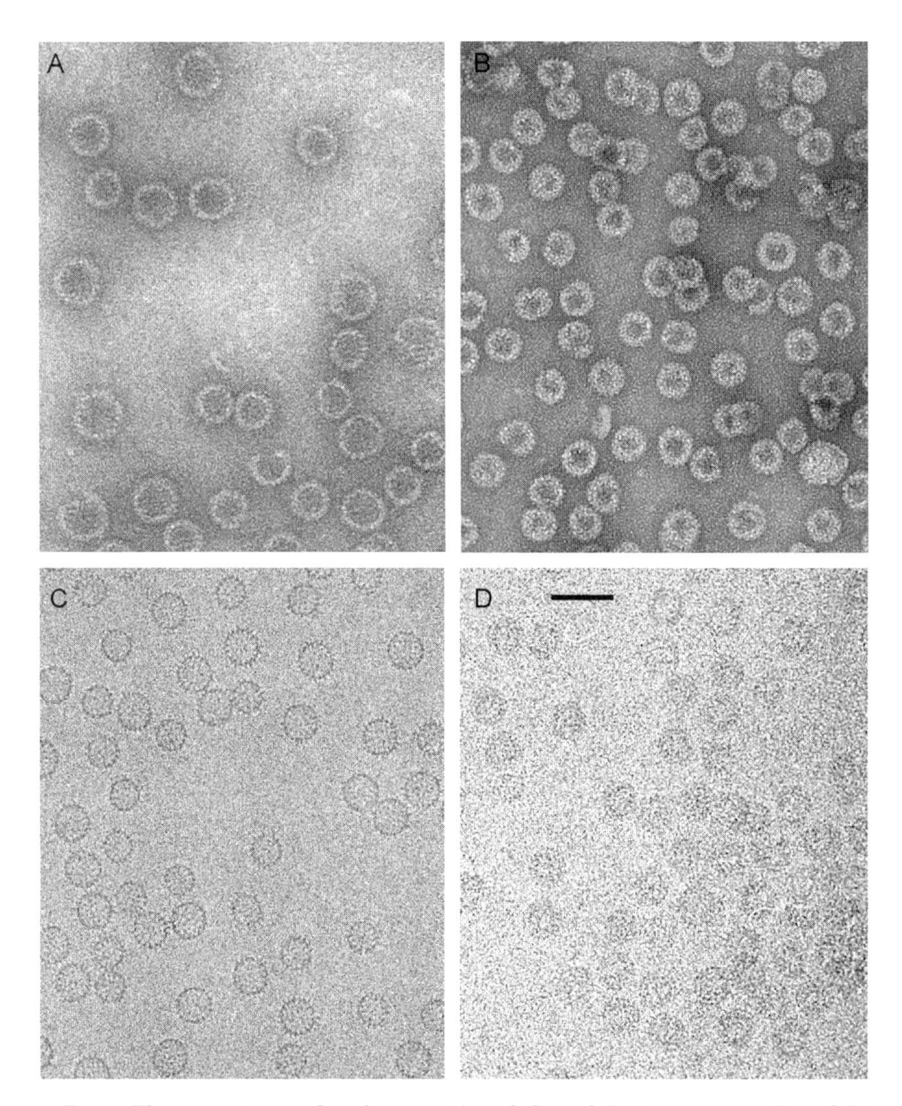


FIG 3. Electron micrographs of empty (A and C) and RNA-containing (B and D) recombinant HBV capsids. Negatively stained preparations are shown in (A) for Cp149 capsids, and in (B) for Cp183 nucleocapsids. Cryoelectron micrographs are shown in (C) for Cp149 capsids, and in (D) for Cp183 nucleocapsids. The two distinct sizes of capsid, $T = 3$ and $T = 4$, are evident in (C) but less so in (D) because almost all Cp183 capsids are the larger, $T = 4$, variant. Negative staining data [e.g., (A) and (B)] also suggest polymorphism but variable amounts of spread/flattening of these specimens, particularly for the empty capsids (A), make it less easy to infer that there are only two size variants. In Cp183 nucleocapsids, the RNA is distributed over the inner surface, producing an inner ring that is discernible in cryomicrographs [e.g., (D)]. In negative staining, the presence

may be extracted from *E. coli* lysates under relatively mild conditions using detergents or chaotropes and purified by conventional chromatographic methods (P. T. Wingfield, unpublished data).

## V. Capsid Structure

### A. *Dimorphism*

It has long been known that HBV capsids are spherical particles of about 300 Å in diameter and, therefore, likely to be icosahedral, but the specific triangulation geometry was determined only relatively recently. Two lines of evidence showed that there are two size variants, corresponding to triangulation numbers $T = 3$ and $T = 4$, respectively. The two size variants were demonstrated by the first cryoelectron microscopy (cryo-EM) reconstructions, at a resolution of ~35 Å (Crowther *et al.*, 1994), which also revealed that the most prominent features of both capsids are external spikes, ~20 Å long, and sizable holes through the capsid wall at the threefold and quasi-sixfold axes, respectively. These data also revealed that the building block is a dimer, consistent with biochemical evidence to this effect (Nassal *et al.*, 1992; Wingfield *et al.*, 1995; Zheng *et al.*, 1992; Zhou and Standring, 1992). Scanning transmission electron microscopy (STEM) mass measurements of Cp149 capsids distinguished two species with masses corresponding to 180 subunits ($T = 3$) and 240 subunits ($T = 4$), respectively; and similar data on Cp183 capsids revealed the same dimorphism and that both size variants contained an additional 30% or so by mass of RNA (Wingfield *et al.*, 1995). Surface lattice diagrams of the packing of dimers and the placement of the seven (i.e., three + four) quasi-equivalent subunits are shown in Fig. 4.

The observation that Cp183 assembled in bacteria consists mainly of $T = 4$ particles suggests that nucleocapsids produced in the course of a natural infection share this property. However, we note that dimorphism has also been detected in nucleocapsids isolated from human liver (Kenney *et al.*, 1995). Conversely, capsid dimorphism should be primarily a property of core domain constructs (see Section VI.B). In them, the morphogenic linker is, in effect, a terminal peptide and

of internalized RNA is manifested more as a thickening of the capsid wall. In suitably focused cryomicrographs [e.g., (C)], externally protruding spikes are discernible around the capsid periphery and are conspicuous when two spikes are coprojected along the line of sight. Scale bar: 500 Å.

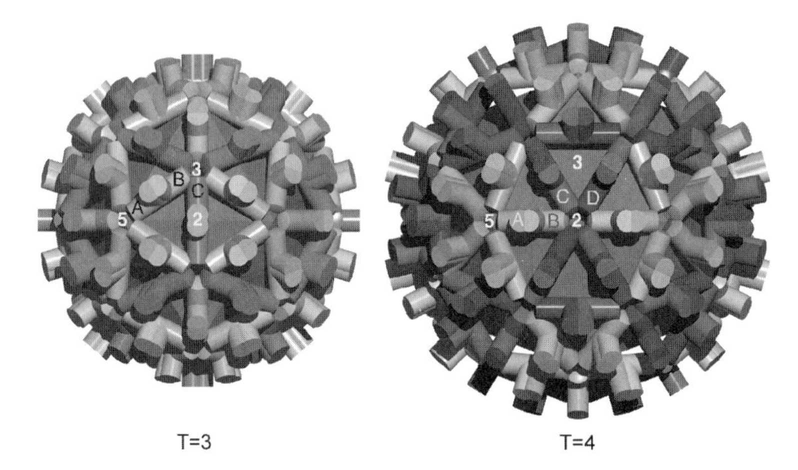


FIG 4. Models of cAg dimer packing in the $T = 3$ and $T = 4$ capsids. The various quasi-equivalent subunits (A, B, C, etc.) are labeled. (See Color Insert.)

examples exist in other viral systems of terminal peptides that influence the size of capsid assembled (e.g., papillomavirus; Chen *et al.*, 2001).

### *B. Progress in Cryoelectron Microscopy*

The next step in resolution, to ~17 Å (Zlotnick *et al.*, 1997), verified the previously described features and showed that there are also smaller holes at the fivefold axes. Taking into consideration the high α-helical content of core domain and the dimensions of the spikes, it was suggested that this feature is a four-helix bundle. This suggestion has been confirmed (see further discussion).

The holes are of considerable functional significance. Because the capsid serves as a compartment for retrotranscription, its design should allow for the diffusive import of nucleotide triphosphates and release of oligonucleotides generated from the pgRNA. The dimensions of the holes—which were confirmed by functional criteria whereby a metal cluster compound with an organic shell having a full diameter of 17 Å may enter capsids (Cheng *et al.*, 1999) whereas another, slightly larger compound of 20 Å diameter may not (Zlotnick *et al.*, 1997)—are appropriate for this purpose. At the same time, the holes are small enough to exclude proteins such as potentially damaging hydrolytic enzymes. Thus the capsid serves as a steric filter.

T=4 T=3

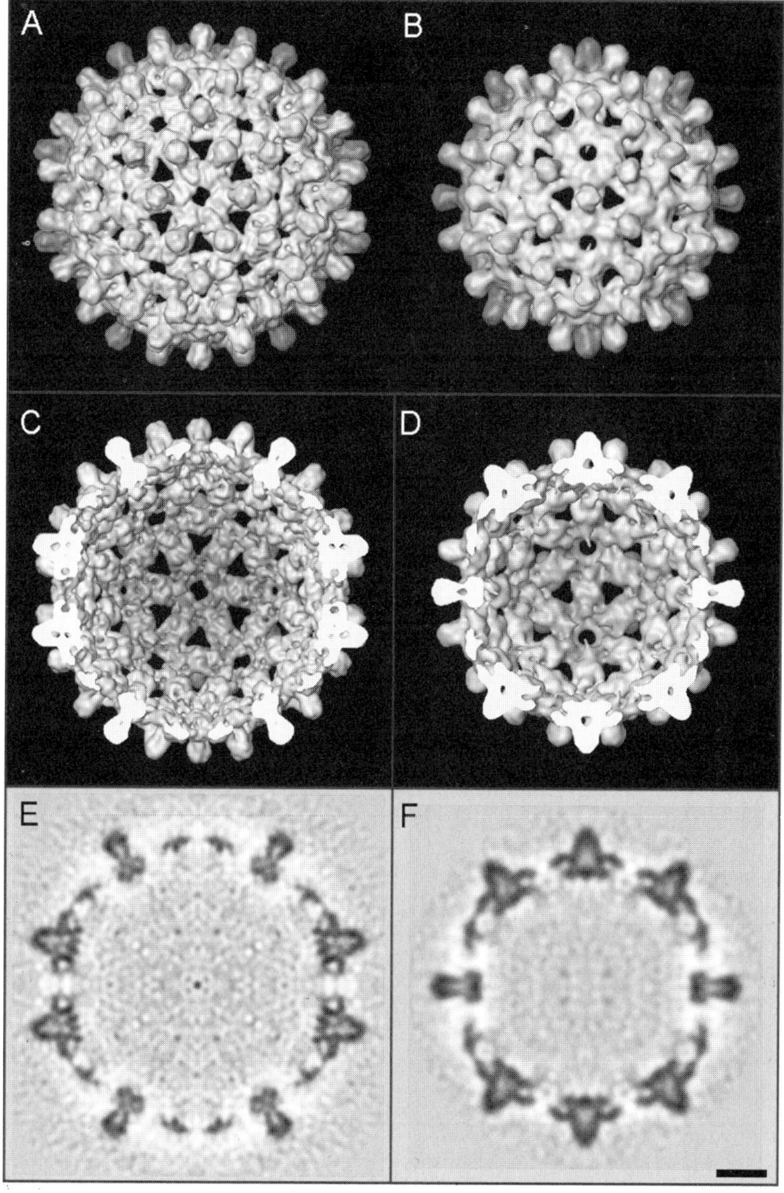

Further extension of the cryo-EM analyses to resolutions of 9 Å (Conway *et al.*, 1997) and 7.3 Å (Böttcher *et al.*, 1997) (Fig. 5) revealed the four-helix bundle in the spike and that these helices extend through the capsid shell to its inner surface. Two approaches were taken to correlating the amino acid sequence with these density maps. We localized the N and C termini by difference mapping with an appended peptide (Conway *et al.*, 1998b) and metal clusters (Cheng *et al.*, 1999; Zlotnick *et al.*, 1997), respectively (Fig. 6), and established that the immunodominant loop (in two copies) occupies its expected position at the tip of the spike by immunolabeling with a monoclonal Fab fragment (Conway *et al.*, 1998a) (see Fig. 13). We also used an EM tilting method to determine the handedness of the capsid, which is ambiguous in cryo-EM analyses because the micrographs are recorded as projection images (Conway *et al.*, 1998a). Böttcher *et al.* (1997) proposed a complete fold, reasoning on the basis of the likely positions of key residues and continuity of density. When the crystal structure was determined (Wynne *et al.*, 1999), this proposal proved correct except for handedness and the location and disposition of the C-terminal portion of the molecule—for instance, residue 149 was assigned an external location, approximately 30 Å removed from its actual location in the capsid interior.

### C. *Crystal Structure: Proteins with Related Folds*

Dimorphism is not conducive to the production of well-ordered crystals. However, using capsids purified by bacterial expression of a Cp149 construct that produces almost exclusively $T = 4$ particles, Wynne *et al.* (1999) obtained well-diffracting crystals that they used to solve the capsid structure. These observations confirmed most of the preceding inferences from cryo-EM, while adding much detail. The HBV capsid protein was the first example of a protein with this fold, which has since been detected in the enzyme Spo0B phosphotransferase (Varughese *et al.*, 1998). A similar topology, albeit with much shorter helices, is exhibited in the C-terminal domain of the human immunodeficiency virus (HIV) capsid protein (Gamble *et al.*, 1997). The three proteins are compared in Fig. 7. The structural

---

FIG 5. Cryo-EM reconstructions of the $T = 4$ and $T = 3$ cAg capsids (Cp147). (A, C, and E) $T = 4$ capsid at a resolution of 9.0 Å; (B, D, and F) $T = 3$ capsid at a resolution of 10 Å. From Conway *et al.* (1997) and Watts *et al.* (2002). (A–D) Outside and inside surface renderings. (E and F) Grayscale central sections of capsids viewed along twofold axes. Protein is dark. Scale bar: 50 Å. (See Color Insert.)

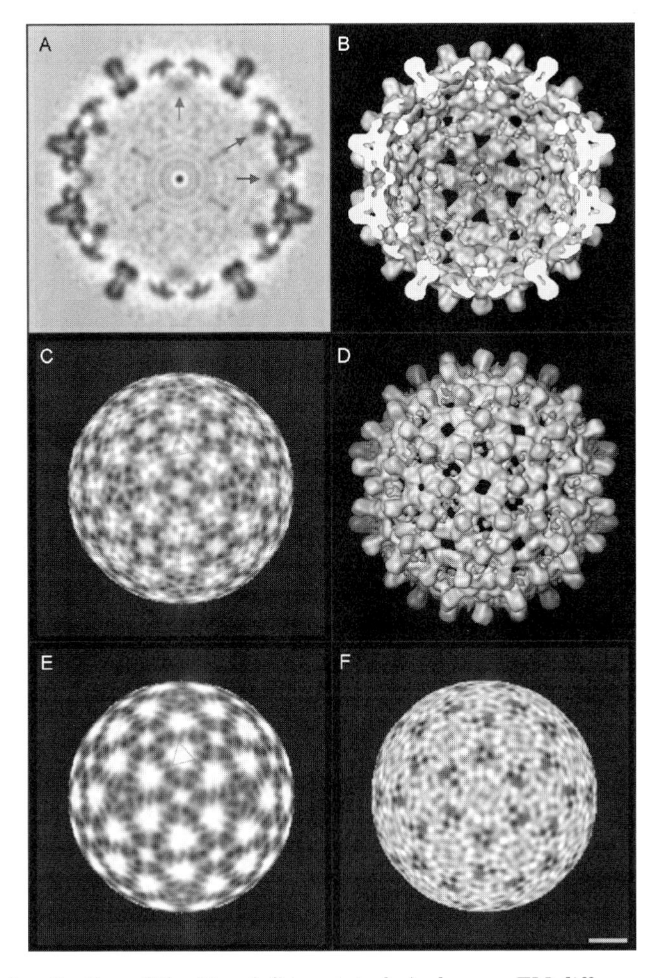


FIG 6. Localization of the N and C termini of cAg by cryo-EM difference mapping. (A and B) A central section and an interior view of the $T = 4$ capsid for a construct that has a cysteine residue appended to Cp149 and a tetrairidium ($Ir_4$) label attached to Cys-150 (Cheng *et al.*, 1999). The extra density from the $Ir_4$ label is indicated with purple arrows (A) or by purple shading (B). The labels underlie symmetry axes. The blue arrow in (A) marks the site of a C-terminal residue (residue 149). In (C–F), the N terminus is marked by the attachment of an extraneous peptide of eight amino acid residues. This difference density is marked in red in (D). (C, E, and F) A comparison of spherical sections at an appropriate radius to sample the appended peptide. (C) is from the peptide-attached capsid; (E) is from the control (no attachment); and (F) is from the difference map. A triplet of peptide-associated densities is marked with a red triangle in each panel. The N terminus (residue 1) lies at the point where the difference density (red in [D]) meets the capsid surface (blue) in (D). Scale bar: 50 Å. (See Color Insert.)

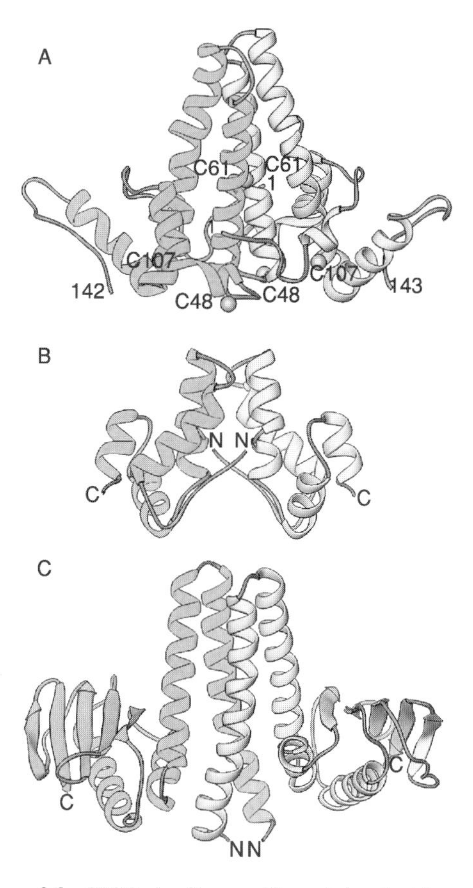


FIG 7. Comparison of the HBV cAg dimer with proteins that have similar structures. (A) cAg (PDB: 1QGT). Cysteine residues C48, C61, and C107 are indicated as ball-and-stick side chains with enlarged gold sulfur atoms. The two C61 residues form an intermolecular disulfide bond within the cAg dimer. (B) C-terminal domain of the HIV capsid protein, CA (PDB: 1A80); and (C) Spo0B phosphotransferase (PDB: 1IXM). All three proteins dimerize through pairing $\alpha$-helical hairpins to create four-helix bundles. The HIV CA dimer has a short spike, whereas those of cAg and Spo0B are similar in length but of opposite hand. Monomers are green and yellow with the N and C termini labeled or numbered. (See Color Insert.)

similarity between cAg and HIV-CA had been anticipated on the basis of (weak) sequence similarities (Zlotnick *et al.*, 1998). This structural relationship is one of several notable points of resemblance between hepadnaviruses and retroviruses (Miller and Robinson, 1986): both are enveloped viruses and they are the only two families to employ retrotranscription in their replication cycles.

No crystal structure has yet been reported for the $T = 3$ capsid. Nevertheless, we have been able to construct a quasi-atomic model (Conway *et al.*, 2003) of it by docking the dimer from the $T = 4$ crystal structure into our cryo-EM map at a resolution of ~10 Å (Conway *et al.*, 1998b). Details of both capsids are shown in Fig. 8. The distributions of charged residues over their inner and outer surfaces are shown in Fig. 9. Both surfaces are predominantly acidic, suggesting that nucleic acid has no particular affinity for the inner surface other than by tethering via the linker and protamine domain.

## VI. Regulation of Core Domain Self-Assembly

### A. *Initiation*

The kinetics of core domain assembly *in vitro* have been studied in depth, both experimentally by monitoring assembly by light scattering and gel filtration under a variety of conditions and by quantitative modeling (Endres and Zlotnick, 2002; Zlotnick *et al.*, 1999a). From analyzing the rate order constants of the initiation and elongation phases of assembly, it was inferred that the initiation complex is a trimer of dimers to which dimers add rapidly and cooperatively to realize the complete capsid (Zlotnick *et al.*, 1999a). Ceres and Zlotnick (2002) used size-exclusion chromatography to study assembly and concluded that it is an entropy-driven process, possibly involving a conformational switch of the dimer to an assembly-active form. Singh and Zlotnick (2003) extended these studies to compare dissociation with association and observed hysteresis effects that they interpreted in terms of a propensity for partially disassembled structures to revert toward higher states of aggregation, with the corollary that uncoating *in vivo* may require some kind of triggering event.

### B. *Linker Peptide as Morphogenic Determinant*

In addition to Cp149 and Cp183, many different core domain constructs have been expressed in *E. coli*. Experiments of this kind revealed that the linker peptide, although not required for assembly, is a morphogenic determinant (Zlotnick *et al.*, 1996). Minimal length core domain, Cp140, produces mostly the smaller $T = 3$ capsids (~80%). Increasing the length of the linker progressively increases the proportion of $T = 4$ capsids to almost 100% with

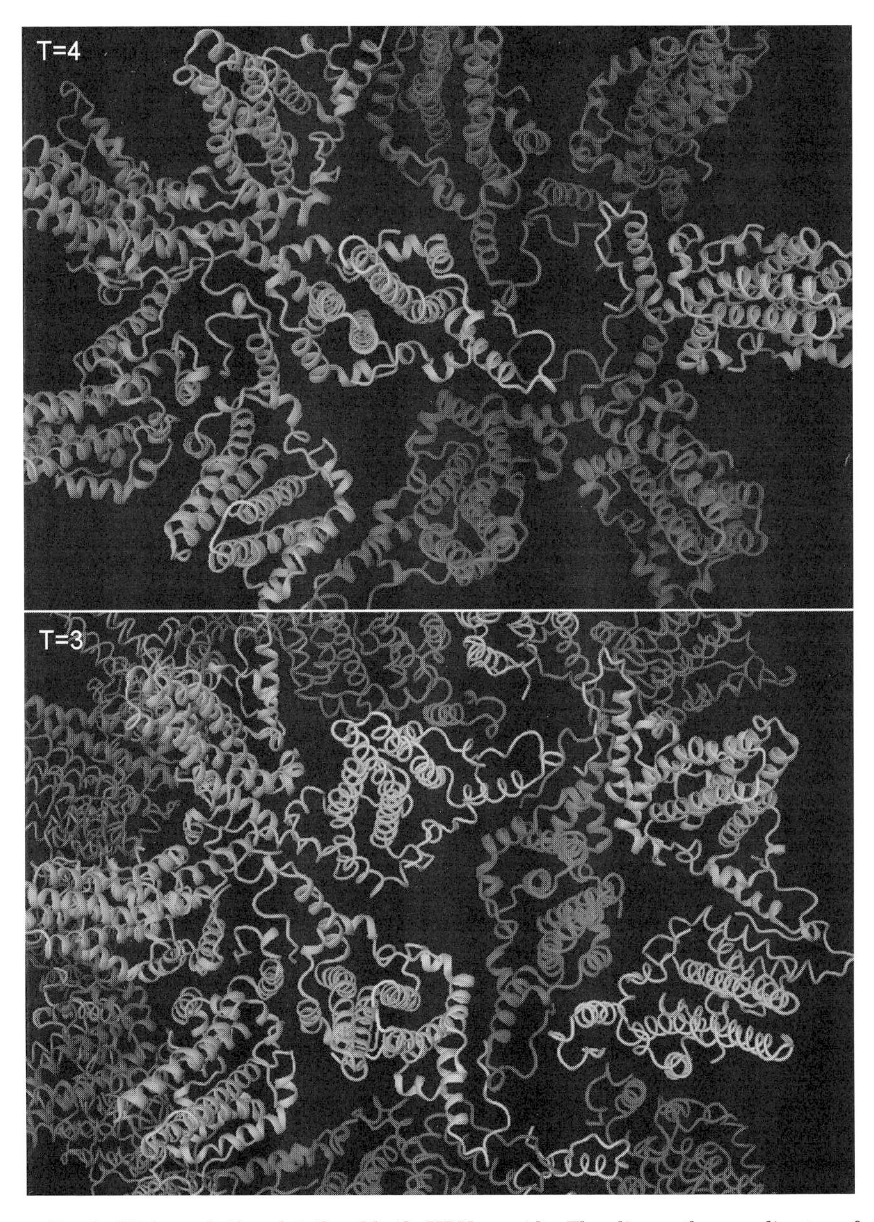


FIG 8. High-resolution details of both HBV capsids. The dimer, the coordinates of which are from PDB (1QGT; Wynne *et al.*, 1999), was docked into cryo-EM reconstructions of the $T = 4$ and the $T = 3$ capsids (Conway *et al.*, 2003). (See Color Insert.)

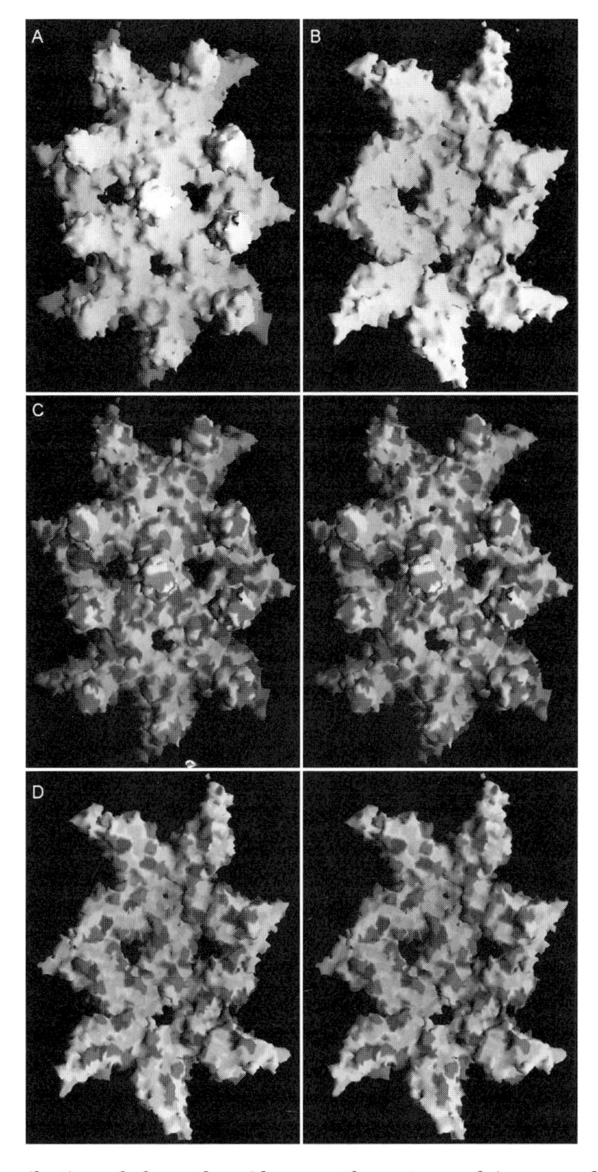


FIG 9. Distribution of charged residues on the outer and inner surfaces of HBV capsids. (A and B) Outer and inner surfaces, respectively, of a fragment of the $T = 4$ capsid comprising five dimers surrounding a fivefold axis (upper part) plus five more dimers around the adjacent twofold axis (bottom part). (C and D) Stereo pairs of A and B, respectively, color coded for charge. Acidic residues are red, basic residues are blue, polar but uncharged residues are yellow, and hydrophobic residues are green. Calculated with GRASP. (See Color Insert.)

full-length linker. Assembly conditions appear also to affect assembly: when full-length core domain is disassembled and reassembled *in vitro*, a larger minority (5–10%) of $T = 3$ capsids is produced.

### *C. Drug-Based Subversion of Capsid Assembly*

Capsid assembly is a viral function that has no evident counterpart in cellular metabolism. As such, it represents a potential target for antiviral drug therapy and some studies have investigated possible effects of low molecular weight compounds on this process. If drugs were to bind to sites on the capsid protein surface that engage in intermolecular interactions, competition for these sites might subvert capsid assembly or—equally effectively—produce malformed capsids and thereby suppress viral replication. Alternatively, allosteric effects of drugs binding to other sites could produce similar effects. In this context, 4,4′-dianilino-1,1′-binaphthyl-5,5′-disulfonic acid (bis-ANS) has attracted interest. This compound binds to hydrophobic surfaces on proteins with a concomitant change in its florescence, providing a convenient assay. Teschke *et al.* (1993) investigated the effects of bis-ANS on bacteriophage P22 capsid assembly *in vitro* and found impairment of capsid assembly at concentrations of 10–50 $\mu$M.

In the HBV system, Zlotnick *et al.* (2002) studied the effects of bis-ANS on the assembly of Cp149 core domains and observed that the drug binds to unassembled dimers at a stoichiometry of one drug molecule per Cp149 dimer with a $K_d$ of $\sim$9 $\mu$M, but not to assembled capsids. When assembly was carried out in the presence of 16–39 $\mu$M bis-ANS, the proportion of assembling material was reduced by approximately half and the assembly products were not closed icosahedral capsids but large aggregates, containing partially and completely assembled capsids embedded in less ordered material (Fig. 10). These authors proposed that the drug acts as a molecular "wedge" that perturbs the interactions sufficiently to derail the process that would normally lead to closure into regular icosahedral particles.

Another class of drugs—HAP compounds (heteroaryldihydropyrimidines)—initially identified as having the ability to suppress HBV replication *in vivo* and *in vitro* (Weber *et al.*, 2002) has been found to act directly on the capsid protein (Deres *et al.*, 2003). The HAP reagent Bay 41-4109 was found to suppress the level of HBV DNA synthesis in stably transfected hepatoma cells, and this effect was accompanied by reduced levels of capsid protein synthesis and capsid assembly. Moreover, evidence was presented that these drugs bind to preassembled capsids *in vitro*.

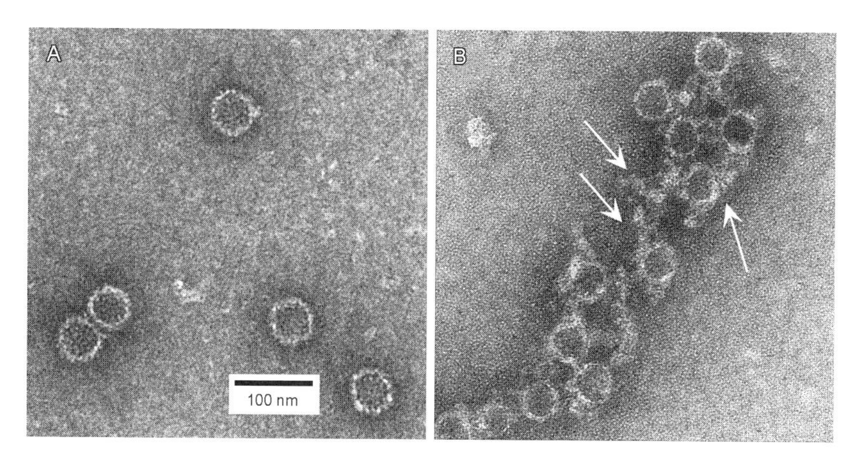


FIG 10. *In vitro* assembly of HBV capsids is misdirected by bis-ANS. (A) A mixture of $T = 4$ and $T = 3$ capsids producing by assembling 10 $\mu$M Cp149 dimers in the presence of 0.25 M NaCl. Association energy is relatively weak under these conditions and only 50% of the protein assembles; the remainder remains as free dimers (see Ceres and Zlotnick, 2002). (B) The same reaction with 23 $\mu$M bis-ANS yields less capsid on the basis of size-exclusion chromatography, but the sample scatters a disproportionate amount of light attributable to noncapsid aggregates. The misassembly effect is even greater at 39 $\mu$M bis-ANS. Under these conditions, less than 20% of the protein assembles into capsid. However, the noncapsid polymer remains correctly folded, according to circular dichroism spectroscopy, and may be induced to assemble into capsids at higher NaCl concentrations. Adapted from Zlotnick *et al.* (2002), with permission.

### *D. Nucleocapsid Assembly: Linker Peptide as Spacer*

Although the linker is not required for assembly of core domain constructs, it is required if the protamine domain is present, as Cp183 deleted for the linker was found not to be assembly-competent (Watts *et al.*, 2002). Residues 142/143 are the last residues to be seen in the crystal structure of Cp149 (Wynne *et al.*, 1999) and these residues have high-temperature factors, suggesting mobility. Nevertheless, the linker peptide, residues 140–149, is sufficiently ordered to have been visualized in a cryo-EM difference map between Cp149 and Cp140 (Watts *et al.*, 2002). Moreover, it is closely similar in sequence to a surface peptide with an extended conformation in cellobiose dehydrogenase (Fig. 11). On the basis of these observations, it has been proposed that the linker serves as a spacer that pivots around a semiflexible hinge at or close to residue 140, connecting the protamine domains and via them, the pregenomic RNA, to the core domain shell. This arrangement would maintain the pgRNA spread over the capsid

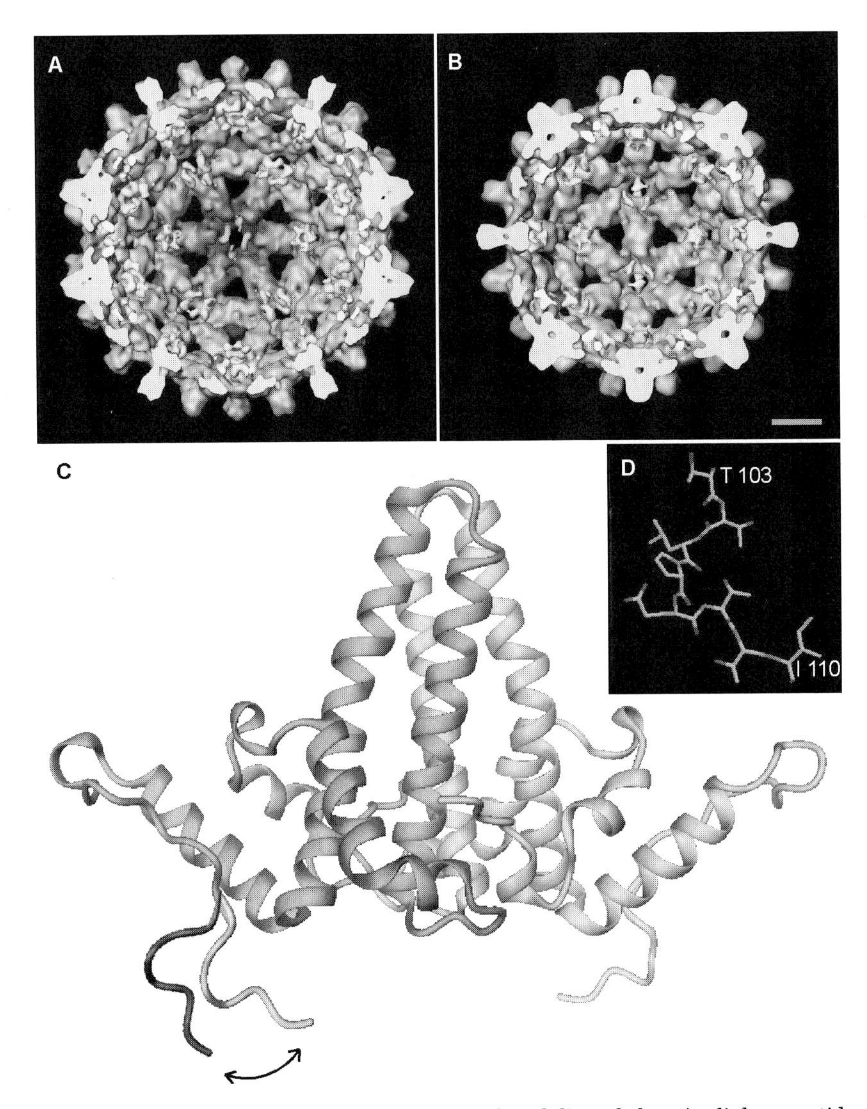


FIG 11. Location, proposed conformation, and mobility of the cAg linker peptide. (A and B) The linker peptide (yellow) was localized on both the $T = 4$ and the $T = 3$ capsids by cryo-EM difference mapping (Watts *et al.*, 2002). (C) Model of Cp142 with linker peptide attached with hinge about residue 142. (D) Its proposed conformation (extended, non-$\alpha$, non-$\beta$) based on that of a surface peptide of cellobiose dehydrogenase (Hallberg *et al.*, 2000), with which it shares nearly complete sequence identity. (See Color Insert.)

inner surface while retaining some mobility, properties that may facilitate reverse transcription (Watts *et al.*, 2002).

## VII. Conformational and Structural Roles of Cysteine Residues: eAg Is Dimeric

Cp183 has cysteines at positions 48, 61, 107, and 183. These residues may be substituted with alanine individually, in combination, or completely, without affecting capsid formation (Nassal *et al.*, 1992; Zheng *et al.*, 1992; Zlotnick *et al.*, 1996). Analysis of Cp183 by sodium dodecyl sulfate–polyacrylamide gel electrophoresis (SDS–PAGE) under nonreducing conditions revealed monomeric, dimeric, and polymeric protein (Nassal *et al.*, 1992; Zheng *et al.*, 1992). Cp183 freshly isolated from *E. coli* runs mainly as a monomer; with air oxidation, there is a slow accumulation of dimer; and after several months of storage, the protein is mostly polymeric. Similar air oxidation of Cp149 leads only to the formation of dimeric protein, indicating that the C-terminal C183 promotes the observed intermolecular cross-linking (dimer to dimer).

In the dimer structure, C61 is at the dyad axis formed by helix 3 (Fig. 7A) and in close proximity to C61 on the adjacent monomer. Because C61 is buried and solvent-inaccessible, the C61–C61 disulfide bond forms slowly, but it can be catalyzed by trace amounts of $Cu^{2+}$ (Nassal *et al.*, 1992). C48 and C107 are not close enough to form disulfides between monomers in a dimer: for example, the sulfur atoms of the two C48 residues are 19.6 Å apart, whereas the distance between $\alpha$ carbons in disulfide bonds range from 4.6 to 7.4 Å (Thornton, 1981). With Cp183, it is the combination of intermolecular cross-linking of monomers within the dimer via C61 and intermolecular linking of dimers via C183 that leads to the polymers observed by nonreducing SDS–PAGE (Fig. 12). This conclusion is supported by data from limited tryptic digestion of Cp183, which releases the C-terminal domain; gel analysis of the digest (equivalent to Cp149) indicates essentially dimeric protein (our unpublished results).

The effects of disulfide formation on capsid stability have been studied by calorimetry (Wingfield *et al.*, 1995). Thermograms were recorded of capsids (Cp183 and Cp149) and Cp149 dimers (at pH 9.6) in the presence and absence of reductant. The results indicated that disulfide bonds enhance the stability of Cp183 capsids and Cp149 dimers, but have little effect on Cp149 capsids.

*In vivo*, the preC signal sequence targets the eAg protein into the secretory pathway. Using transient transfection of a human liver cell

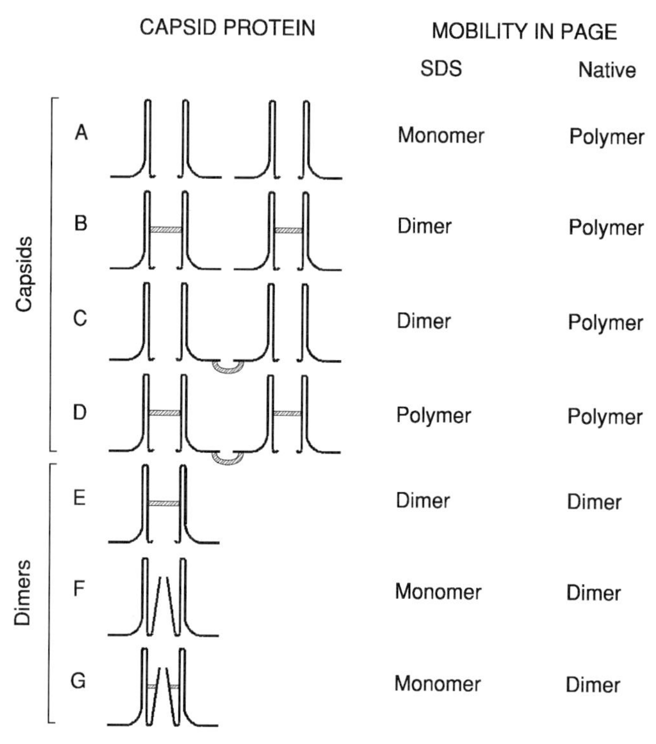


FIG 12. Schematic diagram of disulfide cross-linking and effects on SDS–PAGE mobility for various HBV capsid protein constructs. (A–D) Capsids of Cp149 or Cp183; (E) oxidized Cp149 dimers; (F and G) to Cp(−10) 149 in the reduced or oxidized state. Fully reduced capsids are stable (A). Disulfides form between monomers of the assembly dimer. These intermolecular disulfides connect the two C61 residues at the dimer interface (B). Disulfides formed by intermolecular cross-linking between monomers in adjacent dimers (dimer-to-dimer linking) are mediated mainly by C183 (C). Intermolecular disulfides within and between dimers form a network of cross-links that stabilize the capsids (D). In Cp149 dimers, the intermolecular disulfide between the two C61 residues forms readily (E). In dimeric Cp(−10)149, two intramolecular disulfides between C7 and C61 are formed (G). This dimer is stable in the reduced state (F). All HBV capsid protein variants are monomeric when analyzed by SDS–PAGE under reducing conditions, and are either high molecular weight polymers or dimers under native conditions. SDS–PAGE under nonreducing conditions (no reductant in sample buffers and free sulfhydryls capped with alkylating reagent) gives rise to monomers, dimers, or polymers, as indicated. The monomeric species observed with oxidized Cp(−10)149 (G) migrates anomalously on SDS–PAGE (lower apparent molecular weight than expected) because of intramolecular cross-linking.

line with Cp(−10)149 and various mutations, it was shown that C7 forms an intramolecular disulfide bond with C61 (Nassal and Rieger, 1993; Wasenauer *et al.*, 1993). C61 appeared to be critical for secretion (Wasenauer *et al.*, 1993). The following models were proposed (Nassal and Rieger, 1993): (1) eAg is a secretion-competent monomer containing an intramolecular disulfide between C7 and C61, and exhibits only e antigenicity; or (2) when C7 is absent and C61 is present, a secretion-competent dimer stabilized by a C61 intermolecular disulfide bond, is formed and exhibits both e and c antigenicities. According to these proposals, the main determinant of e antigenicity is related to a monomeric state of the protein and that of c antigenicity to a dimeric state.

One problem in earlier work on disulfide assignments was reliance on SDS–PAGE analysis; purified proteins were not available for more detailed analysis. The crystal structure of Cp149 indicates that the dimer axis is formed by the pairing of helix 3, which is amphipathic (Wynne *et al.*, 1999). As discussed previously, it has been widely held that eAg is monomeric (e.g., Nassal *et al.*, 1992; Schlicht and Wasenauer, 1991). The dimer structure (Fig. 7A) suggests that monomers would be unstable and/or insoluble, exposing extensive hydrophobic surfaces to solvent. This proposition—that there is unlikely to be a folded monomeric state—is supported by the observation that purified Cp(−10)149 is a stable dimer (our unpublished results). Furthermore, denaturation studies with Cp149 dimers show a single unfolding transition with no evidence of separation into monomers before global unfolding (our unpublished results). The dimeric nature of Cp(−10)149 appears to be independent of the presence of the intramolecular disulfide between C7 and C61.

An important question is why, in dimeric Cp(−10)149, is the C7-to-C61 intramolecular linkage formed in preference to the intermolecular C61-to-C61 disulfide? Residue 1 is only ~18 Å from C61 (Fig. 7A) and it would be quite possible for an eight-residue peptide to reach between them, as illustrated by the position of an exogenous peptide appended to the N terminus that we visualized in a cryo-EM difference imaging experiment (Fig. 6D). Because the N-terminal extension of Cp(−10) 149 is hydrophobic, we conjecture that it may intercalate into the dimer interface to promote formation of the C7-to-C61 cross-links. In this scenario, the overall structure of the dimer is maintained but it is locally perturbed in this way. Such perturbation might also contribute to the antigenic distinction between cAg and eAg (see Section VIII.D). Alternatively, assembly of dimers into capsids might be accompanied by conformational changes, affecting in particular the orientation of C61 and favoring the C7-to-C61 interaction over the C61-to-C61 interaction.

In both Cp149 and Cp(−10)149, cysteine residues do not affect assembly; stable capsids or dimers form, respectively. As discussed above, cysteines stabilize the capsid by intra- and intermolecular cross-linking. The physiological significance of this effect is not clear as the cytoplasmic milieu is reducing: on the other hand, secreted eAg is subjected to oxidizing conditions that would allow disulfides to form. Thus, we perceive Cp(−10)149, and probably also eAg, to be a dimeric molecule containing two intramolecular disulfides between C7 and C61. The C7-to-C61 cross-link has been suggested to prevent assembly of eAg into capsids (Schödel *et al.*, 1993). However, we have observed that Cp(−10)149 may be assembled into particles in the presence of high salt and other conditions, albeit with less alacrity than other constructs (our unpublished results).

In summary, C48 and C107 are not strategically placed for the formation of intra- or interdimer disulfides. C61, C183, and C7, when present, control these properties, as summarized schematically in Fig. 12.

## VIII. Antigenicity

Considerable effort has been devoted to investigating the antigenic properties of HBV capsid protein and the basis for the distinction between cAg and eAg. Although the two proteins have different terminal peptides, their epitopes have been assigned to their shared core domain sequence (Seifer and Standring, 1995) (Fig. 2). The cAg determinant is conformational and there has been some debate as to whether there are one or two such epitopes. For eAg, it has been generally agreed that there are two epitopes: e1, which is conformational; and e2, which is linear. The immunodominant loop has been implicated in both the cAg and the e1 epitopes, implying that the distinction between them must have its basis in conformational or accessibility properties. e2 has been assigned to a sequence near the end of the core domain (Seifer and Standring, 1995). In cAg capsids, e2 is cryptic, that is, its epitope is inaccessible to antibodies.

### A. *Mapping Conformational Epitopes*

For a conformational epitope to be recognized by antibodies, the antigen must assume its native three-dimensional structure. Most such epitopes are discontinuous in that they contain peptides from different parts of the polypeptide chain(s) juxtaposed by the three-dimensional fold. Mapping conformational epitopes by immunochemical methods is problematic because the antibodies do not recognize individual peptides in Pepscan assays or (unfolded) protein fragments in Western blots. Moreover, mappings based on competition binding experiments with antibodies that recognize known linear epitopes are difficult to interpret because of steric blocking effects: once bound, an antibody may occlude other epitopes considerably removed from its own. However, cryo-EM of Fab-labeled capsids affords a direct method of elucidating conformational epitopes. Because the crystal structure of the capsid is known, molecular modeling (Wang *et al.*, 1992) may be performed with generic Fab structures from the Protein Data Bank to interpret cryo-EM density maps in greater detail than their nominal resolution. At resolutions of 10–12 Å, we found that the reproducibility with which different Fabs may be docked into a given map, or a given docking accomplished by manual or automated means, to be at the level of ±1 Å in each dimension (Conway *et al.*, 2003). In this way it is possible to specify the point(s) of contact with the antibody and thus identify the peptides on the capsid surface that make up the epitope.

We have used this approach to elucidate four cAg epitopes (Figs. 13–15); monoclonal antibody (mAb) 312 recognizes a linear epitope that was mapped to residues 77 to 82 by Pepscan assays (Sällberg *et al.*, 1991). From cryo-EM, its Fab was observed to bind to the loop on one subunit at the tip of the spike (Conway *et al.*, 1998a), explaining its linear character. In the first instance, this correlation was used to localize these residues to this site. When the crystal structure became available, it was possible to interpret the cryo-EM map in greater detail, whence we concluded that the Fab-binding sequence covers residues 74 to 80 (Belnap *et al.*, 2003). We suspect that the small discrepancy between the two mappings represents conformational effects reflecting how the epitope is presented in the respective contexts.

### B. *Quasi-Equivalent Variations in Binding Affinity*

The other three epitopes that have been characterized are strictly conformational. mAb 3120 (Takahashi *et al.*, 1983) does not bind to the spike and displays pronounced variations in binding affinity between different quasi-equivalent sites (Conway *et al.*, 2003). There are four quasi-equivalent subunits on the $T = 4$ capsid and three on the $T = 3$ capsid, each present in 60 copies (Fig. 3). The $T = 4$ subunits are designated A, B, C, D and the $T = 3$ subunits are designated A, B, C, although they do not necessarily have the same conformations and interactions as the similarly named $T = 4$ subunits. The 3120 epitope

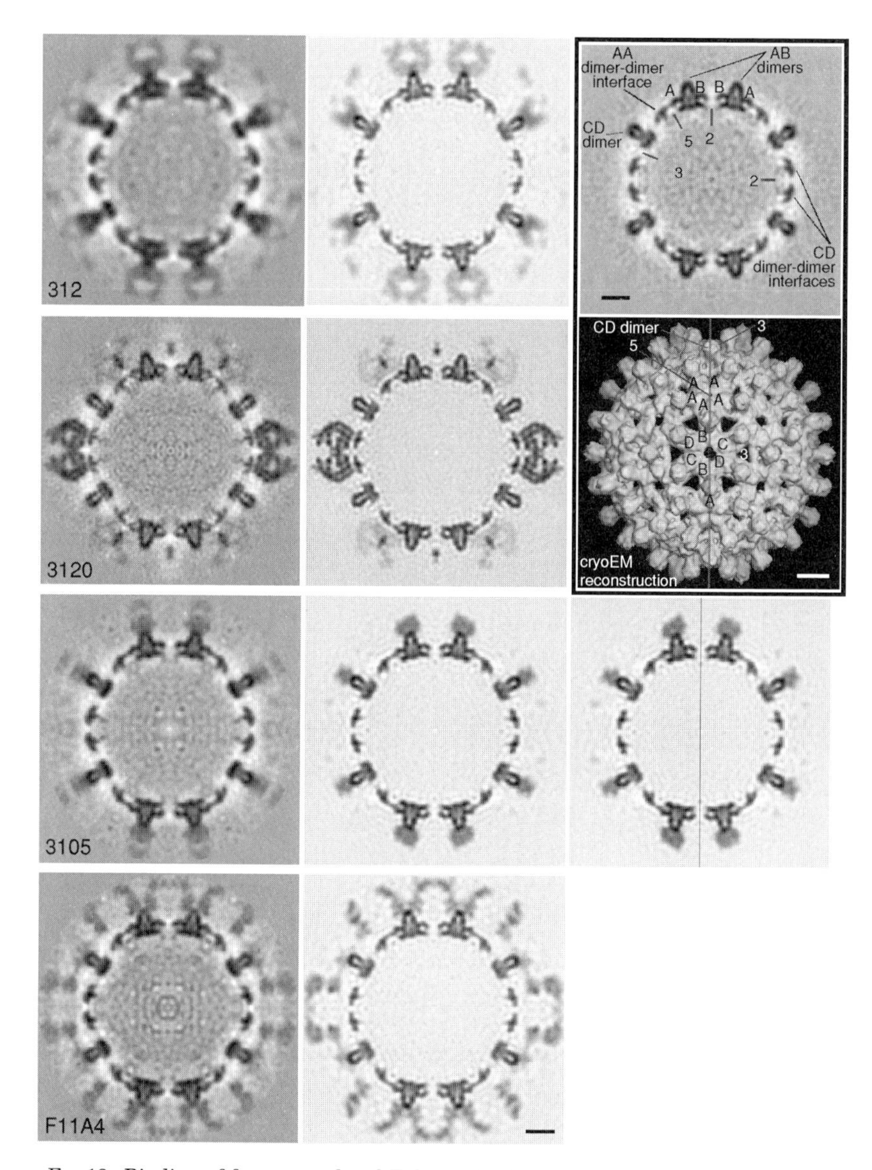


FIG 13. Binding of four monoclonal Fabs to the $T = 4$ HBV capsid. In each case, the central section of a cryo-EM map of the Fab-labeled capsid is shown in the first column. The second column shows the corresponding models obtained by docking a generic Fab structure onto the capsid in the cryo-EM density map of the complex, and then band-limiting the resulting model of capsid plus bound Fab to the same resolution as the original cryo-EM map. Considerable Fab substructure is visualized in the cryo-EM density maps (and models) with Fabs 3120 and F11A4, for which there is relatively little

does not lie on the spike but on an exposed surface around the symmetry axes, where it straddles the interface between subunits on adjacent dimers. Of the seven quasi-equivalent variants of this epitope, only three bind Fab 3120: the C–D epitope on the $T = 4$ capsid, and the A–A epitopes on both the $T = 3$ and $T = 4$ capsids (Fig. 13). Steric interference limits binding to one Fab per fivefold vertex on the $T = 4$ capsid and two Fabs per vertex on the $T = 3$ capsid, reflecting subtle conformational differences between the respective vertices. There are three pairs of epitopes around the twofold axis on the $T = 4$ capsid: C–D, D–B, and B–C. Of these, the C–D epitopes are fully occupied and the other two are vacant, again reflecting conformational nuances deriving from quasi-equivalence. There are two triplets of epitopes around the threefold axis on the $T = 3$ capsid: B–C and C–B. Both are vacant, indicating little or no affinity for this antibody.

### *C. Diversity of cAg Epitopes*

Two more epitopes—for mAbs 3105 (Takahashi *et al.*, 1983) and F11A4 (Ferns and Tedder, 1986)—are also near the tip of the spike and both have contributions from the loops on both subunits, but they differ markedly in both epitope structure and Fab binding disposition (Belnap *et al.*, 2003) (Figs. 13 and 14). As noted previously, steric blocking effects can undermine the interpretation of competition experiments. They can also complicate interpretation of cryo-EM density maps. If two copies of a given epitope are in mutual proximity, binding of a Fab to one of them occludes the other. As a result, the Fab-related density in a cryo-EM map represents the average of Fabs bound at the two sites. This density will have the shape not of a unitary Fab molecule but of a merged and partially overlapping composite. This poses a significant complication because docking experiments usually involve guiding the molecule in question into its envelope as represented by an isodensity contouring of the three-dimensional density map. Nevertheless, modeling may still be accomplished by placing the

overlap between neighboring Fabs. 312 is less detailed because it is at somewhat lower resolution. The Fab 3105 density is less detailed, not because of lower resolution but because there is almost total overlap between two Fabs, respectively, bound on either side of the dyad axis at the spike tip. Their densities are smeared by averaging in the reconstruction. The third panel on the right (row 3) compares the 3105 modeling with a single Fab on one side (*left half*) and on the other side (*right half*) of the dyad axis. First and second panels on the right mark reference points in these sections. From Conway *et al.* (1998a, 2003) and Belnap *et al.* (2003). Scale bars: 50 Å.

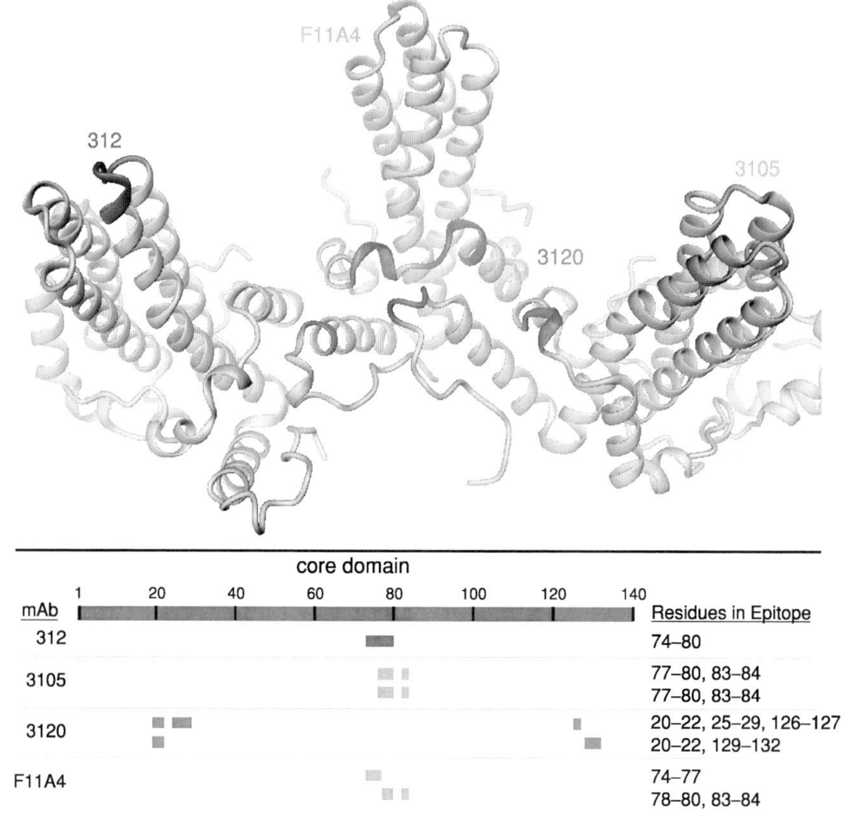


FIG 14. Mapping the epitopes of four monoclonal anti-cAg IgGs. *Top*: The configurations of peptides involved in each case are marked on a ribbon diagram of three adjacent dimers. The epitope for mAb 3120 involves five peptides from two different subunits on neighboring dimers: on the $T = 4$ capsid, only two of four quasi-equivalent variants of the epitope (C–D and A–A) bind the Fab; on the $T = 3$ capsid, only one of three variants (A–A) binds it (Conway *et al.*, 2003). mAbs 3105 and F11A4 each have contributions from both subunits within a dimer. mAb 312 binds to the immunodominant loop on a single subunit. *Bottom*: Mapping of the positions of these peptides on the core domain sequence. Reproduced from Belnap *et al.* (2003). (See Color Insert.)

Fab on the footprint and pivoting it about this point, and similarly at symmetry-related positions, to achieve an optimal match of the averaged modeled density with the observed density. This is illustrated in Fig. 13 for Fab 3105, whose two epitopes at the spike tip are so close together that Fabs bound to them overlap almost completely. In contrast, two F11A4 Fabs may bind to the same spike but they would

FIG 15. Mapping the epitopes of four monoclonal IgGs on the $T = 4$ HBV capsid. The epitopes are marked as yellow patches on a surface rendering of the capsid surface. The binding aspects of the corresponding four Fabs are shown: orange, mAb 312; magenta, mAb 3120; green, mAb 3105; cyan, mAb F11A4. (See Color Insert.)

both occlude an epitope on a neighboring spike so that the maximum occupancy is also 50% in this case.

In summary, the steric barrier posed by the capsid spikes restricts antibody binding to two general regions on the outer surface of the capsid: the outer portion of the spike and the region around the symmetry axes (Figs. 14 and 15). On the spike, there is considerable latitude for epitope diversity: the three epitopes characterized to date are all quite distinct and the full range is likely to be more extensive. So far, we have characterized one epitope (3120) around the symmetry axes and we anticipate a similar microheterogeneity of epitopes in this region.

All three conformational epitopes straddle intersubunit interfaces: two of them (3105 and F11A4) overlie the dimer interface on a given spike; the other, 3120, overlies the interdimer interface. Of the three, 3120 is the only one to exhibit quasi-equivalent variations in binding affinity, and it represents an extreme form of conformational epitope, that is, four of the seven variants have their native folds but do not bind the antibody. A tentative conclusion to be drawn from this observation is that, on HBV capsids, quasi-equivalent variations in conformation are expressed primarily at the interdimer interfaces and the dimer structure is relatively invariant.

### *D. Distinction Between cAg and eAg*

These observations suggest possible explanations for the differing immunologic specificities of cAg and eAg. In particular, they have demonstrated that the two copies of the loop at the spike tip, together with surrounding structural elements, are capable of generating a considerable variety of conformational epitopes. mAbs 312, 3105, and F11A4 all bind to this region, but they differ markedly in their binding aspects and in their epitopes. Thus it appears likely that cAg and e1-Ag represent distinct (sets of) epitopes of this kind. This distinction may relate to their origins, cAg having been identified as eliciting a T cell-independent immune response and eAg, a T cell-dependent response (Milich *et al.*, 1988), or it may reflect intrinsic conformational differences affecting the spike.

Another source of difference relates directly to the state of assembly. Conformational antibodies such as mAb 3120 bind to the interface between adjacent dimers and would not recognize unassembled dimers, that is, they should recognize cAg but not eAg. Conversely, there are sites that are Fab-accessible on unassembled dimers—for instance, at the base of the spike or on its underside—that are inaccessible on capsids (Belnap *et al.*, 2003; Seifer and Standring, 1995).

## IX. Role of Capsid Protein in Envelopment

Envelopment necessarily involves interaction between the capsid and the envelope glycoprotein sAg. Strategically, this step should be deferred until retrotranscription of the pregenome inside the nucleocapsid is complete or nearly so, because once enveloped, the transcription machinery is separated from the source of nucleotides needed to complete this reaction. It has long been suspected that the signal for nucleocapsid envelopment may lie in structural changes of the capsid shell, induced by conversion of the pgRNA into DNA (see Section II). For instance, mutations disrupting RT activity were found to block virion formation (Gerelsaikhan *et al.*, 1996).

To investigate the role of capsid protein, two mutational approaches have been taken. One, by Bruss and co-workers, has been to characterize mutants that remain competent to assemble capsids but are affected in secretion of progeny virions from a transfected cell line. Of 110 insertions and deletions tested, 38 supported capsid formation in *E. coli*. These mutants mapped to loop regions of the capsid protein (Koschel *et al.*, 1999) (Fig. 16A). To test for envelopment competence, eukaryotic expression vectors for the mutants were cotransfected with a capsid-negative HBV genome vector into human hepatoma cells and assayed for secretion of infectious virus. Only one deletion, of P79–S81, secreted at wild-type levels. Because this locus is near the tip of the spike, this region was inferred not to make contact with the cytoplasmic domain of sAg (Koschel *et al.*, 2000). A subsequent study effected alanine substitutions in residues known from the crystal structure to be solvent accessible. Envelopment was evaluated with a system similar to that described previously (Ponsel and Bruss, 2003). Substitutions in 11 sites allowed capsid formation but blocked virion formation. These residues form a ring-like groove at the base of the spike (Fig. 16A) that was suggested to be involved in interactions with sAg.

The other approach, by Shih and co-workers, has been to isolate natural, that is, patient-derived, mutants with altered secretory properties. Analysis of cAg sequences isolated from chronic HBV carriers revealed mutational hot spots (Hosono *et al.*, 1995). The most frequent mutations occur at residues P5, V13, Y38, I59, L60, N87, I97, and P130 (Fig. 16B). When secretion was evaluated for these variants in a eukaryotic coexpression system similar to that outlined previously, some were found to enhance secretion whereas others decreased secretion. For example, a frequent mutation is F97L or I97L, which increases the amount of secreted particles with immature genomes (Yuan *et al.*, 1999). Because a mutation in the preS1 region of sAg (A119F) restores a wild-type secretion phenotype to the F97L mutant, it was suggested that the increased secretion of immature genomes is due to aberrant interaction between sAg and the capsid (Le Pogam and Shih, 2002). Conversely, reduced levels of secretion were observed for mutants P5T and L60V (Le Pogam *et al.*, 2000).

The secretion-affected mutants of both kinds, that is, the *in vitro* and *in vivo* mutations, are mapped on the capsid protein in Fig. 16B. It is noteworthy that the majority cluster around the base of the molecule and are excluded from the protruding helices. Thus, if conformational adjustments in the capsid protein are the signal for secretion, it appears that such adjustments may be affected by mutations in other parts of the protein.

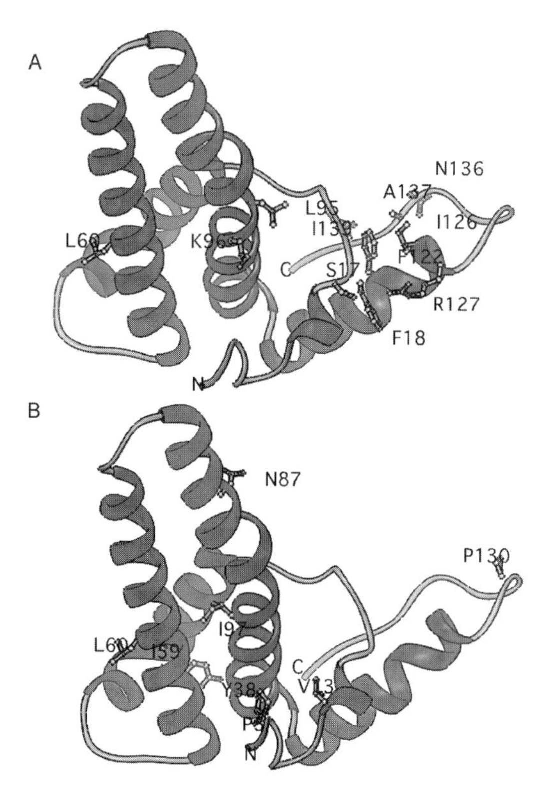


FIG 16. cAg monomers with residues marked whose mutations disrupt HBV secretion. The mutants were identified by (A) *in vitro* mutagenesis or (B) from *in vivo* natural infections. (A) When mutated to alanine, these residues allow nucleocapsid formation but block envelopment and virion formation. These sites—S17, F18, L60, L95, K96, F122, I126, R127, N136, A137, and I139—form a groove around the base of the spike on the portion of the molecule that faces outward on the assembled capsid. This region has been suggested to interact with sAg during envelopment. (B) The most frequent mutations isolated from chronic carriers of HBV. Residues affected include P5, V13, Y38, I59, L60, N87, I97, and P130. Mutation of I97 to L97 increases the amount of immature genome in secreted particles, whereas low-level secretion of virions is caused by mutations P5 to T5 and L60 to V60. In both cases, the residues affected cluster around the base of the spike, not on it. Residues are represented by side chains, which are green ball-and-stick models. N and C termini along with residues are labeled. Based on work by Bruss and co-workers and Shih and co-workers (see text for references). (See Color Insert.)

## X. FUTURE DIRECTIONS

Despite progress, several fundamental questions pertaining to the structure and function of HBV capsids remain to be answered. What is the full range of antigenic determinants? Which epitopes are immunodominant in a clinical setting? How does the nucleocapsid interact with the envelope and what is the nucleocapsid maturation signal? And how, if at all, do capsid shells assembled in infected hepatocytes, with the incorporation of RT and other proteins, differ from assembly products of recombinant proteins *in vitro* and in bacteria?

## Acknowledgments

We thank Adam Zlotnick for much creative discussion and Fig. 10, Vijay Reddy for help with Fig. 9, and Ira Palmer and Joshua Kaufman for expert assistance. This work was supported in part by the NIH Intramural AIDS Targeted Antiviral Program.

## References

Acs, G., Sells, M. A., Purcell, R. H., Price, P., Engle, R., Shapiro, M., and Popper, H. (1987). Hepatitis B virus produced by transfected Hep G2 cells causes hepatitis in chimpanzees. *Proc. Natl. Acad. Sci. USA* **84:**4641–4644.

Baumeister, M. A., Medina-Selby, A., Coit, D., Nguyen, S., George-Nascimento, C., Gyenes, A., Valenzuela, P., Kuo, G., and Chien, D. Y. (2000). Hepatitis B virus e antigen specific epitopes and limitations of commercial anti-HBe immunoassays. *J. Med. Virol.* **60:**256–263.

Belnap, D. M., Watts, N. R., Conway, J. F., Cheng, N., Stahl, S. J., Wingfield, P. T., and Steven, A. C. (2003). Diversity of core antigen epitopes of hepatitis B virus. *Proc. Natl. Acad. Sci. USA* **100:**10884–10889.

Birnbaum, F., and Nassal, M. (1990). Hepatitis B virus nucleocapsid assembly: Primary structure requirements in the core protein. *J. Virol.* **64:**3319–3330.

Blumberg, B. S. (1997). Hepatitis B virus, the vaccine, and the control of primary cancer of the liver. *Proc. Natl. Acad. Sci. USA* **94:**7121–7125.

Booy, F. P., Trus, B. L., Davison, A. J., and Steven, A. C. (1996). The capsid architecture of channel catfish virus, an evolutionarily distant herpesvirus, is largely conserved in the absence of discernible sequence homology with herpes simplex virus. *Virology* **215:**134–141.

Böttcher, B., Wynne, S. A., and Crowther, R. A. (1997). Determination of the fold of the core protein of hepatitis B virus by electron cryomicroscopy. *Nature* **386:**88–91.

Bringas, R. (1997). Folding and assembly of hepatitis B virus core protein: A new model proposal. *J. Struct. Biol.* **118:**189–196.

Ceres, P., and Zlotnick, A. (2002). Weak protein-protein interactions are sufficient to drive assembly of hepatitis B virus capsids. *Biochemistry* **41:**11525–11531.

Chang, C. M., Jeng, K. S., Hu, C. P., Lo, S. J., Su, T. S., Ting, L. P., Chou, C. K., Han, S. H., Pfaff, E., Salfeld, J., and Schaller, H. (1987). Production of hepatitis B virus *in vitro* by transient expression of cloned HBV DNA in a hepatoma cell line. *EMBO J.* **6:**675–680.

Chapman, M.S., and Liljas, L. (2003). Structural folds of viral proteins. *Adv. Protein Chem.* **64:**125–196.

Chen, X. S., Casini, G., Harrison, S. C., and Garcea, R. L. (2001). Papillomavirus capsid protein expression in *Escherichia coli*: Purification and assembly of HPV11 and HPV16 L1. *J. Mol. Biol.* **307:**173–182.

Cheng, N., Conway, J. F., Watts, N. R., Hainfeld, J. F., Joshi, V., Powell, R. D., Stahl, S. J., Wingfield, P. T., and Steven, A.C. (1999). Tetrairidium, a four-atom cluster, is readily visible as a density label in three-dimensional cryo-EM maps of proteins at 10–25 Å resolution. *J. Struct. Biol.* **127:**169–176.

Cheng, R. H., Caston, J. R., Wang, G. J., Gu, F., Smith, T. J., Baker, T. S., Bozarth, R. F., Trus, B. L., Cheng, N., Wickner, R. B., and Steven, A. C. (1994). Fungal virus capsids, cytoplasmic compartments for the replication of double-stranded RNA, formed as icosahedral shells of asymmetric Gag dimers. *J. Mol. Biol.* **244:**255–258.

Cohen, B. J., and Richmond, J. E. (1982). Electron microscopy of hepatitis B core antigen synthesized in *E. coli.*. *Nature* **296:**677–679.

Conway, J. F., Cheng, N., Zlotnick, A., Stahl, S. J., Wingfield, P. T., Belnap, D. M., Kanngiesser, U., Noah, M., and Steven, A. C. (1998a). Hepatitis B virus capsid: Localization of the putative immunodominant loop (residues 78 to 83) on the capsid surface, and implications for the distinction between c and e-antigens. *J. Mol. Biol.* **279:**1111–1121.

Conway, J. F., Cheng, N., Zlotnick, A., Stahl, S. J., Wingfield, P. T., and Steven, A. C. (1998b). Localization of the N terminus of hepatitis B virus capsid protein by peptide-based difference mapping from cryoelectron microscopy. *Proc. Natl. Acad. Sci. USA* **95:**14622–14627.

Conway, J. F., Cheng, N., Zlotnick, A., Wingfield, P. T., Stahl, S. J., and Steven, A. C. (1997). Visualization of a 4-helix bundle in the hepatitis B virus capsid by cryo-electron microscopy. *Nature* **386:**91–94.

Conway, J. F., Watts, N. R., Belnap, D. M., Cheng, N., Stahl, S. J., Wingfield, P. T., and Steven, A. C. (2003). Characterization of a conformational epitope on hepatitis B virus core antigen and quasi-equivalent variations in antibody binding. *J. Virol.* **77:**6466–6473.

Crowther, R. A., Kiselev, N. A., Böttcher, B., Berriman, J. A., Borisova, G. P., Ose, V., and Pumpens, P. (1994). Three-dimensional structure of hepatitis B virus core particles determined by electron cryomicroscopy. *Cell* **77:**943–950.

Daub, H., Blencke, S., Habenberger, P., Kurtenbach, A., Dennenmoser, J., Wisssing, J., Ullrich, A., and Cotton, M. (2002). Identification of SRPK1 and SRPK2 as the major cellular protein kinases phosphorylating hepatitis B virus core protein. *J. Virol.* **76:**8124–8137.

Deres, K., Schroder, C. H., Paessens, A., Goldmann, S., Hacker, H. J., Weber, O., Kramer, T., Niewohner, U., Pleiss, U., Stoltefuss, J., Graef, E., Koletzki, D., Masantschek, R.N., Reimann, A., Jaeger, R., Gross, R., Beckermann, B., Schlemmer, K. H., Haebich, D., and Rubsamen-Waigmann, H. (2003). Inhibition of hepatitis B virus replication by drug-induced depletion of nucleocapsids. *Science* **299:**893–896.

Driedonks, R. A., Krijgsman, P. C., and Mellema, J. E. (1978). Characterization of alfalfa-mosaic-virus protein polymerization in the presence of nucleic acid. *Eur. J. Biochem.* **82:**405–417.

Endres, D., and Zlotnick, A. (2002). Model-based analysis of assembly kinetics for virus capsids or other spherical polymers. *Biophys. J.* **83:**1217–1230.

Ferns, R. B., and Tedder, R. S. (1986). Human and monoclonal antibodies to hepatitis B core antigen recognise a single immunodominant epitope. *J. Med. Virol.* **19:**193–203.

Gamble, T. R., Yoo, S., Vajdoos, F. F., von Schwedler, U. K., Worthylake, D. K., Wang, H., McCutcheon, J. P., Sundquist, W. I., and Hill, C. P. (1997). Structure of the carboxy-terminal dimerization domain of the HIV-1 capsid protein. *Science* **278:**849–853.

Ganem, D. (1996). Hepadnaviridae and their replication. *In* "Field's Virology" (N. F. Fields, D. M. Knipe, and P. M. Howley, eds.), Vol. 2. pp. 2703–2737. Lippincott-Raven, Philadelphia, PA.

Gerelsaikhan, T., Tavis, J. E., and Bruss, V. (1996). Hepatitis B virus nucleocapsid envelopement does not occur without genomic DNA synthesis. *J. Virol.* **70:**4269–4274.

Gerlich, W. H., Godmann, U., Muller, R., Stibbe, W., and Wolff, W. (1982). Specificity and localization of the hepatitis B virus-associated protein kinase. *J. Virol.* **42:**761–766.

Hallberg, B. M., Bergfors, T., Backbro, K., Pettersson, G., Henriksson, G., and Divne, C. (2000). A new scaffold for binding haem in the cytochrome domain of the extracellular flavocytochrome cellobiose dehydrogenase. *Struct. Fold. Des.* **8:**79–88.

Hitzeman, R. A., Chen, C. Y., Hagie, F. E., Patzer, E. J., Liu, C. C., Estell, D. A., Miller, J. V., Yaffe, A., Kleid, D. G., Levinson, A. D., and Oppermann, H. (1983). Expression of hepatitis B virus surface antigen in yeast. *Nucleic Acids Res.* **11:**2745–2763.

Hollinger, F. B. (1996). Hepatitis B virus. *In* "Field's Virology" (N. F. Fields, D. M. Knipe, and P. M. Howley, eds.), Vol. 2. pp. 2738–2808. Lippincott-Raven, Philadelphia, PA.

Honkoop, P., and De Man, R. A. (2003). Entecavir: A potent new antiviral drug for hepatitis B. *Expert Opin. Investig. Drugs* **12:**683–688.

Hosono, S., Tai, P . C., Wang, W., Ambrose, M., Hwang, D. G., Yuan, T. T., Peng, B. H., Yang, C. S., Lee, C. S., and Shih, C. (1995). Core antigen mutations of human hepatitis B virus in hepatomas accumulate in MHC class II-restricted T cell epitopes. *Virology* **212:**151–162.

Hu, J., Toft, D. O., and Seeger, C. (1997). Hepadnavirus assembly and reverse transcription require a multi-component chaperone complex which is incorporated into nucleocapsids. *EMBO J.* **16:**59–68.

Kenney, J. M., von Bonsdorff, C. H., Nassal, M., and Fuller, S. D. (1995). Evolutionary conservation in the hepatitis B virus core structure: Comparison of human and duck cores. *Structure* **3:**1009–1019.

Kniskern, P. J., Hagopian, A., Montgomery, D. L., Burke, P., Dunn, N. R., Hofmann, K. J., Miller, W. J., and Ellis, R. W. (1986). Unusually high-level expression of a foreign gene (hepatitis B virus core antigen) in *Saccharomyces cerevisiae*. *Gene* **46:**135–141.

Koschel, M., Oed, D., Gerelsaikhan, T., Thomssen, R., and Bruss, V. (2000). Hepatitis B virus core gene mutations which block nucleocapsid envelopment. *J. Virol.* **74:**1–7.

Koschel, M., Thomssen, R., and Bruss, V. (1999). Extensive mutagenesis of the hepatitis B virus core gene and mapping of mutations that allow capsid formation. *J. Virol.* **73:**2153–2160.

Kratz, P. A., Böttcher, B., and Nassal, M. (1999). Native display of complete foreign protein domains on the surface of hepatitis B virus capsids. *Proc. Natl. Acad. Sci. USA* **96:**1915–1920.

Lan, Y. T., Li, J., Liao, W., and Ou, J. (1999). Roles of the three major phosphorylation sites of hepatitis B virus core protein in viral replication. *Virology* **259:**342–348.

Le Pogam, S., and Shih, C. (2002). Influence of a putative intermolecular interaction between core and the pre-S1 domain of the large envelope protein on hepatitis B virus secretion. *J. Virol.* **76:**6510–6517.

Le Pogam, S., Yuan, T. T., Sahu, G. K., Chatterjee, S., and Shih, C. (2000). Low-level secretion of human hepatitis B virus virions caused by two independent, naturally occurring mutations (P5T and L60V) in the capsid protein. *J. Virol.* **74:**9099–9105.

Machida, A., Ohnuma, H., Tsuda, F., Yoshikawa, A., Hoshi, Y., Tanaka, T., Kishimoto, S., Akahane, Y., Miyakawa, Y., and Mayumi, M. (1991). Phosphorylation in the carboxyl-terminal domain of the capsid protein of hepatitis B virus: Evaluation with a monoclonal antibody. *J. Virol.* **65:**6024–6030.

Metzger, K., and Bringas, R. (1998). Proline-138 is essential for the assembly of hepatitis B virus core protein. *J. Gen. Virol.* **79:**587–590.

Milich, D. R., Hughes, J., Jones, J., Sällberg, M., and Phillips, T. R. (2001). Conversion of poorly immunogenic malaria repeat sequences into a highly immunogenic vaccine candidate. *Vaccine* **20:**771–788.

Milich, D. R., McLachlan, A., Stahl, S., Wingfield, P., Thornton, G. B., Hughes, J. L., and Jones, J. E. (1988). Comparative immunogenicity of hepatitis B virus core and E antigens. *J. Immunol.* **141:**3617–3624.

Miller, H., and Robinson, W. S. (1986). Common evolutionary origin of hepatitis B virus and retroviruses. *Proc. Natl. Acad. Sci. USA* **83:**2531–2535.

Nassal, M. (1996). Hepatitis B virus morphogenesis. *Curr. Top. Microbiol. Immunol.* **214:**297–337.

Nassal, M., and Rieger, A. (1993). An intramolecular disulfide bridge between Cys-7 and Cys-61 determines the structure of the secretory core gene product (e antigen) of hepatitis B virus. *J. Virol.* **67:**4307–4315.

Nassal, M., Rieger, A., and Steinau, O. (1992). Topological analysis of the hepatitis B virus core particle by cysteine–cysteine cross-linking. *J. Mol. Biol.* **225:**1013–1025.

Nassal, M., and Schaller, H. (1993). Hepatitis B virus nucleocapsid assembly. *In* "Virus Strategies" (W. Doefler and P. Bohm, eds.), pp. 41–75. VCH Publishers, Weinheim, Germany.

Ou, J. H. (1997). Molecular biology of hepatitis B virus e antigen. *J. Gastroenterol. Hepatol.* **12:**S178–S187.

Ponsel, D., and Bruss, V. (2003). Mapping of amino acid side chains on the surface of hepatitis B virus capsids required for envelopment and virion formation. *J. Virol.* **77:**416–422.

Pumpens, P., Borisova, G. P., Crowther, R. A., and Grens, E. (1995). Hepatitis B virus core particles as epitope carriers. *Intervirology* **38:**63–74.

Pumpens, P., and Grens, E. (2001). HBV core particles as a carrier for B cell/T cell epitopes. *Intervirology* **44:**98–114.

Ryu, C. J., Cho, D. Y., Gripon, P., Kim, H. S., Guguen-Guillouzo, C., and Hong, H. J. (2000). An 80-kilodalton protein that binds to the pre-S1 domain of hepatitis B virus. *J. Virol.* **74:**110–116.

Salfeld, J., Pfaff, E., Noah, M., and Schaller, H. (1989). Antigenic determinants and functional domains in core antigen and e antigen from hepatitis B virus. *J. Virol.* **63:**798–808.

Sällberg, M., Ruden, U., Magnius, L. O., Harthus, H. P., Noah, M., and Wahren, B. (1991). Characterisation of a linear binding site for a monoclonal antibody to hepatitis B core antigen. *J. Med. Virol.* **33:**248–252.

Schlicht, H. J., and Wasenauer, G. (1991). The quaternary structure, antigenicity, and aggregational behavior of the secretory core protein of human hepatitis B virus are determined by its signal sequence. *J. Virol.* **65:**6817–6825.

Schödel, F., Moriarty, A. M., Peterson, D. L., Zheng, J. A., Hughes, J. L., Will, H., Leturcq, D. J., McGee, J. S., and Milich, D. R. (1992). The position of heterologous epitopes inserted in hepatitis B virus core particles determines their immunogenicity. *J. Virol.* **66:**106–114 [published erratum appears in *J. Virol.* **66:** 3977].

Schödel, F., Peterson, D., Zheng, J., Jones, J. E., Hughes, J. L., and Milich, D. R. (1993). Structure of hepatitis B virus core and e-antigen: A single precore amino acid prevents nucleocapsid assembly. *J. Biol. Chem.* **268:**1332–1337.

Seeger, C., and Mason, W. S. (2000). Hepatitis B virus biology. *Microbiol. Mol. Biol. Rev.* **64:**51–68.

Seifer, M., and Standring, D. N. (1995). Assembly and antigenicity of hepatitis B virus core particles. *Intervirology* **38:**47–62.
Singh, S., and Zlotnick, A. (2003). Observed hysteresis of virus capsid disassembly is implicit in kinetic models of assembly. *J. Biol. Chem.* **278:**18249–18255.
Stahl, S., MacKay, P., Magazin, M., Bruce, S. A., and Murray, K. (1982). Hepatitis B virus core antigen: Synthesis in *Escherichia coli* and application in diagnosis. *Proc. Natl. Acad. Sci. USA* **79:**1606–1610.
Sureau, C., Romet-Lemonne, J. L., Mullins, J. I., and Essex, M. (1986). Production of hepatitis B virus by a differentiated human hepatoma cell line after transfection with cloned circular HBV DNA. *Cell* **47:**37–47.
Takahashi, K., Machida, A., Funatsu, G., Nomura, M., Usuda, S., Aoyagi, S., Tachibana, K., Miyamoto, H., Imai, M., Nakamura, T., Miyakawa, Y., and Mayumi, M. (1983). Immunochemical structure of hepatitis B e-antigen in the serum. *J. Immunol.* **130:**2903–2907.
Takehara, K., Ireland, D., and Bishop, D. H. (1988). Co-expression of the hepatitis B surface and core antigens using baculovirus multiple expression vectors. *J. Gen. Virol.* **69:**2763–2777.
Teschke, C. M., King, J., and Prevelige, P. E., Jr. (1993). Inhibition of viral capsid assembly by 1,1′-bi(4-anilinonaphthalene-5-sulfonic acid). *Biochemistry* **32:**10658–10665.
Thornton, J. M. (1981). Disulphide bridges in globular proteins. *J. Mol. Biol.* **151:**261–287.
Valenzuela, P., Medina, A., Rutter, W. J., Ammerer, G., and Hall, B. D. (1982). Synthesis and assembly of hepatitis B virus surface antigen particles in yeast. *Nature* **298:**347–350.
Vanlandschoot, P., Cao, T., and Leroux-Roels, G. (2003). The nucleocapsid of the hepatitis B virus: A remarkable immunogenic structure. *Antiviral Res.* **60:**67–74.
Varughese, K. I., Madhusudan, Zhou, X. Z., Whiteley, J. M., and Hoch, J. A. (1998). Formation of a novel four-helix bundle and molecular recognition sites by dimerization of a response regulator phosphotransferase. *Mol. Cell* **2:**485–493.
Wang, G. J., Porta, C., Chen, Z. G., Baker, T. S., and Johnson, J. E. (1992). Identification of a Fab interaction footprint site on an icosahedral virus by cryoelectron microscopy and X-ray crystallography. *Nature* **355:**275–278.
Wasenauer, G., Kock, J., and Schlicht, H. J. (1993). A cysteine and a hydrophobic sequence in the noncleaved portion of the pre-C leader peptide determine the biophysical properties of the secretory core protein (HBe protein) of human hepatitis B virus. *J. Virol.* **66:**5338–5346.
Watts, N. R., Conway, J. F., Cheng, N., Stahl, S. J., Belnap, D. M., Steven, A. C., and Wingfield, P. T. (2002). The morphogenic linker peptide of HBV capsid protein forms a mobile array on the interior surface. *EMBO J.* **21:**876–884.
Weber, O., Schlemmer, K. H., Hartmann, E., Hagelschuer, I., Paessens, A., Graef, E., Deres, K., Goldmann, S., Niewoehner, U., Stoltefuss, J., Haebich, D., Ruebsamen-Waigmann, H., and Wohlfeil, S. (2002). Inhibition of human hepatitis B virus (HBV) by a novel non-nucleosidic compound in a transgenic mouse model. *Antiviral Res.* **54:**69–78.
Wingfield, P. T., Stahl, S. J., Williams, R. W., and Steven, A. C. (1995). Hepatitis core antigen produced in *Escherichia coli*: Subunit composition, conformational analysis, and *in vitro* capsid assembly. *Biochemistry* **34:**4919–4932.
Wynne, S. A., Crowther, R. A., and Leslie, A. G. (1999). The crystal structure of the human hepatitis B virus capsid. *Mol. Cell* **3:**771–780.
Yuan, T. T-T., Gautam, K. S., Whitehead, W. E., Greenberg, R., and Shih, C. (1999). The mechanism of an immature secretion phenotype of a highly frequent naturally occur-

ring missense mutation at codon 97 of human hepatitis B virus core antigen. *J. Virol.* **73:**5731–5740.

Zheng, J., Schödel, F., and Peterson, D. L. (1992). The structure of hepadnaviral core antigens: Identification of free thiols and determination of the disulfide bonding pattern. *J. Biol. Chem.* **267:**9422–9429.

Zhou, S., and Standring, D. N. (1992). Cys residues of the hepatitis B virus capsid protein are not essential for the assembly of viral core particles but can influence their stability. *J. Virol.* **66:**5393–5398.

Zlotnick, A., Ceres, P., Singh, S., and Johnson, J. M. (2002). A small molecule inhibits and misdirects assembly of hepatitis B virus capsids. *J. Virol.* **76:**4848–4854.

Zlotnick, A., Cheng, N., Conway, J. F., Booy, F. P., Steven, A. C., Stahl, S. J., and Wingfield, P. T. (1996). Dimorphism of hepatitis B virus capsids is strongly influenced by the C-terminus of the capsid protein. *Biochemistry* **35:**7412–7421.

Zlotnick, A., Cheng, N., Stahl, S. J., Conway, J. F., Steven, A. C., and Wingfield, P. T. (1997). Localization of the C terminus of the assembly domain of hepatitis B virus capsid protein: Implications for morphogenesis and organization of encapsidated RNA. *Proc. Natl. Acad. Sci. USA* **94:**9556–9561.

Zlotnick, A., Johnson, J. M., Wingfield, P. W., Stahl, S. J., and Endres, D. (1999a). A theoretical model successfully identifies features of hepatitis B virus capsid assembly. *Biochemistry* **38:**14644–14652.

Zlotnick, A., Palmer, I., Kaufman, J. D., Stahl, S. J., Steven, A. C., and Wingfield, P. T. (1999b). Separation and crystallization of $T = 3$ and $T = 4$ icosahedral complexes of the hepatitis B virus core protein. *Acta Crystallogr. D.* **55:**717–720.

Zlotnick, A., Stahl, S. J., Wingfield, P. T., Conway, J. F., Cheng, N., and Steven, A. C. (1998). Shared motifs of the capsid proteins of hepadnaviruses and retroviruses suggest a common evolutionary origin. *FEBS Lett.* **431:**301–304.

ADVANCES IN VIRUS RESEARCH, VOL 64

# MOLECULAR INTERACTIONS IN THE ASSEMBLY OF CORONAVIRUSES

Cornelis A. M. de Haan and Peter J. M. Rottier

Virology Division, Department of Infectious Diseases and Immunology
Faculty of Veterinary Medicine, Utrecht University
3584 CL Utrecht, The Netherlands



## I. Introduction

Viruses are multimolecular assemblies that range from small, regular, and simple to large, pleiomorphic, and complex. They consist of virus-specified proteins and nucleic acids and, in the case of enveloped viruses, of host-derived lipids. In infected cells the assembly of these different components into virions occurs with high precision amidst a huge background of tens of thousands of host compounds. Two key factors determine the efficiency of the assembly process: intracellular transport and molecular interactions.

Directional transport ensures the swift and accurate delivery of the virion components to the cellular compartment(s) where they must meet and form (sub)structures. Some viruses achieve this goal relatively simply when genome production occurs in close proximity




0065-3527/05 $35.00
DOI: 10.1016/S0065-3527(05)64006-7

to the virion assembly site (e.g., picornaviruses). Many viruses, however, have evolved more elaborate strategies. This is illustrated, for instance, by the α-herpesviruses. Assembly of these viruses starts in the nucleus by the encapsidation of viral DNA, using cytoplasmically synthesized capsid proteins; nucleocapsids then migrate to the cytosol, by budding at the inner nuclear membrane followed by deenvelopment, to pick up the tegument proteins. Subsequently, the tegumented capsids obtain their final envelope by budding into vesicles of the *trans*-Golgi network (TGN), where the viral envelope proteins have congregated after their synthesis in the endoplasmic reticulum; the assembled viral particles are finally released by fusion of the virion-containing vesicles with the plasma membrane. To achieve their transport goals viruses provide their components with address labels that can be read by the transport machinery of the cell. Once brought together, formation of the viral (sub)structures is governed and driven by their interactions. Whereas the assembly of nonenveloped viruses is generally restricted to the cell cytoplasm, although often in association with membranes, that of enveloped viruses involves multiple cellular compartments, as exemplified already for herpesviruses.

This review deals with the assembly of coronaviruses. We first describe what is known about the structure of the coronavirion and about the relevant properties of the structural components. We summarize the limited ultrastructural information about coronavirus assembly and budding. The main body of the review describes the interactions between the different structural components of the viruses and discusses their relevance for the process of virion formation. This review has a limited scope; for further information about other aspects of coronavirus biology the reader is referred to other reviews (de Vries *et al.*, 1997; Enjuanes *et al.*, 2001; Gallagher and Buchmeier, 2001; Holmes, 2001; Holmes *et al.*, 2001; Lai, 1997; Lai and Cavanagh, 1997; Lai *et al.*, 1994; Masters, 1999; Perlman, 1998; Rossen *et al.*, 1995; Sawicki and Sawicki, 1998; Siddell, 1995; Ziebuhr *et al.*, 2000).

## II. Structure of the Coronavirion and Its Components

Coronaviruses are a group of enveloped, plus-stranded RNA viruses presently classified as a genus, which, together with the genus *Torovirus*, constitutes the family *Coronaviridae*. These viruses are grouped with two other families, the *Arteriviridae* and the *Roniviridae*, into the order *Nidovirales*. This classification is not based on structural similarities—in fact, structure and composition of the viruses from the different families differ significantly—but on common features of genome organization and gene expression (de Vries *et al.*, 1997; Lai and Cavanagh, 1997).

Coronaviruses infect a wide variety of mammals as well as avian species (Table I). In general they cause respiratory or intestinal infections, but some coronaviruses can also infect other organs (liver, kidney, and brain). Until recently, these viruses were mainly of veterinary importance. This situation has changed quite dramatically because of the emergence of severe acute respiratory syndrome-

TABLE I
CORONAVIRUS GROUPS, THEIR MAIN REPRESENTATIVES, HOSTS, AND PRINCIPAL ASSOCIATED DISEASES

| Group | Virus | Host | Disease |
|---|---|---|---|
| 1 | Feline coronavirus (FCoV) | Cat | Respiratory infection/enteritis/ peritonitis/systemic enteritis |
| | Canine coronavirus (CCoV) | Dog | Enteritis |
| | Transmissible gastroenteritis virus (TGEV) | Pig | Enteritis |
| | Porcine epidemic diarrhea virus (PEDV) | Pig | Enteritis |
| | Porcine respiratory coronavirus (PRCoV) | Pig | Respiratory infection |
| | Human coronavirus (HCoV)-NL63 | Human | Respiratory infection |
| | Human coronavirus (HCoV)-229E | Human | Respiratory infection |
| 2 | Murine hepatitis virus (MHV) | Mouse | Respiratory infection/enteritis/ hepatitis/encephalitis |
| | Rat coronavirus (RCoV) | Rat | Respiratory infection |
| | Bovine coronavirus (BCoV) | Cow | Respiratory infection/enteritis |
| | Hemagglutinating encephalomyelitis virus (HEV) | Pig | Enteritis |
| | Human coronavirus (HCoV)-OC43 | Human | Respiratory infection |
| 3 | Infectious bronchitis virus (IBV) | Chicken | Respiratory infection/enteritis |
| | Turkey coronavirus (TCoV) | Turkey | Enteritis |
| ? | Severe acute respiratory syndrome-associated coronavirus (SARS-CoV) | Human | Respiratory infection/enteritis |

associated coronavirus (SARS-CoV) in late 2002, which emphasized the potential relevance of coronaviruses for humans. On the basis of antigenic and genetic relationships the coronaviruses have been subdivided into three groups (Table I); the taxonomic position of SARS-CoV has not been formally assigned.

### A. *Coronavirion*

Coronavirus particles have a typical appearance under the electron microscope. By the characteristic, approximately 20-nm-long spikes that emanate from their envelope the viruses acquire the solar image to which they owe their name (Fig. 1). The 80- to 120-nm virions have a pleiomorphic appearance that, whether artifact or real, reflects a pliable constellation, a feature that has severely hampered the ultrastructural analysis of these viruses. Hence, our knowledge about the structure of coronaviruses is still rudimentary.

The schematic representation of the current model of the coronavirion drawn in Fig. 1 is based on morphological and biochemical

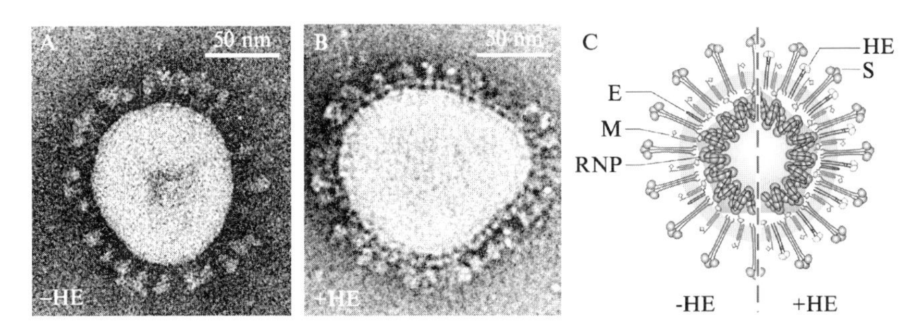


FIG 1. Electron micrographs of mouse hepatitis virus strain A59 (MHV-A59) virions without (A) and with (B) the hemagglutinin-esterase (HE) envelope protein (viruses kindly provided by R. de Groot, Virology Division, Utrecht University, The Netherlands; image courtesy of J. Lepault, VMS-CNRS, Gif-sur-Yvette, France). Large, club-shaped protrusions consisting of spike (S) protein trimers give the viruses their corona solis-like appearance. Viruses containing the HE protein display a second, shorter fringe of surface projections in addition to the spikes. (C) Schematic representation of the coronavirion. The viral RNA is encapsidated by the nucleocapsid (N) protein forming a helical ribonucleoprotein (RNP), which is in turn part of a structure with spherical, probably icosahedral, configuration. The nucleocapsid is surrounded by a lipid bilayer in which the S protein, the membrane glycoprotein (M), and the envelope protein (E) are anchored. In addition, some group 2 coronaviruses contain the HE protein in their lipid envelope as illustrated on the right side of the particle. (See Color Insert.)

observations. As this picture illustrates, the particle consists of a nucleocapsid or core structure that is surrounded by a lipid envelope. Anchored in this envelope are the three canonical coronavirus membrane proteins: the membrane (M) protein, the envelope (E) protein, and the spike (S) protein. Viruses from group 2 have an additional, fourth membrane protein, the hemagglutinin-esterase (HE) protein. As a consequence these viruses display a second, shorter (5 nm) fringe of surface projections in addition to the spikes (Fig. 1B) (Bridger *et al.*, 1978; King *et al.*, 1985; Sugiyama and Amano, 1981).

The ribonucleoprotein (RNP) core contains one copy of the viral genomic RNA. This RNA is packaged into a helical structure by multiple copies of nucleocapsid protein (N). Size estimations of the flexible cylindrical structures varied quite considerably, ranging between 7 and 16 nm in diameter and up to 0.32 $\mu$m in length (see Laude and Masters, 1995). The ribonucleoprotein helix appears in turn to be contained within a spherical, probably icosahedral, configuration as indicated by various ultrastructural approaches using purified transmissible gastroenteritis virus (TGEV) and mouse hepatitis virus (MHV) (Risco *et al.*, 1996, 1998).

The molar ratio of the major structural proteins, S:N:M, has been variously estimated to be approximately 1:8:16 (Sturman *et al.*, 1980), 1:6:15 (Cavanagh, 1983a), 1:8:8 (Hogue and Brian, 1986), and 1:11:10 (Liu and Inglis, 1991), although an M:N molar ratio of 3 has also been reported (Escors *et al.*, 2001a). The S:HE molar ratio was estimated to be 4 (Hogue and Brian, 1986). The E protein is only a minor virion component and was calculated to occur in infectious bronchitis virus (IBV), TGEV, and MHV virions at a rate of approximately 100, 20, and 10 molecules per particle, respectively (Godet *et al.*, 1992; Liu and Inglis, 1991; Vennema *et al.*, 1996).

The lipid composition of coronaviral envelopes has been studied only to a limited extent. Comparison of the phospholipid composition of MHV with that of its host cell showed increased levels of sphingomyelin, phosphatidylserine, and phosphatidylinositol and a decrease in the level of phosphatidylethanolamine (van Genderen *et al.*, 1995). Whether the lipid composition of MHV is an accurate reflection of its budding compartment or whether certain lipids become enriched in the virus during budding is not known.

What follows is a general description of the individual virion components and their properties. This description is by no means complete as it is restricted to the information that is of relevance to the main topic of this review. For a schematic representation of the coronavirus life cycle see Fig. 2.

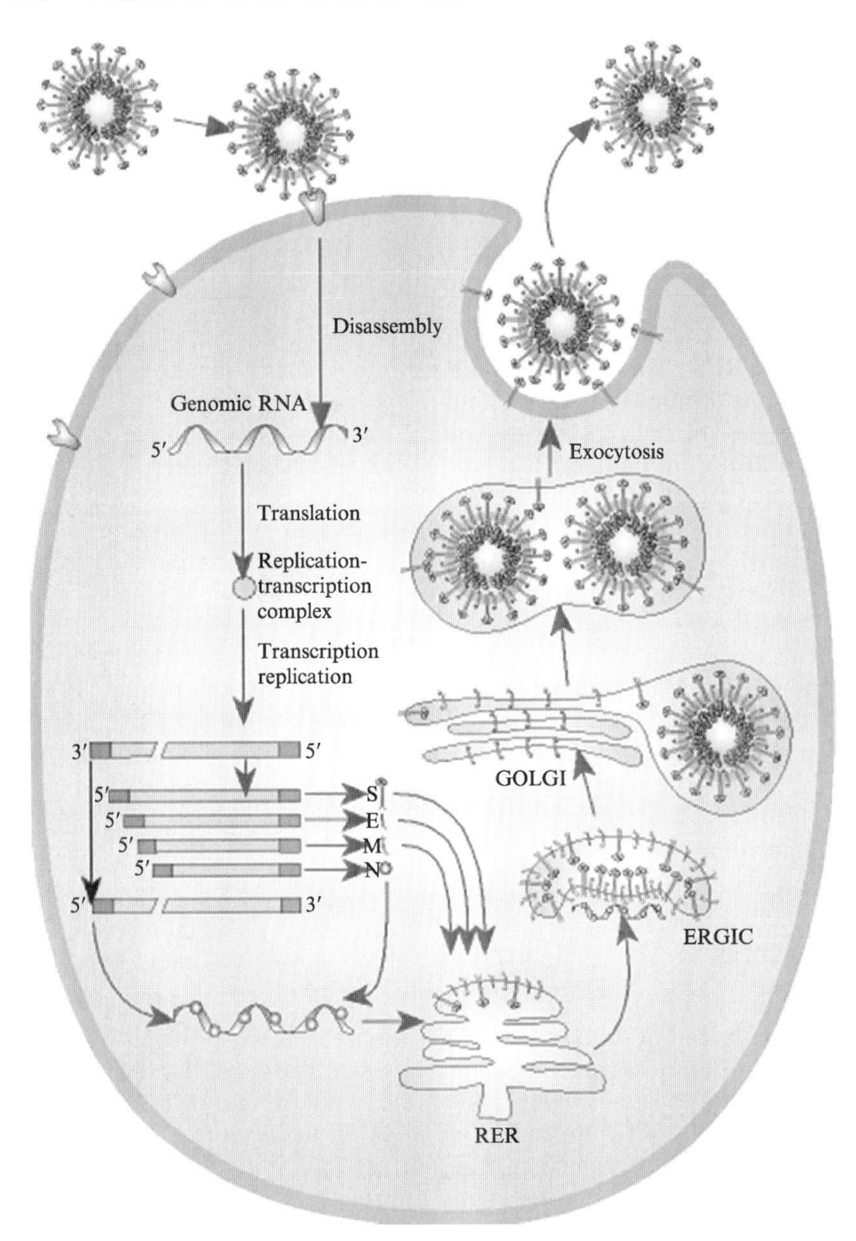


FIG 2. The coronavirus life cycle. The replication cycle starts with attachment of the virion by its S protein, that is, through the S1 subunit thereof, to the receptors on the host cell. This interaction leads to fusion of the virus envelope with a cellular membrane,

### *B. Viral Genome*

Coronaviruses contain a single-stranded positive-sense RNA genome of some 27 to 31 kilobases, the largest nonsegmented viral RNA genomes known. The RNA has a 5′-terminal cap and a 3′-terminal poly(A) tract. Both genomic termini contain untranslated regions (UTRs) of some 200–500 nucleotides that harbor several *cis*-acting sequences and structural elements functioning in viral replication and transcription. Coronaviruses have a typical genome organization characterized by the occurrence of a distinctive set of genes that are essential for viability and occur in a fixed order: 5′-polymerase (*pol*)-S-E-M-N-3′ (Fig. 3). The *pol* gene comprises approximately two-thirds of the genome, from which it is translated directly. It encodes two large precursors (Pol1a and Pol1ab), the many functional cleavage products of which are collectively responsible for RNA replication and transcription (for reviews on coronavirus transcription and replication see de Vries *et al.*, 1997; Lai, 1997; Lai and Cavanagh, 1997; Lai *et al.*, 1994; Sawicki and Sawicki, 1998; Ziebuhr *et al.*, 2000). The more downstream *pol1b* gene is translated by translational readthrough,

for which the S2 subunit is responsible. From the genomic RNA that is released by disassembly of the incoming particle the *pol1a* and *pol1b* genes are translated, resulting in the production of two large precursors (Pol1a and Pol1ab), the many cleavage products of which collectively constitute the functional replication–transcription complex. Genes located downstream of the *pol1b* gene are expressed from a 3′-coterminal nested set of subgenomic (sg) mRNAs, each of which additionally contains a short 5′ leader sequence derived from the 5′ end of the genome (shown in red). Transcription regulatory sequences (TRSs) located upstream of each gene serve as signals for the transcription of the sgRNAs. The leader sequence is joined at a TRS to all genomic sequence distal to that TRS by discontinuous transcription, most likely during the synthesis of negative-strand sgRNAs. In most cases, only the 5′-most gene of each sgRNA is translated. Multiple copies of the N protein package the genomic RNA into a helical structure in the cytoplasm. The structural proteins S, M, and E are inserted into the membrane of the rough endoplasmic reticulum (RER), from where they are transported to the ER-to-Golgi intermediate compartment (ERGIC) to meet the nucleocapsid and assemble into particles by budding. The M protein plays a central role in this process through interactions with all viral assembly partners. It gives rise to the formation of the basic matrix of the viral envelope generated by homotypic, lateral interactions between M molecules, and it interacts with the envelope proteins E, S, and HE (if present), as well as with the nucleocapsid, thereby directing the assembly of the virion. Virions are transported through the constitutive secretory pathway out of the cell—the glycoproteins on their way being modified in their sugar moieties, whereas the S proteins of some but not all coronaviruses are cleaved into two subunits by furin-like enzymes (see text for references). (See Color Insert.)

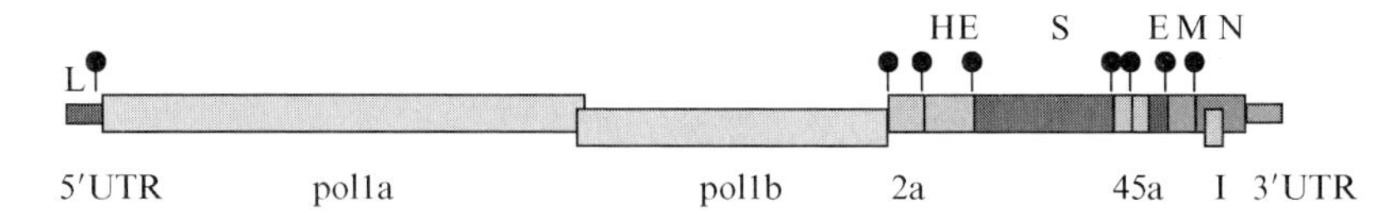


FIG 3. Coronavirus genome organization as illustrated for the group 2 virus MHV. The single-stranded, positive-sense RNA genome contains 5′- and 3′-terminal untranslated regions (UTRs) with a 5′-terminal cap and a 3′-terminal poly(A) tract. The leader sequence (L) in the 5′ UTR is indicated. All coronaviruses have their essential genes in the order 5′-*pol*-S-E-M-N-3′. The *pol1a* and *pol1b* genes comprise approximately two-thirds of the genome. The more downstream *pol1b* gene is translated by translational readthrough, using a ribosomal frameshift mechanism. Transcription regulatory sequences (TRSs) located upstream of each gene, which serve as signals for the transcription of the subgenomic (sg) RNAs, are indicated by circles. The genes encoding the structural proteins HE, S, E, M, and N are specified. Gray boxes indicate the accessory, group-specific genes, in the case of group 2 coronaviruses genes 2a, HE, 4, 5a, and I. (See Color Insert.)

using a ribosomal frameshift mechanism for which a "slippery" sequence and a pseudoknot structure are required.

The genes located downstream of *pol1b* are expressed from a 3′-coterminal nested set of subgenomic (sg) RNAs, each of which additionally contains a short 5′ leader sequence derived from the 5′ end of the genome. Transcription regulatory sequences (TRSs) located upstream of each gene serve as signals for transcription of the sgRNAs. The leader sequence is joined at a TRS to all genomic sequence distal to that TRS by discontinuous transcription, most likely during the synthesis of negative-strand sgRNAs (Sawicki and Sawicki, 1998).

Besides the characteristic genes encoding the replicative and structural functions, coronaviruses have a more variable collection of additional genes that are located in two clusters in the 3′-terminal one-third of the genome. The genes differ distinctly in their nature and genomic position among the coronavirus groups, but they are specific for each group. These so-called group-specific genes appear not to be essential as shown by the occurrence of natural mutants defective in some of them (Brown and Brierley, 1995; Herrewegh *et al.*, 1995; Kennedy *et al.*, 2001; Luytjes, 1995; Shen *et al.*, 2003; Vennema, 1999; Vennema *et al.*, 1998; Woods, 2001) and by the observed viability of engineered deletion mutants lacking some or all of these genes (de Haan *et al.*, 2002b; Fischer *et al.*, 1997; Haijema *et al.*, 2004; Ortego *et al.*, 2003; Sola *et al.*, 2001). Except for the group 2-specific HE protein and, possibly, the poorly characterized I protein (Fischer *et al.*, 1997; Senanayake *et al.*, 1992), the latter encoded by an open reading frame completely contained within the N gene, the group-specific proteins do not appear to occur in virions. Although their functions have not yet been resolved, mutant studies indicate that they play important roles in the interaction of coronaviruses with their host (de Haan *et al.*, 2002b; Fischer *et al.*, 1997; Haijema *et al.*, 2004; Ortego *et al.*, 2003).

### *C. N Protein*

The N protein is the most abundantly expressed viral protein in infected cells (for a review, see Laude and Masters, 1995). Its size varies considerably between viruses from different groups (377–455 amino acids, i.e., molecular masses ranging between 45 and 60 kDa), N proteins from group 2 coronaviruses (Table I) being the largest. Whereas the amino acid sequences of N proteins are quite similar within the groups, the homology between proteins from different coronavirus groups is rather limited (30–35%). An exception is a region spanning about 50 residues within the amino-terminal one-third of the N molecule, where high sequence identity has been conserved across the different groups.

Despite the overall sequence variation the N proteins have a number of common characteristics. Consistent with their role as nucleic acid-binding proteins they are all highly basic because of the abundance of arginine and lysine residues. These are clustered mainly in two nearby regions in the middle of the molecules. The abundance of basic residues is reflected in the calculated overall isoelectric points of the N proteins, the values of which are in the range of 9.7–10.1. These numbers are the more significant in view of the acidic nature of the very carboxy-terminal domain; p*I* values ranging from 4.3 to 5.5 were calculated for the terminal 45 residues (Parker and Masters, 1990). Another general characteristic of the N proteins is their high content (7–11%) of serine residues, which are potential targets for phosphorylation. Although these residues occur all over the N molecule, their relative abundance within the first of the two basic regions is notable.

Little is known about the three-dimensional structure of the N protein. Of the SARS-CoV N protein the amino-terminal domain (residues 45–181) was analyzed by nuclear magnetic resonance spectroscopy. It appeared to consist of a five-stranded $\beta$ sheet with a folding distinct from that of other RNA-binding proteins (Huang *et al.*, 2004).

In coronavirus-infected cells the N protein can often be detected as one major and several minor forms, the latter polypeptides having a slightly lower molecular weight. The major species appeared to comigrate in gels with the N protein observed in virions, indicating that only the full-length N species is incorporated into particles. How the minor N species arise and whether they are of particular significance for infection is unclear. They are most likely derived by proteolytic processing from the major N species. This is supported by studies from Eleouet *et al.* (2000), who showed the TGEV N protein to be cleaved by caspases. Caspase cleavage sites were also predicted in the carboxy terminus of several other coronavirus N proteins (Eleouet *et al.*, 2000; Ying *et al.*, 2004). These features are in agreement with observations showing that antibodies directed against the carboxy terminus of the MHV and TGEV N proteins were not reactive with the faster migrating electrophoretic forms. Furthermore, these smaller N protein forms appeared to be derived from the major species as judged from pulse–chase analyses (for a review see Laude and Masters, 1995).

The N protein is the only coronavirus structural protein known to become phosphorylated (for references see Laude and Masters, 1995). Both the major and minor N species appear to be phosphorylated as shown for MHV-A59 in Sac(−) cells (Rottier *et al.*, 1981b) and for TGEV in LLC-PK1 cells (Garwes *et al.*, 1984). Of the many potential target serines only a few are actually modified in the case of MHV (Stohlman and Lai, 1979; Wilbur *et al.*, 1986). N protein phosphorylation does not seem to play a critical role in the regulation of virus assembly. In contrast, it has been hypothesized that dephosphorylation of the protein might facilitate disassembly during MHV cell entry (Kalicharran *et al.*, 1996; Mohandas and Dales, 1991).

Immunofluorescence microscopy has shown the N protein to be localized in a particulate manner throughout the cytoplasm of coronavirus-infected cells. Although the protein lacks a membrane-spanning domain it was found in association with membranes (Anderson and Wong, 1993; Sims *et al.*, 2000; Stohlman *et al.*, 1983). For MHV, the N protein was found to colocalize partly with the membrane-associated viral replication complexes (Denison *et al.*, 1999; van der Meer *et al.*, 1999). In addition to its cytoplasmic localization, the N proteins of IBV, MHV, and TGEV have also been demonstrated to localize to the nucleolus both in coronavirus-infected cells and when expressed independently (Hiscox *et al.*, 2001; Wurm *et al.*, 2001). Putative nuclear localization signals were identified in these proteins. The IBV N protein was found to interact with nucleolar antigens, which appeared to occur more efficiently when the N protein was phosphorylated, and to affect the cell cycle (Chen *et al.*, 2002). However, because MHV is able to replicate in enucleated cells (Brayton *et al.*, 1981; Wilhelmsen *et al.*, 1981) the nucleolar localization of the N protein does not appear an essential step during infection.

Although the primary function of the N protein is the formation of the viral ribonucleoprotein complex, several studies indicate the protein to be multifunctional. As indicated by its intracellular localization, the N protein is a likely component of the coronavirus replication and transcription complex. Its presence is not an absolute requirement for replication and transcription because a human coronavirus (HCoV) RNA vector containing the complete *pol1ab* gene appeared to be functional in the absence of the N protein (Thiel *et al.*, 2003). However, the efficiency of the system was much enhanced when the protein was present. Furthermore, using an *in vitro* system, it was demonstrated that antibodies to the N protein, but not those against the S and M proteins, inhibited viral RNA synthesis by 90% (Compton *et al.*, 1987). Interactions that have been observed between the N protein and leader/TRS sequences (Baric *et al.*, 1988; Nelson *et al.*, 2000; Stohlman *et al.*, 1988) and between N protein and the 3′ UTR (Zhou *et al.*, 1996) suggest a role for the N protein in the discontinuous transcription process. Furthermore, the N protein was also shown to interact with cellular proteins that play a role in coronavirus RNA replication and transcription (Choi *et al.*, 2002; Shi *et al.*, 2000). In addition, the N protein was reported to function as a translational enhancer of MHV sgRNAs (Tahara *et al.*, 1998).

### *D. M Protein*

The M protein (previously known as E1 protein) is the most abundant envelope protein. It is the "building block" of the coronavirion and has been shown to interact with virtually every other virion component, as detailed in Section IV. The M protein is 221–230 residues in length, except for the group 1 M proteins, of which the amino terminus is about 30 residues longer. Despite large differences in primary sequences between M proteins from different antigenic groups, their hydropathicity profiles are remarkably similar. The M protein is highly hydrophobic. It has three hydrophobic domains alternating with short hydrophilic regions in the amino-terminal half of the protein, with the exception of the aforementioned group 1 M proteins, which have at their amino terminus a fourth hydrophobic domain that functions as a cleavable signal peptide. The carboxy-terminal half of the protein is amphipathic, with a short hydrophilic domain at the carboxy-terminal end (Fig. 4). In the center of the protein, directly adjacent to the third hydrophobic domain, is a stretch of eight amino acids that is well conserved (SWWSFNPE). The conservation of the overall chemical features suggests that there are rigid structural constraints on the

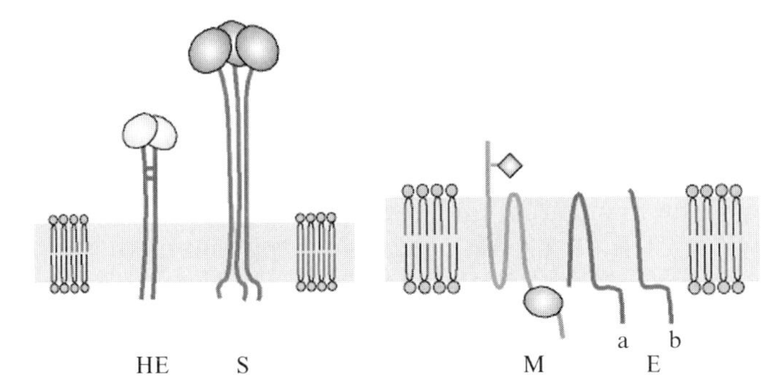


FIG 4. Membrane topology of the coronavirus envelope proteins. The HE and S proteins are both type I membrane proteins, with short carboxy-terminal cytoplasmic tails. The HE protein forms disulfide-linked homodimers, whereas the S protein forms noncovalently linked homotrimers. The S1 subunits presumably constitute the globular head, whereas the S2 subunits form the stalk-like region of the spike. The M protein spans the lipid bilayer three times, leaving a small amino-terminal domain in the lumen of intracellular organelles (or on the outside of the virion), whereas the carboxy-terminal half of the protein is located on the cytoplasmic side of the membrane (or inside the virion). In TGEV virions some of the M proteins have their cytoplasmic tail exposed on the outside (not shown). The M protein is glycosylated at its amino terminus (indicated by a diamond). The amphipathic domain of the M protein is represented by an oval. The hydrophilic carboxy terminus of the E protein is exposed on the cytoplasmic side of cellular membranes or on the inside of the virion. The E protein may span the bilayer once (b) or twice (a). (See Color Insert.)

M protein as a result of functional requirements (for a review on the M protein, see Rottier, 1995).

Biochemical and theoretical studies led to a topological model for the MHV M protein (Armstrong *et al.*, 1984; Rottier *et al.*, 1984, 1986), in which the polypeptide spans the lipid bilayer three times, leaving a small amino-terminal domain (15–35 residues) in the lumen of intracellular organelles (or outside the virus), whereas the carboxy-terminal half of the protein is located on the cytoplasmic side of the membrane (or inside the virion). The lumenal domain and the hydrophilic carboxy terminus are susceptible to protease digestion and are thus exposed. The bulk of the carboxy-terminal half of the M protein is protease resistant, indicating that the amphipathic part of the protein is either folded tightly or embedded in the polar surface of the membrane. Indeed, a mutant lacking all three transmembrane domains was found to be associated with membranes (Mayer *et al.*, 1988). The model for the disposition of the M protein in the membrane was confirmed for IBV (Cavanagh *et al.*, 1986). Interestingly, the

M protein of TGEV was shown to adopt an additional conformation. In virions about one-third of the M molecules have their carboxy terminus exposed on the virus surface rather than buried inside the particle (Escors *et al.*, 2001a; Risco *et al.*, 1995). This appears to have immunological consequences (Risco *et al.*, 1995) but the real significance of the dual topology is unclear. In MHV the M protein was found to assume only one defined membrane topology (Raamsman *et al.*, 2000).

The coronavirus M protein is almost invariably glycosylated in its exposed amino-terminal domain. This provides the virion with a diffuse, hydrophilic cover on its outer surface. Whereas the group 1 and 3 coronaviruses and SARS-CoV all contain M proteins with only N-linked sugars, the M proteins of group 2 coronaviruses are O-glycosylated (for a review see Rottier, 1995). An exception is MHV-2, the M protein of which carries both O- and N-linked sugars (Yamada *et al.*, 2000). N-Glycosylation is initiated in the endoplasmic reticulum by the cotranslational linkage of a large oligosaccharide structure to the polypeptide at asparagine residues within the consensus sequence NXS/T (where X is any amino acid). In contrast, mucin-type O-glycosylation starts posttranslationally with the addition of an *N*-acetylgalactosamine (GalNAc) monosaccharide to a hydroxylamino acid. O-Glycosylation is subsequently completed by stepwise addition of other monosaccharides such as galactose, *N*-acetylglucosamine, fucose, and sialic acid. MHV M proteins carry a well-conserved SS(X)TTXXP sequence at their extreme amino terminus. Despite the apparent presence of multiple hydroxylamino acids as potential oligosaccharide acceptor sites the M protein of MHV-A59 was found to be modified by the addition of only a single oligosaccharide side chain (de Haan *et al.*, 1998b). This side chain, when studied in OST7-1 cells, appeared to be attached to the threonine at position 5. Mutation studies, however, revealed that alternative acceptor sites can also be used. No unique sequence motifs for O-glycosylation of MHV M could be identified, which is probably related to the occurrence in cells of multiple GalNAc transferases (de Haan *et al.*, 1998b). As the expression of these enzymes varies in cells, conservation of the SS(X)TTXXP motif in MHV M protein may serve to increase opportunities for the protein to become glycosylated in different cell types.

The distinct conservation of N- and O-glycosylation among the M proteins of the different groups of coronaviruses suggests that the presence and the particular type of carbohydrates are somehow beneficial to the virus, most likely in its interaction with the host. Glycosylation of the M protein appeared not to be required for envelope assembly (de Haan *et al.*, 1998a,) or for interaction with the S protein (de Haan *et al.*, 1999), nor did it influence virus replication *in vitro* (de Haan *et al.*, 2002a, 2003; Laude *et al.*, 1992). Coronaviruses are able to induce interferon α (IFN-α) by their glycoproteins (Baudoux *et al.*, 1998a,b). For TGEV (Charley and Laude, 1988; Laude *et al.*, 1992) and MHV (de Haan *et al.*, 2003), the oligosaccharides linked to the M protein were demonstrated to be important for efficient IFN induction *in vitro*. The glycosylation status of the MHV M protein was found to influence the ability of the virus to replicate in the liver but not in the brain (de Haan *et al.*, 2003). Thus, viruses with N-glycosylated M proteins replicated to a significantly higher extent in liver than otherwise identical viruses carrying O-glycosylated M proteins. MHV with unglycosylated M proteins replicated to the lowest extent. The mechanism behind these observations remains to be elucidated.

When expressed in cells independently from the other viral proteins, the M proteins of MHV, IBV, TGEV, and feline coronavirus (FCoV) accumulate in the Golgi compartment, that is, beyond the site of virus budding (Klumperman *et al.*, 1994; Locker *et al.*, 1992; Machamer and Rose, 1987; Machamer *et al.*, 1990; Rottier and Rose, 1987), which is the intermediate compartment between the ER and the Golgi (ERGIC). The fine localization of the different proteins is, however, not the same. For instance, whereas the MHV M protein is concentrated in the *trans*-most Golgi compartments, the IBV M protein localizes to the *cis* side of the Golgi complex. Signals for localization appear to reside in the hydrophilic part of the cytoplasmic tail and in the transmembrane domains. The extreme carboxy-terminal tail of MHV M was shown to be necessary, although not sufficient, for Golgi localization (Armstrong and Patel, 1991; Locker *et al.*, 1994). Mutant proteins lacking this domain were transported to the plasma membrane. Also, mutation of a single tyrosine in this domain, which occurs in the context of a potential internalization signal, resulted in plasma membrane localization (C. A. M. de Haan and P. J. M. Rottier, unpublished results). The first transmembrane domain of the IBV M protein was shown to be required and sufficient for localization to the *cis*-Golgi region (Machamer and Rose, 1987; Machamer *et al.*, 1990, 1993; Swift and Machamer, 1991). This is not the case for the MHV M protein of which mutants with only the first transmembrane domain did not leave the ER (Armstrong *et al.*, 1990; Locker *et al.*, 1994; Rottier *et al.*, 1990). Moreover, insertion of the first transmembrane domain of MHV M into a reporter protein resulted in a chimeric protein that was transported to the cell surface (Armstrong and Patel, 1991; Machamer *et al.*, 1993), unlike a similar chimeric protein containing the first transmembrane domain of IBV that was retained in the Golgi compartment (Machamer *et al.*, 1993). Other MHV M mutant proteins lacking the first and second transmembrane domains were also not efficiently retained in the Golgi compartment and were diverted to endosomal structures (Armstrong *et al.*, 1990; Locker *et al.*, 1994). The mechanism by which Golgi retention of M proteins is regulated has not yet been resolved. However, oligomerization of the proteins, mediated by the transmembrane domains, seems to play an important role, perhaps in combination with retrieval mechanisms (de Haan *et al.*, 2000; Maceyka and Machamer, 1997). Formation of oligomeric complexes has been demonstrated to correlate with Golgi retention of a reporter protein containing the first transmembrane domain of IBV M (Weisz *et al.*, 1993) while also the Golgi-resident MHV M protein was found to occur in large, homomeric complexes (Locker *et al.*, 1995). The lumenal domain of the M protein does not appear to contribute to localization; its deletion from MHV M did not affect the intracellular destination of the protein (Mayer *et al.*, 1988; Rottier *et al.*, 1990).

In infected cells the M proteins of IBV and MHV were observed to occur in the membranes of the budding compartment as well as in the Golgi compartment. Under these conditions their *cis–trans* distribution in the Golgi compartment was the same as when these proteins were expressed independently (Klumperman *et al.*, 1994; Machamer *et al.*, 1990).

### *E. E Protein*

The E protein (previously known as sM protein) is a small protein (76–109 residues) and a minor component of the coronaviral envelope. Although the primary structures of E proteins are quite conserved within the different coronavirus groups, they share little homology between the groups. However, the proteins have several structural features in common. The E protein contains a relatively large hydrophobic region in its amino-terminal half, followed by a cysteine-rich region, an absolutely conserved proline residue, and a hydrophilic tail. E is an integral membrane protein, which is assembled in membranes without the involvement of a cleaved signal peptide (Raamsman *et al.*, 2000). Its membrane topology has not been firmly established. Although the opposite was proposed initially for the TGEV E protein (Godet *et al.*, 1992), there seems to be consensus about the hydrophilic carboxy terminus being exposed on the cytoplasmic side in cells or on the inside of the virion (Corse and Machamer, 2000; Raamsman *et al.*, 2000). The amino terminus of the MHV E protein was not detectably present on the virion outside (Raamsman *et al.*, 2000) but appeared to be exposed cytoplasmically when it was extended with an amino-terminal epitope tag (Maeda *et al.*, 2001), consistent with a topological model in which the hydrophobic domain spans the bilayer twice. For the IBV E protein evidence was provided indicating that the amino terminus is exposed lumenally in cells, consistent with a single spanning topology (Corse and Machamer, 2000) (Fig. 4).

The E protein is not glycosylated but appears to become palmitoylated. This was shown most convincingly for the IBV E protein by labeling with [$^3$H]palmitate (Corse and Machamer, 2002), both in IBV-infected cells and when the protein was expressed. Mutagenesis revealed that one or both of the two conserved cysteines became modified. The result is consistent with the observed increase in electrophoretic mobility of the MHV E protein after treatment with hydroxylamine, an agent that cleaves thioester-linked acyl chains (Yu *et al.*, 1994). Others, however, were not able to confirm this post-translational modification (Godet *et al.*, 1992; Raamsman *et al.*, 2000).

In coronavirus-infected cells the E protein has been observed by immunofluorescence studies to occur at intracellular membranes as well as at the cell surface (Godet *et al.*, 1992; Smith *et al.*, 1990; Tung *et al.*, 1992; Yu *et al.*, 1994). When expressed exogenously from cDNA, the E protein was detected only in intracellular organelles, although at different locations. The MHV E protein localized to pre-Golgi membrane compartments, as was demonstrated by its colocalization with rab-1, a marker for the endoplasmic reticulum and the ERGIC, by electron microscopy (Raamsman *et al.*, 2000). The IBV E protein, tagged at its amino terminus with an epitope, was also localized to pre-Golgi compartments (Lim and Liu, 2001). In another study, however, the IBV E protein was shown to accumulate in the Golgi apparatus, being distributed throughout the complex (Corse and Machamer, 2000, 2002). While the former study identified an ER-targeting signal in the extreme carboxy terminus of the E protein (Lim and Liu, 2001), the latter studies, using carboxy-terminal truncations, mapped the Golgi-targeting information to a region between tail residues 13 and 63 (Corse and Machamer, 2002). In addition, these authors showed the IBV E cytoplasmic tail to be necessary and sufficient for Golgi targeting.

The E protein was identified as a virion component relatively late, due to its low abundance and its small size (Godet *et al.*, 1992; Liu and Inglis, 1991; Yu *et al.*, 1994). It was estimated to occur in IBV, TGEV, and MHV virions at a rate of about 100, 20, and 10 molecules per particle, respectively (Godet *et al.*, 1992; Liu and Inglis, 1991; Vennema *et al.*, 1996). Because of its low abundance the E protein may not have a genuine structural function in the virion envelope. Rather, it may have a morphogenetic function by taking strategic positions within the M protein lattice to generate the required membrane curvature. Alternatively, it may serve to close the neck of the budding particle as it pinches off the membrane (Vennema *et al.*, 1996). Expression of the E protein alone induced the formation of characteristic membrane structures also observed in infected cells, which apparently consist of masses of tubular, smooth convoluted membranes (David-Ferreira and Manaker, 1965; Raamsman *et al.*, 2000). In addition, it resulted in the formation of vesicles containing the E protein, shown to be released from the cells (Corse and Machamer, 2000; Maeda *et al.*, 1999).

MHV infection induces caspase-dependent apoptosis in some, but not all, cells. By expressing the viral structural proteins separately in cells, the activity could be attributed to the E protein (An *et al.*, 1999). Apoptosis induction has not been reported for E proteins from other coronaviruses.

Coronavirus E proteins share structural similarities with small hydrophobic membrane proteins found in other enveloped viruses. Examples are the Vpu protein of HIV-1, the 6K protein of alphaviruses, and the M2 protein of influenza virus. These proteins, also known as viroporins (Gonzalez and Carrasco, 2003), were demonstrated to modify membrane permeability and to help the efficient release of progeny virus.

### *F. S Protein*

The S protein (previously known as E2) constitutes the spikes, the hallmark of coronaviruses under the electron microscope. It is the major determinant of host range, tissue tropism, pathogenesis, and virulence. It is a relatively large, 1160- to 1452-amino acid-long type I glycoprotein with a cleavable N-terminal signal sequence and a membrane-anchoring sequence followed by a short hydrophilic carboxy-terminal tail of about 30 residues (Fig. 4). When comparing primary sequences, the S protein shows two faces: an amino-terminal half with hardly any sequence similarities and a carboxy-terminal half in which regions with significant conservation can be observed (de Groot *et al.*, 1987a,b; for a review see Cavanagh, 1995), consistent with the distinctive functions of these domains (see later).

The S protein is synthesized as a heavily glycosylated polypeptide as demonstrated by the susceptibility of the glycans to endoglycosidases and by the dramatic effect of the N-glycosylation inhibitor tunicamycin.

The number of potential N-glycosylation sites ranges from 21 (MHV) to 35 [feline infectious peritonitis virus (FIPV)]. The S protein has not been reported to contain O-linked sugars. Cotranslational N-glycosylation is an essential requirement for proper folding, oligomerization, and transport of the S protein, as has also been shown for other (viral) glycoproteins (Doms *et al.*, 1993). Growth of coronaviruses in the presence of tunicamycin resulted in the production of spikeless, noninfectious particles (Holmes *et al.*, 1981; Mounir and Talbot, 1992; Rottier *et al.*, 1981a; Stern and Sefton, 1982). These particles were devoid of S protein, which was found to aggregate in the endoplasmic reticulum when glycosylation was inhibited (Delmas and Laude, 1990).

Folding of the S protein is a relatively slow process. Besides the addition of oligosaccharides it involves the formation and rearrangement of many intramolecular disulfide bonds. For the S protein of MHV-A59, the lumenal domain of which contains 42 cysteine residues, the major conformational events appear to take about 20 min during which the protein passes through a continuous spectrum of folding intermediates (Opstelten *et al.*, 1993a). Folding of S is probably the rate-limiting step in the process of oligomerization. Sufficiently folded S protein monomers associate in the endoplasmic reticulum to form trimers (Delmas and Laude, 1990; Lin *et al.*, 2004), with a half-time of approximately 1 h (Delmas and Laude, 1990; Vennema *et al.*, 1990a,b). Trimerization is likely to be required for export out of the endoplasmic reticulum. In infected cells S protein trimers interact with M protein (Opstelten *et al.*, 1995) and perhaps also with E protein, and migrate to the virus assembly site. A fraction of the S protein is transported to the plasma membrane where it can cause cell–cell fusion, a feature formally attributed to the S protein by its individual expression in cells (de Groot *et al.*, 1989; Pfleiderer *et al.*, 1990). Under such expression conditions the bulk of the S protein remains intracellularly (Vennema *et al.*, 1990a) in the endoplasmic reticulum (Opstelten *et al.*, 1995). Retrieval signals have been identified in the cytoplasmic tail of the S proteins from coronavirus groups 1 and 3 as well as in the tail of the SARS-CoV S protein, but not in the group 1 MHV S protein (Lontok *et al.*, 2004).

During its transport to the cell surface, either alone or as part of virions, the S protein undergoes further modifications. The N-linked sugars are modified and become mature during passage through the Golgi complex. The MHV S protein was shown to become palmitoylated, a modification that may already take place in the endoplasmic reticulum (van Berlo *et al.*, 1987). As a late step the S protein can be cleaved. A basic amino acid sequence resembling the furin consensus sequence motif (RXR/KR) occurs approximately in the middle of the protein and was shown to be the target of a furin-like enzyme in the case of MHV-A59 (de Haan *et al.*, 2004). Cleavage has been demonstrated for S proteins from coronavirus groups 2 and 3, but not for S proteins from group 1 viruses (Cavanagh, 1995) or from SARS-CoV (Bisht *et al.*, 2004). The resulting amino-terminal S1 subunit and the membrane-anchored S2 subunit remain noncovalently linked. It has been suggested that the S1 subunit constitutes the globular head, whereas the S2 subunit forms the stalk-like region of the spike (Cavanagh, 1983b; de Groot *et al.*, 1987a,b).

The coronavirus S protein has two functions, which appear to be spatially separated. The S1 subunit (or the equivalent part in viruses with uncleaved S protein) is responsible for receptor binding, and the S2 subunit is responsible for membrane fusion. For several coronaviruses the receptor-binding site in S1 has been mapped. For MHV strain JHM (MHV-JHM), for instance, it was located in the domain composed of the amino-terminal 330 residues of the S molecule (Kubo *et al.*, 1994), residues 62–65 and 214–216 being particularly important (Saeki *et al.*, 1997; Suzuki and Taguchi, 1996). This amino-terminal domain also determined CEACAM1 receptor specificity of various MHV strains (Tsai *et al.*, 2003). For TGEV (Godet *et al.*, 1994), HCoV-229E (Bonavia *et al.*, 2003; Breslin *et al.*, 2003), and SARS-CoV (Babcock *et al.*, 2004; Wong *et al.*, 2004) the receptor-binding domains have also been mapped to the S1 subunit, although in different regions. In several cases neutralizing antibodies were demonstrated to bind the receptor-binding domains and to prevent the interaction with the receptor (Godet *et al.*, 1994; Kubo *et al.*, 1994; Sui *et al.*, 2004).

The interaction between the S protein and its receptor is the major determinant for virus entry and host range restriction. Nonpermissive cell lines can be rendered susceptible by making them express the receptor (see later references). Coronaviruses can also be retargeted to specific cells by exchanging the ectodomain of the S protein for that of an appropriate other coronavirus, as was demonstrated for MHV (Kuo *et al.*, 2000) and FIPV (Haijema *et al.*, 2003). Receptors have so far been identified for the group 2 coronavirus MHV (CEACAM; Dveksler *et al.*, 1991, 1993; Williams *et al.*, 1991); the group 1 coronaviruses TGEV and porcine respiratory coronavirus (PRCoV) (pAPN; Delmas *et al.*, 1992, 1993), FIPV (fAPN; Tresnan *et al.*, 1996), and HCoV-229E (hAPN; Yeager *et al.*, 1992); and for SARS-CoV (ACE2; Li *et al.*, 2003). The S proteins of the group 2 coronaviruses have been observed to exhibit hemagglutinating activities. Although for bovine coronavirus (BCoV), HCoV-OC43, and hemagglutinating encephalomyelitis virus (HEV) 9-O-acetylated sialic acids were identified as a receptor determinant (Krempl *et al.*, 1995; Kunkel and Herrler, 1993; Schultze and Herrler, 1992; Schultze *et al.*, 1991a; Vlasak *et al.*, 1988b), specific receptors for these viruses have not been identified. Also, the MHV S protein appears to bind sialic acid derivatives in addition to its specific receptor CEACAM, which may suggest that sialic acids function as an additional receptor determinant for MHV-like coronaviruses (Wurzer *et al.*, 2002).

The ectodomain of the S2 subunit, which is involved in the fusion process, contains two heptad repeat (HR) regions (de Groot *et al.*, 1987a,b), a sequence motif characteristic of coiled coils. Mutations in the first (i.e., membrane-distal) HR region of the MHV S protein resulted in fusion-negative phenotypes (Luo and Weiss, 1998) or in a low-pH dependence for fusion (Gallagher *et al.*, 1991), whereas mutations in the second HR region caused defects in S protein oligomerization and fusion ability (Luo *et al.*, 1999). A fusion peptide has not yet been identified in any of the coronavirus spike proteins, but is predicted to be located at (Bosch *et al.*, 2004b; Chambers *et al.*, 1990) or within (Luo and Weiss, 1998) the amino terminus of the first HR region. Binding of the S1 subunit to the (soluble) receptor, or exposure to 37 °C and an elevated pH, has been shown to trigger conformational changes that are supposed to facilitate virus entry by activation of the fusion function of the S2 subunit (Breslin *et al.*, 2003; Gallagher, 1997; Lewicki and Gallagher, 2002; Matsuyama and Taguchi, 2002; Miura *et al.*, 2004; Sturman *et al.*, 1990; Taguchi and Matsuyama, 2002; Zelus *et al.*, 2003). This conformational change is thought to lead to exposure of the fusion peptide and its interaction with the target membrane, further changes resulting in the formation of a heterotrimeric six-helix bundle, characteristic of class I viral fusion proteins, during the membrane fusion process. Indeed, peptides corresponding to the HR regions of MHV (Bosch *et al.*, 2003; Xu *et al.*, 2004) and SARS-CoV (Bosch *et al.*, 2004b; Ingallinella *et al.*, 2004; Liu *et al.*, 2004; Tripet *et al.*, 2004; Zhu *et al.*, 2004) were found to assemble into stable oligomeric complexes in an antiparallel manner, which in the natural situation would result in the close colocation of the fusion peptide and the transmembrane domain. These peptides were further shown to be inhibitors for viral entry (Bosch *et al.*, 2003, 2004b; Liu *et al.*, 2004; Yuan *et al.*, 2004; Zhu *et al.*, 2004).

Besides the HR regions, other parts of the S protein are also likely to be important for the fusion process. All coronavirus S proteins contain a highly conserved region (de Groot *et al.*, 1987a), rich in aromatic residues, downstream of the second HR region, part of which may form the start of the transmembrane domain. The function of this domain is unknown, but a similar region in the HIV-1 Env protein was demonstrated to be important for viral fusion and Env incorporation into virions (Salzwedel *et al.*, 1999). Immediately downstream of the transmembrane domain all S proteins contain a cysteine-rich region (de Groot *et al.*, 1987a). Using a mutational approach including deletions, insertions, and substitutions, both the transmembrane domain and the cysteine-rich region immediately downstream thereof, but not the carboxy-terminal part of the cytoplasmic tail, were shown to be important for MHV S protein-induced cell–cell fusion (Bos *et al.*, 1995; Chang and Gombold, 2001; Chang *et al.*, 2000) (B. J. Bosch, C. A. M. de Haan, and P. J. M. Rottier, unpublished results).

The cleavage requirements of the S proteins for the biological activities of the coronavirus spike remain enigmatic. Whereas the S proteins of group 1 coronaviruses, such as FIPV (Vennema *et al.*, 1990a), are not cleaved, those of other coronaviruses, particularly of groups 2 and 3, are cleaved to variable extents, depending on the viral strain and the cell type in which the viruses are grown (Frana *et al.*, 1985; reviewed by Cavanagh, 1995). Cleavage of the S proteins is not required to expose the internal fusion peptide. Whereas cleavage of the MHV S protein generally correlates strongly with cell–cell fusion (Cavanagh, 1995), virus–cell fusion appeared not to be affected by preventing S protein cleavage, indicating that these fusion events have different requirements (de Haan *et al.*, 2004). Similarly, whereas trypsin activation of SARS-CoV S protein was required for cell–cell fusion, it did not enhance the infectivity of cell-free pseudovirions (Simmons *et al.*, 2004). For MHV-4, the spikes of which are able to initiate fusion without prior interaction with the primary MHV receptor (Gallagher *et al.*, 1992), the stability of the S1–S2 heterodimers after S protein cleavage is low, allowing receptor-independent fusion. During cell culture adaptation, however, selected mutant viruses carried deletions in the S1 subunit, downstream of the receptor-binding domain, which resulted in stabilized S1–S2 heterodimers and receptor-dependent fusion activity (Krueger *et al.*, 2001).

### *G. HE Protein*

Virions of group 2 coronaviruses generally contain a fringe of shorter surface projections in addition to the characteristic spikes (Bridger *et al.*, 1978; King *et al.*, 1985; Sugiyama and Amano, 1981). These viruses express and incorporate into their particles an additional membrane protein, HE (for a review see Brian *et al.*, 1995). Although all group 2 viruses contain an HE gene, the protein is not expressed by all MHV strains (Luytjes *et al.*, 1988; Yokomori *et al.*, 1991), indicating that HE is a nonessential protein also in these viruses.

The HE gene encodes a type I membrane protein of 424–439 residues that contains a cleavable signal peptide at its amino terminus (Hogue *et al.*, 1989; Kienzle *et al.*, 1990) and a transmembrane domain close to its carboxy terminus, leaving a short cytoplasmic tail of about 10 residues (Fig. 4). The ectodomain contains 8–10 putative N-linked glycosylation sites. The putative esterase active site (FGDS) is located near the (signal-cleaved) HE amino terminus. The coronavirus HE protein has 30% amino acid identity with the HE-1 subunit of the HE fusion protein of influenza C virus and the HE protein of torovirus (Cornelissen *et al.*, 1997). It has been suggested that coronaviruses have captured their HE module from influenza C virus or a related virus (Luytjes *et al.*, 1988). However, influenza C virus, toroviruses, and coronaviruses may well have acquired their HE sequences independently, not from each other but from yet another source (Cornelissen *et al.*, 1997). The HE protein becomes cotranslationally N-glycosylated when expressed in cells, giving rise to a polypeptide of approximately 60–65 kDa that rapidly forms disulfide-linked dimers (Hogue *et al.*, 1989; Kienzle *et al.*, 1990; King *et al.*, 1985; Parker *et al.*, 1989; Yokomori *et al.*, 1989; Yoo *et al.*, 1992). The HE dimers (or a higher order structure thereof) become incorporated into virions, while a proportion is transported to the cell surface (Kienzle *et al.*, 1990; Pfleiderer *et al.*, 1991).

Little is still known about the function(s) of the coronavirus HE protein. The protein contains hemagglutinin and acetyl esterase activities (Brian *et al.*, 1995). While the HE proteins of BCoV, HEV, and HCoV-OC43 hydrolyze the 9-*O*-acetyl group of sialic acid and therefore appear to function as receptor-destroying enzymes (Schultze *et al.*, 1991b; Vlasak *et al.*, 1988a), the HE proteins of MHV-like coronaviruses function as sialate-4-*O*-acetylesterases (Klausegger *et al.*, 1999; Regl *et al.*, 1999; Wurzer *et al.*, 2002). Although inhibition of the esterase activity of BCoV resulted in a 100- to 400-fold reduction in viral infectivity (Vlasak *et al.*, 1988a), it was shown both for BCoV and for an MHV strain expressing an HE gene that the S protein is required and sufficient for infection (Gagneten *et al.*, 1995; Popova and Zhang, 2002). In view of these results it has been proposed that the HE protein might play a role at an even earlier step and may mediate viral adherence to the intestinal wall through the specific yet reversible binding to mucopolysaccharides. The process of binding to sialic acid receptors followed by cleavage and rebinding to intact receptors could theoretically result in virus motility and even allow migration through the mucus layer covering the epithelial target cells in the respiratory and enteric tracts (Cornelissen *et al.*, 1997).

Several studies have indicated the HE protein to play a role in pathogenicity. The HE protein of BCoV (Deregt and Babiuk, 1987), but not that of MHV, was able to induce neutralizing antibodies. However, passive immunization of mice with nonneutralizing, MHV HE-specific antibodies protected the animals against a lethal MHV infection (Yokomori *et al.*, 1992). Furthermore, intracerebral expression of the HE protein in mice was found to affect the neuropathogenicity of MHV (Yokomori *et al.*, 1995; Zhang *et al.*, 1998). Strikingly, HE protein-defective MHV mutants were rapidly selected during viral infection in the mouse brain (Yokomori *et al.*, 1993), which may suggest that the HE protein plays a more critical role during the infection of other tissues.

## III. Ultrastructural Observations of Coronavirus Morphogenesis

### A. *Viral Budding*

Early electron microscopic studies demonstrated that coronavirus morphogenesis takes place at intracellular membranes and identified the cisternae of the endoplasmic reticulum as the site of budding of IBV and HCoV-229E (Becker *et al.*, 1967; Chasey and Alexander, 1976; Hamre *et al.*, 1967; Oshiro *et al.*, 1971). Later studies revealed that early in infection particle formation occurs predominantly at smooth-walled, tubulovesicular membranes located intermediately between the rough endoplasmic reticulum and the Golgi complex (ERGIC). This so-called intermediate compartment was shown to be used as the early budding compartment by MHV, IBV, FIPV, TGEV, and SARS-CoV (Goldsmith *et al.*, 2004; Klumperman *et al.*, 1994; Tooze *et al.*, 1984). At later times during infection the rough endoplasmic reticulum was seen to gradually become the major site of MHV budding in fibroblasts (Tooze *et al.*, 1984).

As already mentioned, ultrastructural studies localized the MHV and IBV M proteins in the budding compartment(s) but also in the Golgi complex, that is, beyond the site of budding (Klumperman *et al.*, 1994; Tooze *et al.*, 1984). Apparently, accumulation of M protein alone is not sufficient to determine the site of budding; other viral and/or cellular factors are required as well. For MHV the E protein, which was found to localize to the intermediate compartment by immunoelectron microscopy (immuno-EM), was suggested to be such a candidate (Raamsman *et al.*, 2000), but more players are likely to be involved. Whether the helical nucleocapsids, visible as electron-dense cytoplasmic elements adjacent to budding profiles (David-Ferreira and Manaker, 1965; Dubois-Dalcq *et al.*, 1982; Massalski *et al.*, 1982; Risco *et al.*, 1998; and references given previously), are a determining factor is unclear. In this respect, knowledge about the budding location of coronavirus-like particles (see later) might be informative.

### *B. Postassembly Maturation of Virions*

Coronavirions are subject to an intracellular postbudding maturation process that occurs while they are on their way through the constitutive exocytic pathway by which they are exported out of the cell (Risco *et al.*, 1998; Salanueva *et al.*, 1999; Tooze *et al.*, 1987). Indications of this had already been noticed in early morphological studies with HCoV-229E (Becker *et al.*, 1967; Chasey and Alexander, 1976; Hamre *et al.*, 1967; Oshiro *et al.*, 1971) and MHV (Holmes and Behnke, 1981; Holmes *et al.*, 1981), but were described in somewhat more detail for MHV by Tooze and coworkers (1987). The pictures show "immature" virions in pre-Golgi compartments and Golgi cisternae that appear as spherical structures with the ribonucleoprotein core immediately below the viral envelope and with an "empty" center. By contrast, virions in the *trans*-Golgi network and beyond have the mature morphology showing a fairly uniform, high internal electron density.

An extensive analysis of the structural maturation of coronavirions was reported for TGEV (Risco *et al.*, 1998; Salanueva *et al.*, 1999). Budding was shown to yield relatively large virions with an annular, electron-dense internal periphery and a clear central area. Smaller particles, with the characteristic morphology of extracellular virions, that is, having a compact, dense inner core with polygonal contours, were seen to accumulate in secretory vesicles in the periphery of the infected cell. Both types of particles appeared to coexist in the Golgi complex (Fig. 5). Obviously, the larger particles are the precursors of the smaller mature virions (Salanueva *et al.*, 1999) and probably undergo their morphological maturation during their transport through the Golgi complex. The reorganization of the particle gives

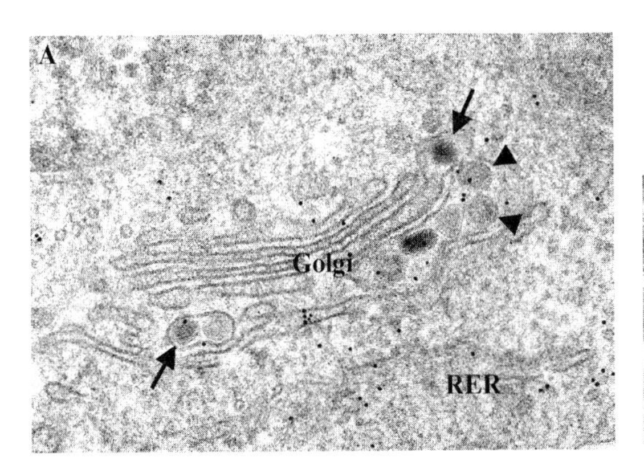


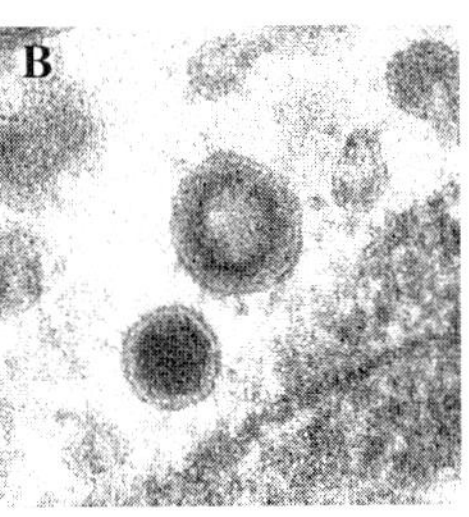


FIG 5. Structural maturation of coronavirus particles. Two types of virion-related particles were detected in TGEV-infected cells. Although large virions with an electron-dense internal periphery and a clear central area are abundant at perinuclear regions, smaller viral particles, with the characteristic morphology of extracellular virions, accumulate inside secretory vesicles that reach the plasma membrane. (A) Large virions (arrowheads) and small dense viral particles (arrows) coexist within the Golgi complex of infected cells (Risco *et al.*, 1998). (B) For a direct comparison of size and morphology a small, dense particle and a large particle are shown (Salanueva *et al.*, 1999). Pictures were kindly provided by C. Risco.

rise to the supposedly icosahedral core shell and is accompanied by a dramatic, approximately 50% reduction of the particle volume.

It is presently unknown what triggers the morphological reorganization in the Golgi complex. Application of drugs affecting the state of the organelle did not give clues. As virions encounter an increasingly acidic pH on passage through the Golgi stack, studies addressing this parameter were done with lysosomotropic agents. Thus, chloroquine and $NH_4Cl$ were applied to MHV-infected cells to elevate the pH at the *trans* side of the Golgi complex, but no effect on the maturation of MHV was observed (Tooze *et al.*, 1987). Monensin, a drug that reversibly disorganizes the Golgi complex and blocks transport along the exocytic pathway, led to the accumulation of the large, annular TGEV virions; after reversal of the blockade formation of the small, compact particles was again restored (Salanueva *et al.*, 1999). These observations confirm that the Golgi complex is necessary for TGEV structural transformation. Nocodazole treatment of cells causes a reversible fragmentation of the Golgi complex. Under these conditions TGEV virions were still able to undergo normal structural maturation. In contrast, still another Golgi-disrupting compound, brefeldin A, prevented their maturation (Risco *et al.*, 1998). This compound leads to a redistribution of Golgi membranes to the ER, leaving no definable Golgi system. Interestingly, MHV particles accumulated in infected cells under these conditions appeared to be infectious when liberated by sonication (J. Meertens and P. J. M. Rottier, unpublished results), leaving us with an intriguing question about the function of the maturation process.

## IV. Molecular Interactions in Assembly of the Coronavirion

### A. *Nucleocapsid Assembly*

#### 1. *Nucleocapsids in Infected Cells and Virions*

Helical nucleocapsids are assembled in the cytoplasm of coronavirus-infected cells. They have been recognized by their tubular appearance in electron microscopy studies with several viruses including IBV, HCoV-229E, TGEV, and MHV (Becker *et al.*, 1967; Chasey and Alexander, 1976; David-Ferreira and Manaker, 1965; Dubois-Dalcq *et al.*, 1982; Hamre *et al.*, 1967; Massalski *et al.*, 1982; Oshiro *et al.*, 1971; Risco *et al.*, 1998). Large inclusions of nucleocapsids were seen to accumulate late in the infection of cells with HCoV (Caul and Egglestone, 1977) and MHV-JHM (Dubois-Dalcq *et al.*, 1982).

The structure of the nucleocapsid as it occurs in infected cells has not been studied in any detail. Ribonucleoprotein particles supposed to represent nucleocapsids have been isolated from MHV-infected cells and were shown to consist of genomic RNA and N protein (Perlman *et al.*, 1986; Robb and Bond, 1979; Spaan *et al.*, 1981). The particles sedimented as EDTA-resistant structures of 200–230S in sucrose gradients. During the active phase of viral replication the majority (90%) of the intracellular genome-size RNA was found in these structures (Spaan *et al.*, 1981).

Ultrastructural studies of nucleocapsids derived from purified virion preparations have shown quite a variety of helical structures, depending on the virus and the experimental conditions used. The overall feature, however, was that of a thread-like coil, sometimes appearing to be hollow, with a diameter varying between 9 and 16 nm and a length ranging from 0.32 $\mu$m up to 6 $\mu$m (Caul *et al.*, 1979; Davies *et al.*, 1981; Kennedy and Johnson-Lussenburg, 1975; Macnaughton and Davies, 1978)

Biochemical analysis of nucleocapsids prepared by detergent disruption of purified coronaviruses revealed the presence of genomic RNA and N protein. Interestingly, however, particles obtained by treatment of virions with Nonidet P-40 appeared to be spherical when viewed under the electron microscope and, in addition, to contain M protein, as was observed with TGEV (Garwes *et al.*, 1976), HEV (Pocock and Garwes, 1977), and MHV-JHM (Wege *et al.*, 1979). The presence of the M protein was found for MHV-A59 to depend on the preparation conditions. Whereas the protein was absent when the virions had been disrupted with Nonidet P-40 at 4°C, solubilization at 37 °C resulted in copurification of M protein with the nucleocapsid (Sturman *et al.*, 1980). The higher temperature was found to cause a conformational change in the M protein, leading to its aggregation and association with the viral RNA in the nucleocapsid. Similarly, nucleocapsid structures essentially lacking the M protein were also reported for IBV when virions were treated with detergent at low temperature (Davies *et al.*, 1981).

Although there is no direct evidence yet, it seems reasonable to assume that the helical nucleocapsids seen accumulating in the cytosol of infected cells constitute the reservoir that feeds into the viral budding system. The location where these nucleocapsids are assembled has not been defined. Their production may take place either free in the cytoplasm, where the N protein is synthesized, or, alternatively, in association with the membrane-bound structures where genomic RNA is produced. The observed colocalization of N protein with the replication complexes (Bost *et al.*, 2000, 2001; Denison *et al.*, 1999; van der Meer *et al.*, 1999) is consistent with the latter possibility. Coronavirus replication appears to occur on double-membrane vesicles (Gosert *et al.*, 2002), which utilize components of the cellular autophagy pathway (Prentice *et al.*, 2004). Whereas early in infection the replication complexes were shown to be almost entirely discrete from sites of M protein accumulation, at later times of infection helicase and N proteins appeared to colocalize with the M protein (Bost *et al.*, 2000, 2001). It was proposed that the translocation of helicase–N protein complexes to sites of virus assembly may serve as a mechanism to deliver the newly synthesized RNA and nucleocapsids and to facilitate the retention of the M protein in the intermediate compartment.

### 2. *Packaging Signals*

Encapsidation of genomic RNA into a nucleocapsid is presumably initiated by an interaction of the N protein with a specific nucleotide sequence, the packaging signal, which is subsequently followed by the polymerization of N proteins around the RNA molecule in a non-sequence-specific manner. The selective incorporation of genomic RNA into virions would predict the packaging signal to be located in sequences unique to this RNA, that is, within the approximately 20-kb region comprising the 5′ UTR and open reading frame 1 (ORF1) with the exception of the leader sequence. Although the data obtained so far support this prediction, no consistent picture has emerged yet.

The approach generally used to map the packaging signal involved the study of helper virus-assisted encapsidation of natural and artificially obtained defective RNA genomes. Thus, a 650-nucleotide region located at the 3′ end of the *pol1b* gene was initially identified for MHV (van der Most *et al.*, 1991), which was subsequently narrowed to an area of 190 nucleotides (Fosmire *et al.*, 1992). Within this area a stable stem–loop of 69 nucleotides was predicted. Mutation studies revealed that the integrity of this secondary structure was important and that the sequence of the packaging signal could be trimmed further to a minimum stretch of 61 nucleotides (Fosmire *et al.*, 1992). The signal appeared to be sufficient for RNA packaging as its inclusion allowed a synthetic subgenomic mRNA of MHV-A59 to be packaged specifically; the encapsidation efficiency of the mRNA was, however, significantly lower than that of the defective genomic RNA from which it was transcribed (Bos *et al.*, 1997). Even a nonviral RNA was found to be packaged into MHV particles when provided with the packaging signal (Woo *et al.*, 1997). Buoyant density analysis of the particles revealed that the RNA was not assembled separately but copackaged with helper virus RNA.

Studies of the corresponding *pol1b* region of another group 2 coronavirus, BCoV, indicated that within this group the packaging signal is structurally and functionally conserved. A 69-nucleotide sequence with significant homology (74%) to that of MHV was identified within a cloned 291-nucleotide segment sharing 72% homology overall (Cologna and Hogue, 2000). When this segment was fused to a noncoronavirus reporter gene sequence, the resulting RNA appeared to be packaged not only by the homologous helper virus BCoV but also by MHV. Conversely, when the MHV packaging signal was fused to the reporter gene sequence, the RNA was found to be encapsidated also in the context of a BCoV-infected cell (Cologna and Hogue, 2000).

Mapping studies of packaging signals in the genomes of group 1 and group 3 coronaviruses have yielded quite different results. For IBV, deletion mutagenesis of a defective RNA led to the conclusion that only the sequences in the 5′ UTR and/or a region of the 3′ UTR were specifically required for packaging, although parts of the *pol1b* sequence, but not any part in particular, also enhanced the efficiency (Dalton *et al.*, 2001). Somewhat similar conclusions could be drawn from a study with TGEV (Izeta *et al.*, 1999). By comparing packaging efficiencies of different defective genomes it was inferred that information for packaging was present both at the genomic 5′ end (about 1.0 kb) and in parts of *pol1b*. A packaging signal was subsequently mapped to a fragment representing the 5′ terminal 649 nucleotides of the genome by inserting a series of overlapping segments covering a stretch of about 2300 nucleotides from the 5′ end of the genome into an mRNA reporter expression construct contained within a defective genome (Escors *et al.*, 2003); only the 5′ terminal sequence conferred to the mRNA the ability to become packaged by helper virus.

It is too early to conclude that the apparently contrasting results reflect true, fundamental differences in encapsidation strategies between the different (groups of) coronaviruses. Although the overview may suggest the existence of different *cis*-acting signals, the data still allow a scenario in which multiple domains in the genome are involved cooperatively, each one contributing differently to (the efficiency of) the encapsidation process. Such contributions would not necessarily concern N protein binding only; the exceptional complexity of the coronaviral genome might call for additional provisions, related perhaps to the structuring of the encapsidation complex. Several observations indeed imply the involvement of multiple domains. The efficient rescue, for instance, of a BCoV defective genome (Drep) that completely lacks the putative 69-nucleotide packaging signal entails the participation of additional sequence(s) (Chang and Brian, 1996; Cologna and Hogue, 2000). Another example is the strongly increased rescue of an otherwise poorly packaged defective TGEV genome (M22) due to the presence of about 4.1 kb of sequences derived from the *pol1b* gene (M62) (Izeta *et al.*, 1999).

*3. N–RNA Interactions*

There is no direct evidence yet for the actual functioning of the presumed packaging signals in the initiation of nucleocapsid assembly. Binding of N protein to these signals, the first step in the process, has so far been addressed only for the 69-nucleotide sequence of MHV. Specific binding to RNA transcripts containing this sequence was indeed demonstrated biochemically with MHV N protein derived from infected cells, from virions, and from cells expressing the protein (Molenkamp and Spaan, 1997). The binding efficiency, however, appeared to be relatively weak as was shown by comparing N protein binding to different parts of a packageable defective genome (MIDI-C) (Cologna *et al.*, 2000). The highest binding efficiency was observed with an RNA transcript representing about 1 kb from the 5′ end of *pol1a*. Remarkably, not even removal of the packaging signal from

MIDI-C RNA affected binding to the N protein as measured in the filter-binding assay used. The observations indicate that the domain containing the 69-nucleotide sequence does not function as a packaging signal in the conventional sense, adding further support to the notion that the intricacies of coronaviral nucleocapsid assembly are complex.

Apart from the studies mentioned, the occurrence of N–RNA interactions has been amply documented. This is not surprising as the N protein has been implicated in several other processes that involve interaction with RNA, such as replication, transcription, and translation (Lai and Cavanagh, 1997; see also Section II.C). As the relevance of these interactions for viral assembly is generally unclear, a brief survey of the available information is included here. In addition, an overview of data on the mapping of RNA interactions on the N polypeptide is schematically presented in Fig. 6.

A high-affinity interaction between the N protein and the 5′ leader was demonstrated for MHV-A59 (Stohlman *et al.*, 1988). Using an RNA overlay protein blot assay and various *in vitro* RNA transcripts, the binding of N protein was localized to a stretch of nucleotides (nucleotides 56–65) at the 3′ end of the leader (Stohlman *et al.*, 1988). The stretch included the pentanucleotide repeat UCUAA now known to be critical for transcription. Biochemical analyses of the interaction measured a dissociation constant ($K_d$) of 14 n$M$ for bacterially expressed MHV N protein to the leader RNA (Nelson *et al.*, 2000). Consistent with the presence of a leader at the 5′ end of all viral RNAs, an N protein-specific monoclonal antibody coimmunoprecipitated genomic RNA as well as the subgenomic RNAs from MHV-infected cells (Baric *et al.*, 1988). Similar observations were made for BCoV (Cologna *et al.*, 2000). Packaging of subgenomic RNAs has been reported for TGEV (Sethna *et al.*, 1991), BCoV (Hofmann *et al.*, 1990), and IBV (Zhao *et al.*, 1993) but not for MHV, suggesting that their incorporation is not mediated by N protein–leader interaction. Whereas for TGEV and IBV the relative packaging of subgenomic RNAs was found to be inefficient, genomic RNA appearing in virions at a more than 10-fold molar excess over any subgenomic RNA species, the BCoV subgenomic N and M mRNAs appeared to be packaged as abundantly as the genome (Hofmann *et al.*, 1990). A reevaluation for TGEV revealed that the detection of subgenomic RNAs in virions was related to the purity of virus preparations, indicating that mRNAs were not specifically encapsidated (Escors *et al.*, 2003).

Besides the leader, the N protein has been found to bind to other parts of the coronaviral genome. In addition to the binding site in the MHV

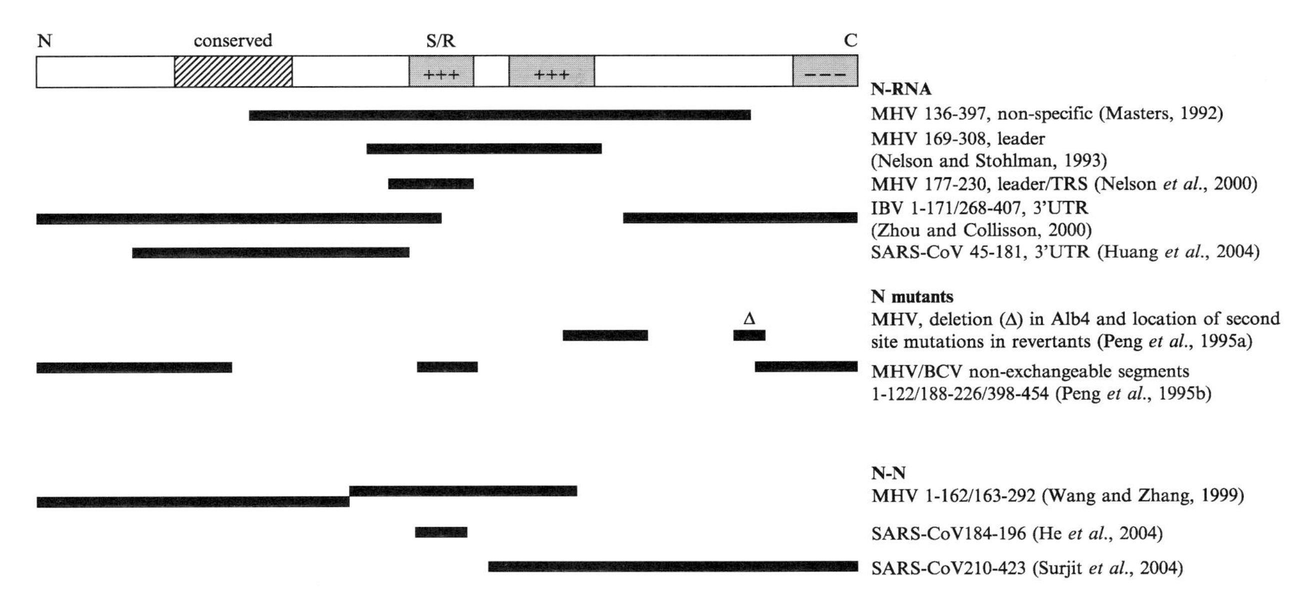


FIG 6. Structural organization of the coronavirus N protein. Common features and their distribution along the polypeptide chain are shown schematically. The hatched box indicates the most conserved part of the N protein, with a high proportion of aromatic residues. The N protein contains many basic residues throughout the polypeptide, but with particular clustering in two regions (+++). The upstream cluster contains a serine/arginine-rich region (S/R). The carboxy terminus, which contains a high proportion of acidic residues, is also indicated (−−−). The bars indicate parts of the N protein that have been implicated in N–N and N–RNA interaction. Furthermore, the location of the deletion in the MHV-A59 mutant Alb4 (Δ) is indicated as well as the domain where second-site mutations in revertant viruses of Alb4 are mapped. Finally, the parts of the N protein that could not be transferred from BCV into the MHV genome are marked. References are included in the figure.

*pol1a* gene mentioned previously, a high-efficiency binding site was identified by the same authors in the 3′ half of the N gene of both MHV and BCoV (Cologna *et al.*, 2000). The IBV N protein was shown to bind sequences in the 3′ terminal UTR of the genome (Zhou *et al.*, 1996).

Coronavirus N proteins do not contain sequence motifs typically found in other RNA-binding proteins. They appear to bind RNA both nonspecifically (Masters, 1992; Robbins *et al.*, 1986) and in a sequence-specific way (Cologna *et al.*, 2000; Nelson and Stohlman, 1993; Nelson *et al.*, 2000; Stohlman *et al.*, 1992). Non-sequence-specific RNA binding has been mapped to a large central domain of the MHV N molecule (Fig. 6) (Masters, 1992). Also, the leader-binding property was assigned to this domain; this activity was initially mapped to the area containing the two highly basic regions (Nelson and Stohlman, 1993), but was later narrowed to a 55-residue segment containing the serine/arginine-rich basic region (Nelson *et al.*, 2000). Interestingly, this particular region could not be interchanged with its BCoV counterpart in a study on the functional equivalence of the N proteins from these related viruses (Peng *et al.*, 1995b). Another domain in the MHV N protein implicated in viral RNA binding was mapped to an area that partly overlaps with the second basic region. The assignment was based on an analysis of second-site revertants of MHV mutant Alb4, the virions of which are extremely thermolabile because of a 29-residue deletion located between the central and carboxy-terminal domain of the N protein (Koetzner *et al.*, 1992). The reverting mutations correlated with restoration of the disturbed RNA-binding capacity of the MHV N protein and were found clustered close to the basic region some 80 residues on the amino side of the deletion (Fig. 6) (Peng *et al.*, 1995a). Although all these studies consistently attribute a major role in RNA binding to the central portion of the coronaviral N protein, the interaction of the IBV N protein with the 3′ UTR of IBV RNA mentioned previously was mapped to the amino- and carboxy-terminal domains of the molecule (Zhou and Collisson, 2000). 3′ UTR RNA-binding activity was also assigned to the amino-terminal domain of the SARS-CoV N protein on the basis of studies using nuclear magnetic resonance spectroscopy (Huang *et al.*, 2004).

### 4. *N–N Interactions*

It is obvious that the wrapping of the 30-kb coronaviral genome into the compact helical nucleocapsid is largely driven by N protein interactions. As there are no indications for packaging of the RNA into a preformed capsid, these interactions can be described by the following model. Packaging is initiated by binding of the N protein, either as a monomer or in a multimeric form, to the RNA. By analogy to other RNA viruses, this sequence-specific interaction may induce a conformational change in the N protein, thereby creating a nucleation site for the cooperative stacking of N protein units along the entire length of the RNA, now in a non-sequence-specific way. These N units can again be monomeric or consist of defined multimers. Finally, helix formation is driven by interactions between N molecules separated along the ribonucleoprotein chain but that become adjacent in neighboring helices. This model predicts multiple nonequivalent interactions between N molecules.

N–N interactions have been experimentally demonstrated for MHV, BCoV, and HCoV-OC43. High molecular weight species of the N protein, possibly trimers, were detected by sodium dodecyl sulfate–polyacrylamide gel electrophoresis (SDS–PAGE) of virion preparations under nonreducing (but not under reducing) conditions, which is indicative of intermolecular disulfide bonds between the N subunits (Hogue *et al.*, 1984; Narayanan *et al.*, 2003b; Robbins *et al.*, 1986). These complexes are likely to be additionally stabilized by noncovalent interactions as coronavirus N protein cysteines are not well conserved, the SARS-CoV N protein even lacking any cysteine residues. Both monomeric and oligomeric N species were able to bind RNA (Robbins *et al.*, 1986). Multimeric forms of the N protein were also found in association with intracellular genomic RNA in MHV-infected cells as shown after the selective isolation of this ribonucleoprotein through coimmunoprecipitation with the M protein (Narayanan *et al.*, 2003b). High molecular weight forms of the N protein corresponding to dimers and trimers were also demonstrated *in vitro* after ultraviolet (UV) cross-linking of BCoV N protein to RNAs (Cologna *et al.*, 2000).

Few studies have addressed the identification of the N–N interaction domains. The results so far are inconsistent (Fig. 6), but this might as well reflect the predicted occurrence of nonequivalent interactions. Interaction sites were mapped to the amino-terminal part of the MHV N protein (Wang and Zhang, 1999). Using an *in vitro* binding assay in which the full-length N protein was incubated with bacterially expressed fusion proteins containing different segments of the N protein, interaction was observed with a polypeptide derived from the amino-terminal one-third (residues 1–162) of the protein and with a polypeptide representing the central part (residues 163–292). The latter domain contains the serine/arginine-rich region implicated in the binding to the leader/TRS-specific sequences (Nelson *et al.*, 2000). This domain could not be replaced by the corresponding domain from BCoV without loss of viral viability, from which the authors indeed inferred an involvement in protein–protein interactions (Peng *et al.*, 1995b). The same domain was also shown to be essential for the homotypic association of the SARS-CoV N protein (He *et al.*, 2004). Using a mammalian two-hybrid approach, the N–N interaction appeared to be abolished completely when the serine/arginine-rich region had been deleted. However, in another study using the yeast two-hybrid system this interaction was not confirmed. A polypeptide consisting of the amino-terminal two-thirds of the SARS-CoV N protein, that contains the serine/arginine-rich region, exhibited no association with the full-length protein (Surjit *et al.*, 2004). Rather, self-association was attributed to the carboxy-terminal 209 residues of the molecule, which lacks the motif.

## *B. Envelope Assembly*

### *1. Formation of Virus-Like Particles: M–E Interactions*

Unlike most other enveloped viruses, coronaviruses have the remarkable feature of being able to independently assemble their envelope. Indications for this were already noticed in early electron microscopy studies of viral preparations and infected cells showing the occurrence of apparently "empty" particles (Afzelius, 1994; Chasey and Alexander, 1976; Macnaughton and Davies, 1980). Incomplete virions with the typical coronavirus morphology but lacking the N protein and the genome could indeed be separated from normal IBV particles by their lower density in sucrose gradients (Macnaughton and Davies, 1980). The definition of virus-like particles (VLPs) and the requirements for their formation were established by the coexpression of the coronaviral structural proteins in mammalian cells (Vennema *et al.*, 1996). Membranous particles were assembled when the MHV envelope proteins M, E, and S were coexpressed, without the need for an N protein or genomic RNA. The particles were released from the cells and, when examined under the electron microscope, appeared to be morphologically indistinguishable from authentic virions, that is, they had the characteristic shape and dimensions of normal virions. Also, their membrane protein composition was similar to that of MHV, with a high abundance of M protein and only trace amounts of E protein. Quite surprisingly, only the M and E proteins were required for particle assembly. Both S and N proteins were dispensable for particle formation but, whereas the S protein became incorporated when present, this was not the case for the N protein (Vennema *et al.*, 1996), except in combination with (defective) genomic RNA, in which case a nucleocapsid was coassembled (Bos *et al.*, 1996; Kim *et al.*, 1997). Individual expression of the M or the E protein in cells did not give rise to formation of VLPs although E protein synthesis by itself led to the secretion of E-containing vesicles (Corse and Machamer, 2000; Maeda *et al.*, 1999). The nature of these particles has not been characterized in much detail; the vesicles induced by MHV E sedimented slightly slower than virions (Maeda *et al.*, 1999) whereas the particles obtained with IBV E had about the same density as virions (Corse and Machamer, 2000).

Besides for MHV and IBV, VLPs have so far been described for BCoV (Baudoux *et al.*, 1998b), FIPV (Godeke *et al.*, 2000), and TGEV (Baudoux *et al.*, 1998b). The observations demonstrate the unique budding mechanism of coronaviruses, which is dependent solely on the envelope proteins M and E but independent of a nucleocapsid. Somewhat similar observations have been described for the flavivirus tick-borne encephalitis virus and for hepatitis B virus, which also produce proteolipid particles on expression of their envelope proteins preM and E (Allison *et al.*, 1995; Mason *et al.*, 1991) and S (Patzer *et al.*, 1986; Simon *et al.*, 1988), respectively, but these particles are much smaller than the corresponding virions. In contrast, particles with the typical, large size of coronaviruses are acquired by the concerted action of just the proteins M and E (Baudoux *et al.*, 1998b; Vennema *et al.*, 1996). Budding of enveloped viruses generally requires a nucleocapsid (for a review see Garoff *et al.*, 1998). For retroviruses the Gag protein, the precursor to the nucleocapsid, is all that is needed to obtain particles resembling immature virions; the Env protein is dispensable. Budding of alphaviruses, on the other hand, requires both the envelope proteins and the nucleocapsid. Interestingly, the same appears to hold true for arteriviruses (R. Wieringa, A. A. F. de Vries, and P. J. M. Rottier, unpublished observations), which are closely related to the coronaviruses, share with them a triple-spanning envelope protein, and bud into early membranes of the secretory pathway, like coronaviruses but unlike alphaviruses.

It is unknown how the coronavirus M and E proteins cooperate in budding. As the extensive electron microscopy work with M proteins from various coronaviruses gave no indications that this protein causes membrane bulging by itself, it is believed that the function of the E protein in coronavirus budding is in the induction of curvature in the M protein lattice (see later) and the subsequent budding of the membrane (Vennema *et al.*, 1996). By its low abundance in the virion, the E protein does not seem to serve a genuine structural function in that it occupies frequent, regular positions in the M protein framework. Consistent with an important role of the E protein in particle morphogenesis, mutations in its hydrophilic carboxy-terminal part, introduced by targeted recombination into the MHV genome, yielded thermolabile viruses one of which showed aberrant virion morphology with pinched and elongated shapes when viewed in the electron microscope (Fischer *et al.*, 1998). Revertant analyses revealed that a single second-site amino acid change within the E protein was able to reverse the phenotypic effect of the original mutations, providing support for possible interactions between E protein monomers during budding (Fischer *et al.*, 1998). Unexpectedly, complete deletion of the E gene from the coronaviral genome does not abolish virion formation, demonstrating that the protein is not essential for budding. Whereas this deletion dramatically (at least 1000-fold) reduced the release of infectivity from infected cells in the case of MHV-A59 (Kuo and Masters, 2003), knockout of the TGEV E gene resulted in a lethal phenotype (Curtis *et al.*, 2002; Ortego *et al.*, 2002). Remarkably, however, in the latter case virions still assembled but these appeared to be unable to leave the cells (J. Ortego and L. Enjuanes, personal communication).

The VLP system offers a convenient assay to study many aspects of coronavirus envelope assembly. It was thus used to analyze the primary structure requirements of the M and E proteins for particle formation. For the M protein such studies demonstrated each of its different domains to be important. In general, mutations (deletions, insertions, and point mutations) in the lumenal domain, the transmembrane domains, the amphiphilic domain, or the carboxy-terminal domain of the MHV M protein strongly affected its ability to form VLPs (de Haan *et al.*, 1998a). The assembly process was particularly sensitive to changes in the carboxy terminus of the protein. Truncation by only one residue reduced the efficiency severely whereas removal of two residues fully abolished particle formation. These effects appeared to be less severe in the context of a normal coronaviral infection, probably because additional interactions can compensate. The single-residue deletion, when introduced into the MHV genome, was without measurable phenotype and also a mutant virus with a truncation of two residues could be obtained, although with difficulty, as it was severely affected in its growth (de Haan *et al.*, 1998a; Kuo and Masters, 2002). The importance of the M protein cytoplasmic and transmembrane domains was confirmed by VLP studies in the IBV system; mutant proteins lacking portions of either of these domains were unable to support particle assembly (Corse and Machamer, 2003).

Studies of the primary structure requirements of the E protein for VLP formation revealed that the sequence of its hydrophobic domain was not critical. The assembly capacity of the protein was maintained when its transmembrane region was partly or completely replaced by the corresponding domain of the vesicular stomatitis virus (VSV) G protein (Corse and Machamer, 2003). However, when its small amino-terminal ectodomain was additionally replaced by the large VSV G protein counterpart, the chimeric protein became nonfunctional. Deletion of the transmembrane, but not the amino-terminal, domain rendered the E protein essentially assembly incompetent (Lim and Liu, 2001). Deletions in the cytoplasmic carboxy-terminal half of the E protein mapped the cysteine-rich region as the most important part for VLP assembly (Lim and Liu, 2001).

Although the interaction between M and E proteins is amply demonstrated by their interdependence for VLP formation, direct evidence for their interaction has actually been provided only for the IBV proteins (Corse and Machamer, 2003; Lim and Liu, 2001). The two proteins could be cross-linked to each other in IBV-infected cells and in cells coexpressing the M and E genes (Corse and Machamer, 2003). It appeared that the cytoplasmic tails of both proteins were required, suggesting they are involved in the interaction. In another study M–E interaction was demonstrated by a coimmunoprecipitation assay. Also in this assay the cytoplasmic domain of the E protein, comprising the cysteine-rich region, was found to be important as its deletion affected M–E interaction to the greatest extent when compared with other deletion mutant E proteins (Lim and Liu, 2001). The results from both studies also showed that the ability of mutant E or M proteins to interact did not correlate with their assembly competence. Apparently, other requirements such as homotypic E or M interactions or interactions with host cell components must be met.

The specificity of the interaction between the M and E proteins during particle assembly was further demonstrated by the poorly successful attempts to generate chimeric VLPs. No particles were observed when heterologous combinations of TGEV and BCoV M and E proteins were coexpressed (Baudoux *et al.*, 1998b) and the same was true for heterologous combinations of FIPV and MHV M and E proteins (H. Vennema and P. J. M. Rottier, unpublished results). In both studies chimeric M and E proteins were also tested, demonstrating that, except in one case, exchanges between corresponding domains rendered the proteins assembly incompetent. Only when the TGEV M protein amino-terminal ectodomain was replaced with that of BCoV did the chimeric polypeptide support VLP formation in combination with the TGEV E protein; VLP formations was also supported, but to different extents, with TGEV/BCoV chimeric E proteins and—poorly, however—with the BCoV E protein (Baudoux *et al.*, 1998b). The reciprocal M construct, the BCoV M protein carrying the TGEV ectodomain, was nonfunctional, even in combination with TGEV E protein. Consistently, replacement of the ectodomain with that of FIPV M also abolished the productive partnership of MHV M protein with MHV E (de Haan *et al.*, 1999).

*2. M–M Interactions*

As the disproportionate amounts of M and E proteins in VLPs already imply, homotypic interactions between M molecules must constitute the energetic basis underlying the formation of the coronaviral envelope. In MHV-based VLPs generated by coexpression of M and E proteins, for instance, the sheer excess of M protein—the relative molar presence of E in the particles is less than 1%—is evidence for the strong interactive forces between M molecules. Hence, envelope assembly is thought to be driven primarily by laterally interacting M molecules that form a two-dimensional lattice in intracellular membranes (Opstelten *et al.*, 1993b, 1995). Large multimeric complexes of M protein have indeed been demonstrated biochemically after individual expression of the MHV protein in cells. When the association of the M molecules was maintained by the careful selection of cell lysis conditions, sucrose gradient analysis revealed the existence of large heterogeneous (up to about 40 molecules) complexes, which accumulated in the Golgi compartment (Locker *et al.*, 1995). Somewhat smaller complexes were obtained when the cytoplasmic tail of the protein was removed; these complexes were no longer retained in the Golgi apparatus but transported to the cell surface. Apparently, the tail domain is not essential for the lateral interactions between M proteins, but it is critically required for budding (de Haan *et al.*, 1998a, 2000). Similar higher order complexes of the M protein have also been demonstrated in MHV-infected cells as well as in MHV virions (Opstelten *et al.*, 1993b, 1995).

Further support for the existence of homotypic M protein interactions and additional insight into the domains involved in these interactions came from work with mutant M proteins that are unable to assemble into VLPs. In these studies MHV M proteins with deletions in either the transmembrane regions, the amphipathic domain, or the extreme carboxy terminus or with substitutions of the lumenal domain were tested for their ability to associate with other M proteins and to be rescued into VLPs formed by assembly-competent M proteins (de Haan *et al.*, 1998a, 2000). It appeared that the mutant proteins maintained these biological activities despite the often severe alterations; actually, the only mutant protein that had lost these abilities was one in which all three transmembrane domains had been replaced by a heterologous transmembrane domain. It was concluded that M protein molecules interact with each other through multiple contact sites, particularly at the transmembrane level. It was furthermore hypothesized that the full complement of interactions between the M molecules is required for efficient particle formation; possibly, all these interactions are required to provide the free energy to generate and stabilize the budding envelope. The failure of M protein mutants capable of associating with assembly-competent M protein to assemble into VLPs by themselves (de Haan *et al.*, 1998a, 2000) indicates that additional interactions with viral (E) and/or host proteins is required. In this respect it is of note that the IFN-inducing capacity of the M protein, demonstrated for TGEV and BCoV, also requires the presence of the E protein, which suggests that the induction of IFN is dependent on a specific, probably regularly organized structure of the M protein (Baudoux *et al.*, 1998a,b). It is unclear how the presence of the E protein alters the M protein lattice to achieve this effect.

### 3. *M–S and M–HE Interactions*

Coronavirus envelope assembly is not dependent on the S protein or the HE protein. This is obvious from work with VLPs as well as with viruses showing that bona fide particles were produced when these proteins were either simply absent or unavailable for assembly. Availability can be compromised under conditions in which proper folding of the proteins is affected. Inhibition of N-glycosylation by the drug tunicamycin, for instance, can lead to aggregation and retention of membrane proteins in the endoplasmic reticulum and has been shown to prevent the incorporation into virions of both the S protein (Holmes *et al.*, 1981; Mounir and Talbot, 1992; Rottier *et al.*, 1981a; Stern and Sefton, 1982) and the HE protein (Mounir and Talbot, 1992). The same effect has been observed with temperature-sensitive MHV mutants carrying defects in their S gene, which, when grown at the restrictive temperature, gave rise to spikeless particles (Luytjes *et al.*, 1997; Ricard *et al.*, 1995).

Both S and HE proteins are assembled into the coronaviral envelope through interactions with the M protein. Such interactions have been demonstrated for MHV and BCoV M and S proteins and for BCoV M and HE proteins, in infected cells, in cells coexpressing the proteins, and in virions (Nguyen and Hogue, 1997; Opstelten *et al.*, 1995). Complexes of the proteins were shown by coimmunoprecipitation and cosedimentation analyses as well as by immunofluorescence studies in which the intracellular transport of S and HE proteins to the plasma membrane was found to be inhibited by coexpressed M protein, the proteins being retained in the Golgi apparatus, the natural residence of the M protein.

The kinetics with which the proteins engage in heteromeric complex formation appeared to be different for the different proteins. This effect is due to their different rates of folding and oligomerization. For the S protein these rates are low, involving the formation of multiple intramolecular disulfide bonds and the addition of numerous oligosaccharide side chains (Delmas and Laude, 1990; Opstelten *et al.*, 1993a; Vennema *et al.*, 1990a,b). In contrast, folding of the MHV and BCoV M proteins is independent of disulfide bonds and glycosylation, as a result of which they are, for instance, swiftly transported out of the endoplasmic reticulum (Opstelten *et al.*, 1993a). As a consequence, M molecules enter into M–S and M–HE complexes immediately after their synthesis whereas for newly synthesized S and HE molecules it took 15-30 min before they started to appear in these heterocomplexes (Nguyen and Hogue, 1997; Opstelten *et al.*, 1995). The importance of folding as a major rate-limiting step was illustrated by the inability of the S protein to interact with M protein when its folding had been inhibited by *in vivo* reduction; only completely oxidized S molecules were association competent (Opstelten *et al.*, 1993a, 1995). Whether the M protein interacts with S and HE proteins while they are still in their monomeric form or only after their oligomerization remains to be elucidated. It is, however, clear that the proteins engage in interaction with each other in early compartments, most likely the endoplasmic reticulum, as judged from the oligosaccharide maturation states of freshly formed protein complexes (de Haan *et al.*, 1999; Nguyen and Hogue, 1997; Opstelten *et al.*, 1995). Only dimers of HE were associated with HE–M–S complexes that were observed in BCoV-infected cells; because the appearance of HE in these complexes correlated with the kinetics of HE dimerization it was concluded that proper oligomerization is most likely a requirement for its association (Nguyen and Hogue, 1997). Interestingly, such heterotrimeric complexes were not observed on coexpression of the three proteins in cells. Under these conditions only the heterodimeric M–S and M–HE associations were detected.

The structural domains of M and S proteins that are involved in the formation and stabilization of their complex have been identified. Using the coimmunoprecipitation and colocalization assays referred to previously, the essential domains in the MHV M protein were mapped by a mutagenetic approach (de Haan *et al.*, 1999). It appeared that M–S complex formation was sensitive to changes in all membrane-associated parts of the M molecule. Interactions between M and S proteins were found to occur at the level of the transmembrane domains and of the amphipathic domain, which is located on the cytoplasmic face of cellular membranes. In contrast, neither the lumenally exposed amino terminus nor the hydrophilic cytoplasmic tail of the M protein was required; even the deletion of these parts—known to abrogate the ability of the protein to form VLPs—did not prevent association with the S protein.

Chimeric S proteins were used to show that the large ectodomains of the spikes are not involved in interaction with M proteins. Such chimeric proteins were constructed from the MHV and FIPV S proteins and consisted of the ecto- or lumenal domain from the one and the transmembrane plus endodomain from the other. These proteins, which seemed biologically fit as they were still fusion active, were initially tested in coexpression studies with the M and E proteins from either virus for their ability to be incorporated into VLPs. They were found to assemble only into viral particles of the species from which their carboxy-terminal domain originated (Godeke *et al.*, 2000). The chimeric S genes were subsequently incorporated into the proper coronavirus genomic background, creating the chimeric viruses fMHV and mFIPV, the spike ectodomains of which are from the feline and murine coronavirus, respectively; these studies provided the basis for the development of a novel targeted recombination system for reverse genetics of coronaviruses (Haijema *et al.*, 2003; Kuo *et al.*, 2000).

Further fine mapping of the carboxy-terminal parts of the S protein involved in M–S protein interaction revealed the importance of the cytoplasmic tail. Again using coimmunoprecipitation and VLP incorporation assays, it appeared that increasing truncations gradually abolished the association with the M protein (B. J. Bosch, C. A. M. de Haan, and P. J. M. Rottier, unpublished results). The significance of the tail domain was demonstrated most convincingly by showing the coimmunoprecipitation and VLP assembly of a chimeric VSV G protein the cytoplasmic tail of which had been replaced by that of MHV S. Tail truncations were tolerated in the context of the coronavirus; recombinant MHVs were generated that lacked 12 or 25 (but not 35) residues from the S protein carboxy terminus, but their growth was impaired by about 10- and $10^4$-fold, respectively. Also, tail extensions were tolerated, allowing the construction of a recombinant MHV with a spike protein extended at its carboxy terminus by the green fluorescent protein (GFP), yielding green fluorescent virions (Bosch *et al.*, 2004a). The extension was, however, lost quite rapidly on serial passaging of the virus.

Molecular details of the interaction of M and HE proteins and the requirements of HE for incorporation into viral particles have not been described. One study reported that HE protein mutants lacking part of their ectodomain were not assembled into particles (Liao *et al.*, 1995). Most likely, however, this observation was due to folding or maturation defects of the mutant proteins. In another study the BCoV S and HE proteins were shown to be incorporated into MHV particles when coexpressed in MHV-infected cells. Apparently, homology between the proteins of these related group 2 coronaviruses is sufficiently high for heterologous M–S and M–HE interactions to occur (Popova and Zhang, 2002).

## *C. Virion Assembly*

### *1. Envelope–Nucleocapsid Interactions*

Numerous electron microscopy studies have pictured the process of virion assembly in the coronavirus-infected cell. They show the close apposition of—presumably preassembled—tubular nucleocapsids to intracellular membranes, the appearance of membrane curvature at the contact sites, the "growth" of these buds into particle-sized vesicles, and the ultimate detachment of virions from the membranes by pinching off.

It has become clear that the M protein is the central player, which, through its interactions with every known component of the virion, orchestrates the entire assembly process (see Fig. 7). In the process two levels of interaction can be distinguished. One is the level of the membrane where, as detailed in the previous section, the M protein interacts (1) with itself, to generate the basic molecular framework of the viral envelope, (2) with the E protein, to induce curving and budding of the M protein-modified membrane, and (3) with S and HE, to coassemble these spikes into the viral envelope. The other level at which the M protein operates involves the incorporation of the nucleocapsid into the virion. Here, two types of interactions have been described: interactions of the M protein with the N protein and with the viral genome.

An instrumental role of the M protein in drawing the nucleocapsid into the budding particle is indicated by their demonstrated interaction in studies with virion preparations. The M protein has been shown to remain associated with subviral particles obtained after treatment of virions with detergent that removes the spikes (Escors *et al.*, 2001a,b; Garwes *et al.*, 1976; Lancer and Howard, 1980; Wege *et al.*, 1979).

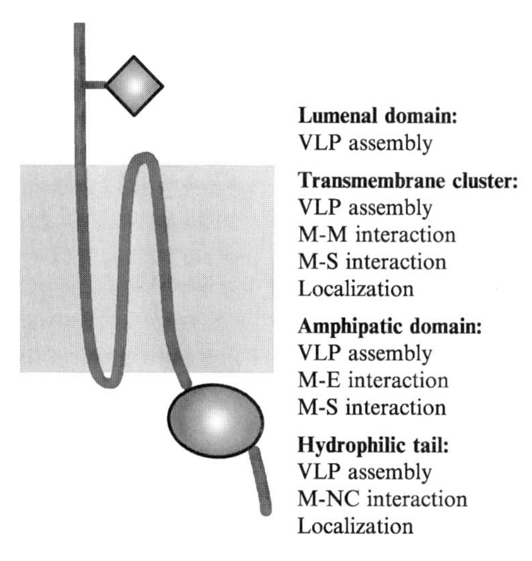


FIG 7. The various domains of the MHV M protein and the processes for which they are important. The amphipathic domain of the M protein is represented by an oval. See text for references. (See Color Insert.)

The association was shown for MHV to be temperature dependent (Sturman *et al.*, 1980). In the case of TGEV the association of M with the spherical structure, termed the core, was stabilized by basic pH and divalent cations but lost at high salt concentration, resulting in disruption of the core structure and release of the helical nucleocapsid (Escors *et al.*, 2001a,b; Risco *et al.*, 1996). In an elegant study of the binding of *in vitro*-translated M polypeptides to purified nucleocapsids the ionic interaction was mapped to a 16-residue sequence in the hydrophilic carboxy-terminal tail domain (Escors *et al.*, 2001b). Also in infected cells, interaction of the M protein with ribonucleoprotein structures, presumably nucleocapsids, has been demonstrated. Using M-specific antibodies, structures containing both N protein and genomic RNA were coimmunoprecipitated with M protein from MHV-infected cell lysates (Narayanan *et al.*, 2000). Conversely, M protein was coprecipitated when an N-specific antibody was used, while in this case all viral mRNAs copurified because of their known leader-mediated affinity for the N protein. These interactions did not require an S or an E protein (Narayanan *et al.*, 2000).

Although interactions between the M and N proteins might intuitively be expected to drive the process of attachment of the nucleocapsid to the intracellular target membrane, direct experimental evidence for this interaction is strikingly lacking. Significantly, MHV M and N proteins coexpressed in cells were found not to interact (Narayanan *et al.*, 2000) nor did purified TGEV N protein interact with *in vitro*-synthesized M protein (Escors *et al.*, 2001b). Because the N protein occurs in coronavirus-infected cells in various configurations—as a free protein and in association with an array of partners including the viral genome—it is obvious that a selection mechanism must act to ensure that only nucleocapsids are assembled into particles. Consistently, coexpressed N protein is not incorporated into VLPs, but its inclusion depends on the presence of viral RNA (Bos *et al.*, 1996; Vennema *et al.*, 1996). Thus, unless the selection process is performed by a mechanism not involving the N protein, its association with genomic RNA to form a nucleocapsid seems required to generate the unique conformation that enables it to interact with the M protein. The only evidence to date for an interaction between M and N proteins is indirect and comes from genetic studies. Analysis of second-site revertants of a constructed MHV-A59 mutant virus lacking the two carboxy-terminal residues of its M protein revealed that the highly defective growth phenotype of this virus could be restored, among others, by mutations in the carboxy-terminal domain of the N protein (Kuo and Masters, 2002). In two independently obtained revertants the N protein had lost 15 residues of this—among different strains of MHV highly conserved—domain because of a frameshifting 10-nucleotide deletion. The results argue strongly for a direct cooperation of the carboxy-terminal regions of the M and N proteins during virion formation. Other indications supporting the occurrence of M–N interactions come from studies of complexes of M protein with ribonucleoprotein from MHV-infected cells and with TGEV cores (Escors *et al.*, 2001b; Narayanan *et al.*, 2000). When such complexes were treated with RNase the association of M and N proteins was not destroyed, suggesting a direct interaction. However, the presence of short RNAs inaccessible to the RNase but sufficient to bridge the M–N interaction could not be excluded.

The most unusual interaction that the coronaviral M protein seems to engage in involves genomic RNA. This interaction has so far been reported only for MHV, by Makino and co-workers. These workers had shown earlier that the 69-nucleotide packaging signal located in the *pol1b* gene could mediate the incorporation into virions of RNAs of even noncoronaviral origin (Woo *et al.*, 1997). They subsequently showed that this incorporation is most likely effected by a direct and specific interaction of the signal with the M protein. When defective genomic RNAs or nonviral RNAs were introduced into helper MHV-infected cells, they could subsequently be isolated as ribonucleoproteins from lysates of the cells by immunoprecipitation with an M-specific antibody, but only if the RNAs contained the packaging signal (Narayanan and Makino, 2001). Coexpression experiments using noncoronaviral vectors showed the interaction to be independent of the N protein. A reporter gene transcript generated in cells expressing the M protein could be coimmunoprecipitated with an anti-M monoclonal antibody provided that the RNA carried the packaging signal (Narayanan *et al.*, 2003a). Moreover, when the E protein was additionally coexpressed, the signal-containing RNA—but not an identical RNA lacking this sequence—was found to be coincorporated into VLPs, irrespective of the presence of the N protein. Altogether these observations reveal a hitherto unknown type of interaction between a viral envelope protein and genomic RNA. Although its significance remains to be further established the M–RNA interaction seems to provide additional selectivity to the assembly of the coronaviral nucleocapsid.

### 2. *Specificity and Flexibility*

Assembly of viruses is a process of generally high specificity. Directed by specific targeting signals, the viral structural components colocalize to distinct places in the cell where unique and complex molecular interactions control their assembly. These rules hold particularly for naked viruses and many of the smaller enveloped viruses; there are, however, many examples where the process is considerably less selective and where "nonself" (host or viral) components are coassembled (see, e.g., Garoff *et al.*, 1998). Interestingly, formation of the large, pleiomorphic coronaviruses appears to combine aspects of both great selectivity and extreme flexibility.

With the M and E proteins as the fixed minimal requirement, coronaviral particles appear to tolerate the presence of all other viral components in practically every possible combination. A nucleocapsid is not required but, if available, it can take almost any length as defective (including chimeric) genomes of largely varying sizes have been accommodated. RNAs need not necessarily be packaged into a nucleocapsid; whether of viral or nonviral origin, if provided with the proper packaging signal they can be taken in even in the absence of an N protein.

Also in the composition of their viral envelope these viruses are highly flexible. Spikes seem to be incorporated in variable numbers depending on availability. They tolerate severe manipulation, both of their ectodomain and of their endodomain. Thus, swapping of ectodomains between unrelated coronaviruses (i.e., from different groups) creates viable chimeric viruses whereas a foreign protein such as GFP appended to the S protein endodomain is accommodated in the particle, although reluctantly. Some coronaviruses have an HE protein but continue growing well if for any reason the gene is not (properly) expressed as happens among different MHVs. Consistently, S and HE proteins are incorporated independent from each other. Direct interactions between S and HE were not observed when the proteins were coexpressed in cells (Nguyen and Hogue, 1997). This result is consistent with the idea that these proteins are separately drawn into the M protein lattice by their distinctive interactions with M molecules. It is also consistent with the concept that these proteins assume different positions within this lattice, a hypothesis based on the presumed different geometric requirements for the incorporation of trimeric and di- or tetrameric S and HE complexes, respectively.

How, in the face of this enormous flexibility in accommodating all these various numbers and combinations of viral components, do coronaviruses manage to maintain specificity? Host proteins have not been noticed to occur in virions, although this may simply not have been looked at carefully enough. By probing the specificity, using viral and nonviral membrane proteins, it appeared that foreign proteins are effectively excluded from coronaviral particles (de Haan *et al.*, 2000). However, some missorting was found to occur, consistent with earlier observations (Yoshikura and Taguchi, 1978).

The picture of coronaviral envelope formation is one that is directed entirely by lateral interactions between the envelope proteins. In infected cells, membrane proteins—viral and cellular—are sampled for fit into the lattice formed by M molecules. The specificity of the molecular interactions acts as a quality control system to warrant the formation of the two-dimensional assemblies that contain the full complement of viral membrane proteins but from which cellular proteins are segregated. For each cellular protein the efficiency of this exclusion process is determined by its lack of interaction with the M protein, its lack of fit in the M protein framework, and its success in competing with the S and HE oligomers for the (geometrically different) vacancies within this framework.

### 3. *Localization of Budding*

The precise location of coronavirus budding and the factors that govern it have not been established. Although it is clear that particle formation occurs at membranes early in the secretory pathway, up to the *cis*-Golgi compartment, the precise site has not been identified for any coronavirus. Several considerations may explain this lack of knowledge. One is that these early compartments are themselves rather complex and highly dynamic and have hence been difficult to define structurally. Another is the possible alteration of the structural integrity of these compartments by infection; studies of these effects have not been described. A third complication may be that coronaviruses do not behave uniformly, different viruses possibly preferring different membranes for budding. In this respect it may be of note that differences have been observed, for instance, in the intrinsic localization of the M proteins from IBV and MHV; they appeared to accumulate on the *cis* and *trans* side of the Golgi apparatus, respectively (Klumperman *et al.*, 1994; Machamer *et al.*, 1990).

It has long been assumed that the M protein determines the site of coronavirus budding. When, however, this protein appeared to localize beyond this site, the idea became attractive that the association of the envelope proteins may create the novel targeting signals that direct these multimeric complexes to the budding site. In support of such a notion is the fact that the S and HE proteins, when coexpressed with the M protein, are retained in the Golgi apparatus rather than being transported to the plasma membrane. The critical question now is whether and how the E protein affects the localization of the M protein. As the E protein by itself does not seem to localize to the virion budding site it will be of great interest to determine the membranes at which VLPs assemble.

As the envelope proteins can direct particle formation by themselves, it may seem that the nucleocapsid is not leading the assembly process. Still, besides giving rise to virions rather than VLPs, its involvement in assembly might have important consequences. First of all, nucleocapsids may enhance the efficiency of the budding process. The physical yields of VLPs obtained by coexpression of the envelope proteins in cells are generally poor. Although there may be many reasons for this, in infected cells the availability of nucleocapsids is likely to facilitate particle production. Empty particles, considered to be VLPs, have nevertheless been observed during natural infection (Afzelius, 1994; Macnaughton and Davies, 1980). Their formation might simply serve as a means to dispose of excess viral membrane proteins from infected cells if required.

Another effect of the nucleocapsid could involve the localization of budding. It is conceivable that, unless a defined budding station is created by a specific interplay between viral and host proteins (for which no indications yet exist), preassembled nucleocapsids dock at those intracellular membrane sites where sufficiently sized patches of M protein-based envelope structure have accumulated. Early in infection such patches might start to form only after the envelope proteins have left the endoplasmic reticulum and become concentrated in intermediate membranes on their way to the Golgi complex. Later, when viral protein synthesis increases, this density might be reached earlier, perhaps explaining the observed late budding in the endoplasmic reticulum (Goldsmith *et al.*, 2004; Klumperman *et al.*, 1994; Tooze *et al.*, 1984). It will again be interesting to learn where VLPs independently bud and how this relates to the local density of the envelope proteins because this information will shed light on the role of the nucleocapsid in the localization of virion assembly.

## V. Perspectives

The picture of coronavirus assembly that the available literature allows us to draw in this review is still a rough draft. We know the identity and some characteristics of the key elements of the picture, we know the relative positions and orientations of most of them, but we are unable to fit them all into a sensible composition.

Although this may seem like a discontented retrospective, it certainly is not. In the 25 years that the senior author has been in coronavirology research, enormous progress has been made in practically all its aspects including virion assembly. However, the rewarding act of compiling and ordering the available information and trying to abstract from it actual knowledge was at the same time a sharp and recurrent confrontation with the unknown. We want to conclude this work by summarizing what in our opinion will be the main issues for the near future.

With the obstacle of reverse genetics technology solved, "structure" will be the dominating issue of the next decade. Biology has taught that molecular insight into processes will eventually depend critically on detailed structural information. For coronavirus assembly this means data on the individual structural components and, particularly, on the virion. With respect to the former, this will be most challenging for the membrane proteins, notably M and E. Virions, by their apparent elasticity, have eluded structural analysis. Here, despite the still limited resolution to be expected, cryoelectron microscopy should provide the urgently required insight into the structural organization of the particle.

Another issue will be the cell biology of assembly. This actually refers to a number of poignant problems at every stage of the process. Starting with nucleocapsid formation we must admit that we know practically nothing. By which interactions and where on the genome the packaging is initiated, how the wrapping of the RNA proceeds, how the condensation of the ribonucleoprotein structure takes place, and where in the cell these activities take place are all unresolved questions. Although we seem to know more about the budding process, several fundamental issues are still unresolved. Obvious issues are the site of budding and the determinants of its location, and the inclusion of the nucleocapsid into the budding particle. An intriguing issue is the budding mechanism itself: how is membrane curvature generated and, particularly, how is the directionality determined. Coronaviruses, like other intracellularly budding viruses, direct their particles out of the cytoplasm into the organelles, that is, opposite to the natural direction of cellular vesicle budding. Once again, simply nothing is known about the governing principles.

A third field of research that has yet to open is the contribution of host cellular factors to the assembly process. Work has so far been concentrating on the viral components and their interplay. Although there have been incidental indications, studies on the specific involvement of host proteins apparently had to await the development of appropriate technologies and these are now becoming available.

Although the serious health threat caused by the 2002–2003 epidemic of SARS apparently has waned, the coronavirological community has welcomed the consequent increased interest in this family of viruses. The boost that the research in this field has since been experiencing warrants an exciting future and accelerated progress with the elucidation of the fascinating process of coronavirus assembly.

## Acknowledgments

We thank Henk Halsema for making most of the drawings. We are grateful to Cristina Risco for providing the electron micrographs demonstrating the structural maturation of virions; to Raoul de Groot, Berend Jan Bosch, and Jean Lepault for their combined efforts in providing the electron micrographs of the purified virions; and to Raoul de Groot for critical reading of the manuscript.

## References

Afzelius, B. A. (1994). Ultrastructure of human nasal epithelium during an episode of coronavirus infection. *Virchows Arch.* **424:**295–300.

Allison, S. L., Stadler, K., Mandl, C. W., Kunz, C., and Heinz, F. X. (1995). Synthesis and secretion of recombinant tick-borne encephalitis virus protein E in soluble and particulate form. *J. Virol.* **69:**5816–5820.

An, S., Chen, C. J., Yu, X., Leibowitz, J. L., and Makino, S. (1999). Induction of apoptosis in murine coronavirus-infected cultured cells and demonstration of E protein as an apoptosis inducer. *J. Virol.* **73:**7853–7859.

Anderson, R., and Wong, F. (1993). Membrane and phospholipid binding by murine coronaviral nucleocapsid N protein. *Virology* **194:**224–232.

Armstrong, J., and Patel, S. (1991). The Golgi sorting domain of coronavirus E1 protein. *J. Cell Sci.* **98:**567–575.

Armstrong, J., Niemann, H., Smeekens, S., Rottier, P., and Warren, G. (1984). Sequence and topology of a model intracellular membrane protein, E1 glycoprotein, from a coronavirus. *Nature* **308:**751–752.

Armstrong, J., Patel, S., and Riddle, P. (1990). Lysosomal sorting mutants of coronavirus E1 protein, a Golgi membrane protein. *J. Cell Sci.* **95:**191–197.

Babcock, G. J., Esshaki, D. J., Thomas, W. D., and Ambrosino, D. M. (2004). Amino acids 270 to 510 of the severe acute respiratory syndrome coronavirus spike protein are required for interaction with receptor. *J. Virol.* **78:**4552–4560.

Baric, R. S., Nelson, G. W., Fleming, J. O., Deans, R. J., Keck, J. G., Casteel, N., and Stohlman, S. A. (1988). Interactions between coronavirus nucleocapsid protein and viral RNAs: Implications for viral transcription. *J. Virol.* **62:**4280–4287.

Baudoux, P., Besnardeau, L., Carrat, C., Rottier, P., Charley, B., and Laude, H. (1998a). Interferon $\alpha$ inducing property of coronavirus particles and pseudoparticles. *Adv. Exp. Med. Biol.* **440:**377–386.

Baudoux, P., Carrat, C., Besnardeau, L., Charley, B., and Laude, H. (1998b). Coronavirus pseudoparticles formed with recombinant M and E proteins induce $\alpha$ interferon synthesis by leukocytes. *J. Virol.* **72:**8636–8643.

Becker, W. B., McIntosh, K., Dees, J. H., and Chanock, R. M. (1967). Morphogenesis of avian infectious bronchitis virus and a related human virus (strain 229E). *J. Virol.* **1:**1019–1027.

Bisht, H., Roberts, A., Vogel, L., Bukreyev, A., Collins, P. L., Murphy, B. R., Subbarao, K., and Moss, B. (2004). Severe acute respiratory syndrome coronavirus spike protein expressed by attenuated vaccinia virus protectively immunizes mice. *Proc. Natl. Acad. Sci. USA* **101:**6641–6646.

Bonavia, A., Zelus, B. D., Wentworth, D. E., Talbot, P. J., and Holmes, K. V. (2003). Identification of a receptor-binding domain of the spike glycoprotein of human coronavirus HCoV-229E. *J. Virol.* **77:**2530–2538.

Bos, E. C., Dobbe, J. C., Luytjes, W., and Spaan, W. J. (1997). A subgenomic mRNA transcript of the coronavirus mouse hepatitis virus strain A59 defective interfering (DI) RNA is packaged when it contains the DI packaging signal. *J. Virol.* **71:**5684–5687.

Bos, E. C., Heijnen, L., Luytjes, W., and Spaan, W. J. (1995). Mutational analysis of the murine coronavirus spike protein: Effect on cell-to-cell fusion. *Virology* **214:**453–463.

Bos, E. C., Luytjes, W., van der Meulen, H. V., Koerten, H. K., and Spaan, W. J. (1996). The production of recombinant infectious DI-particles of a murine coronavirus in the absence of helper virus. *Virology* **218:**52–60.

Bosch, B. J., de Haan, C. A. M., and Rottier, P. J. M. (2004a). The coronavirus spike glycoprotein, extended carboxy terminally with the green fluorescent protein, is assembly competent. *J. Virol.* **78:**7369–7378.

Bosch, B. J., Martina, B. E., Van Der Zee, R., Lepault, J., Haijema, B. J., Versluis, C., Heck, A. J., De Groot, R., Osterhaus, A. D., and Rottier, P. J. (2004b). Severe acute respiratory syndrome coronavirus (SARS-CoV) infection inhibition using spike protein heptad repeat-derived peptides. *Proc. Natl. Acad. Sci. USA* **101:**8455–8460.

Bosch, B. J., van der Zee, R., de Haan, C. A., and Rottier, P. J. (2003). The coronavirus spike protein is a class I virus fusion protein: Structural and functional characterization of the fusion core complex. *J. Virol.* **77:**8801–8811.

Bost, A. G., Carnahan, R. H., Lu, X. T., and Denison, M. R. (2000). Four proteins processed from the replicase gene polyprotein of mouse hepatitis virus colocalize in the cell periphery and adjacent to sites of virion assembly. *J. Virol.* **74:**3379–3387.

Bost, A. G., Prentice, E., and Denison, M. R. (2001). Mouse hepatitis virus replicase protein complexes are translocated to sites of M protein accumulation in the ERGIC at late times of infection. *Virology* **285:**21–29.

Brayton, P. R., Ganges, R. G., and Stohlman, S. A. (1981). Host cell nuclear function and murine hepatitis virus replication. *J. Gen. Virol.* **56:**457–460.

Breslin, J. J., Mork, I., Smith, M. K., Vogel, L. K., Hemmila, E. M., Bonavia, A., Talbot, P. J., Sjostrom, H., Noren, O., and Holmes, K. V. (2003). Human coronavirus 229E: Receptor binding domain and neutralization by soluble receptor at 37 °C. *J. Virol.* **77:**4435–4438.

Brian, D. A., Hague, B. G., and Kienzle, T. E. (1995). The coronavirus hemagglutinin esterase glycoprotein. *In* "The *Coronaviridae*" (S. G. Siddell, ed.), pp. 165–179. Plenum Press, New York.

Bridger, J. C., Caul, E. O., and Egglestone, S. I. (1978). Replication of an enteric bovine coronavirus in intestinal organ cultures. *Arch. Virol.* **57:**43–51.

Brown, T. D. K., and Brierley, I. (1995). The coronavirus nonstructural proteins. *In* "The *Coronaviridae*" (S. G. Siddell, ed.), pp. 191–217. Plenum Press, New York.

Caul, E. O., Ashley, C. R., Ferguson, N. M., and Egglestone, S. I. (1979). Preliminary studies on the isolation of coronavirus 229E nucleocapsids. *FEMS Microbiol. Lett.* **48:**193.

Caul, E. O., and Egglestone, S. I. (1977). Further studies on human enteric coronaviruses. *Arch. Virol.* **54:**107–117.

Cavanagh, D. (1983a). Coronavirus IBV: Further evidence that the surface projections are associated with two glycopolypeptides. *J. Gen. Virol.* **64:**1787–1791.

Cavanagh, D. (1983b). Coronavirus IBV: Structural characterization of the spike protein. *J. Gen. Virol.* **64:**2577–2583.

Cavanagh, D. (1995). The coronavirus surface glycoprotein. *In* "The *Coronaviridae*" (S. G. Siddell, ed.), pp. 73–113. Plenum Press, New York.

Cavanagh, D., Davis, P. J., and Pappin, D. J. (1986). Coronavirus IBV glycopolypeptides: Locational studies using proteases and saponin, a membrane permeablizer. *Virus Res.* **4:**145–156.

Chambers, P., Pringle, C. R., and Easton, A. J. (1990). Heptad repeat sequences are located adjacent to hydrophobic regions in several types of virus fusion glycoproteins. *J. Gen. Virol.* **71:**3075–3080.

Chang, K. W., and Gombold, J. L. (2001). Effects of amino acid insertions in the cysteine-rich domain of the MHV-A59 spike protein on cell fusion. *Adv. Exp. Med. Biol.* **494:**205–211.

Chang, K. W., Sheng, Y., and Gombold, J. L. (2000). Coronavirus-induced membrane fusion requires the cysteine-rich domain in the spike protein. *Virology* **269:**212–224.

Chang, R. Y., and Brian, D. A. (1996). *Cis* requirement for N-specific protein sequence in bovine coronavirus defective interfering RNA replication. *J. Virol.* **70:**2201–2207.

Charley, B., and Laude, H. (1988). Induction of α interferon by transmissible gastroenteritis coronavirus: Role of transmembrane glycoprotein E1. *J. Virol.* **62:**8–11.

Chasey, D., and Alexander, D. J. (1976). Morphogenesis of avian infectious bronchitis virus in primary chick kidney cells. *Arch. Virol.* **52:**101–111.

Chen, H., Wurm, T., Britton, P., Brooks, G., and Hiscox, J. A. (2002). Interaction of the coronavirus nucleoprotein with nucleolar antigens and the host cell. *J. Virol.* **76:**5233–5250.

Choi, K. S., Huang, P., and Lai, M. M. (2002). Polypyrimidine-tract-binding protein affects transcription but not translation of mouse hepatitis virus RNA. *Virology* **303:**58–68.

Cologna, R., and Hogue, B. G. (2000). Identification of a bovine coronavirus packaging signal. *J. Virol.* **74:**580–583.

Cologna, R., Spagnolo, J. F., and Hogue, B. G. (2000). Identification of nucleocapsid binding sites within coronavirus-defective genomes. *Virology* **277:**235–249.

Compton, S. R., Rogers, D. B., Holmes, K. V., Fertsch, D., Remenick, J., and McGowan, J. J. (1987). *In vitro* replication of mouse hepatitis virus strain A59. *J. Virol.* **61:**1814–1820.

Cornelissen, L. A. H. M., Wierda, C. M. H., van der Meer, F. J., Herrewegh, A. A. P. M., Horzinek, M. C., Egberink, H. F., and de Groot, R. J. (1997). Hemagglutinin-esterase, a novel structural protein of torovirus. *J. Virol.* **71:**5277–5286.

Corse, E., and Machamer, C. E. (2000). Infectious bronchitis virus E protein is targeted to the Golgi complex and directs release of virus-like particles. *J. Virol.* **74:**4319–4326.

Corse, E., and Machamer, C. E. (2002). The cytoplasmic tail of infectious bronchitis virus E protein directs Golgi targeting. *J. Virol.* **76:**1273–1284.

Corse, E., and Machamer, C. E. (2003). The cytoplasmic tails of infectious bronchitis virus E and M proteins mediate their interaction. *Virology* **312:**25–34.

Curtis, K. M., Yount, B., and Baric, R. S. (2002). Heterologous gene expression from transmissible gastroenteritis virus replicon particles. *J. Virol.* **76:**1422–1434.

Dalton, K., Casais, R., Shaw, K., Stirrups, K., Evans, S., Britton, P., Brown, T. D., and Cavanagh, D. (2001). *Cis*-acting sequences required for coronavirus infectious bronchitis virus defective-RNA replication and packaging. *J. Virol.* **75:**125–133.

David-Ferreira, J. F., and Manaker, R. A. (1965). An electron microscopic study of the development of a mouse hepatitis virus in tissue culture cells. *J. Cell Biol.* **24:**57–78.

Davies, H. A., Dourmashkin, R. R., and Macnaughton, M. R. (1981). Ribonucleoprotein of avian infectious bronchitis virus. *J. Gen. Virol.* **53:**67–74.

de Groot, R. J., Lenstra, J. A., Luytjes, W., Niesters, H. G., Horzinek, M. C., van der Zeijst, B. A., and Spaan, W. J. (1987a). Sequence and structure of the coronavirus peplomer protein. *Adv. Exp. Med. Biol.* **218:**31–38.

de Groot, R. J., Luytjes, W., Horzinek, M. C., van der Zeist, B. A., Spaan, W. J., and Lenstra, J. A. (1987b). Evidence for a coiled-coil structure in the spike proteins of coronaviruses. *J. Mol. Biol.* **196:**963–966.

de Groot, R. J., van Leen, R. W., Dalderup, M. J., Vennema, H., Horzinek, M. C., and Spaan, W. J. (1989). Stably expressed FIPV peplomer protein induces cell fusion and elicits antibodies in mice. *Virology* **171:**493.

de Haan, C. A., de Wit, M., Kuo, L., Montalto-Morrison, C., Haagmans, B. L., Weiss, S. R., Masters, P. S., and Rottier, P. J. (2003). The glycosylation status of the murine hepatitis coronavirus M protein affects the interferogenic capacity of the virus *in vitro* and its ability to replicate in the liver but not the brain. *Virology* **312:**395–406.

de Haan, C. A., Kuo, L., Masters, P. S., Vennema, H., and Rottier, P. J. (1998a). Coronavirus particle assembly: Primary structure requirements of the membrane protein. *J. Virol.* **72:**6838–6850.

de Haan, C. A., Masters, P. S., Shen, X., Weiss, S., and Rottier, P. J. (2002b). The group-specific murine coronavirus genes are not essential, but their deletion, by reverse genetics, is attenuating in the natural host. *Virology* **296:**177–189.

de Haan, C. A., Roestenberg, P., de Wit, M., de Vries, A. A., Nilsson, T., Vennema, H., and Rottier, P. J. (1998b). Structural requirements for O-glycosylation of the mouse hepatitis virus membrane protein. *J. Biol. Chem.* **273:**29905–29914.

de Haan, C. A., Smeets, M., Vernooij, F., Vennema, H., and Rottier, P. J. (1999). Mapping of the coronavirus membrane protein domains involved in interaction with the spike protein. *J. Virol.* **73:**7441–7452.

de Haan, C. A., Vennema, H., and Rottier, P. J. (2000). Assembly of the coronavirus envelope: Homotypic interactions between the M proteins. *J. Virol.* **74:**4967–4978.

de Haan, C. A., de Wit, M., Kuo, L., Montalto, C., Masters, P. S., Weiss, S. R., and Rottier, P. J. (2002a). O-glycosylation of the mouse hepatitis coronavirus membrane protein. *Virus Res.* **82:**77–81.

de Haan, C. A. M., Stadler, K., Godeke, G.-J., Bosch, B. J., and Rottier, P. J. M. (2004). Cleavage inhibition of the murine coronavirus spike protein by a furin-like enzyme affects cell–cell but not virus–cell fusion. *J. Virol.* **78:**6048–6054.

de Vries, A. A., Horzinek, M. C., Rottier, P. J., and de Groot, R. J. (1997). The genome organization of the nidovirales: Similarities and differences between arteri-, toro-, and coronaviruses. *Semin. Virol.* **8:**33–47.

Delmas, B., Gelfi, J., L'Haridon, R., Vogel, L. K., Sjostrom, H., Noren, O., and Laude, H. (1992). Aminopeptidase N is a major receptor for the entero-pathogenic coronavirus TGEV. *Nature* **357:**417–420.

Delmas, B., Gelfi, J., Sjostrom, H., Noren, O., and Laude, H. (1993). Further characterization of aminopeptidase-N as a receptor for coronaviruses. *Adv. Exp. Med. Biol.* **342:**293–298.

Delmas, B., and Laude, H. (1990). Assembly of coronavirus spike protein into trimers and its role in epitope expression. *J. Virol.* **64:**5367–5375.

Denison, M. R., Spaan, W. J., van der Meer, Y., Gibson, C. A., Sims, A. C., Prentice, E., and Lu, X. T. (1999). The putative helicase of the coronavirus mouse hepatitis virus is processed from the replicase gene polyprotein and localizes in complexes that are active in viral RNA synthesis. *J. Virol.* **73:**6862–6871.

Deregt, D., and Babiuk, L. A. (1987). Monoclonal antibodies to bovine coronavirus: Characteristics and topographical mapping of neutralizing epitopes on the E2 and E3 glycoproteins. *Virology* **161:**410–420.

Doms, R., Lamb, R. A., Rose, J. K., and Helenius, A. (1993). Folding and assembly of viral membrane proteins. *Virology* **193:**545–562.

Dubois-Dalcq, M. E., Doller, E. W., Haspel, M. V., and Holmes, K. V. (1982). Cell tropism and expression of mouse hepatitis viruses (MHV) in mouse spinal cord cultures. *Virology* **119:**317–331.

Dveksler, G. S., Dieffenbach, C. W., Cardellichio, C. B., McCuaig, K., Pensiero, M. N., Jiang, G. S., Beauchemin, N., and Holmes, K. V. (1993). Several members of the mouse carcinoembryonic antigen-related glycoprotein family are functional receptors for the coronavirus mouse hepatitis virus-A59. *J. Virol.* **67:**1–8.

Dveksler, G. S., Pensiero, M. N., Cardellichio, C. B., Williams, R. K., Jiang, G. S., Holmes, K. V., and Dieffenbach, C. W. (1991). Cloning of the mouse hepatitis virus (MHV) receptor: Expression in human and hamster cell lines confers susceptibility to MHV. *J. Virol.* **65:**6881–6891.

Eleouet, J. F., Slee, E. A., Saurini, F., Castagne, N., Poncet, D., Garrido, C., Solary, E., and Martin, S. J. (2000). The viral nucleocapsid protein of transmissible gastroenteritis coronavirus (TGEV) is cleaved by caspase-6 and -7 during TGEV-induced apoptosis. *J. Virol.* **74:**3975–3983.

Enjuanes, L., Sola, I., Almazan, F., Ortego, J., Izeta, A., Gonzalez, J. M., Alonso, S., Sanchez, J. M., Escors, D., Calvo, E., Riquelme, C., and Sanchez, C. (2001). Coronavirus derived expression systems. *J. Biotechnol.* **88:**183–204.

Escors, D., Camafeita, E., Ortego, J., Laude, H., and Enjuanes, L. (2001a). Organization of two transmissible gastroenteritis coronavirus membrane protein topologies within the virion and core. *J. Virol.* **75:**12228–12240.

Escors, D., Izeta, A., Capiscol, C., and Enjuanes, L. (2003). Transmissible gastroenteritis coronavirus packaging signal is located at the 5′ end of the virus genome. *J. Virol.* **77:**7890–7902.

Escors, D., Ortego, J., Laude, H., and Enjuanes, L. (2001b). The membrane M protein carboxy terminus binds to transmissible gastroenteritis coronavirus core and contributes to core stability. *J. Virol.* **75:**1312–1324.

Fischer, F., Peng, D., Hingley, S. T., Weiss, S. R., and Masters, P. S. (1997). The internal open reading frame within the nucleocapsid gene of mouse hepatitis virus encodes a structural protein that is not essential for viral replication. *J. Virol.* **71:**996–1003.

Fischer, F., Stegen, C. F., Masters, P. S., and Samsonoff, W. A. (1998). Analysis of constructed E gene mutants of mouse hepatitis virus confirms a pivotal role for E protein in coronavirus assembly. *J. Virol.* **72:**7885–7894.

Fosmire, J. A., Hwang, K., and Makino, S. (1992). Identification and characterization of a coronavirus packaging signal. *J. Virol.* **66:**3522–3530.

Frana, M. F., Behnke, J. N., Sturman, L. S., and Holmes, K. V. (1985). Proteolytic cleavage of the E2 glycoprotein of murine coronavirus: Host-dependent differences in proteolytic cleavage and cell fusion. *J. Virol.* **56:**912–920.

Gagneten, S., Gout, O., Dubois-Dalcq, M., Rottier, P., Rossen, J., and Holmes, K. V. (1995). Interaction of mouse hepatitis virus (MHV) spike glycoprotein with receptor glycoprotein MHVR is required for infection with an MHV strain that expresses the hemagglutinin-esterase glycoprotein. *J. Virol.* **69:**889–895.

Gallagher, T. M. (1997). A role for naturally occurring variation of the murine coronavirus spike protein in stabilizing association with the cellular receptor. *J. Virol.* **71:**3129–3137.

Gallagher, T. M., and Buchmeier, M. J. (2001). Coronavirus spike proteins in viral entry and pathogenesis. *Virology* **279:**371–374.

Gallagher, T. M., Buchmeier, M. J., and Perlman, S. (1992). Cell receptor-independent infection by a neurotropic murine coronavirus. *Virology* **191:**517–522.

Gallagher, T. M., Escarmis, C., and Buchmeier, M. J. (1991). Alteration of the pH dependence of coronavirus-induced cell fusion: Effect of mutations in the spike glycoprotein. *J. Virol.* **65:**1916–1928.

Garoff, H., Hewson, R., and Opstelten, D. J. (1998). Virus maturation by budding. *Microbiol. Mol. Biol. Rev.* **62:**1171–1190.

Garwes, D. J., Bountiff, L., Millson, G. C., and Elleman, C. J. (1984). Defective replication of porcine transmissible gastroenteritis virus in a continuous cell line. *Adv. Exp. Med. Biol.* **173:**79–93.

Garwes, D. J., Pocock, D. H., and Pike, B. V. (1976). Isolation of subviral components from transmissible gastroenteritis virus. *J. Gen. Virol.* **32:**283–294.

Godeke, G. J., de Haan, C. A., Rossen, J. W., Vennema, H., and Rottier, P. J. (2000). Assembly of spikes into coronavirus particles is mediated by the carboxy-terminal domain of the spike protein. *J. Virol.* **74:**1566–1571.

Godet, M., Grosclaude, J., Delmas, B., and Laude, H. (1994). Major receptor-binding and neutralization determinants are located within the same domain of the transmissible gastroenteritis virus (coronavirus) spike protein. *J. Virol.* **68:**8008–8016.

Godet, M., L'Haridon, R., Vautherot, J. F., and Laude, H. (1992). TGEV corona virus ORF4 encodes a membrane protein that is incorporated into virions. *Virology* **188:**666–675.

Goldsmith, C., Tatti, K., Ksiazek, T., Rollin, P., Comer, J., Lee, W., Rota, P., Bankamp, B., Bellini, W., and Zaki, S. (2004). Ultrastructural characterization of SARS coronavirus. *Emerg. Infect. Dis.* **10:**320–326.

Gonzalez, M. E., and Carrasco, L. (2003). Viroporins. *FEBS Lett.* **552:**28–34.

Gosert, R., Kanjanahaluethai, A., Egger, D., Bienz, K., and Baker, S. C. (2002). RNA replication of mouse hepatitis virus takes place at double-membrane vesicles. *J. Virol.* **76:**3697–3708.

Haijema, B. J., Volders, H., and Rottier, P. J. (2003). Switching species tropism: An effective way to manipulate the feline coronavirus genome. *J. Virol.* **77:**4528–4538.

Haijema, B. J., Volders, H., and Rottier, P. J. (2004). Live, attenuated coronavirus vaccines through the directed deletion of group-specific genes provide protection against feline infectious peritonitis. *J. Virol.* **78:**3863–3871.

Hamre, D., Kindig, D. A., and Mann, J. (1967). Growth and intracellular development of a new respiratory virus. *J. Virol.* **1:**810–816.

He, R., Dobie, F., Ballantine, M., Leeson, A., Li, Y., Bastien, N., Cutts, T., Andonov, A., Cao, J., Booth, T. F., Plummer, F. A., Tyler, S., Baker, L., and Li, X. (2004). Analysis of multimerization of the SARS coronavirus nucleocapsid protein. *Biochem. Biophys. Res. Commun.* **316:**476–483.

Herrewegh, A. A., Vennema, H., Horzinek, M. C., Rottier, P. J., and de Groot, R. J. (1995). The molecular genetics of feline coronaviruses: Comparative sequence analysis of the ORF7a/7b transcription unit of different biotypes. *Virology* **212:**622–631.

Hiscox, J. A., Wurm, T., Wilson, L., Britton, P., Cavanagh, D., and Brooks, G. (2001). The coronavirus infectious bronchitis virus nucleoprotein localizes to the nucleolus. *J. Virol.* **75:**506–512.

Hofmann, M. A., Sethna, P. B., and Brian, D. A. (1990). Bovine coronavirus mRNA replication continues throughout persistent infection in cell culture. *J. Virol.* **64:**4108–4114.

Hogue, B. G., and Brian, D. A. (1986). Structural proteins of human respiratory coronavirus OC43. *Virus Res.* **5:**131–144.

Hogue, B. G., Kienzle, T. E., and Brian, D. A. (1989). Synthesis and processing of the bovine enteric coronavirus haemagglutinin protein. *J. Gen. Virol.* **70:**345–352.

Hogue, B. G., King, B., and Brian, D. A. (1984). Antigenic relationships among proteins of bovine coronavirus, human respiratory coronavirus OC43, and mouse hepatitis coronavirus A59. *J. Virol.* **51:**384–388.

Holmes, K. V. (2001). Enteric infections with coronaviruses and toroviruses. *Novartis Found. Symp.* **238:**258–269; discussion 269–275.

Holmes, K. V., and Behnke, J. N. (1981). Evolution of a coronavirus during persistent infection *in vitro*. *Adv. Exp. Med. Biol.* **142:**287–299.

Holmes, K. V., Doller, E. W., and Sturman, L. S. (1981). Tunicamycin resistant glycosylation of coronavirus glycoprotein: Demonstration of a novel type of viral glycoprotein. *Virology* **115:**334–344.

Holmes, K. V., Zelus, B. D., Schickli, J. H., and Weiss, S. R. (2001). Receptor specificity and receptor-induced conformational changes in mouse hepatitis virus spike glycoprotein. *Adv. Exp. Med. Biol.* **494:**173–181.

Huang, Q., Yu, L., Petros, A. M., Gunasekera, A., Liu, Z., Xu, N., Hajduk, P., Mack, J., Fesik, S. W., and Olejniczak, E. T. (2004). Structure of the N-terminal RNA-binding domain of the SARS CoV nucleocapsid protein. *Biochemistry* **43:**6059–6063.

Ingallinella, P., Bianchi, E., Finotto, M., Cantoni, G., Eckert, D. M., Supekar, V. M., Bruckmann, C., Carfi, A., and Pessi, A. (2004). Structural characterization of the fusion-active complex of severe acute respiratory syndrome (SARS) coronavirus. *Proc. Natl. Acad. Sci. USA* **101:**8709–8714.

Izeta, A., Smerdou, C., Alonso, S., Penzes, Z., Mendez, A., Plana-Duran, J., and Enjuanes, L. (1999). Replication and packaging of transmissible gastroenteritis coronavirus-derived synthetic minigenomes. *J. Virol.* **73:**1535–1545.

Kalicharran, K., Mohandas, D., Wilson, G., and Dales, S. (1996). Regulation of the initiation of coronavirus JHM infection in primary oligodendrocytes and L-2 fibroblasts. *Virology* **225:**33–43.

Kennedy, D. A., and Johnson-Lussenburg, C. M. (1975). Isolation and morphology of the internal component of human coronavirus, strain 229E. *InterVirology* **6:**197–206.

Kennedy, M., Boedeker, N., Gibbs, P., and Kania, S. (2001). Deletions in the 7a ORF of feline coronavirus associated with an epidemic of feline infectious peritonitis. *Vet. Microbiol.* **81:**227–234.

Kienzle, T. E., Abraham, S., Hogue, B. G., and Brian, D. A. (1990). Structure and orientation of expressed bovine coronavirus hemagglutinin-esterase protein. *J. Virol.* **64:**1834–1838.

Kim, K. H., Narayanan, K., and Makino, S. (1997). Assembled coronavirus from complementation of two defective interfering RNAs. *J. Virol.* **71:**3922–3931.

King, B., Potts, B. J., and Brian, D. A. (1985). Bovine coronavirus hemagglutinin protein. *Virus Res.* **2:**53–59.

Klausegger, A., Strobl, B., Regl, G., Kaser, A., Luytjes, W., and Vlasak, R. (1999). Identification of a coronavirus hemagglutinin-esterase with a substrate specificity different from those of influenza C virus and bovine coronavirus. *J. Virol.* **73:**3737–3743.

Klumperman, J., Locker, J. K., Meijer, A., Horzinek, M. C., Geuze, H. J., and Rottier, P. J. (1994). Coronavirus M proteins accumulate in the Golgi complex beyond the site of virion budding. *J. Virol.* **68:**6523–6534.

Koetzner, C. A., Parker, M. M., Ricard, C. S., Sturman, L. S., and Masters, P. S. (1992). Repair and mutagenesis of the genome of a deletion mutant of the coronavirus mouse hepatitis virus by targeted RNA recombination. *J. Virol.* **66:**1841–1848.

Krempl, C., Schultze, B., and Herrler, G. (1995). Analysis of cellular receptors for human coronavirus OC43. *Adv. Exp. Med. Biol.* **380:**371–374.

Krueger, D. K., Kelly, S. M., Lewicki, D. N., Ruffolo, R., and Gallagher, T. M. (2001). Variations in disparate regions of the murine coronavirus spike protein impact the initiation of membrane fusion. *J. Virol.* **75:**2792–2802.

Kubo, H., Yamada, Y. K., and Taguchi, F. (1994). Localization of neutralizing epitopes and the receptor-binding site within the amino-terminal 330 amino acids of the murine coronavirus spike protein. *J. Virol.* **68:**5403–5410.

Kunkel, F., and Herrler, G. (1993). Structural and functional analysis of the surface protein of human coronavirus OC43. *Virology* **195:**195–202.

Kuo, L., and Masters, P. S. (2002). Genetic evidence for a structural interaction between the carboxy termini of the membrane and nucleocapsid proteins of mouse hepatitis virus. *J. Virol.* **76:**4987–4999.

Kuo, L., Godeke, G. J., Raamsman, M. J., Masters, P. S., and Rottier, P. J. (2000). Retargeting of coronavirus by substitution of the spike glycoprotein ectodomain: Crossing the host cell species barrier. *J. Virol.* **74:**1393–1406.

Kuo, L., and Masters, P. S. (2003). The small envelope protein E is not essential for murine coronavirus replication. *J. Virol.* **77:**4597–4608.

Lai, M. M. (1997). RNA–protein interactions in the regulation of coronavirus RNA replication and transcription. *Biol. Chem.* **378:**477–481.

Lai, M. M., and Cavanagh, D. (1997). The molecular biology of coronaviruses. *Adv. Virus Res.* **48:**1–100.

Lai, M. M., Liao, C. L., Lin, Y. J., and Zhang, X. (1994). Coronavirus: How a large RNA viral genome is replicated and transcribed. *Infect. Agents Dis.* **3:**98–105.

Lancer, J. A., and Howard, C. R. (1980). The disruption of infectious bronchitis virus (IBV-41 strain) with Triton X-100 detergent. *J. Virol. Methods* **1:**121–131.

Laude, H., Gelfi, J., Lavenant, L., and Charley, B. (1992). Single amino acid changes in the viral glycoprotein M affect induction of α interferon by the coronavirus transmissible gastroenteritis virus. *J. Virol.* **66:**743–749.

Laude, H., and Masters, P. S. (1995). The coronavirus nucleocapsid protein. *In* "The *Coronaviridae*" (S. G. Siddell, ed.), pp. 141–163. Plenum Press, New York.

Lewicki, D. N., and Gallagher, T. M. (2002). Quaternary structure of coronavirus spikes in complex with carcinoembryonic antigen-related cell adhesion molecule cellular receptors. *J. Biol. Chem.* **277:**19727–19734.

Li, W., Moore, M. J., Vasilieva, N., Sui, J., Wong, S. K., Berne, M. A., Somasundaran, M., Sullivan, J. L., Luzuriaga, K., Greenough, T. C., Choe, H., and Farzan, M. (2003). Angiotensin-converting enzyme 2 is a functional receptor for the SARS coronavirus. *Nature* **426:**450–454.

Liao, C. L., Zhang, X., and Lai, M. M. (1995). Coronavirus defective-interfering RNA as an expression vector: The generation of a pseudorecombinant mouse hepatitis virus expressing hemagglutinin-esterase. *Virology* **208:**319–327.

Lim, K. P., and Liu, D. X. (2001). The missing link in coronavirus assembly. Retention of the avian coronavirus infectious bronchitis virus envelope protein in the pre-Golgi compartments and physical interaction between the envelope and membrane proteins. *J. Biol. Chem.* **276:**17515–17523.

Lin, Y., Yan, X., Cao, W., Wang, C., Feng, J., Duan, J., and Xie, S. (2004). Probing the structure of the SARS coronavirus using scanning electron microscopy. *Antiviral Ther.* **9:**287–289.

Liu, D. X., and Inglis, S. C. (1991). Association of the infectious bronchitis virus 3c protein with the virion envelope. *Virology* **185:**911–917.

Liu, S., Xiao, G., Chen, Y., He, Y., Niu, J., Escalante, C. R., Xiong, H., Farmar, J., Debnath, A. K., Tien, P., and Jiang, S. (2004). Interaction between heptad repeat 1 and 2 regions in spike protein of SARS-associated coronavirus: Implications for virus fusogenic mechanism and identification of fusion inhibitors. *Lancet* **363:**938–947.

Locker, J. K., Griffiths, G., Horzinek, M. C., and Rottier, P. J. (1992). O-glycosylation of the coronavirus M protein. Differential localization of sialyltransferases in N- and O-linked glycosylation. *J. Biol. Chem.* **267:**14094–14101.

Locker, J. K., Klumperman, J., Oorschot, V., Horzinek, M. C., Geuze, H. J., and Rottier, P. J. (1994). The cytoplasmic tail of mouse hepatitis virus M protein is essential but not sufficient for its retention in the Golgi complex. *J. Biol. Chem.* **269:**28263–28269.

Locker, J. K., Opstelten, D. J., Ericsson, M., Horzinek, M. C., and Rottier, P. J. (1995). Oligomerization of a *trans*-Golgi/*trans*-Golgi network retained protein occurs in the Golgi complex and may be part of its retention. *J. Biol. Chem.* **270:**8815–8821.

Lontok, E., Corse, E., and Machamer, C. E. (2004). Intracellular targeting signals contribute to localization of coronavirus spike proteins near the virus assembly site. *J. Virol.* **78:**5913–5922.

Luo, Z., Matthews, A. M., and Weiss, S. R. (1999). Amino acid substitutions within the leucine zipper domain of the murine coronavirus spike protein cause defects in oligomerization and the ability to induce cell-to-cell fusion. *J. Virol.* **73:**8152–8159.

Luo, Z., and Weiss, S. R. (1998). Roles in cell-to-cell fusion of two conserved hydrophobic regions in the murine coronavirus spike protein. *Virology* **244:**483–494.

Luytjes, W. (1995). Coronavirus gene expression. *In* "The *Coronaviridae*" (S. G. Siddell, ed.), pp. 33–54. Plenum Press, New York.

Luytjes, W., Bredenbeek, P. J., Noten, A. F., Horzinek, M. C., and Spaan, W. J. (1988). Sequence of mouse hepatitis virus A59 mRNA 2: Indications for RNA recombination between coronaviruses and influenza C virus. *Virology* **166:**415–422.

Luytjes, W., Gerritsma, H., Bos, E., and Spaan, W. (1997). Characterization of two temperature-sensitive mutants of coronavirus mouse hepatitis virus strain A59 with maturation defects in the spike protein. *J. Virol.* **71:**949–955.

Maceyka, M., and Machamer, C. E. (1997). Ceramide accumulation uncovers a cycling pathway for the *cis*-Golgi network marker, infectious bronchitis virus M protein. *J. Cell Biol.* **139:**1411–1418.

Machamer, C. E., Grim, M. G., Esquela, A., Chung, S. W., Rolls, M., Ryan, K., and Swift, A. M. (1993). Retention of a *cis* Golgi protein requires polar residues on one face of a predicted α-helix in the transmembrane domain. *Mol. Biol. Cell* **4:**695–704.

Machamer, C. E., Mentone, S. A., Rose, J. K., and Farquhar, M. G. (1990). The E1 glycoprotein of an avian coronavirus is targeted to the *cis* Golgi complex. *Proc. Natl. Acad. Sci. USA* **87:**6944–6948.

Machamer, C. E., and Rose, J. K. (1987). A specific transmembrane domain of a coronavirus E1 glycoprotein is required for its retention in the Golgi region. *J. Cell Biol.* **105:**1205–1214.

Macnaughton, M. R., and Davies, H. A. (1978). Ribonucleoprotein-like structures from coronavirus particles. *J. Gen. Virol.* **39:**545–549.

Macnaughton, M. R., and Davies, H. A. (1980). Two particle types of avian infectious bronchitis virus. *J. Gen. Virol.* **47:**365–372.

Maeda, J., Maeda, A., and Makino, S. (1999). Release of coronavirus E protein in membrane vesicles from virus-infected cells and E protein-expressing cells. *Virology* **263:**265–272.

Maeda, J., Repass, J. F., Maeda, A., and Makino, S. (2001). Membrane topology of coronavirus E protein. *Virology* **281:**163–169.

Mason, P. W., Pincus, S., Fournier, M. J., Mason, T. L., Shope, R. E., and Paoletti, E. (1991). Japanese encephalitis virus–vaccinia recombinants produce particulate forms of the structural membrane proteins and induce high levels of protection against lethal JEV infection. *Virology* **180:**294–305.

Massalski, A., Coulter-Mackie, M., Knobler, R. L., Buchmeier, M. J., and Dales, S. (1982). *In vivo* and *in vitro* models of demyelinating diseases. V. Comparison of the assembly of mouse hepatitis virus, strain JHM, in two murine cell lines. *InterVirology* **18:**135–146.

Masters, P. S. (1992). Localization of an RNA-binding domain in the nucleocapsid protein of the coronavirus mouse hepatitis virus. *Arch. Virol.* **125:**141–160.

Masters, P. S. (1999). Reverse genetics of the largest RNA viruses. *Adv. Virus Res.* **53:**245–264.

Matsuyama, S., and Taguchi, F. (2002). Receptor-induced conformational changes of murine coronavirus spike protein. *J. Virol.* **76:**11819–11826.

Mayer, T., Tamura, T., Falk, M., and Niemann, H. (1988). Membrane integration and intracellular transport of the coronavirus glycoprotein E1, a class III membrane glycoprotein. *J. Biol. Chem.* **263:**14956–14963.

Miura, H. S., Nakagaki, K., and Taguchi, F. (2004). N-terminal domain of the murine coronavirus receptor CEACAM1 is responsible for fusogenic activation and conformational changes of the spike protein. *J. Virol.* **78:**216–223.

Mohandas, D. V., and Dales, S. (1991). Endosomal association of a protein phosphatase with high dephosphorylating activity against a coronavirus nucleocapsid protein. *FEBS Lett.* **282:**419–424.

Molenkamp, R., and Spaan, W. J. (1997). Identification of a specific interaction between the coronavirus mouse hepatitis virus A59 nucleocapsid protein and packaging signal. *Virology* **239:**78–86.

Mounir, S., and Talbot, P. J. (1992). Sequence analysis of the membrane protein gene of human coronavirus OC43 and evidence for O-glycosylation. *J. Gen. Virol.* **73:**2731–2736.

Narayanan, K., Chen, C. J., Maeda, J., and Makino, S. (2003a). Nucleocapsid-independent specific viral RNA packaging via viral envelope protein and viral RNA signal. *J. Virol.* **77:**2922–2927.

Narayanan, K., Kim, K. H., and Makino, S. (2003b). Characterization of N protein self-association in coronavirus ribonucleoprotein complexes. *Virus Res.* **98:**131–140.

Narayanan, K., Maeda, A., Maeda, J., and Makino, S. (2000). Characterization of the coronavirus M protein and nucleocapsid interaction in infected cells. *J. Virol.* **74:**8127–8134.

Narayanan, K., and Makino, S. (2001). Cooperation of an RNA packaging signal and a viral envelope protein in coronavirus RNA packaging. *J. Virol.* **75:**9059–9067.

Nelson, G. W., and Stohlman, S. A. (1993). Localization of the RNA-binding domain of mouse hepatitis virus nucleocapsid protein. *J. Gen. Virol.* **74:**1975–1979.

Nelson, G. W., Stohlman, S. A., and Tahara, S. M. (2000). High affinity interaction between nucleocapsid protein and leader/intergenic sequence of mouse hepatitis virus RNA. *J. Gen. Virol.* **81:**181–188.

Nguyen, V. P., and Hogue, B. G. (1997). Protein interactions during coronavirus assembly. *J. Virol.* **71:**9278–9284.

Opstelten, D. J., de Groote, P., Horzinek, M. C., Vennema, H., and Rottier, P. J. (1993a). Disulfide bonds in folding and transport of mouse hepatitis coronavirus glycoproteins. *J. Virol.* **67:**7394–7401.

Opstelten, D. J., Horzinek, M. C., and Rottier, P. J. (1993b). Complex formation between the spike protein and the membrane protein during mouse hepatitis virus assembly. *Adv. Exp. Med. Biol.* **342:**189–195.

Opstelten, D. J., Raamsman, M. J., Wolfs, K., Horzinek, M. C., and Rottier, P. J. (1995). Envelope glycoprotein interactions in coronavirus assembly. *J. Cell Biol.* **131:**339–349.

Ortego, J., Escors, D., Laude, H., and Enjuanes, L. (2002). Generation of a replication-competent, propagation-deficient virus vector based on the transmissible gastroenteritis coronavirus genome. *J. Virol.* **76:**11518–11529.

Ortego, J., Sola, I., Almazan, F., Ceriani, J. E., Riquelme, C., Balasch, M., Plana, J., and Enjuanes, L. (2003). Transmissible gastroenteritis coronavirus gene 7 is not essential but influences *in vivo* virus replication and virulence. *Virology* **308:**13–22.

Oshiro, L. S., Schieble, J. H., and Lennette, E. H. (1971). Electron microscopic studies of coronavirus. *J. Gen. Virol.* **12:**161–168.

Parker, M. D., Cox, G. J., Deregt, D., Fitzpatrick, D. R., and Babiuk, L. A. (1989). Cloning and *in vitro* expression of the gene for the E3 haemagglutinin glycoprotein of bovine coronavirus. *J. Gen. Virol.* **70:**155–164.

Parker, M. M., and Masters, P. S. (1990). Sequence comparison of the N genes of five strains of the coronavirus mouse hepatitis virus suggests a three domain structure for the nucleocapsid protein. *Virology* **179:**463–468.
Patzer, E. J., Nakamura, G. R., Simonsen, C. C., Levinson, A. D., and Brands, R. (1986). Intracellular assembly and packaging of hepatitis B surface antigen particles occur in the endoplasmic reticulum. *J. Virol.* **58:**884–892.
Peng, D., Koetzner, C. A., and Masters, P. S. (1995a). Analysis of second-site revertants of a murine coronavirus nucleocapsid protein deletion mutant and construction of nucleocapsid protein mutants by targeted RNA recombination. *J. Virol.* **69:**3449–3457.
Peng, D., Koetzner, C. A., McMahon, T., Zhu, Y., and Masters, P. S. (1995b). Construction of murine coronavirus mutants containing interspecies chimeric nucleocapsid proteins. *J. Virol.* **69:**5475–5484.
Perlman, S. (1998). Pathogenesis of coronavirus-induced infections: Review of pathological and immunological aspects. *Adv. Exp. Med. Biol.* **440:**503–513.
Perlman, S., Ries, D., Bolger, E., Chang, L. J., and Stoltzfus, C. M. (1986). MHV nucleocapsid synthesis in the presence of cycloheximide and accumulation of negative strand MHV RNA. *Virus Res.* **6:**261–272.
Pfleiderer, M., Routledge, E., Herrler, G., and Siddell, S. G. (1991). High level transient expression of the murine coronavirus haemagglutinin-esterase. *J. Gen. Virol.* **72:**1309–1315.
Pfleiderer, M., Routledge, E., and Siddell, S. G. (1990). Functional analysis of the coronavirus MHV-JHM surface glycoproteins in vaccinia virus recombinants. *Adv. Exp. Med. Biol.* **276:**21–31.
Pocock, D. H., and Garwes, D. J. (1977). The polypeptides of haemagglutinating encephalomyelitis virus and isolated subviral particles. *J. Gen. Virol.* **37:**487–499.
Popova, R., and Zhang, X. (2002). The spike but not the hemagglutinin/esterase protein of bovine coronavirus is necessary and sufficient for viral infection. *Virology* **294:**222–236.
Prentice, E., Jerome, W. G., Yoshimori, T., Mizushima, N., and Denison, M. R. (2004). Coronavirus replication complex formation utilizes components of cellular autophagy. *J. Biol. Chem.* **279:**10136–10141.
Raamsman, M. J., Locker, J. K., de Hooge, A., de Vries, A. A., Griffiths, G., Vennema, H., and Rottier, P. J. (2000). Characterization of the coronavirus mouse hepatitis virus strain A59 small membrane protein E. *J. Virol.* **74:**2333–2342.
Regl, G., Kaser, A., Iwersen, M., Schmid, H., Kohla, G., Strobl, B., Vilas, U., Schauer, R., and Vlasak, R. (1999). The hemagglutinin-esterase of mouse hepatitis virus strain S is a sialate-4-*O*-acetylesterase. *J. Virol.* **73:**4721–4727.
Ricard, C. S., Koetzner, C. A., Sturman, L. S., and Masters, P. S. (1995). A conditional-lethal murine coronavirus mutant that fails to incorporate the spike glycoprotein into assembled virions. *Virus Res.* **39:**261–276.
Risco, C., Anton, I. M., Sune, C., Pedregosa, A. M., Martin-Alonso, J. M., Parra, F., Carrascosa, J. L., and Enjuanes, L. (1995). Membrane protein molecules of transmissible gastroenteritis coronavirus also expose the carboxy-terminal region on the external surface of the virion. *J. Virol.* **69:**5269–5277.
Risco, C., Anton, I. M., Enjuanes, L., and Carrascosa, J. L. (1996). The transmissible gastroenteritis coronavirus contains a spherical core shell consisting of M and N proteins. *J. Virol.* **70:**4773–4777.
Risco, C., Muntion, M., Enjuanes, L., and Carrascosa, J. L. (1998). Two types of virus-related particles are found during transmissible gastroenteritis virus morphogenesis. *J. Virol.* **72:**4022–4031.

Robb, J. A., and Bond, C. W. (1979). Pathogenic murine coronaviruses. I. Characterization of biological behavior *in vitro* and virus-specific intracellular RNA of strongly neurotropic JHMV and weakly neurotropic A59V viruses. *Virology* **94:**352–370.

Robbins, S. G., Frana, M. F., McGowan, J. J., Boyle, J. F., and Holmes, K. V. (1986). RNA-binding proteins of coronavirus MHV: Detection of monomeric and multimeric N protein with an RNA overlay-protein blot assay. *Virology* **150:**402–410.

Rossen, J. W., Horzinek, M. C., and Rottier, P. J. (1995). Coronavirus infection of polarized epithelial cells. *Trends Microbiol.* **3:**486–490.

Rottier, P. J. M. (1995). The coronavirus membrane glycoprotein. *In* "The *Coronaviridae*" (S. G. Siddell, ed.), pp. 115–139. Plenum Press, New York.

Rottier, P. J. M., Brandenburg, D., Armstrong, J., van der Zeijst, B., and Warren, G. (1984). Assembly *in vitro* of a spanning membrane protein of the endoplasmic reticulum: The E1 glycoprotein of coronavirus mouse hepatitis virus A59. *Proc. Natl. Acad. Sci. USA* **81:**1421–1425.

Rottier, P. J. M., Horzinek, M. C., and van der Zeijst, B. A. (1981a). Viral protein synthesis in mouse hepatitis virus strain A59-infected cells: Effect of tunicamycin. *J. Virol.* **40:**350–357.

Rottier, P. J. M., Locker, J. K., Horzinek, M. C., and Spaan, W. J. (1990). Expression of MHV-A59 M glycoprotein: Effects of deletions on membrane integration and intracellular transport. *Adv. Exp. Med. Biol.* **276:**127–135.

Rottier, P. J. M., and Rose, J. K. (1987). Coronavirus E1 glycoprotein expressed from cloned cDNA localizes in the Golgi region. *J. Virol.* **61:**2042–2045.

Rottier, P. J. M., Spaan, W. J., Horzinek, M. C., and van der Zeijst, B. A. (1981b). Translation of three mouse hepatitis virus strain A59 subgenomic RNAs in *Xenopus laevis* oocytes. *J. Virol.* **38:**20–26.

Rottier, P. J. M., Welling, G. W., Welling-Wester, S., Niesters, H. G., Lenstra, J. A., and Van der Zeijst, B. A. (1986). Predicted membrane topology of the coronavirus protein E1. *Biochemistry* **25:**1335–1339.

Saeki, K., Ohtsuka, N., and Taguchi, F. (1997). Identification of spike protein residues of murine coronavirus responsible for receptor-binding activity by use of soluble receptor-resistant mutants. *J. Virol.* **71:**9024–9031.

Salanueva, I. J., Carrasco, J. L., and Risco, C. (1999). Structural maturation of the transmissible gastroenteritis coronavirus. *J. Virol.* **73:**7952–7964.

Salzwedel, K., West, J. T., and Hunter, E. (1999). A conserved tryptophan-rich motif in the membrane-proximal region of the human immunodeficiency virus type 1 gp41 ectodomain is important for Env-mediated fusion and virus infectivity. *J. Virol.* **73:**2469–2480.

Sawicki, S. G., and Sawicki, D. L. (1998). A new model for coronavirus transcription. *Adv. Exp. Med. Biol.* **440:**215–219.

Schultze, B., Gross, H. J., Brossmer, R., and Herrler, G. (1991a). The S protein of bovine coronavirus is a hemagglutinin recognizing 9-O-acetylated sialic acid as a receptor determinant. *J. Virol.* **65:**6232–6237.

Schultze, B., and Herrler, G. (1992). Bovine coronavirus uses *N*-acetyl-9-*O*-acetylneuraminic acid as a receptor determinant to initiate the infection of cultured cells. *J. Gen. Virol.* **73:**901–906.

Schultze, B., Wahn, K., Klenk, H. D., and Herrler, G. (1991b). Isolated HE-protein from hemagglutinating encephalomyelitis virus and bovine coronavirus has receptor-destroying and receptor-binding activity. *Virology* **180:**221–228.

Senanayake, S. D., Hofmann, M. A., Maki, J. L., and Brian, D. A. (1992). The nucleocapsid protein gene of bovine coronavirus is bicistronic. *J. Virol.* **66:**5277–5283.

Sethna, P. B., Hofmann, M. A., and Brian, D. A. (1991). Minus-strand copies of replicating coronavirus mRNAs contain antileaders. *J. Virol.* **65:**320–325.
Shen, S., Wen, Z. L., and Liu, D. X. (2003). Emergence of a coronavirus infectious bronchitis virus mutant with a truncated 3b gene: Functional characterization of the 3b protein in pathogenesis and replication. *Virology* **311:**16–27.
Shi, S. T., Huang, P., Li, H. P., and Lai, M. M. (2000). Heterogeneous nuclear ribonucleoprotein A1 regulates RNA synthesis of a cytoplasmic virus. *EMBO J.* **19:**4701–4711.
Siddell, S. G. (ed.) (1995). "The *Coronaviridae*." Plenum Press, New York.
Simmons, G., Reeves, J. D., Rennekamp, A. J., Amberg, S. M., Piefer, A. J., and Bates, P. (2004). Characterization of severe acute respiratory syndrome-associated coronavirus (SARS-CoV) spike glycoprotein-mediated viral entry. *Proc. Natl. Acad. Sci. USA* **101:**4240–4245.
Simon, K., Lingappa, V. R., and Ganem, D. (1988). Secreted hepatitis B surface antigen polypeptides are derived from a transmembrane precursor. *J. Cell Biol.* **107:**2163–2168.
Sims, A. C., Ostermann, J., and Denison, M. R. (2000). Mouse hepatitis virus replicase proteins associate with two distinct populations of intracellular membranes. *J. Virol.* **74:**5647–5654.
Smith, A. R., Boursnell, M. E., Binns, M. M., Brown, T. D., and Inglis, S. C. (1990). Identification of a new membrane-associated polypeptide specified by the coronavirus infectious bronchitis virus. *J. Gen. Virol.* **71:**3–11.
Sola, I., Alonso, S., Sanchez, C., Sanchez-Morgado, J. M., and Enjuanes, L. (2001). Expression of transcriptional units using transmissible gastroenteritis coronavirus derived minigenomes and full-length cDNA clones. *Adv. Exp. Med. Biol.* **494:**447–451.
Spaan, W. J., Rottier, P. J., Horzinek, M. C., and van der Zeijst, B. A. (1981). Isolation and identification of virus-specific mRNAs in cells infected with mouse hepatitis virus (MHV-A59). *Virology* **108:**424–434.
Stern, D. F., and Sefton, B. M. (1982). Coronavirus proteins: Structure and function of the oligosaccharides of the avian infectious bronchitis virus glycoproteins. *J. Virol.* **44:**804–812.
Stohlman, S. A., Baric, R. S., Nelson, G. N., Soe, L. H., Welter, L. M., and Deans, R. J. (1988). Specific interaction between coronavirus leader RNA and nucleocapsid protein. *J. Virol.* **62:**4288–4295.
Stohlman, S. A., Fleming, J. O., Patton, C. D., and Lai, M. M. (1983). Synthesis and subcellular localization of the murine coronavirus nucleocapsid protein. *Virology* **130:**527–532.
Stohlman, S. A., Kyuwa, S., Cohen, M., Bergmann, C., Polo, J. M., Yeh, J., Anthony, R., and Keck, J. G. (1992). Mouse hepatitis virus nucleocapsid protein-specific cytotoxic T lymphocytes are Ld restricted and specific for the carboxy terminus. *Virology* **189:**217–224.
Stohlman, S. A., and Lai, M. M. (1979). Phosphoproteins of murine hepatitis viruses. *J. Virol.* **32:**672–675.
Sturman, L. S., Holmes, K. V., and Behnke, J. (1980). Isolation of coronavirus envelope glycoproteins and interaction with the viral nucleocapsid. *J. Virol.* **33:**449–462.
Sturman, L. S., Ricard, C. S., and Holmes, K. V. (1990). Conformational change of the coronavirus peplomer glycoprotein at pH 8.0 and 37 °C correlates with virus aggregation and virus-induced cell fusion. *J. Virol.* **64:**3042–3050.
Sugiyama, K., and Amano, Y. (1981). Morphological and biological properties of a new coronavirus associated with diarrhea in infant mice. *Arch. Virol.* **67:**241–251.
Sui, J., Li, W., Murakami, A., Tamin, A., Matthews, L. J., Wong, S. K., Moore, M. J., Tallarico, A. S., Olurinde, M., Choe, H., Anderson, L. J., Bellini, W. J., Farzan, M., and

Marasco, W. A. (2004). Potent neutralization of severe acute respiratory syndrome (SARS) coronavirus by a human mAb to S1 protein that blocks receptor association. *Proc. Natl. Acad. Sci. USA* **101:**2536–2541.

Surjit, M., Liu, B., Kumar, P., Chow, V. T. K., and Lal, S. K. (2004). The nucleocapsid protein of the SARS coronavirus is capable of self-association through a C-terminal 209 amino acid interaction domain. *Biochem. Biophys. Res. Commun.* **317:**1030–1036.

Suzuki, H., and Taguchi, F. (1996). Analysis of the receptor-binding site of murine coronavirus spike protein. *J. Virol.* **70:**2632–2636.

Swift, A. M., and Machamer, C. E. (1991). A Golgi retention signal in a membrane-spanning domain of coronavirus E1 protein. *J. Cell Biol.* **115:**19–30.

Taguchi, F., and Matsuyama, S. (2002). Soluble receptor potentiates receptor-independent infection by murine coronavirus. *J. Virol.* **76:**950–958.

Tahara, S. M., Dietlin, T. A., Nelson, G. W., Stohlman, S. A., and Manno, D. J. (1998). Mouse hepatitis virus nucleocapsid protein as a translational effector of viral mRNAs. *Adv. Exp. Med. Biol.* **440:**313–318.

Thiel, V., Karl, N., Schelle, B., Disterer, P., Klagge, I., and Siddell, S. G. (2003). Multigene RNA vector based on coronavirus transcription. *J. Virol.* **77:**9790–9798.

Tooze, J., Tooze, S., and Warren, G. (1984). Replication of coronavirus MHV-A59 in sac-cells: Determination of the first site of budding of progeny virions. *Eur. J. Cell Biol.* **33:**281–293.

Tooze, J., Tooze, S. A., and Fuller, S. D. (1987). Sorting of progeny coronavirus from condensed secretory proteins at the exit from the *trans*-Golgi network of AtT20 cells. *J. Cell Biol.* **105:**1215–1226.

Tresnan, D. B., Levis, R., and Holmes, K. V. (1996). Feline aminopeptidase N serves as a receptor for feline, canine, porcine, and human coronaviruses in serogroup I. *J. Virol.* **70:**8669–8674.

Tripet, B., Howard, M. W., Jobling, M., Holmes, R. K., Holmes, K. V., and Hodges, R. S. (2004). Structural characterization of the SARS-coronavirus spike S fusion protein core. *J. Biol. Chem.* **279:**20836–20849.

Tsai, J. C., Zelus, B. D., Holmes, K. V., and Weiss, S. R. (2003). The N-terminal domain of the murine coronavirus spike glycoprotein determines the CEACAM1 receptor specificity of the virus strain. *J. Virol.* **77:**841–850.

Tung, F. Y., Abraham, S., Sethna, M., Hung, S. L., Sethna, P., Hogue, B. G., and Brian, D. A. (1992). The 9-kDa hydrophobic protein encoded at the 3′ end of the porcine transmissible gastroenteritis coronavirus genome is membrane-associated. *Virology* **186:**676–683.

van Berlo, M. F., van den Brink, W. J., Horzinek, M. C., and van der Zeijst, B. A. (1987). Fatty acid acylation of viral proteins in murine hepatitis virus-infected cells: Brief report. *Arch. Virol.* **95:**123–128.

van der Meer, Y., Snijder, E. J., Dobbe, J. C., Schleich, S., Denison, M. R., Spaan, W. J., and Locker, J. K. (1999). Localization of mouse hepatitis virus nonstructural proteins and RNA synthesis indicates a role for late endosomes in viral replication. *J. Virol.* **73:**7641–7657.

van der Most, R. G., Bredenbeek, P. J., and Spaan, W. J. (1991). A domain at the 3′ end of the polymerase gene is essential for encapsidation of coronavirus defective interfering RNAs. *J. Virol.* **65:**3219–3226.

van Genderen, I. L., Godeke, G. J., Rottier, P. J., and van Meer, G. (1995). The phospholipid composition of enveloped viruses depends on the intracellular membrane through which they bud. *Biochem. Soc. Trans.* **23:**523–526.

Vennema, H. (1999). Genetic drift and genetic shift during feline coronavirus evolution. *Vet. Microbiol.* **69:**139–141.
Vennema, H., Godeke, G. J., Rossen, J. W., Voorhout, W. F., Horzinek, M. C., Opstelten, D. J., and Rottier, P. J. (1996). Nucleocapsid-independent assembly of coronavirus-like particles by co-expression of viral envelope protein genes. *EMBO J.* **15:**2020–2028.
Vennema, H., Heijnen, L., Zijderveld, A., Horzinek, M. C., and Spaan, W. J. (1990a). Intracellular transport of recombinant coronavirus spike proteins: Implications for virus assembly. *J. Virol.* **64:**339–346.
Vennema, H., Poland, A., Foley, J., and Pedersen, N. C. (1998). Feline infectious peritonitis viruses arise by mutation from endemic feline enteric coronaviruses. *Virology* **243:**150–157.
Vennema, H., Rottier, P. J., Heijnen, L., Godeke, G. J., Horzinek, M. C., and Spaan, W. J. (1990b). Biosynthesis and function of the coronavirus spike protein. *Adv. Exp. Med. Biol.* **276:**9–19.
Vlasak, R., Luytjes, W., Leider, J., Spaan, W., and Palese, P. (1988a). The E3 protein of bovine coronavirus is a receptor-destroying enzyme with acetylesterase activity. *J. Virol.* **62:**4686–4690.
Vlasak, R., Luytjes, W., Spaan, W., and Palese, P. (1988b). Human and bovine coronaviruses recognize sialic acid-containing receptors similar to those of influenza C viruses. *Proc. Natl. Acad. Sci. USA* **85:**4526–4529.
Wang, Y., and Zhang, X. (1999). The nucleocapsid protein of coronavirus mouse hepatitis virus interacts with the cellular heterogeneous nuclear ribonucleoprotein A1 *in vitro* and *in vivo*. *Virology* **265:**96–109.
Wege, H., Nagashima, K., and ter Meulen, V. (1979). Structural polypeptides of the murine coronavirus JHM. *J. Gen. Virol.* **42:**37–47.
Weisz, O. A., Swift, A. M., and Machamer, C. E. (1993). Oligomerization of a membrane protein correlates with its retention in the Golgi complex. *J. Cell Biol.* **122:**1185–1196.
Wilbur, S. M., Nelson, G. W., Lai, M. M., McMillan, M., and Stohlman, S. A. (1986). Phosphorylation of the mouse hepatitis virus nucleocapsid protein. *Biochem. Biophys. Res. Commun.* **141:**7–12.
Wilhelmsen, K. C., Leibowitz, J. L., Bond, C. W., and Robb, J. A. (1981). The replication of murine coronaviruses in enucleated cells. *Virology* **110:**225–230.
Williams, R. K., Jiang, G. S., and Holmes, K. V. (1991). Receptor for mouse hepatitis virus is a member of the carcinoembryonic antigen family of glycoproteins. *Proc. Natl. Acad. Sci. USA* **88:**5533–5536.
Wong, S. K., Li, W., Moore, M. J., Choe, H., and Farzan, M. (2004). A 193-amino acid fragment of the SARS coronavirus S protein efficiently binds angiotensin-converting enzyme 2. *J. Biol. Chem.* **279:**3197–3201.
Woo, K., Joo, M., Narayanan, K., Kim, K. H., and Makino, S. (1997). Murine coronavirus packaging signal confers packaging to nonviral RNA. *J. Virol.* **71:**824–827.
Woods, R. D. (2001). Efficacy of a transmissible gastroenteritis coronavirus with an altered ORF-3 gene. *Can. J. Vet. Res.* **65:**28–32.
Wurm, T., Chen, H., Hodgson, T., Britton, P., Brooks, G., and Hiscox, J. A. (2001). Localization to the nucleolus is a common feature of coronavirus nucleoproteins, and the protein may disrupt host cell division. *J. Virol.* **75:**9345–9356.
Wurzer, W. J., Obojes, K., and Vlasak, R. (2002). The sialate-4-*O*-acetylesterases of coronaviruses related to mouse hepatitis virus: A proposal to reorganize group 2 *Coronaviridae*. *J. Gen. Virol.* **83:**395–402.

Xu, Y., Liu, Y., Lou, Z., Qin, L., Li, X., Bai, Z., Pang, H., Tien, P., Gao, G. F., and Rao, Z. (2004). Structural basis for coronavirus-mediated membrane fusion: Crystal structure of MHV spike protein fusion core. *J. Biol. Chem.* **279:**30514–30522.

Yamada, Y. K., Yabe, M., Ohtsuki, T., and Taguchi, F. (2000). Unique N-linked glycosylation of murine coronavirus MHV-2 membrane protein at the conserved O-linked glycosylation site. *Virus Res.* **66:**149–154.

Yeager, C. L., Ashmun, R. A., Williams, R. K., Cardellichio, C. B., Shapiro, L. H., Look, A. T., and Holmes, K. V. (1992). Human aminopeptidase N is a receptor for human coronavirus 229E. *Nature* **357:**420–422.

Ying, W., Hao, Y., Zhang, Y., Peng, W., Qin, E., Cai, Y., Wei, K., Wang, J., Chang, G., Sun, W., Dai, S., Li, X., Zhu, Y., Li, J., Wu, S., Guo, L., Dai, J., Wan, P., Chen, T., Du, C., Li, D., Wan, J., Kuai, X., Li, W., Shi, R., Wei, H., Cao, C., Yu, M., Liu, H., Dong, F., Wang, D., Zhang, X., Qian, X., Zhu, Q., and He, F. (2004). Proteomic analysis on structural proteins of severe acute respiratory syndrome coronavirus. *Proteomics* **4:**492–504.

Yokomori, K., Asanaka, M., Stohlman, S. A., Makino, S., Shubin, R. A., Gilmore, W., Weiner, L. P., Wang, F. I., and Lai, M. M. (1995). Neuropathogenicity of mouse hepatitis virus JHM isolates differing in hemagglutinin-esterase protein expression. *J Neurovirol.* **1:**330–339.

Yokomori, K., Baker, S. C., Stohlman, S. A., and Lai, M. M. C. (1992). Hemagglutinin-esterase specific monoclonalantibodies alter the neuropathogenicity of mouse hepatitis virus. *J. Virol.* **66:**2865–2874.

Yokomori, K., Banner, L. R., and Lai, M. M. (1991). Heterogeneity of gene expression of the hemagglutinin-esterase (HE) protein of murine coronaviruses. *Virology* **183:**647–657.

Yokomori, K., La Monica, N., Makino, S., Shieh, C. K., and Lai, M. M. (1989). Biosynthesis, structure, and biological activities of envelope protein gp65 of murine coronavirus. *Virology* **173:**683–691.

Yokomori, K., Stohlman, S. A., and Lai, M. M. (1993). The detection and characterization of multiple hemagglutinin-esterase (HE)-defective viruses in the mouse brain during subacute demyelination induced by mouse hepatitis virus. *Virology* **192:**170–178.

Yoo, D., Graham, F. L., Prevec, L., Parker, M. D., Benko, M., Zamb, T., and Babiuk, L. A. (1992). Synthesis and processing of the haemagglutinin-esterase glycoprotein of bovine coronavirus encoded in the E3 region of adenovirus. *J. Gen. Virol.* **73:**2591–2600.

Yoshikura, H., and Taguchi, F. (1978). Mouse hepatitis virus strain MHV-S: Formation of pseudotypes with a murine leukemia virus envelope. *InterVirology* **10:**132–136.

Yu, X., Bi, W., Weiss, S. R., and Leibowitz, J. L. (1994). Mouse hepatitis virus gene 5b protein is a new virion envelope protein. *Virology* **202:**1018–1023.

Yuan, K., Yi, L., Chen, J., Qu, X., Qing, T., Rao, X., Jiang, P., Hu, J., Xiong, Z., Nie, Y., Shi, X., Wang, W., Ling, C., Yin, X., Fan, K., Lai, L., Ding, M., and Deng, H. (2004). Suppression of SARS-CoV entry by peptides corresponding to heptad regions on spike glycoprotein. *Biochem. Biophys. Res. Commun.* **319:**746–752.

Zelus, B. D., Schickli, J. H., Blau, D. M., Weiss, S. R., and Holmes, K. V. (2003). Conformational changes in the spike glycoprotein of murine coronavirus are induced at 37 °C either by soluble murine CEACAM1 receptors or by pH 8. *J. Virol.* **77:**830–840.

Zhang, X., Hinton, D. R., Park, S., Parra, B., Liao, C. L., Lai, M. M., and Stohlman, S. A. (1998). Expression of hemagglutinin/esterase by a mouse hepatitis virus coronavirus defective-interfering RNA alters viral pathogenesis. *Virology* **242:**170–183.

Zhao, X., Shaw, K., and Cavanagh, D. (1993). Presence of subgenomic mRNAs in virions of coronavirus IBV. *Virology* **196:**172–178.

Zhou, M., and Collisson, E. W. (2000). The amino and carboxyl domains of the infectious bronchitis virus nucleocapsid protein interact with 3′ genomic RNA. *Virus Res.* **67:**31–39.

Zhou, M., Williams, A. K., Chung, S. I., Wang, L., and Collisson, E. W. (1996). The infectious bronchitis virus nucleocapsid protein binds RNA sequences in the 3′ terminus of the genome. *Virology* **217:**191–199.

Zhu, J., Xiao, G., Xu, Y., Yuan, F., Zheng, C., Liu, Y., Yan, H., Cole, D. K., Bell, J. I., Rao, Z., Tien, P., and Gao, G. F. (2004). Following the rule: Formation of the 6-helix bundle of the fusion core from severe acute respiratory syndrome coronavirus spike protein and identification of potent peptide inhibitors. *Biochem. Biophys. Res. Commun.* **319:**283–288.

Ziebuhr, J., Snijder, E. J., and Gorbalenya, A. E. (2000). Virus-encoded proteinases and proteolytic processing in the Nidovirales. *J. Gen. Virol.* **81:**853–879.

ADVANCES IN VIRUS RESEARCH, VOL 64

# MECHANISM OF MEMBRANE FUSION BY VIRAL ENVELOPE PROTEINS

Stephen C. Harrison

Children's Hospital, Harvard Medical School, and Howard Hughes Medical Institute
Boston, Massachusetts 02115



## I. Introduction

Enveloped viruses enter cells by fusing their lipid bilayer membrane with a cellular membrane. They bear on their surface oligomers of a fusion protein, often part of a polypeptide that performs other functions, such as receptor binding. Most viral fusion proteins require priming by proteolytic processing, either of the fusion protein itself or of an accompanying protein. The priming step, which often occurs during transport of the fusion protein to the cell surface but may also occur extracellularly, then prepares the fusion protein for triggering by events that accompany attachment and uptake. For example, proton binding is frequently such a trigger, which provides the virus with a mechanism for detecting that it has arrived in the low-pH milieu of an endosome.

Two classes of viral fusion proteins have been identified so far by structural studies. Later, discuss what is known about fusion by members of each of those two classes. The fusion of two bilayers that these proteins catalyze is likely to proceed by the same pathway in both




0065-3527/05 $35.00
DOI: 10.1016/S0065-3527(05)64007-9

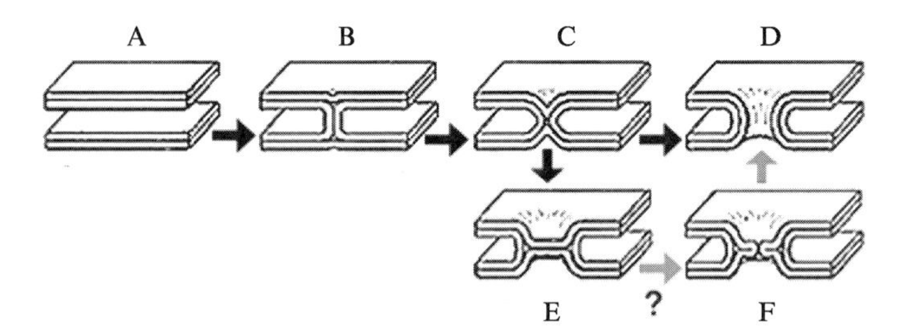


FIG 1. Fusion of two lipid bilayers. (A) Two parallel bilayer membranes. There is a substantial barrier to close approach. (B) Hemifusion stalk. (C) Proposed transition structure. (D) Fusion pore (before lateral expansion). (E) Hemifusion diaphragm. (F) Some models include perforation of the hemifusion diaphragm as a productive step toward fusion pore formation. Adapted from Jahn *et al.* (2003); see also Cohen *et al.* (2002).

cases. That is, these proteins are like enzymes that have different structures but that still catalyze the same chemical reaction.

The bilayer fusion reaction common to all the enveloped viral entry pathways is shown schematically in Fig. 1 (for a review, see Cohen *et al.*, 2002). It is believed to pass through an intermediate known as a "hemifusion stalk" (Fig. 1B) (Markin *et al.*, 1984; Siegel, 1993). In this intermediate, the two apposed leaflets have fused, but not the distal ones. Hemifused bilayers can proceed to form a "fusion pore" (Fig. 1D) or form a structure in which the two distal leaflets create a single bilayer. This state, which can spread laterally, is called a "hemifusion diaphragm" (Fig. 1E). Bilayers do not fuse spontaneously (e.g., concentrated liposomes are quite stable), because the reaction in Fig. 1 has a high activation barrier, both at the step between the precursor bilayers and the hemifusion stalk and at the step between the hemifusion stalk and the fusion pore. A newly opened pore appears to revert frequently to a hemifusion structure ("flickering"), and the largest kinetic barrier may be for the step in which the pore dilates rather than reverts. Once the fusion pore has dilated, the fused stucture is stable with respect to the initial, unfused structure.

## II. Class I Viral Fusion Proteins

The fusion proteins of myxo- and paramyxoviruses, retroviruses, filoviruses, and at least some coronaviruses have sufficient common characteristics to classify and describe them together (Skehel and

Wiley, 1998). They are trimers, with a large, N-terminal ectodomain and a C-proximal transmembrane anchor. The subunit is synthesized as a precursor chain (e.g., $HA_0$, gp160), which is processed, usually late in the secretory pathway, by a proteolytic cleavage, which generates two associated chains (e.g., $HA_1$ and $HA_2$, gp120 and gp41). The carboxy-terminal fragment is the actual fusion effector, and it bears a relatively hydrophobic, glycine-rich "fusion peptide" at or near the cleavage site.

Cleavage of the precursor chain primes the fusion protein, but in HA and gp160 it does not produce major conformational alterations. By contrast, molecular interactions associated with attachment and uptake trigger massive rearrangements. Only for the influenza virus hemagglutinin (HA) do we currently have pre- and postfusion structures, so the extent of rearrangement must be extrapolated from that one example (Skehel and Wiley, 2000). But indirect lines of evidence suggest that the extrapolation is reasonable. The detailed descriptions that follow are restricted to influenza HA and human immunodeficiency virus/simian immunodeficiency virus (HIV/SIV) Env—the two for which the most extensive structural and biochemical data are available at the time of this writing.

### A. *Influenza A Hemagglutinin*

The precursor trimer, $HA_0$, and the primed $HA_1$–$HA_2$ are similar in structure (Fig. 2A,B). The receptor-binding domain, the core of $HA_1$, is borne on a stalk formed not only by $HA_2$ but also by the N- and C-terminal segments of $HA_1$. The loop of $HA_0$ destined to be cleaved by furin is exposed and partly disordered. After cleavage, the N terminus of $HA_2$ (the fusion peptide) tucks between the long helices that cluster around the threefold axis (for many references on influenza virus HA before 2000, see Skehel and Wiley, 2000).

Sialic acid, linked to complex glycans on either glycoproteins or glycolipids, is the receptor for influenza virus. Receptor binding does not induce any significant conformational changes in HA. It merely attaches the virus to the cell surface and allows capture by endocytic vesicles. The trigger for a fusion-inducing conformational change is the binding of one or more protons, as the pH of the endosome becomes progressively lower. Most strains of influenza A have a critical pH of about 5–5.5, corresponding to the pH of a relatively late endosomal compartment. No particular titrating residue accounts for the transition; rather, when the charge on the protein destabilizes it sufficiently,

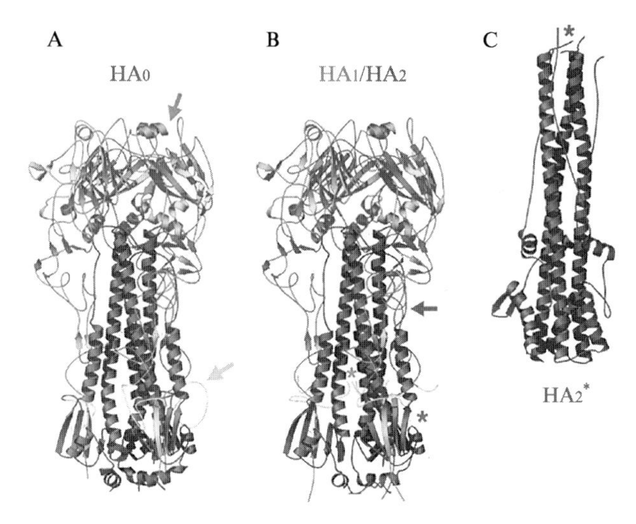


FIG 2. The influenza virus hemagglutinin. (A) $HA_0$, before cleavage between $HA_1$ and $HA_2$. The $HA_1$ part of the protein is blue; the $HA_2$ part, is purple. The fusion peptide, looped out before cleavage, is yellow. The sialic acid-binding site and the cleavage point are shown by blue and yellow arrows, respectively. (B) The mature HA after cleavage but before low-pH triggering. The only change from the structure in (A) is insertion of the fusion peptide (the N terminus of $HA_2$, in yellow) into a crevice along the three-fold axis (dark yellow asterisk). A purple arrow points to the loop between the shorter, N-proximal helix and the longer, central helix. A purple asterisk indicates the position of residues that will move to the top of the molecule during the low pH-induced transition. (C) $HA_2$ after exposure to low pH. The same structure can be obtained by refolding $HA_2$ expressed in *E. coli*. It is the minimal free-energy state of $HA_2$ unconstrained by covalent association with $HA_1$. The long loop in the prefusion structure (purple arrow in [B]) has now become helical, elevating the N terminus of the protein (the fusion peptide itself is not included in this structure) to the top of the molecule (purple asterisk). A break and reversal of direction in the central $\alpha$-helix of the prefusion trimer likewise projects the C terminus of the protein to the top. The figure is aligned with respect to (B) so this break is roughly at the same height in both panels. In the actual transition, the chain reversal is likely to occur by melting and rezipping of the C-terminal helical segment, as shown in Fig. 3C. For detailed references, see Skehel and Wiley (2000). (See Color Insert.)

a cooperative conformational rearrangement ensues (Daniels *et al.*, 1985).

The conformational change triggered by proton binding (Fig. 2B,C, and Fig. 3) has the following characteristics (Skehel and Wiley, 2000). (1) The $HA_1$ receptor-binding domains separate from each other. Because $HA_1$ is linked to $HA_2$ by a disulfide bond, it cannot dissociate completely, but the final conformation of $HA_2$ requires that $HA_1$

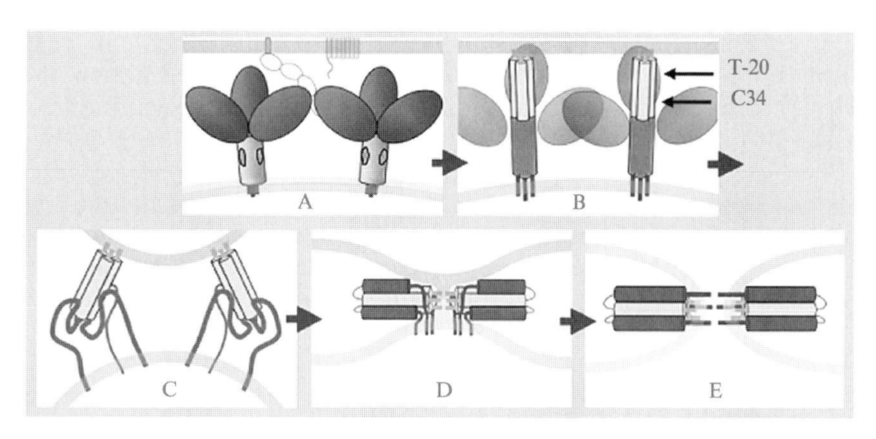


FIG 3. Diagram of membrane fusion mediated by class I viral fusion proteins. (A) Receptor binding (shown here for HIV or SIV Env, where the schematic receptor and coreceptor symbolize CD4 and CXCR4 or CCR5). (B) Dissociation of the receptor-binding domain (in the case of HA, a disulfide bond prevents complete dissociation, but the structural rearrangement requires that $HA_1$ move away from the threefold axis) and projection of the fusion peptide toward the target cell membrane. This state is known as the "prefusion intermediate" (or, sometimes, the "prehairpin intermediate"). The arrows show the positions along the core helical bundle at which, on HIV-1 gp41, the inhibitor peptides T-20 and C34 are expected to bind. (C) Folding back of the C-terminal part of the molecule. This zipping up of a segment of the fusion protein sometimes designated "helical region 2" (HR2); helical region 1 (HR1) forms the inner core of the postfusion trimer) draws the two membranes together as the N-terminal fusion peptide, inserted into the target cell membrane, and the C-terminal transmembrane segment, which is anchored in the viral membrane, are forced by the refolding to approach each other. (D) Hemifusion stalk formation. Provided that they insert only into the outer leaflet of the bilayer, the fusion peptides can migrate into the stalk, as proposed here, and stabilize it. (E) Fusion pore formation. Hemifusion structures flicker transiently into unstable pores, which reseal. The pore can be trapped by the final refolding step, in which the three transmembrane segments snap into place around the inserted fusion peptides. In the case of HA, this step may be driven by formation of a set of interactions that cap the inner core helices. (See Color Insert.)

move away from the threefold axis to allow $HA_2$ to form new threefold contacts. (2) $HA_2$ straightens by a loop-to-helix transition of a segment previously tucked beneath $HA_1$ (Bullough *et al.*, 1994; Carr and Kim, 1993). This rearrangement thrusts the N-terminal fusion peptide up (using the coordinates of Fig. 2 as a reference for "up" and "down"), where it can encounter the membrane of the target cell. The extended intermediate, with fusion peptides inserted into the target cell membrane, may have a lifetime of seconds to minutes. (3) The long helix of $HA_2$ breaks and reverses direction, and the distal part of the $HA_2$ chain zips up along the stem formed by long central helix, which now runs from the fusion peptide to the breakpoint. In its final conformation, the $HA_2$ ectodomain has its N terminus, which bears the fusion peptide, adjacent to its C terminus, which leads to the transmembrane anchor (Chen *et al.*, 1999). In other words, the two membrane-interacting segments on each chain are next to each other. Thus, assuming that all three fusion peptides insert into the target membrane and that the C-terminal transmembrane segments remain anchored in the viral membrane, the zipping up of the distal part of $HA_2$ will drag the two membranes together (Fig. 3).

Only the inital (Fig. 2B) and final (Fig. 2C) conformations of $HA_2$ can be studied easily by structural methods. Evidence for the pathway described previously and shown diagrammatically in Fig. 3 comes from biophysical studies of HA-mediated fusion and from the effects of various mutations on the fusion process (reviewed extensively elsewhere and outlined only in summary here).

1. *The low-pH trigger.* The threshold pH for fusion can be affected by a variety of amino acid substitutions, and some that increase the threshold pH can be selected by isolating virus resistant to amantadine (which raises the pH of endosomes). These mutations map in a variety of locations, which have in common that they appear to contribute to interactions that hold together the HA trimer (Daniels *et al.*, 1985). Thus, there appears not to be a single, localized trigger, but rather a cumulative effect of various groups that bind protons at pH 5 and above.

2. *Fusion peptide emergence and insertion.* The amino acid sequence of the loop in $HA_2$ has a particularly strong helical coiled-coil signature, and peptides with this sequence form stable coiled coils in solution (Carr and Kim, 1993). Release of the constraints imposed on the conformation of this peptide by $HA_1$ is therefore likely to lead to coiled-coil formation and translocation of the fusion peptide away from the base of the trimer. The phrase "spring-loaded" has been used to describe this mechanism (Carr and Kim, 1993). Experiments with target membranes containing photoactivatable cross-linking reagents demonstrate insertion of the fusion peptide into the target cell membrane, probably before fusion (Tsurudome *et al.*, 1992). A peptide having the amino acid sequence of the first 20 residues of $HA_2$ forms a gently kinked, amphipathic helix in the presence of detergent micelles (Han *et al.*, 2001). On a membrane, this structure would lie partially embedded in the outer monolayer, with the two ends dipping somewhat more deeply into the bilayer outer leaflet than the apex of

the kink. That is, the fusion peptide probably inserts only into the proximal leaflet of a membrane lipid bilayer. The effects of mutations in the fusion peptide vary with position (Qiao *et al.*, 1999; Steinhauer *et al.*, 1995). They do not correlate in any simple way with the structure just described, but it may be difficult to deconvolute the requirements for its conformation in the mature $HA_1$–$HA_2$ trimer from those for productive membrane insertion.

3. *Fold-back*. The best evidence for an extended "prefusion intermediate," with the fusion peptides inserted into the target cell membrane but with the C-terminal parts of the fusion protein still at the opposite end of the molecule, comes from work on HIV-1 gp41 (Furuta *et al.*, 1998; Rimsky *et al.*, 1998) (see later). In the mature HA trimer at neutral pH, the long axial helices diverge from each other at about the position at which the folding back occurs (Fig. 2B,C), and it is plausible that in the extended intermediate, the "lower" part of the molecule is unstable and locally unfolded. There has been some confusion in the literature about the folding back process. Although the expressions "fold back" and "jack-knifing" have been used to describe the conformational change, the only way it can plausibly occur in practice is by melting at least some of the C-terminal segments of the $HA_2$ ectodomain and zipping up these segments along the central core (Fig. 3). If the links between the fusion peptides and the central core are flexible, then the zipping process will necessarily displace the position of the chain reversal (the fold-back point) laterally, away from the constriction. During the transition from the extended intermediate to the folded-back conformation, threefold symmetry is broken; it is restored only in the final, postfusion structure.

The free energy required to deform the two membranes, squeeze them together, and initiate fusion is probably more than can be obtained from refolding of a single trimer, and it is likely that several HA trimers cooperate to form a fusion pore. Evidence for this statement includes the observation that the lag time between acidification and fusion of cells expressing HA on their plasma membranes depends on the HA surface density (Danieli *et al.*, 1996). One picture for the process would invoke formation of a structurally defined ring of extended intermediates, surrounding the position at which fusion will occur. An alternative, more stochastic model, is also possible, as described in more detail later, in Section IV.A.

4. *Hemifusion*. The fold-back step brings the two membranes close together. Are there further structural features that accelerate hemifusion? Distortion of one or both membranes by introduction of positive curvature would be one way to do so; stabilization of a hemifusion stalk

once formed, thereby preventing reversal, would be another. We need to know more about the conformation and properties of an inserted fusion peptide before we can assess these possibilities. The discovery, that class II fusion peptides insert only into the outer part of the target membrane, suggests a possible mechanism for hemifusion stalk stabilization (see later).

5. *Fusion pore formation.* A noteworthy feature of the rearranged structure of $HA_2$ ectodomain is a tight "cap," formed by the way the C-terminal part of the polypeptide chain associates with the N-terminal end of the central, three-helix bundle (Fig. 2C) (Chen *et al.*, 1999). This cap is likely to snap the folded-back conformation into place. In the model for the fusion reaction diagrammed in Fig. 3, the final step involves stabilization of the fusion pore by interposition of the transmembrane anchors of $HA_2$. Formation of the $HA_2$ cap may be a critical feature of this step. When HA is linked to a glycophosphatidylinositol (GPI) anchor or to a truncated transmembrane anchor that does not cross the bilayer, it catalyzes hemifusion quite readily, but it promotes fusion poorly, if at all (Kemble *et al.*, 1994; Melikyan *et al.*, 1995). Catalyzing fusion does not require a specific structure or amino acid sequence on the inside of the membrane; the transmembrane anchor must simply traverse the bilayer completely (Armstrong *et al.*, 2000). Truncation of the SV5 fusion protein produces a similar result (Dutch and Lamb, 2001). Snapping the $HA_2$ cap into place and dragging the cytosolic end of the transmembrane anchor into a newly opened pore will prevent the pore from resealing (which it will otherwise tend to do) and force the fusion reaction to proceed to completion.

### *B. Influenza C HEF*

The fusion protein of influenza C, known as HEF (for hemagglutinin, esterase, fusion protein) contains, in addition to the receptor-binding (hemagglutinating) and fusion activities, an esterase domain, which hydrolyzes the 9-*O*-acetyl-sialic acid receptor, to promote viral escape from the cell surface. With the exception of the added esterase module, the protein is recognizably similar to the HA of influenza A (Rosenthal *et al.*, 1998) (Fig. 4). The esterase domain resembles in structure other acetyl hydrolases. In the amino acid sequence, it falls in two parts—one at the N-terminal end of the sialic acid-binding domain and the other at the C-terminal end. That is, the receptor-binding domain is an insert into the esterase, which in turn is an insert between the N- and C-terminal segments of $HEF_1$. One can thus imagine that

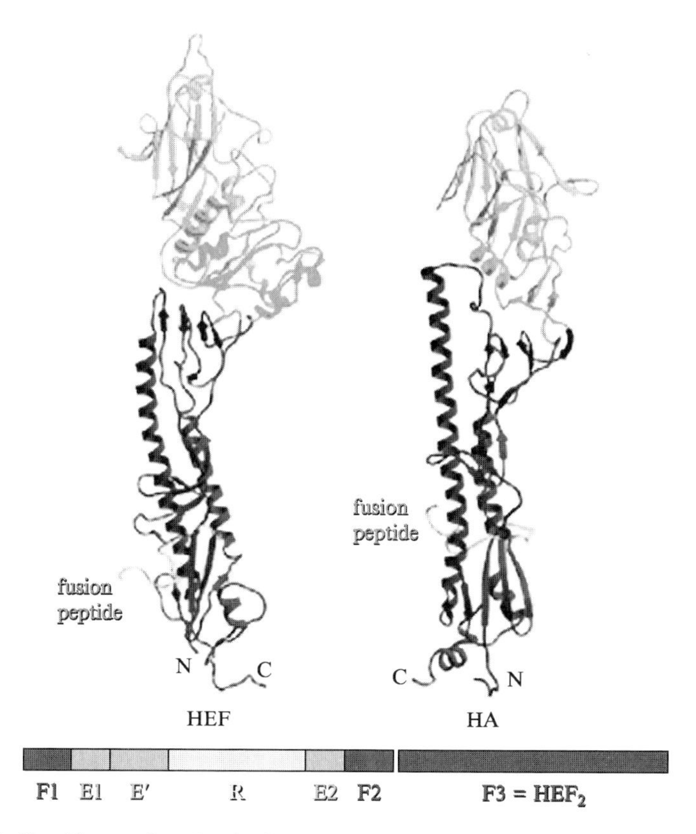


FIG 4. *Top*: The ectodomain of influenza virus C HEF and its structural and apparent evolutionary relationship to influenza A HA. *Bottom*: A key to the color scheme for the HEF polypeptide chain. The N- and C- terminal segments of $HEF_1$ and all of $HEF_2$ (except for the fusion peptide) are in red. The acetylase enzymatic domain (E1 + E″ + E2) is in green; the receptor-binding domain (R) is in blue. Think of the red fragments as an elementary fusion protein and the green and blue fragments as insertions. In HA, most of the acetylase domain has been deleted, except for the E′ fragment, which becomes an adaptor to connect the elementary fusion protein with the receptor-binding domain. Adapted from Rosenthal *et al.* (1998). (See Color Insert.)

the N- and C-terminal segments of $HEF_1$ (or $HA_1$) and the entirety of $HEF_2$ (or $HA_2$) constitute an elementary fusion module, into which other functions have been inserted in the course of evolution (Rosenthal *et al.*, 1998; Skehel and Wiley, 2000).

### *C. HIV and SIV Env*

The HIV and SIV envelope glycoproteins have two essential functions in viral entry. They attach the virus to target cells and they catalyze membrane fusion. The single-chain envelope glycoprotein precursor, gp160, is cleaved after trimerization by a furin-like protease in a late compartment of the export pathway (Allan *et al.*, 1985; Robey *et al.*, 1985; Veronese *et al.*, 1985). The two fragments, gp120 and gp41, remain associated noncovalently, but the contact is relatively weak, and gp120 tends to dissociate ("shed") from mature virions (Moore *et al.*, 1990).

Fusion occurs at the cell surface and does not require a change in pH. Rather, infection of suitable cell types is determined by a requirement for both a primary receptor, CD4 (Dalgleish *et al.*, 1984), and a coreceptor, one of several members of the chemokine receptor family (Feng *et al.*, 1996). Binding of CD4 induces a conformational change in gp120/gp41, detected by altered antigenic properties, enhanced proteolytic sensitivity of the gp120 moiety, and enhanced shedding (Sattentau and Moore, 1991; Sattentau *et al.*, 1993). The change increases affinity of gp120/gp41 for the coreceptor, probably by organizing the site for coreceptor binding (Trkola *et al.*, 1996; Wu *et al.*, 1996). Coreceptor attachment leads to fusion, probably because the coreceptor induces gp120 dissociation, fusion peptide exposure, and gp41 refolding (Chan *et al.*, 1997; Weissenhorn *et al.*, 1997). The diagram in Fig. 5 summarizes current thinking.

Direct data are available for only two of the postulated structures in the complete envelope protein transformation. These show gp120 complexed with CD4 and the Fab from monoclonal antibody (mAb) 17b (Kwong *et al.*, 1998) and the rearranged form of the gp41 ectodomain (Chan *et al.*, 1997; Weissenhorn *et al.*, 1997). Both structures correspond to postfusion states (or at least to posttriggering states). Both structures are also of molecules truncated in important ways to facilitate structural studies. Thus, the gp120 "core" that has been crystallized (in complex with CD4 and the Fab from mAb 17b) is simply the receptor- and coreceptor-binding element, stripped of nonessential variable loops and lacking N- and C-terminal segments. These terminal segments of gp120 can be thought of as part of the basic fusion module, by analogy with influenza A HA and influenza C HEF (Pollard *et al.*, 1992) (Fig. 6). The soluble form of the gp41 ectodomain (Blacklow *et al.*, 1995; Lu *et al.*, 1995), studied in several crystal forms (Chan *et al.*, 1997; Tan *et al.*, 1997; Weissenhorn *et al.*, 1997) and also by nuclear magnetic resonance (NMR) (Caffrey *et al.*, 1998), lacks the

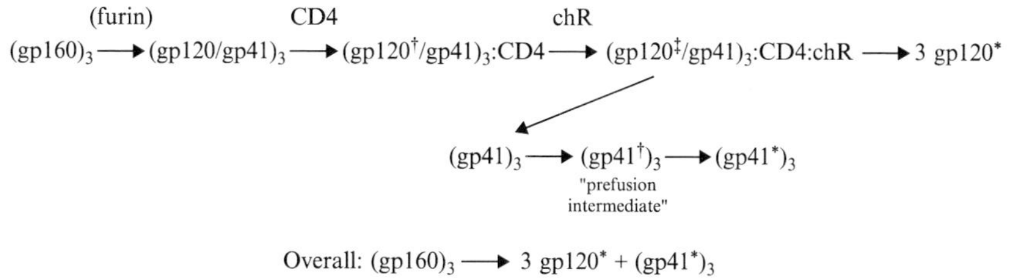

$$(gp160)_3 \xrightarrow{(furin)} (gp120/gp41)_3 \xrightarrow{CD4} (gp120^{\dagger}/gp41)_3{:}CD4 \xrightarrow{chR} (gp120^{\ddagger}/gp41)_3{:}CD4{:}chR \longrightarrow 3\ gp120^*$$

$$(gp41)_3 \longrightarrow (gp41^{\dagger})_3 \longrightarrow (gp41^*)_3$$

"prefusion intermediate"

$$\text{Overall: } (gp160)_3 \longrightarrow 3\ gp120^* + (gp41^*)_3$$

FIG 5. States of the HIV-1 envelope protein, as detected by biochemical, immunological, physicochemical, and structural analyses. The trimeric gp160 precursor is cleaved by a furin-like protease to gp120 and gp41, which retain their threefold association. Binding of CD4 (one per trimer may be sufficient, but the degree of cooperativity is not determined) induces a conformational change in gp120 ($gp120^{\dagger}$); binding of a suitable chemokine receptor (chR) may induce or lock in a further change ($gp120^{\ddagger}$). Ultimately gp120 is probably shed from the gp41 stem ($gp120^*$). Once liberated, the gp41 trimer undergoes a two-stage, fusion-inducing conformational change: a transition to an extended, prefusion intermediate, $(gp41^{\dagger})_3$, with the fusion peptide inserted into the target cell membrane, followed by a folding back to form the final, trimer-of-hairpins structure $(gp41^*)_3$, generating membrane fusion in the process.

first 30 residues of gp41, that is (the fusion peptide plus some residues that follow it) and (in most cases) a loop in the middle.

The structure of the gp120 core bound with CD4 and the Fab of mAb 17b has been determined for two different HIV-1 isolates (Kwong *et al.*, 1998, 2000) (Fig. 7). The affinity of gp120 for mAb 17b is enhanced by CD4 binding, and the antibody footprint is thought to mark the coreceptor site (Rizzuto *et al.*, 1998). Thus, the crystal structure probably shows the released form of gp120 (Fig. 5).

In what is generally agreed to be its final, postfusion conformation, the ectodomain of gp41 is a trimer of $\alpha$-helical hairpins, as shown in Fig. 8. The crystal structures do not contain the residues immediately proximal to either membrane (or to the single, fused membrane at the end of the fusion process), including a number of residues critical for inhibition by the peptide drug DP178/T-20/enfuvirtide (Wild *et al.*, 1994). The prefusion conformation of gp41 and the unliganded structure of gp120 remains undetermined, and we therefore know rather little about how they might fit together into the gp160 trimer.

From the available data, what can we deduce about the conformational changes that lead to fusion? The rearrangements in gp120 that initiate the process are probably quite extensive (Myszka *et al.*, 2000). The receptor-binding core folds into two domains, termed "inner" and "outer" (Fig. 7). One hairpin loop from each of these

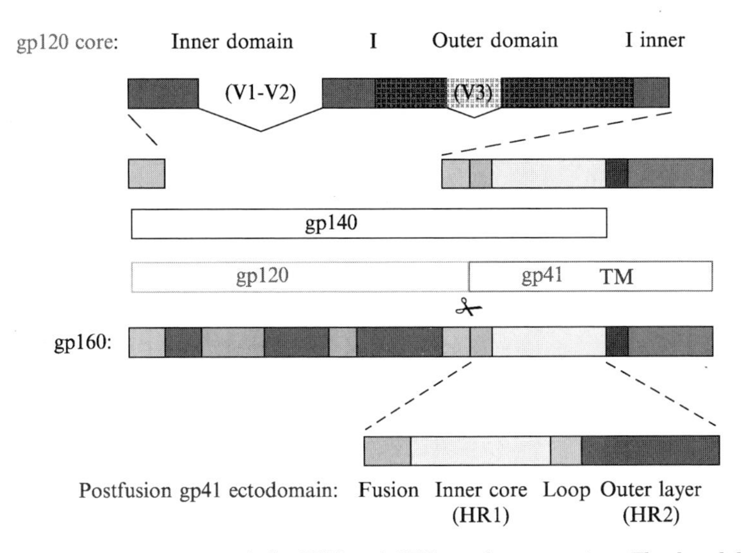


FIG 6. Primary structure of the HIV and SIV envelope proteins. The bar labeled "gp160" represents, schematically, the regions of the envelope precursor. The parts of gp160 corresponding to gp120, gp41, and gp140 (the gp160 ectodomain) are shown as open bars. The scissors symbol shows the furin cleavage point. The gp120 core can be considered a receptor-binding insertion into an elementary fusion protein, as diagrammed in the top two bars. Compare with Fig. 4—the gp120 core is the analog of the R and E′ regions of influenza HA. The principal elements of the gp41 ectodomain are diagrammed in the bottom bar. (See Color Insert.)

domains contributes to the "bridging sheet," a $\beta$ sheet that locks the relative positions and orientations of the two domains and that creates the site for coreceptor interaction. It has been suggested that the bridging sheet is not present before encounter with CD4 and that receptor binding locks the sheet into place (Kwong *et al.*, 1998). The inner and outer domains could thus have quite different relative orientations in the precursor conformation. The inner domain contains both N and C termini of the gp120 core. They are essentially adjacent to each other in the structure, consistent with the notion that the receptor-binding element is an insert into an elementary fusion module (compare with Fig. 4).

The conformational changes induced by CD4 binding weaken interactions with the rest of the trimer and lead to gp120 dissocation. This event in turn liberates gp41 to rearrange. The likely rearrangement, leading to the final trimer of hairpins, follows the mechanism already outlined for influenza HA and shown in Fig. 3. The N-terminal helices

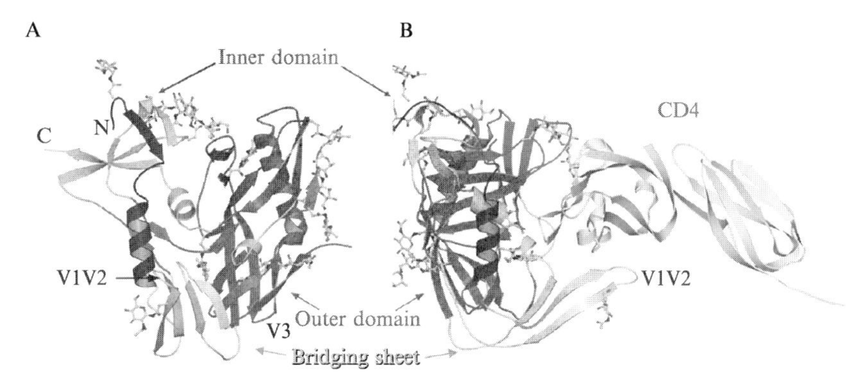


FIG 7. Structure of the gp120 core in the CD4-bound state (Kwong *et al.*, 1998). (A) Ribbon representation, showing the inner and outer domains, linked by a bridging sheet. The locations of the V1–V2 and V3 loops, deleted from the core construct (compare with Fig. 6), are shown. The locations of carbohydrate chains are shown by molecular ball-and-stick representations of those sugars found to be well-ordered in the crystal structure. The view is into the CD4-binding pocket. N and C termini of the core are labeled. (B) Side view of the same structure, with the first two immunoglobulin-like domains of CD4 also shown. (See Color Insert.)

of each hairpin cluster into a three-helix coiled coil, forming the inner core of the rearranged gp41. The fusion peptides project from the N-terminal end of this inner core. The C-terminal helix of each hairpin runs along the outside of the core, so that the transmembrane anchor at which it terminates lies adjacent to a fusion peptide. Although missing from the crystal structures, the loop between inner core helix and outer layer helix probably has a structure that resembles the conformation seen in fragments of the fusion proteins from Moloney murine leukemia virus (Fass *et al.*, 1996) and Ebola virus (Weissenhorn *et al.*, 1998), both of which have a conserved disulfide bond in this region, in common with HIV and SIV Env.

Peptides derived from the outer layer of rearranged gp41 inhibit fusion and viral infectivity by targeting the intermediate shown in Fig. 3B (Rimsky *et al.*, 1998). Evidence for the properties of this intermediate comes from kinetic studies, in which binding of outer layer-derived peptides has been analyzed by detecting their effects on fusion (Furuta *et al.*, 1998; Jones *et al.*, 1998; Munoz-Barroso *et al.*, 1999). The intermediate has a measurable lifetime. For example, in cell–cell fusion assays, conformational changes in gp120/gp41 begin within 1 min of mixing donor and target cells, but a peptide corresponding to residues in the outer layer helix inhibits fusion even if added 15 min later (Munoz-Barroso *et al.*, 1999). Two classes of outer layer peptide

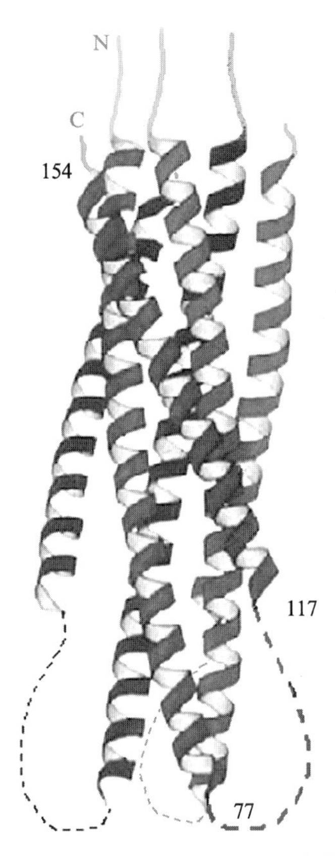


FIG 8. The inner core (HR1) and outer layer (HR2) of the gp41 ectodomain in the postfusion state of the gp41 trimer (Weissenhorn *et al.*, 1997). The structure as determined crystallographically contains six peptides—three inner and three outer. The dashed lines show covalent connectivities that would be present in the intact gp41 trimer. The residue numbers (for gp41, counting from the N terminus of the fusion peptide) for the beginning and end of the inner and outer layer helices are shown for the red subunit. (See Color Insert.)

have been studied. One includes residues in the stretch just preceding the transmembrane segment (e.g., DP178/T-20/enfuvirtide, now a licensed antiretroviral drug [Kilby and Eron, 2003; Wild *et al.*, 1994]). The other includes residues toward the N terminus of the outer layer helix (e.g., a peptide called C34; Chan *et al.*, 1998). These two classes may block fusion at slightly different stages. There is some evidence that C34-like peptides block both hemifusion (lipid mixing) and full fusion (content mixing), whereas T-20-like peptides block only the latter (Kliger *et al.*, 2001). Figure 3 is consistent with this difference. According to Fig. 3, early stages in zipping up the outer layer, blocked by C34, would be required to bring together the two membranes and hence for the initial hemifusion step; later stages would probably be required primarily to complete the process and to lock the open fusion pore in place.

Peptides from the inner core of rearranged gp41 also inhibit fusion and viral infectivity, probably by trimerizing and capturing outer layer segments before the zipping-up step (Jiang *et al.*, 1993; Lu *et al.*, 1995; Wild *et al.*, 1992). The inner core is quite insoluble, and a more efficient way to study this mode of inhibition is to cover two of its three grooves with outer layer peptides, allowing one groove free to capture an outer layer segment from gp41 as it refolds. A single-chain version of such a structure (in which three inner layer segments, interspersed with two outer layer segments, have been concatenated with suitable short linkers) is indeed an effective inhibitor of gp41-mediated fusion and of HIV-1 infectivity (Root *et al.*, 2001).

Inspection of the trimer of inner core helices shows that there is a particularly striking hydrophobic pocket on each of the three symmetric surfaces of the three-helix bundle, near the end distal to the fusion peptide (Chan *et al.*, 1997). This pocket is occupied in the final, postfusion structure by Trp-117 and adjacent residues from an outer layer helix. A 16-residue cyclic peptide synthesized from D-amino acids can bind tightly at this site and in so doing inhibit fusion and HIV infectivity (Eckert *et al.*, 1999). This pocket is also the target of a small molecule selected from a biased combinatorial library, synthesized onto the N terminus of a partial outer layer peptide (Ferrer *et al.*, 1999; Zhou *et al.*, 2000).

## III. Class II Viral Fusion Proteins

The fusion machinery of flaviviruses and alphaviruses resembles that of the class I fusion apparatus in certain broad respects: it requires activation by proteolytic cleavage and it uses the reduced pH of an endosome as a trigger for conformational change. The cleavage does not, however, modify the fusion protein itself, but rather a second, "protector" protein, which in its uncleaved state blocks the conformational rearrangement. Moreover, the fusion peptide is a loop within the folded polypeptide chain, and we refer to it here as the "fusion loop."

Until structures were determined (Lescar *et al.*, 2001; Rey *et al.*, 1995), the similarity of alphavirus and flavivirus surface proteins was not evident. During alphavirus infection, proteins pE2 (precursor of E2) and E1 are synthesized, and pE2 is cleaved to E2 plus E3 during transport of a heterodimer to the cell surface. Budding occurs at the plasma membrane. Some alphaviruses retain E3; others shed it. During flavivirus infection, proteins prM (presursor of M) and E are incorporated into immature virions, which bud into the endoplasmic reticulum. Processing of prM to M (the C-terminal fragment dissociates) occurs late in the export pathway. The structures of E1 and E are very similar; those of pE2 and prM are not yet known. E1 and E2 assemble into a $T = 4$ icosahedral lattice on the surface of an alphavirus particle; E also forms an icosahedral array, with 180 subunits (90 dimers), but not one with a quasi-equivalent design (Kuhn *et al.*, 2002).

### A. *Flaviviruses*

The envelope protein, E, does not project from the viral membrane like the "spikes" of many viruses; instead, it forms a relatively thin, tightly packed layer on the virion surface (Fig. 9A; note that sequence similarities among all flavivirus envelope proteins allow us to apply structural results from one virus to analysis of another) (Kuhn *et al.*, 2002; Rey *et al.*, 1995). A central, $\beta$-sandwich domain (domain I) organizes its folded structure (Fig. 9B) (Modis *et al.*, 2003; Rey *et al.*, 1995). Two long and elaborate loops, stabilized by disulfide bridges, emanate from this central domain, forming a distinct subdomain termed domain II. At the tip of domain II is a hydrophobic sequence, identified from its conservation among all flaviviruses as the fusion loop (Allison *et al.*, 2001). A third domain (domain III) follows the others in the polypeptide-chain sequence; it has an immunoglobulin-like fold, and various lines of evidence indicate that it binds receptors (at least in the case of several, well-studied flaviviruses). Between domain III and the transmembrane anchor are about 50 amino acid residues—a segment that has been called the "stem" (Allison *et al.*, 1999). Although not a part of the protein that crystallizes, its conformation on the virion has been deduced from cryo-EM image reconstructions. It is a pair of $\alpha$ helices with an intervening loop, sandwiched between the main part of the E protein and the outer surface of the lipid bilayer (W. Zhang *et al.*, 2003). The transmembrane anchor is also a helical hairpin, as the virus relies on signal protease (in the endoplasmic reticulum [ER] lumen) to cleave E from a precursor polyprotein.

E is a dimer on the surface of a mature virion (Heinz *et al.*, 1991; Kuhn *et al.*, 2002). Its stable (stemless) ectodomain ("soluble E" or sE, comprising domains I, II, and III—approximately 395 of the 450

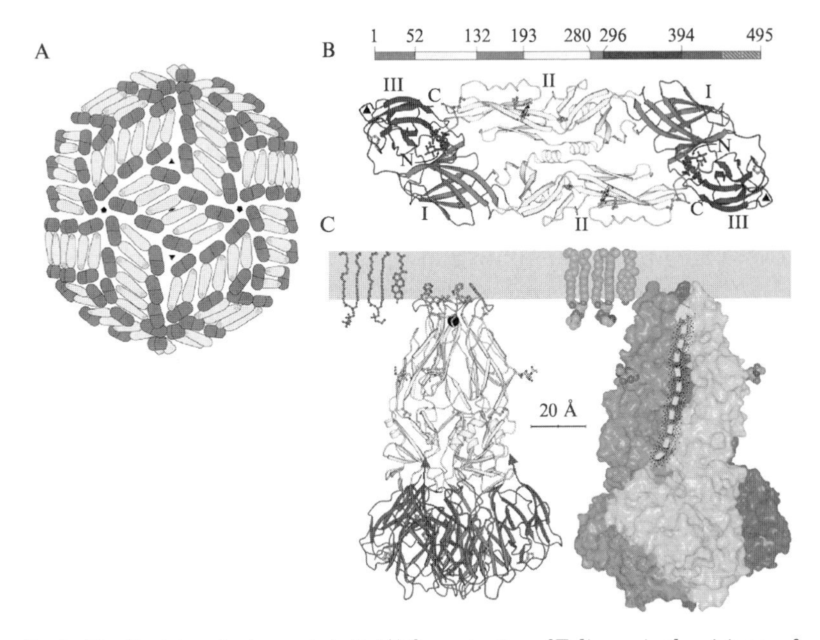


FIG 9. The flavivirus fusion protein E. (A) Organization of E dimers in the virion surface. Each subunit is shown in three colors: domain I in red, domain II in yellow, and domain III in blue—based on the structure described by Kuhn *et al.*, (2002). (B) The soluble ectodomain, sE of dengue virus type 2, in the dimeric prefusion conformation found on mature virions (Modis *et al.*, 2003). The domain colors are as in (A). The bar above the ribbon diagram shows the relationship of domains to primary structure. The "stem" segment between residue 394 and the transmembrane anchor is not included in the three-dimensional structure. (C) The sE trimer (Modis *et al.*, 2004). The proteins are shown in relation to a schematic lipid bilayer to illustrate the likely degree of penetration of the fusion loops (*top*) into the membrane. The ribbon diagram (*left*) is colored as in (A) and (B). Arrows at the C terminus of the polypeptide chain suggest its presumed continuing direction. The surface rendering (*right*) includes a dashed arrow to show the proposed course of the stem peptide, which would lead to the transmembrane anchor. (See Color Insert.)

residues that lie outside the viral membrane) is also a dimer in solution at reasonable concentrations (Heinz *et al.*, 1991). The fusion loop at the tip of domain II lies at the dimer interface, sequestered from potential membrane insertions until lowered pH triggers rearrangement (Rey *et al.*, 1995). On the immature virion, E and prM form heterodimers, which in turn form somewhat asymmetric trimer clusters, and prM (rather than another E) appears to protect the fusion loop from exposure (Y. Zhang *et al.*, 2003). Cleavage of prM releases all but 40 ectodomain residues, leading to dimerization of E and formation of the tightly packed surface array (Heinz *et al.*, 1994).

When exposed to lowered pH, E rearranges yet again into trimers (Allison *et al.*, 1995). The rearranged, presumably postfusion, structure of sE has been determined [for dengue virus type 2 (Modis *et al.*, 2004) and for TBE virus (Bressanelli *et al.*, 2004)] by taking advantage of the observation that sE dissociates reversibly to monomers in solution at low pH, but trimerizes irreversibly if liposomes are present (Stiasny *et al.*, 2002). The conformational change involves reorientations of the three domains with respect to each other and probably an extension of the stem (not present in the sE trimer that crystallizes; Fig. 9C). The clustered domains II present their fusion loops at one end of the trimer. The loops themselves have essentially the same conformation as they do in the prefusion dimer, and the extent of their joint hydrophobic surface leads to the conclusion that they probably interact only with the outer leaflet of the target membrane bilayer and that they can dip by no more than 5 to 10-Å into its hydrophobic region. The reorientation of domain II with respect to domain I is a rotation of about 20°. The reorientation of domain III is substantially greater; it flips over in such a way that its C terminus now projects back toward the fusion loops. The structure leads directly to the proposal that in the fusion rearrangement of intact E, the three stems extend along the lateral faces of the clustered domains II, forcing the C-terminal transmembrane segments to approach the fusion loops. This feature of the conformational transition accomplishes essentially the same result as the fold-back of class I viral fusion proteins: it brings together the two membrane-associated parts of the protein and thus can cause the target cell membrane, into which the fusion loops are presumably inserted, to approach the viral membrane, into which the transmembrane anchor penetrates (Fig. 10).

Structural analysis of the rearrangements that E undergoes during fusion suggests at least two distinct strategies for discovering fusion inhibitors. One follows from the analogy with class I viral fusion and the approach embodied in the development of T-20/enfuvirtide: use of a peptide containing sequences derived from the stem. Provided that the stem interacts as strongly with the threefold clustered domains II as the outer layer helices of gp41 interact with the trimeric core, then stem-derived peptides would be expected to frustrate the zipping-up step (Fig. 10D), just as T-20 or C34 frustrate the zipping up of gp41. A second strategy follows from the observation that a detergent molecule can occupy the hydrophobic pocket at the domain I–domain II interface (Modis *et al.*, 2003). This pocket disappears during the fusion rearrangement, so small molecules that lodge there would probably retard refolding.

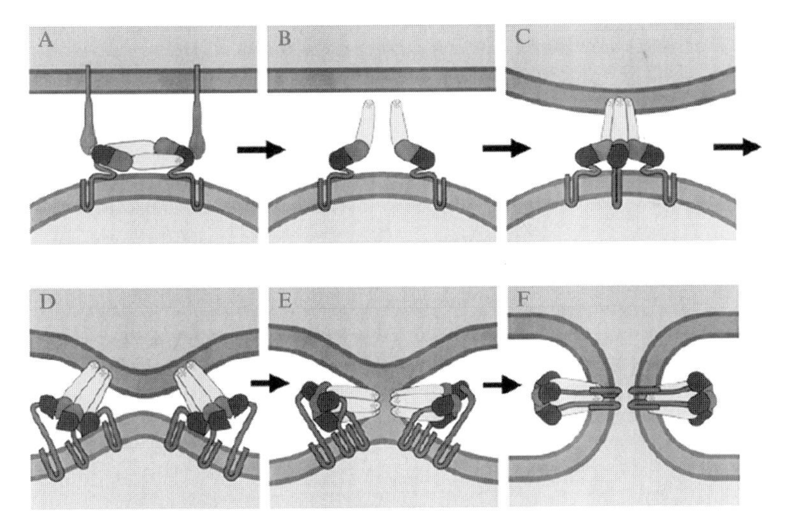


FIG 10. Diagram of membrane fusion by class II viral fusion proteins. (A) Receptor binding through domain III of E (flaviviruses). (B) Lowered pH in an endosome leads to dissociation of the dimer interactions. On release of dimer constraints, monomers can flex outward, presenting their fusion loops to the target cell membrane. (C) Insertion of the fusion loops into the target cell membrane and initial formation of trimer contacts among the projecting domains II. (D) Domain III flips over and the stem zips up along the outside of the trimer. (E) Hemifusion stalk. The diagram shows a proposed role for the inserted fusion loops—stabilization of the hemifusion stalk. (F) Formation of a fusion pore. Completing the zipping up of the stem drives fusion forward, because the cytosolic tails enter the pore and commit it to dilation. (See Color Insert.)

### *B. Alphaviruses*

The structures of two representative alphaviruses—Sindbis virus and Semliki Forest virus (SFV)—have been examined in considerable detail by electron cryomicroscopy (Lescar *et al.*, 2001; Pletnev *et al.*, 2001; Zhang *et al.*, 2002), and crystal structures of the E1 protein from SFV have been obtained, both for the pre- and postfusion forms (Gibbons *et al.*, 2004b; Lescar *et al.*, 2001). E1 has the same architecture as flavivirus E. Each of the domains in E1 resembles closely its E counterpart, including the position and conformation of the fusion loop (Fig. 11). The stem segment of E1 is substantially shorter (30 residues instead of 55). The E1 domain III does not bind receptor; this function resides instead on E2. The soluble ectodomain fragment, E1*, which lacks all but about 10 residues of the stem, is a monomer.

E1 and E2 are a closely associated heterodimer on the surface of the virion (Lescar *et al.*, 2001; Pletnev *et al.*, 2001; Zhang *et al.*, 2002). Their

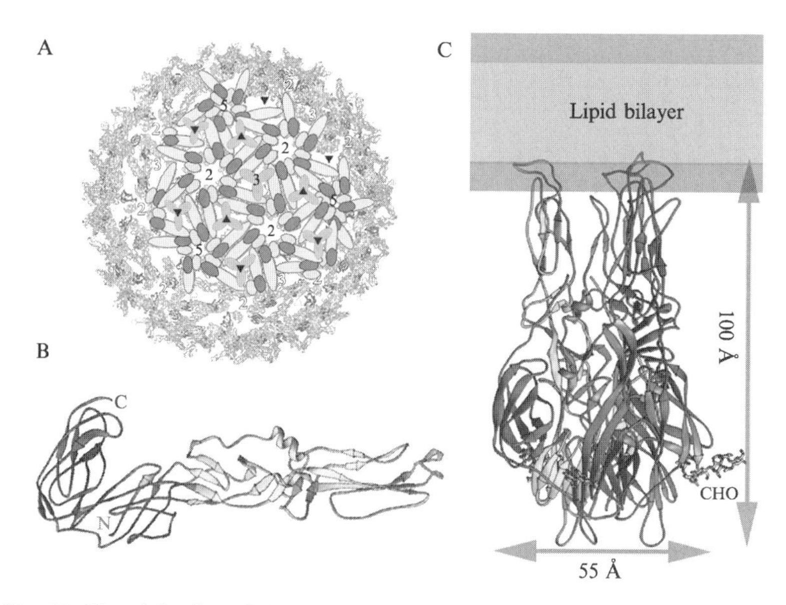


FIG 11. The alphavirus fusion protein E1. (A) Organization of E1 and E2 on the surface of virions. Simplified representations of subunits have been superimposed on a model of the fit of the SFV E1 crystal structure into an image reconstruction of the virion from electron cryomicroscopy (Lescar *et al.*, 2001). E1 is red (domain I), yellow (domain II), and blue (domain III). The E2 trimer (for which only a 9-Å structure is currently known, from electron microscopy) is represented by a green trefoil. It projects outward, capping the fusion loop of E1. Numbers (5, 3, and 2) show the positions of fivefold, threefold, and twofold icosahedral symmetry axes; triangles show the positions of local threefold axes in the T = 4 icosahedral surface lattice. (B) The soluble ectodomain, E1*, of Semliki Forest virus. Domain colors as in (A). (C) The E1* postfusion trimer. Each subunit is a single color. The stem of E1, shorter than the stem of flavivirus E proteins (compare with Fig. 9), would link the C terminus of E1* to the transmembrane anchor. (B) and (C) are adapted from Gibbons *et al.* (2004b). (See Color Insert.)

transmembrane anchors are in contact as they pass through the membrane. E2, for which no high-resolution structure is yet available, projects outward and over E1, covering the E1 fusion loops. Three E2 subunits cluster around the threefold and quasi-threefold axes of the $T = 4$ icosahedral surface lattice, creating a spike like feature in images from electron microscopy; the three associated E1 subunits form a triangular "skirt" around the base of the E2 spike (Fig. 11A).

When alphavirus particles are exposed to lowered pH, protein packing in the icosahedral surface lattice changes substantially (Wahlberg *et al.*, 1989, 1992). The E1–E2 heterodimers dissociate, and E1 trimerizes. Thus, the E2 subunits move away from the threefold axis, permitting E1 to cluster there. As outlined later, the analogy to what happens in influenza HA rearrangement is direct: the receptor-binding fragment, which covers the fusion fragment, loses its threefold interactions and allows the latter to associate along the threefold axis, initially as an extended prefusion intermediate and subsequently as a folded-back, postfusion trimer.

Trimerization and membrane insertion of the soluble E1* from SFV can be induced by lowering the pH in the presence of liposomes (Ahn *et al.*, 2002; Gibbons *et al.*, 2004a; Klimjack *et al.*, 1994). The lipid requirements are the same as those for fusion of intact virions: the target membrane must contain cholesterol and sphingolipid. Electron microscopy of liposomes with inserted E1 or E1* gives similar pictures: arrays of uniform trimers (Gibbons *et al.*, 2003). Thus, the presence of the C-terminal transmembrane anchor does not affect the way the fusion loops insert into the membrane. In the absence of a liposome bilayer into which to insert, E1* remains monomeric at lowered pH, like flavivirus sE. In both cases, the bilayer catalyzes the trimeric association, presumably by concentrating the monomers and organizing them in essentially parallel orientations. The irreversibility of the monomer-to-trimer transition probably resides in both cases in local $\beta$-sheet reogranizations at the two interdomain boundaries.

The crystal structure of the SFV E1* trimer, extracted from liposomes after the procedure just described, shows that the low pH-induced conformational change is essentially just like the one outlined for flavivirus E (Figs. 9–11). The clustering of domain II is less pronounced at the fusion loop tip, so that the three loops are not in contact, as they are in the trimer of dengue sE, and the hydrophobic insertion surface is therefore somewhat different (Fig. 11C). Nonetheless, one can infer the same restricted extent of insertion into the target bilayer, both from an analysis of hydrophobicity at the tip of domain II and from a comparison with the images of membrane-inserted trimers mentioned previously. Why cholesterol should be necessary for membrane insertion of the SFV fusion loop is not immediately evident from the structure. A mutation in E1 that partially alleviates the cholesterol requirement (Vashishtha *et al.*, 1998) lies close to the lipid headgroup region in the model for a membrane-interacting trimer, but the segment of the protein that bears it (see Fig. 11B,C) may still be too far from the membrane to interact with headgroups directly.

In the SFV E1* trimer crystals, there is a lateral interaction between fusion loops of subunits in adjacent trimers. It has been suggested that this contact might represent the way in which a ring of trimers could cooperate in creating a fusion pore (Bressanelli *et al.*, 2004; Gibbons *et al.*, 2004b) (see Section IV.A).

### *C. Class II Fusion Mechanism*

Fusion by class II proteins may proceed with somewhat different kinetics than fusion by HA and other class I proteins. In the case of fusion by SFV E1, once a pore opens, it proceeds to dilate; flickering is infrequent (Samsonov *et al.*, 2002). Thus, folding over of domain III and zipping up of the stem may be strongly concerted processes, so that once a pore has opened, the transmembrane anchors snap firmly into place. It is possible that this property reflects intertrimer cooperativity, as discussed later (Section IV.A).

We can outline five steps in the class II fusion process that parallel the five steps described previously for class I fusion by influenza virus HA.

1. *The low-pH trigger.* Mutations that alter the threshold pH for flavivirus fusion cluster around a hydrophobic pocket at the hinge between domains I and II (Modis *et al.*, 2003). A nonionic detergent molecule inserts into this pocket when crystals of DEN-2 sE are grown in its presence. Several of these mutations change the size of hydrophobic side chains, suggesting that their effect is on the free energy cost of altering the domain I–domain II hinge position, rather than on the p*K* of a titrating residue. Candidates for proton-binding residues are two conserved histidines at the domain I–domain III interface in the prefusion dimer (Bressanelli *et al.*, 2004). One might expect protonation of these histidines to destabilize that interface, inducing dissociation of the dimer as domain III swings away both from its own domain I and from the fusion loop region of the other subunit.

2. *Fusion loop emergence and insertion.* Unlike class I fusion peptides, which when exposed cause the fusion protein to aggregate (e.g., into "rosettes"), class II fusion loops appear to be relatively stable when exposed to an aqueous environment. Flavivirus sE fragments dissociate reversibly at low pH in the absence of liposomes, and the SFV E1 ectodomain monomer crystallizes without any special protection for its fusion loop. Even three clustered fusion loops, as in the DEN-2 sE trimer, do not seem to bind a detergent micelle (Modis *et al.*, 2004). Nonetheless, they clearly insert stably into bilayers, as seen directly by electron microscopy for SFV E1 ectodomain, DEN-2 sE, and tick-borne encephalitis (TBE) sE (Gibbons *et al.*, 2003; Modis *et al.*, 2004; Stiasny *et al.*, 2004). Although the insertion has not been visualized at high resolution, we can infer by comparing the

high-resolution structures with images from negative-stain electron microscopy that the fusion loops penetrate only a short distance into the outer bilayer leaflet. Conserved tryptophans in the loops may lodge at the polar group–hydrocarbon interface, as frequently found in integral membrane proteins.

3. *Fold-back.* Class I fusion proteins are generally trimers both before and after the fusion transition, but trimerization is actually part of the transition in the class II proteins. The structures suggest that the finger-like domains II might initiate the trimer cluster, which could then lock in place as each monomer reverses on itself. The stem segments are likely to be quite flexible, and the description we gave of class I folding back as first a melting out (of the C-terminal part of $HA_2$) and then a zipping up probably applies to class II proteins as well. In this case, the melted-out part would be the stem. Even if the two $\alpha$-helical segments of the flavivirus stem retained their secondary structure during the refolding, the stem could still extend to nearly 100 Å in length, giving more than enough slack for the process diagrammed in Fig. 10D,E.

4. *Hemifusion.* As stated earlier, fusion proteins could catalyze the hemifusion step by promoting appropriate membrane curvature or by stabilizing the hemifusion intermediate (or both). Insertion of the fusion loops only into the outer bilayer leaflet will indeed produce some degree of positive curvature in the target membrane. A simple area calculation suggests, however, that it would take a relatively large number of trimers—more than are likely to contribute to one fusion event—to produce the degree of curvature pictured in some theoretical models of hemifusion (see, e.g., Kuzmin *et al.*, 2001). A more important consequence of the limited insertion may be the possibility for the fusion loops to migrate into the hemifusion stalk itself, thereby stabilizing the stalk and preventing its reveral (fission). This notion is attractive for two reasons. First, it depends on a general property of the fusion segment (limited insertion), and it could apply to class I fusion mechanisms as well. Second, it explains the observation of stable hemifusion (so far, for class I fusion only) when proceeding to full fusion is substantially retarded. This proposed incorporation of fusion loops into the hemifusion stalk is included in the the diagram in Fig. 10.

5. *Fusion pore formation.* Finishing the refolding steps, by bringing the transmembrane segments back onto the threefold axis of the protein trimer, can also lock a fusion pore into place, provided that the fusion loops have not migrated through the hemifusion stalk and into the viral membrane. At its waist, the curvature of the stalk (as seen

from outside the cell by the fusion protein) is positive, and the fusion loops would be expected to stabilize it and perhaps to influence its geometry. Conversely, migration of fusion loops into a hemifusion stalk would be energetically favorable, holding them there until the opening of a pore. These proposals all require testing: the structural data available so far show only the initial and final states of the fusion process.

## IV. Some Questions

### A. *How Many Trimers Are Needed for Fusion?*

Diagrams such as those in Figs. 3 and 10 represent a schematic cross-section through a (presumably) axially symmetric fusion structure. But how many fusion proteins actually surround the hemifusion stalk? Is the number precise, determined by cooperative, lateral interactions among the proteins, or is it variable? What is the minimal number of fusion proteins required to generate a productive fusion event?

The minimal number of fusion proteins required for a productive encounter is set in part by the free energy barrier to fusion and the free energy recovered from the overall refolding reaction. The latter value will obviously vary from case to case. The former has been estimated to be roughly 30–50 kcal/mol (Kuzmin *et al.*, 2001). In principle, this value could be recovered from refolding of one or a few trimers.

Evidence for cooperativity among trimers in HA-mediated fusion supports the notion that more than one trimer participates (Danieli *et al*, 1996). A possible mechanistic basis for cooperativity in class II fusion has been described in connection with analysis of the SFV E1* trimer structure (Gibbons *et al.*, 2004). Lateral contacts between the outer rims of fusion loops from adjacent trimers in the crystals of the low pH-induced form of SFV E1* suggest that these interactions may create a ring of five trimers, angled at roughly 45° to the viral membrane, with their fusion loops (15 in all) forming a hydrophobic "crater" at the tip of a fusion "volcano" (Bressanelli *et al.*, 2004; Gibbons *et al.*, 2004b). These 15 loops might indeed be sufficient to stabilize a nipple-like outpouching of the target cell membrane, a first step toward formation of a hemifusion stalk, and their coordinated participation could promote rapid transit through the hemifusion stage to an open fusion pore, as observed (Samsonov *et al.*, 2002).

Another possible source of cooperativity, even without specific inter-trimer interactions at any step in the fusion process, comes from common anchoring of several participating trimers in both viral and target cell membranes. The prefusion intermediate (Figs. 3B, 10C) presumably flickers toward the zipped-up conformation but proceeds along that direction only when it meets a low enough barrier. If squeezing the two membranes together requires more than one trimer—that is, if the resistance of the membranes to deformation counteracts the tendency of a single extended intermediate to collapse by zipping up into the folded-back structure—then only when several trimers proceed toward the zipped-up conformation in concert will the refolding proceed. In this view, cooperativity is achieved not by defined protein–protein interactions (ring formation), but rather by coupling through the elastic properties of the membranes undergoing fusion. A hemifusion stalk might be generated by as few as one or two trimers or by as many as can fit around the stalk (probably five or six), depending on circumstances, concentrations, and of course the actual free energy recovery from a single trimer. One advantage of this picture is that it does not require additional protein–protein contacts between necessarily flickering and partly disordered intermediate structures.

### *B. What Is the Structure of the Hemifusion Intermediate?*

The detailed structure of the hemifusion state for two bilayers has not yet been determined. Several possible geometries are still consistent with available data. Instead of opening up as a fusion pore, the stalk shown in Fig. 1 could resolve as a hemifusion diaphragm. This state, which can spread laterally, is more likely to be a nonproductive dead end, as rupture of the diaphragm would be required to achieve content mixing and as the fusion protein would probably be relegated to the periphery.

The proposal, embodied in Figs. 3 and 10, that the fusion peptide or fusion loop can migrate into the hemifusion stalk, derives from the conclusion that these segments interact solely with the proximal leaflet of a lipid bilayer and that this migration is feasible. Because hemifusion states can be trapped by suitable mutations in the fusion protein, it will in principle be possible to test this proposal directly.

### *C. Are There Additional Structural Classes of Viral Fusion Proteins?*

The distinction between class I and class II viral fusion proteins is essentially an architectural rather than a mechanistic one. Class I proteins assemble as trimers, are primed by proteolytic cleavage, and rearrange to a postfusion structure based on a trimeric coiled coil. Class II proteins assemble as heterodimers with a protector protein. Cleavage of the latter is the priming event, and the fusion-inducing rearrangement includes trimer formation. The class II proteins studied so far are clearly homologs. The same is not so evident for the class I proteins. Those of the lentiviruses, oncoretroviruses, and filoviruses have some apparent evolutionary kinship, but how they relate to influenza HA (for example) is less obvious. In any case, it is unlikely that the categories just defined exhaust the possible structural solutions to the problem of promoting bilayer fusion. It is probable, however, that fusion proteins of classes yet to be discovered will resemble the ones we know already in how they actually fuse membranes. That is, we can expect them to work through the sequence of fusion peptide exposure, fusion peptide insertion, and overall folding back that we have already analyzed in detail.

### *D. Can Small Molecules Inhibit the Fusion Transition?*

Peptides can inhibit the HIV gp41 fold-back step. How general is this strategy, and can a more drug-like small molecule do the same? An outer layer peptide from Ebola virus fusion protein GP2 blocks entry of VSV pseudotyped with Ebola virus glycoprotein GP1–GP2 (Watanabe *et al.*, 2000). Thus, HIV gp41 is not unique in having an outer layer that binds tightly enough to work as an inhbitor *in trans*. The cyclic peptide that targets a pocket on the inner core of the HIV gp41 trimer has a footprint not much larger than that of a more conventional drug, so even the zipping-up phase of the refolding reaction might be susceptible to small molecule inhibitors. Because HIV fuses at the cell surface, the gp41 fusion transition is more likely to be accessible to a peptide than that of a protein that refolds in an endosome. Small molecules that can penetrate to low-pH compartments may therefore be necessary if the principle demonstrated by the successful use of T-20/enfuvirtide is to be applied to the entry of many other viral pathogens.

## References

Ahn, A., Gibbons, D. L., and Kielian, M. (2002). The fusion peptide of Semliki Forest virus associates with sterol-rich membrane domains. *J. Virol.* **76:**3267–3275.

Allan, J. S., Coligan, J. E., Barin, F., McLane, M. F., Sodroski, J. G., Rosen, C. A., Haseltine, W. A., Lee, T. H., and Essex, M. (1985). Major glycoprotein antigens that induce antibodies in AIDS patients are encoded by HTLV-III. *Science* **228:**1091–1094.

Allison, S. L., Schalich, J., Stiasny, K., Mandl, C. W., and Heinz, F. X. (2001). Mutational evidence for an internal fusion peptide in flavivirus envelope protein E. *J. Virol.* **75:**4268–4275.

Allison, S. L., Schalich, J., Stiasny, K., Mandl, C. W., Kunz, C., and Heinz, F. X. (1995). Oligomeric rearrangement of tick-borne encephalitis virus envelope proteins induced by an acidic pH. *J. Virol.* **69:**695–700.

Allison, S. L., Stiasny, K., Stadler, K., Mandl, C. W., and Heinz, F. X. (1999). Mapping of functional elements in the stem-anchor region of tick-borne encephalitis virus envelope protein E. *J. Virol.* **73:**5605–5612.

Armstrong, R. T., Kushnir, A. S., and White, J. M. (2000). The transmembrane domain of influenza hemagglutinin exhibits a stringent length requirement to support the hemifusion to fusion transition. *J. Cell Biol.* **151:**425–437.

Blacklow, S. C., Lu, M., and Kim, P. S. (1995). A trimeric subdomain of the simian immunodeficiency virus envelope glycoprotein. *Biochemistry* **34:**14955–14962.

Bressanelli, S., Stiasny, K., Allison, S. L., Stura, E. A., Duquerroy, S., Lescar, J., Heinz, F. X., and Rey, F. A. (2004). Structure of a flavivirus envelope glycoprotein in its low-pH-induced membrane fusion conformation. *EMBO J.* **23:**728–738.

Bullough, P. A., Hughson, F. M., Skehel, J. J., and Wiley, D. C. (1994). Structure of influenza haemagglutinin at the pH of membrane fusion. *Nature* **371:**37–43.

Caffrey, M., Cai, M., Kaufman, J., Stahl, S. J., Wingfield, P. T., Covell, D. G., Gronenborn, A. M., and Clore, G. M. (1998). Three-dimensional solution structure of the 44 kDa ectodomain of SIV gp41. *EMBO J.* **17:**4572–4584.

Carr, C. M., and Kim, P. S. (1993). A spring-loaded mechanism for the conformational change of influenza hemagglutinin. *Cell* **73:**823–832.

Chan, D. C., Chutkowski, C. T., and Kim, P. S. (1998). Evidence that a prominent cavity in the coiled coil of HIV type 1 gp41 is an attractive drug target. *Proc. Natl. Acad. Sci. USA* **95:**15613–15617.

Chan, D. C., Fass, D., Berger, J. M., and Kim, P. S. (1997). Core structure of gp41 from the HIV envelope glycoprotein. *Cell* **89:**263–273.

Chen, J., Skehel, J. J., and Wiley, D. C. (1999). N- and C-terminal residues combine in the fusion-pH influenza hemagglutinin $HA_2$ subunit to form an N cap that terminates the triple-stranded coiled coil. *Proc. Natl. Acad. Sci. USA* **96:**8967–8972.

Cohen, F. S., Markosyan, R. M., and Melikyan, G. B. (2002). The process of membrane fusion: Nipples, hemifusion, pores, and pore growth. *Curr. Top. Membr.* **52:**501–529.

Dalgleish, A. G., Beverley, P. C., Clapham, P. R., Crawford, D. H., Greaves, M. F., and Weiss, R. A. (1984). The CD4 (T4) antigen is an essential component of the receptor for the AIDS retrovirus. *Nature* **312:**763–767.

Danieli, T., Pelletier, S. L., Henis, Y. I., and White, J. M. (1996). Membrane fusion mediated by the influenza virus hemagglutinin requires the concerted action of at least three hemagglutinin trimers. *J. Cell Biol.* **133:**559–569.

Daniels, R. S., Downie, J. C., Hay, A. J., Knossow, M., Skehel, J. J., Wang, M. L., and Wiley, D. C. (1985). Fusion mutants of the influenza virus hemagglutinin glycoprotein. *Cell* **40:**431–439.

Dutch, R. E., and Lamb, R. A. (2001). Deletion of the cytoplasmic tail of the fusion protein of the paramyxovirus simian virus 5 affects fusion pore enlargement. *J. Virol.* **75:**5363–5369.

Eckert, D. M., Malashkevich, V. N., Hong, L. H., Carr, P. A., and Kim, P. S. (1999). Inhibiting HIV-1 entry: Discovery of D-peptide inhibitors that target the gp41 coiled-coil pocket. *Cell* **99:**103–115.

Fass, D., Harrison, S. C., and Kim, P. S. (1996). Retrovirus envelope domain at 1.7 Å resolution. *Nat. Struct. Biol.* **3:**465–469.

Feng, Y., Broder, C. C., Kennedy, P. E., and Berger, E. A. (1996). HIV-1 entry cofactor: Functional cDNA cloning of a seven-transmembrane, G protein-coupled receptor. *Science* **272:**872–877.

Ferrer, M., Kapoor, T. M., Strassmaier, T., Weissenhorn, W., Skehel, J. J., Oprian, D., Schreiber, S. L., Wiley, D. C., and Harrison, S. C. (1999). Selection of gp41-mediated HIV-1 cell entry inhibitors from biased combinatorial libraries of non-natural binding elements. *Nat. Struct. Biol.* **6:**953–960.

Furuta, R. A., Wild, C. T., Weng, Y., and Weiss, C. D. (1998). Capture of an early fusion-active conformation of HIV-1 gp41. *Nat. Struct. Biol.* **5:**276–279.

Gibbons, D. L., Ahn, A., Liao, M., Hammar, L., Cheng, R. H., and Kielian, M. (2004a). Multistep regulation of membrane insertion of the fusion peptide of Semliki Forest virus. *J. Virol.* **78:**3312–3318.

Gibbons, D. L., Erk, I., Reilly, B., Navaza, J., Kielian, M., Rey, F. A., and Lepault, J. (2003). Visualization of the target-membrane-inserted fusion protein of Semliki Forest virus by combined electron microscopy and crystallography. *Cell* **114:**573–583.

Gibbons, D. L., Vaney, M. C., Roussel, A., Vigouroux, A., Reilly, B., Lepault, J., Kielian, M., and Rey, F. A. (2004b). Conformational change and protein–protein interactions of the fusion protein of Semliki Forest virus. *Nature* **427:**320–325.

Han, X., Bushweller, J. H., Cafiso, D. S., and Tamm, L. K. (2001). Membrane structure and fusion-triggering conformational change of the fusion domain from influenza hemagglutinin. *Nat. Struct. Biol.* **8:**715–720.

Heinz, F. X., Mandl, C. W., Holzmann, H., Kunz, C., Harris, B. A., Rey, F., and Harrison, S. C. (1991). The flavivirus envelope protein E: Isolation of a soluble form from tick-borne encephalitis virus and its crystallization. *J. Virol.* **65:**5579–5583.

Heinz, F. X., Stiasny, K., Puschner-Auer, G., Holzmann, H., Allison, S. L., Mandl, C. W., and Kunz, C. (1994). Structural changes and functional control of the tick-borne encephalitis virus glycoprotein E by the heterodimeric association with protein prM. *Virology* **198:**109–117.

Jahn, R., Lang, T., and Sudhof, T. C. (2003). Membrane fusion. *Cell* **112:**519–533.

Jiang, S., Lin, K., Strick, N., and Neurath, A. R. (1993). HIV-1 inhibition by a peptide. *Nature* **365:**113.

Jones, P. L., Korte, T., and Blumenthal, R. (1998). Conformational changes in cell surface HIV-1 envelope glycoproteins are triggered by cooperation between cell surface CD4 and co-receptors. *J. Biol. Chem.* **273:**404–409.

Kemble, G. W., Danieli, T., and White, J. M. (1994). Lipid-anchored influenza hemagglutinin promotes hemifusion, not complete fusion. *Cell* **76:**383–391.

Kilby, J. M., and Eron, J. J. (2003). Novel therapies based on mechanisms of HIV-1 cell entry. *N. Engl. J. Med.* **348:**2228–2238.

Kliger, Y., Gallo, S. A., Peisajovich, S. G., Munoz-Barroso, I., Avkin, S., Blumenthal, R., and Shai, Y. (2001). Mode of action of an antiviral peptide from HIV-1: Inhibition at a post-lipid mixing stage. *J. Biol. Chem.* **276:**1391–1397.

Klimjack, M. R., Jeffrey, S., and Kielian, M. (1994). Membrane and protein interactions of a soluble form of the Semliki Forest virus fusion protein. *J. Virol.* **68:**6940–6946.

Kuhn, R. J., Zhang, W., Rossmann, M. G., Pletnev, S. V., Corver, J., Lenches, E., Jones, C. T., Mukhopadhyay, S., Chipman, P. R., Strauss, E. G., Baker, T. S., and Strauss, J. H. (2002). Structure of dengue virus: Implications for flavivirus organization, maturation, and fusion. *Cell* **108:**717–725.

Kuzmin, P. I., Zimmerberg, J., Chizmadzhev, Y. A., and Cohen, F. S. (2001). A quantitative model for membrane fusion based on low-energy intermediates. *Proc. Natl. Acad. Sci. USA* **98:**7235–7240.

Kwong, P. D., Wyatt, R., Majeed, S., Robinson, J., Sweet, R. W., Sodroski, J., and Hendrickson, W. A. (2000). Structures of HIV-1 gp120 envelope glycoproteins from laboratory-adapted and primary isolates. *Struct. Fold. Des.* **8:**1329–1339.

Kwong, P. D., Wyatt, R., Robinson, J., Sweet, R. W., Sodroski, J., and Hendrickson, W. A. (1998). Structure of an HIV gp120 envelope glycoprotein in complex with the CD4 receptor and a neutralizing human antibody. *Nature* **393:**648–659.

Lescar, J., Roussel, A., Wien, M. W., Navaza, J., Fuller, S. D., Wengler, G., and Rey, F. A. (2001). The fusion glycoprotein shell of Semliki Forest virus: An icosahedral assembly primed for fusogenic activation at endosomal pH. *Cell* **105:**137–148.

Lu, M., Blacklow, S. C., and Kim, P. S. (1995). A trimeric structural domain of the HIV-1 transmembrane glycoprotein. *Nat. Struct. Biol.* **2:**1075–1082.

Markin, V. S., Kozlov, M. M., and Borovjagin, V. L. (1984). On the theory of membrane fusion: The stalk mechanism. *Gen. Physiol. Biophys.* **5:**361–377.

Melikyan, G. B., White, J. M., and Cohen, F. S. (1995). GPI-anchored influenza hemagglutinin induces hemifusion to both red blood cell and planar bilayer membranes. *J. Cell Biol.* **131:**679–691.

Modis, Y., Ogata, S., Clements, D., and Harrison, S. C. (2003). A ligand-binding pocket in the dengue virus envelope glycoprotein. *Proc. Natl. Acad. Sci. USA* **100:**6986–6991.

Modis, Y., Ogata, S., Clements, D., and Harrison, S. C. (2004). Structure of the dengue virus envelope protein after membrane fusion. *Nature* **427:**313–319.

Moore, J. P., McKeating, J. A., Weiss, R. A., and Sattentau, Q. J. (1990). Dissociation of gp120 from HIV-1 virions induced by soluble CD4. *Science* **250:**1139–1142.

Munoz-Barroso, I., Salzwedel, K., Hunter, E., and Blumenthal, R. (1999). Role of the membrane-proximal domain in the initial stages of human immunodeficiency virus type 1 envelope glycoprotein-mediated membrane fusion. *J. Virol.* **73:**6089–6092.

Myszka, D. G., Sweet, R. W., Hensley, P., Brigham-Burke, M., Kwong, P. D., Hendrickson, W. A., Wyatt, R., Sodroski, J., and Doyle, M. L. (2000). Energetics of the HIV gp120–CD4 binding reaction. *Proc. Natl. Acad. Sci. USA* **97:**9026–9031.

Pletnev, S. V., Zhang, W., Mukhopadhyay, S., Fisher, B. R., Hernandez, R., Brown, D. T., Baker, T. S., Rossmann, M. G., and Kuhn, R. J. (2001). Locations of carbohydrate sites on alphavirus glycoproteins show that E1 forms an icosahedral scaffold. *Cell* **105:**127–136.

Pollard, S. R., Rosa, M. D., Rosa, J. J., and Wiley, D. C. (1992). Truncated variants of gp120 bind CD4 with high affinity and suggest a minimum CD4 binding region. *EMBO J.* **11:**585–591.

Qiao, H., Armstrong, R. T., Melikyan, G. B., Cohen, F. S., and White, J. M. (1999). A specific point mutant at position 1 of the influenza hemagglutinin fusion peptide displays a hemifusion phenotype. *Mol. Biol. Cell* **10:**2759–2769.

Rey, F. A., Heinz, F. X., Mandl, C., Kunz, C., and Harrison, S. C. (1995). The envelope glycoprotein from tick-borne encephalitis virus at 2 Å resolution. *Nature* **375:**291–298.

Rimsky, L. T., Shugars, D. C., and Matthews, T. J. (1998). Determinants of human immunodeficiency virus type-I resistance to gp41-derived inhibitory peptides. *J. Virol.* **72:**986–993.

Rizzuto, C. D., Wyatt, R., Hernandez-Ramos, N., Sun, Y., Kwong, P. D., Hendrickson, W. A., and Sodroski, J. (1998). A conserved HIV gp120 glycoprotein structure involved in chemokine receptor binding. *Science* **280:**1949–1953.

Robey, W. G., Safai, B., Oroszlan, S., Arthur, L. O., Gonda, M. A., Gallo, R. C., and Fischinger, P. J. (1985). Characterization of envelope and core structural gene products of HTLV-III with sera from AIDS patients. *Science* **228:**593–595.

Root, M. J., Kay, M. S., and Kim, P. S. (2001). Protein design of an HIV-1 entry inhibitor. *Science* **291:**884–888.

Rosenthal, P. B., Zhang, X., Formanowski, F., Fitz, W., Wong, C. H., Meier-Ewert, H., Skehel, J. J., and Wiley, D. C. (1998). Structure of the haemagglutinin–esterase-fusion glycoprotein of influenza C virus. *Nature* **396:**92–96.

Samsonov, A. V., Chatterjee, P. K., Razinkob, V. I., Eng, C. H., Kielian, M., and Cohen, F. S. (2002). Effects of membrane potential and sphingolipid structures on fusion of Semliki Forest virus. *J. Virol.* **76:**12691–12702.

Sattentau, Q. J., and Moore, J. P. (1991). Conformational changes induced in the human immunodeficiency virus envelope glycoprotein by soluble CD4 binding. *J. Exp. Med.* **174:**407–415.

Sattentau, Q. J., Moore, J. P., Vignaux, F., Traincard, F., and Poignard, P. (1993). Conformational changes induced in the envelope glycoproteins of the human and simian immunodeficiency viruses by soluble receptor binding. *J. Virol.* **67:**7383–7393.

Siegel, D. P. (1993). Energetics of intermediates in membrane fusion: Comparison of stalk and inverted micellar intermediate mechanisms. *Biophys. J.* **65:**2124–2140.

Skehel, J. J., and Wiley, D. C. (1998). Coiled coils in both intracellular vesicle and viral membrane fusion. *Cell* **95:**871–874.

Skehel, J. J., and Wiley, D. C. (2000). Receptor binding and membrane fusion in virus entry: The influenza hemagglutinin. *Annu. Rev. Biochem.* **69:**531–569.

Steinhauer, D. A., Wharton, S. A., Skehel, J. J., and Wiley, D. C. (1995). Studies of the membrane fusion activities of fusion peptide mutants of influenza virus hemagglutinin. *J. Virol.* **69:**6643–6651.

Stiasny, K., Allison, S. L., Schalich, J., and Heinz, F. X. (2002). Membrane interactions of the tick-borne encephalitis virus fusion protein E at low pH. *J. Virol.* **76:**3784–3790.

Stiasny, K., Bressanelli, S., Lepault, J., Rey, F. A., and Heinz, F. X. (2004). Characterization of a membrane-associated trimeric low-pH induced form of the class II viral fusion protein E from tick-borne encephalitis virus and its crystallization. *J. Virol.* **78:**3178–3183.

Tan, K., Liu, J., Wang, J., Shen, S., and Lu, M. (1997). Atomic structure of a thermostable subdomain of HIV-1 gp41. *Proc. Natl. Acad. Sci. USA* **94:**12303–12308.

Trkola, A., Dragic, T., Arthos, J., Binley, J. M., Olson, W. C., Allaway, G. P., Cheng-Mayer, C., Robinson, J., Maddon, P. J., and Moore, J. P. (1996). CD4-dependent, antibody-sensitive interactions between HIV-1 and its co-receptor CCR-5. *Nature* **384:**184–187.

Tsurudome, M., Gluck, R., Graf, R., Falchetto, R., Schaller, U., and Brunner, J. (1992). Lipid interactions of the hemagglutinin $HA_2$ $NH_2$-terminal segment during influenza virus-induced membrane fusion. *J. Biol. Chem.* **267:**20225–20232.

Vashishtha, M., Phalen, T., Marquardt, M. T., Ryu, J. S., Ng, A. C., and Kielian, M. (1998). A single point mutation controls the cholesterol dependence of Semliki Forest virus entry and exit. *J. Cell Biol.* **140:**91–99.

Veronese, F. D., DeVico, A. L., Copeland, T. D., Oroszlan, S., Gallo, R. C., and Sarngadharan, M. G. (1985). Characterization of gp41 as the transmembrane protein coded by the HTLV-III/LAV envelope gene. *Science* **229:**1402–1405.

Wahlberg, J. M., Boere, W. A., and Garoff, H. (1989). The heterodimeric association between the membrane proteins of Semliki Forest virus changes its sensitivity to low pH during virus maturation. *J. Virol.* **63:**4991–4997.

Wahlberg, J. M., Bron, R., Wilschut, J., and Garoff, H. (1992). Membrane fusion of Semliki Forest virus involves homotrimers of the fusion protein. *J. Virol.* **66:**7309–7318.

Watanabe, S., Takada, A., Watanabe, T., Ito, H., Kida, H., and Kawaoka, Y. (2000). Functional importance of the coiled-coil of the Ebola virus glycoprotein. *J. Virol.* **74:**10194–10201.

Weissenhorn, W., Carfi, A., Lee, K. H., Skehel, J. J., and Wiley, D. C. (1998). Crystal structure of the Ebola virus membrane fusion subunit, GP2, from the envelope glycoprotein ectodomain. *Mol. Cell* **2:**605–616.

Weissenhorn, W., Dessen, A., Harrison, S. C., Skehel, J. J., and Wiley, D. C. (1997). Atomic structure of the ectodomain from HIV-1 gp41. *Nature* **387:**426–430.

Wild, C., Oas, T., McDanal, C., Bolognesi, D., and Matthews, T. (1992). A synthetic peptide inhibitor of human immunodeficiency virus replication: Correlation between solution structure and viral inhibition. *Proc. Natl. Acad. Sci. USA* **89:**10537–10541.

Wild, C. T., Shugars, D. C., Greenwell, T. K., McDanal, C. B., and Matthews, T. J. (1994). Peptides corresponding to a predictive $\alpha$-helical domain of human immunodeficiency virus type 1 gp41 are potent inhibitors of virus infection. *Proc. Natl. Acad. Sci. USA* **91:**9770–9774.

Wu, L., Gerard, N. P., Wyatt, R., Choe, H., Parolin, C., Ruffing, N., Borsetti, A., Cardoso, A. A., Desjardin, E., Newman, W., Gerard, C., and Sodroski, J. (1996). CD4-induced interaction of primary HIV-1 gp120 glycoproteins with the chemokine receptor CCR-5. *Nature* **384:**179–183.

Zhang, W., Chipman, P. R., Corver, J., Johnson, P. R., Zhang, Y., Mukhopadhyay, S., Baker, T. S., Strauss, J. H., Rossmann, M. G., and Kuhn, R. J. (2003). Visualization of membrane protein domains by cryo-electron microscopy of dengue virus. *Nat. Struct. Biol.* **10:**907–912.

Zhang, Y., Corver, J., Chipman, P. R., Zhang, W., Pletnev, S. V., Sedlak, D., Baker, T. S., Strauss, J. H., Kuhn, R. J., and Rossmann, M. G. (2003). Structures of immature flavivirus particles. *EMBO J.* **22:**2604–2613.

Zhou, G., Ferrer, M., Chopra, R., Kapoor, T. M., Strassmaier, T., Weissenhorn, W., Skehel, J. J., Oprian, D., Schreiber, S. L., Harrison, S. C., and Wiley, D. C. (2000). The structure of an HIV-1 specific cell entry inhibitor in complex with the HIV-1 gp41 trimeric core. *Bioorg. Med. Chem.* **8:**2219–2227.

Zhang, W., Mukhopadhyay, S., Pletnev, S. V., Baker, T. S., Kuhn, R. J., and Rossmann, M. G. (2002). Placement of the structural proteins in Sindbis virus. *J. Virol.* **76:**11645–11658.

ADVANCES IN VIRUS RESEARCH, VOL 64

# STRUCTURE AND ASSEMBLY OF ICOSAHEDRAL ENVELOPED RNA VIRUSES

Richard J. Kuhn and Michael G. Rossmann

Markey Center for Structural Biology, Department of Biological Sciences
Purdue University, West Lafayette, Indiana 47907



## I. Introduction

Considerable progress has been made in deciphering the structure of enveloped viruses and their component proteins. This has been particularly true for enveloped viruses that exhibit icosahedral symmetry in their outer glycoprotein shell. Two groups of enveloped viruses have been shown to have icosahedral symmetry and although they reside in different virus families, they have a number of features that suggest their common evolutionary origins (Strauss and Strauss, 2001). The alphaviruses, belonging to the *Togaviridae* family, and the flaviviruses, belonging to the *Flaviviridae* family, are spherical icosahedral enveloped viruses with plus-strand RNA genomes (Lindenbach and Rice, 2001; Schlesinger and Schlesinger, 2001). Originally they were placed within the same family but genome sequencing and molecular characterization showed differing gene order and differing replication strategies. Whether other members within each family, for example, rubella virus within the *Togaviridae* family or hepatitis




0065-3527/05 $35.00
DOI: 10.1016/S0065-3527(05)64008-0

C virus within *Flaviviridae*, will share the property of icosahedral virions remains to be seen.

Advances in cryoelectron microscopy (cryo-EM) have provided significant insights into the structures of enveloped viruses. Several enveloped virus structures have been reported, using cryo-EM to a resolution better than 10 Å. This is significant, as these lipid-containing viruses have proved recalcitrant in X-ray crystallographic studies. Although the enveloped viruses have been difficult to study by X-ray crystallography, many of the structural proteins of these viruses have been examined at atomic resolution. This has permitted the fitting of atomic resolution structures into cryo-EM density maps, resulting in "pseudo-atomic" structures of enveloped viruses. By analyzing different intermediates in the virus assembly and entry pathways, these dynamic processes can be understood at the molecular level.

## II. General Structural Features of Enveloped Viruses

The defining component for an enveloped virus is a lipid bilayer membrane that is derived from the host (Harrison, 2001). This membrane is acquired by the virus in a process known as budding (Garoff *et al.*, 1998). The driving force behind this process has not been elaborated for these viruses, but it is likely that lateral interactions between the envelope proteins are the predominant determinant. The composition of the viral membrane will be similar if not identical to that found in the cell at the site of budding. However, in the case of the alphaviruses and flaviviruses, there appears to be exclusion of host proteins so that only the viral envelope proteins are present in the budded virion (Strauss, 1978). The tight lateral interactions that are present in both alphavirus and flavivirus envelope proteins prevent host proteins from interfering with the glycoprotein lattice, thus ensuring host protein exclusion.

Viruses that contain a lipid bilayer have distinct advantages in egress and entry over their nonenveloped counterparts. These enveloped viruses have evolved proteins that promote the fusion of their lipid bilayer with that of the host membrane (Heinz and Allison, 2000; Kielian, 1995). This process has been shown to be efficient for both the alphaviruses and flaviviruses and probably contributes to their low particle-to-infectivity ratios (Corver *et al.*, 2000). Viruses that lack membranes must establish a protein pore or disrupt the integrity of a cellular membrane in order to enter a new host cell, and this appears to be a less efficient process than membrane fusion. Enveloped viruses usually exit their host cell without disruption of cellular membranes and death of the cell. Enveloped viruses have been shown to bud from numerous cellular membranes including the nuclear, endoplasmic reticulum (ER), Golgi, and plasma membranes. In contrast, the majority of nonenveloped viruses require cell lysis to exit the host, and thus all viral protein production and replication must be complete at the time of virus release.

Proteins that are incorporated into the transmembrane region of an enveloped virus usually contain a signal sequence that directs translocation into the ER. Once in the lumen of the ER, viral proteins can be posttranslationally processed just like normal cellular proteins destined for the plasma membrane or for secretion. An enveloped virus takes advantage of these processing steps. For example, the ability to add sugar moieties, using the glycosylation apparatus, is necessary to carry out late maturation events involving proteolytic processing necessary to activate viral glycoproteins for their subsequent entry stage. Because the viral proteins are being transported to and processed in the lumen of the ER, and because genome replication occurs in the cytoplasm for plus-strand RNA viruses, an enveloped virion will assemble into distinct layers of protein and membrane. For both alphaviruses and flaviviruses, the outer layer contains a set of envelope proteins that form a regular array anchored to the membrane. The membrane separates this layer from an inner shell of capsid proteins and genome RNA. This multilayer organization is a hallmark of enveloped virus structure.

## III. Alphavirus Structure

### *A. Cryo-electron Microscopy and Image Reconstructions*

The genus *Alphavirus* contains at least 27 distinct viruses, with Sindbis virus being the prototype member (Strauss and Strauss, 1994). Other important members of this group include Semliki Forest, Venezuelan equine encephalitis, eastern equine encephalitis, and Ross River virus. Early electron microscopy studies of the alphaviruses revealed the multilayered disposition and the regular spherical arrangement of the native virus particle (Murphy, 1980). Analysis of the plasma membrane from cells infected with Sindbis virus demonstrated a lattice arrangement of the surface envelope proteins and suggested that these proteins were arranged in a repeating and regular manner (von Bonsdorff and Harrison, 1975). Furthermore, it was clear from electron microscopy images that the virus particle had surface projections, called spikes, which emanated from the membrane surface and decorated the exterior of the virus particle.

Three major proteins are found within the alphavirus particle: the nucleocapsid protein (C) and the two envelope glycoproteins, E1 and E2. A small protein called 6K is also found in the particle, but in much lower amounts than the 240 copies of the 3 major structural proteins (Gaedigk-Nitschko and Schlesinger, 1990). The structural proteins are proteolytically processed from a polyprotein that is translated from a subgenomic mRNA. This use of a separate mRNA allows the virus to differentially regulate the expression of its replication and virion-associated proteins. Early biochemical studies revealed the pathway of alphavirus assembly. The C protein was implicated as a protease and was shown to rapidly associate with ribosomes after its synthesis (Aliperti and Schlesinger, 1978; Söderlund and Ulmanen, 1977). The C protein then interacts with genome RNA and forms a nucleocapsid core (NC) in the cytoplasm of the infected cell. The envelope glycoproteins are translocated into the lumen of the ER and processed by cellular proteases to generate the proteins that ultimately are found in the virion (Strauss and Strauss, 1994). These biochemical studies revealed an association between glycoproteins E1 and PE2, a precursor form of the E2 glycoprotein that is described in greater detail in Section V (Jones *et al.*, 1977; Ziemiecki *et al.*, 1980). The E1–PE2 heterodimer pair is glycosylated, undergoes carbohydrate trimming as it proceeds through the Golgi, and undergoes a maturation cleavage converting PE2 into E2 and E3 just before its arrival at the plasma membrane. At the plasma membrane, the E1–E2 heterodimers oligomerize into E1–E2 trimers, forming spikes. Budding and membrane envelopment occur as the NC and the envelope proteins associate at the plasma membrane (Garoff and Simons, 1974).

Numerous cryo-EM studies have been carried out with alphaviruses and have succeeded in producing structures at increasingly higher resolution (Cheng *et al.*, 1995; Fuller, 1987; Mancini *et al.*, 2000; Paredes *et al.*, 1993, 2001; Vénien-Bryan and Fuller, 1994; Vogel *et al.*, 1986; Zhang *et al.*, 2002a,b). Here, we focus on high-resolution studies that have provided the most detailed images of the virion. Structures of both Semliki Forest virus and Sindbis virus have been reported to resolutions of approximately 9.0 Å, extending and enhancing many of the features that were seen previously at lower resolutions (Mancini *et al.*, 2000; Zhang *et al.*, 2002b). A surface-shaded view of Sindbis virus (Fig. 1A) reveals several characteristic features of the alphavirus virion. Perhaps the most striking feature is the spike-like projections that decorate the surface of the virion. Each spike can be

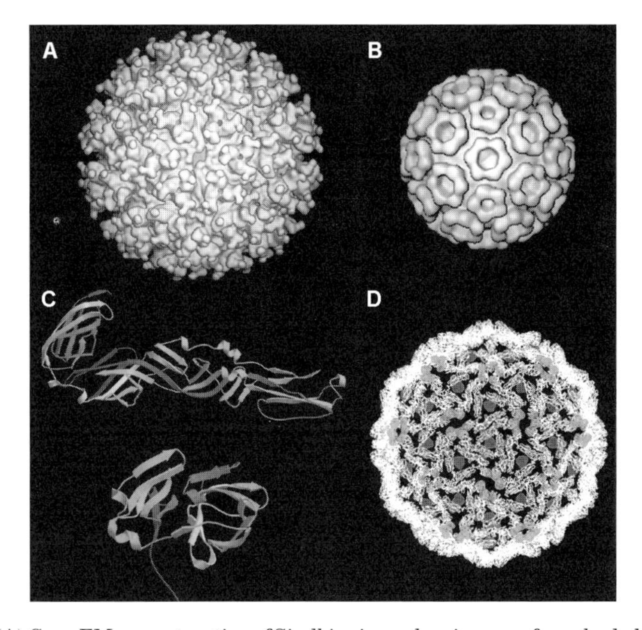


FIG 1. (A) Cryo-EM reconstruction of Sindbis virus, showing a surface-shaded view of the virus as viewed down a twofold axis. The glycoproteins E1 and E2 are collectively colored in blue and the bilayer is green. The resolution of the structure is 20 Å. (B) An alphavirus reconstruction in which the surface features and lipid bilayer were subtracted, leaving a surface-shaded view of the nucleocapsid core. (C) Ribbon diagrams showing atomic structures of the Semliki Forest virus E1 protein ectodomain (residues 1–381) and the Sindbis virus capsid protein proteinase domain (residues 106–264). $\beta$-strands are indicated by the arrows, and $\alpha$-helices are indicated by the coiled ribbons. The color gradation indicates the direction of the $C_\alpha$ backbone as it traverses from the amino terminus (blue) to the carboxy terminus (red) of the protein. (D) Fit of the E1 $C_\alpha$ backbone into the cryo-EM density of Sindbis virus, as viewed down the icosahedral threefold axis. E1 is shown in white, and the glycosylation site at Sindbis E2 residue 318 is shown in the magenta density cage; E1 glycosylation residues 139 and 245 are shown in green and red, respectively. (See Color Insert.)

divided into a set of three petals, suggesting a trimeric organization of the spike. The regular icosahedral arrangement of the spikes is another distinguishing feature of these viruses and is readily apparent in the surface-shaded view. The tight packing of the spikes and the supporting protein shell or skirt provide an extensive but not complete covering of the viral membrane. The two envelope glycoproteins both traverse and penetrate the inner face of the lipid bilayer, with E2 making direct contact with the nucleocapsid core. The nucleocapsid core, like the glycoprotein spikes on the exterior of the virion, is organized into a $T = 4$ icosahedron. A surface-shaded representation of the core

(Fig. 1B) clearly shows the pentameric and hexameric projections or capsomers. However, insight into the molecular organization of the virus particle requires details of the individual protein components.

### *B. Structures of E1 Glycoprotein and Capsid Protein*

The atomic structure of the E1 glycoprotein ectodomain from Semliki Forest virus was determined by X-ray crystallography (Lescar *et al.*, 2001). This protein was obtained by detergent treatment of purified virus followed by proteolysis, using subtilisin, to release the transmembrane components and PNGaseF to disrupt the E1–E2 heterodimer (Wengler *et al.*, 1999). The structure of the E1 ectodomain (Fig. 1C, top) can be divided into three domains: domain I is the central domain, an eight-stranded $\beta$ barrel, connecting dimerization domain II to the immunoglobulin-like domain III. The distal end of domain II contains the fusion peptide, which has been shown to play an important role in the process of membrane fusion (Levy-Mintz and Kielian, 1991). The structure of the E1 glycoprotein revealed a striking structural similarity to the previously solved atomic structure of the flavivirus E protein, despite having little sequence similarity (Rey *et al.*, 1995). Although the E1 ectodomain is a monomer in solution, it crystallized as a dimer. The dimer interface is head to tail with the monomers facing back to back, leaving the fusion peptide exposed. There is an angle of about 20° between the two monomers with the fusion peptide pointing up.

The structures of the C-terminal domain (residues 114 through 264 in Sindbis virus) of the C proteins of both Sindbis and Semliki Forest viruses were solved by X-ray crystallography (Choi *et al.*, 1991, 1997). This conserved domain of the C protein has a chymotrypsin-like fold, composed of two Greek key $\beta$-barrel subdomains with the substrate-binding site located between them (Fig. 1C, bottom). The structure supports the claims of several laboratories that the C protein had proteolytic activity (Aliperti and Schlesinger, 1978; Simmons and Strauss, 1974). The crystal structure revealed that the specificity pocket was occupied by Trp-264, the final residue of the C protein, as a result of autocatalysis. This suggested that the proteinase was incapable of subsequent *trans* cleavages, supporting previous evidence for a single *cis* cleavage (Hahn and Strauss, 1990; Melancon and Garoff, 1987). Further crystallographic studies have identified a hydrophobic pocket located on an outer surface of the C protein that has been postulated to bind C protein during the early nucleocapsid core assembly process and shown to bind the glycoprotein E2 during the later stages of virus assembly (Lee *et al.*, 1996; Skoging *et al.*, 1996).

### *C. Pseudo-atomic Structure of an Alphavirus*

The availability of both X-ray structures of viral components and high-resolution cryo-EM reconstructions allowed for the fitting of the former into the latter. Previous studies had fitted the Sindbis C protein into the Ross River virus reconstruction. However, that result was limited by the low (25 Å) resolution of the Ross River reconstruction, which failed to provide sufficient density features to allow an unambiguous fit of the protein (Cheng *et al.*, 1995). Pletnev *et al.* used a different strategy to establish an acceptable fit of the E1 and E2 glycoproteins into cryo-EM density (Pletnev *et al.*, 2001). By introducing single amino acid substitutions that abrogated glycosylation sites, followed by cryo-EM reconstructions of mutant and wild-type viruses, difference maps were produced that could be used as positional markers for the glycosylation sites. A surprising observation was made on the basis of the location of glycosylation sites for the E1 and E2 glycoproteins. The two sites present in the E2 glycoprotein were roughly perpendicular to the plane of the membrane and separated by 77 Å. However, the two sites of the E1 glycoprotein were roughly at the same radius from the virion center and were separated by a distance of 70 Å. Given the size of the E1 glycoprotein, it was suggested that the bulk of the E1 ectodomain was parallel to the viral membrane, analogous with a previous hypothesis from the E protein of flaviviruses (see Section IV). At the same time, Rey and collaborators had established the structure of the Semliki Forest E1 ectodomain and demonstrated the similarity in structure between alphavirus and flavivirus fusion proteins (Lescar *et al.*, 2001). Thus, it was reasonable to use the glycosylation sites as restraints to dock the glycoprotein structure from tick-borne encephalitis E protein into the Sindbis map and refine the four independent positions with the program EMfit (Rossmann *et al.*, 2001). This produced an ordered arrangement of the glycoprotein into an icosahedral scaffold for the virus and it made up the majority of the region under the spike and covering the membrane, a region previously referred to as the skirt (Fig. 1d). The E1 glycoprotein was found to be at an angle of about 35° rather than parallel to the membrane. Three E1 molecules were arranged in a triangular array around the perimeter of a spike. The dimer arrangement seen in the crystal structure identified the lateral interactions of E1 between adjacent spikes. Using their newly solved atomic structure, Lescar *et al.* were able to directly fit the Semliki Forest E1 ectodomain into the cryo-EM structure of Semliki Forest virus determined independently by Mancini *et al.* and demonstrated similar results (Lescar *et al.*, 2001).

The 9.0-Å resolution map of Sindbis virus and the Mancini *et al.* map both showed density across the 40-Å lipid bilayer, indicating the helical nature of the E1–E2 transmembrane regions (Mancini *et al.*, 2000; Zhang *et al.*, 2002b). The density in the Sindbis structure was of sufficient quality to permit the fitting of the unrelated heterodimeric coiled coil of GCN4 into the transmembrane region (Zhang *et al.*, 2002b). Furthermore, it was possible to trace the density into the inner side of the lipid bilayer and follow the E2 cytoplasmic domain. E2 has 33 amino acids that penetrate the bilayer and numerous experiments have suggested that direct contact between the C protein and E2 is required for budding (Garoff *et al.*, 1998; Metsikkö and Garoff, 1990). The unique density features present in the reported 9.0-Å structure of Sindbis permitted the fitting of the Sindbis C protein into the nucleocapsid core density. The fitting, using the EMfit program, provided a quantitatively unique orientation that was different than that previously reported (Rossmann *et al.*, 2001). The new orientation of the C protein placed the N terminus toward the inner face of the core, where it might be expected to interact with genomic RNA. The density for the E2 glycoprotein could be monitored as it entered the hydrophobic pocket of the C protein, thus confirming the previous prediction as to site of the E2 cytoplasmic domain–C protein interaction (Lee *et al.*, 1996; Skoging *et al.*, 1996).

The structure of the E2 glycoprotein has not yet been elucidated by X-ray crystallography. However, after docking the E1 structure into the Sindbis cryo-EM density, a difference density map was calculated to show the remaining density that represents the E2 protein (Zhang *et al.*, 2002b). The E2 glycoprotein is a long and thin molecule, somewhat similar in shape to and twined around E1. It sits on top of E1, capping and protecting the fusion peptide, and forms the external, large leaf-like domain of the alphavirus spike. Three E1–E2 heterodimers lean against each other, making a three-start, right-handed helix in the formation of each spike.

## IV. Flavivirus Structure

### A. *Structure of E Glycoprotein*

Flaviviruses comprise a genus within the family *Flaviviridae* and represent a group of extremely important human pathogens (Lindenbach and Rice, 2001). Examples of the more than 70 viruses that are found within the genus are yellow fever virus, dengue virus, West Nile virus, Japanese encephalitis virus, and tick-borne encephalitis virus. Like alphaviruses, they contain a lipid bilayer membrane that surrounds the plus-strand genomic RNA and multiple copies of the C protein. A protein shell outside the bilayer contains two transmembrane proteins. The major surface protein is approximately 500 amino acids in size and is called the E, or envelope, protein. This protein is responsible for both receptor binding and for membrane fusion. A small membrane or M protein of about 75 amino acids is also found in the virion, although its function in the mature virion is unknown. The M protein is proteolytically processed during the maturation of the virus from a larger precursor (prM) and is discussed in Section V.

Treatment of purified tick-borne encephalitis virus particles with trypsin resulted in the release of the E protein ectodomain in the form of dimers (Heinz *et al.*, 1991). The protein was crystallized and its structure was solved to a resolution of 2.0 Å (Rey *et al.*, 1995). The protein has a three-domain structure that is homologous to the alphavirus E1 structure described previously (Fig. 2C). Domain III contains the IgC-like motif and has been implicated in receptor attachment, a significant difference in functional role from the alphavirus E1 domain III. Domain I is the central domain and domain II, containing the fusion peptide, acts as the dimerization domain. The dimer is organized differently from the alphavirus E1 dimer, with its interface in a head-to-tail orientation and the monomers facing front to front. This places the fusion peptide on the interior of the dimer, protected from the aqueous environment, unlike the E1 dimer. The atomic structure for the dengue virus E protein has also been solved by X-ray crystallography (Modis *et al.*, 2003; Y. Zhang, S. Ogata, R. J. Kuhn, and M. G. Rossmann, unpublished data). As expected, on the basis of 37% sequence identity, the dengue E protein structure is homologous to the tick-borne encephalitis E protein. Two crystal forms of the dengue protein were identified, one with, and one without, the presence of the detergent *n*-octyl-β-D-glucoside in a hydrophobic pocket (Modis *et al.*, 2003). This pocket is lined with residues that influence the pH threshold of fusion and was found in two conformations depending on the presence or absence of detergent. It has been postulated that this conformational rearrangement may serve as a pivot point between domains I and II during the fusion process.

### *B. Organization of Flavivirus Particles*

Early electron microscopy studies of flaviviruses showed small particles with a dense outer layer but relatively few features to permit a clear prediction as to icosahedral symmetry (Murphy, 1980). The first structural insight into the icosahedral organization of a flavivirus

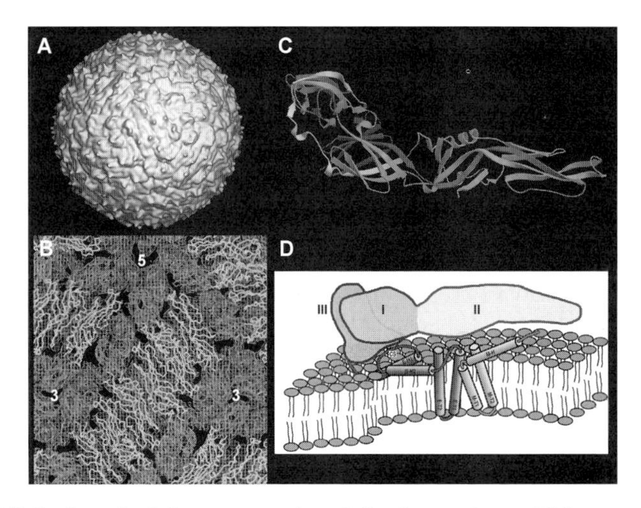


FIG 2. (A) Surface-shaded representation of the dengue 2 cryo-EM reconstruction at a resolution of 12 Å, as viewed down a twofold axis. The blue envelope proteins completely obscure the underlying membrane. (B) The fit of E monomers into the dengue structure is shown with domains I, II, and III in red, yellow, and blue, respectively. The limits of one icosahedral asymmetric unit can be defined by connecting the fivefold and threefold axes. (C) Ribbon diagram of the dengue E protein ectodomain (residues 1–395). $\beta$-Strands are indicated by the arrows and $\alpha$-helices by the coiled ribbons. The color gradation indicates the direction of the $C_\alpha$ backbone as it traverses from the amino terminus (blue) to the carboxy terminus (red) of the protein. (D) Diagrammatic drawing of the dengue virus ectodomain and transmembrane domain proteins. The volume occupied by the ectodomain of an E monomer is shown in pink (domain I), yellow (domain II), and lilac (domain III). The stem and anchor helices of E and M are shown in blue and orange, respectively. Helices are identified by the nomenclature described in text. CS represents the linker region between E-H1 and E-H2. (See Color Insert.)

came from the cryo-EM structure determination of a recombinant subviral particle (RSP) (Ferlenghi *et al.*, 2001). In many flavivirus infections, virus-specific particles are released from infected cells that are substantially smaller than mature virions and lack infectivity. These particles lack both capsid protein and genome RNA but retain hemagglutinating and fusion activity, and can be used as an immunogen with protective capability (Corver *et al.*, 2000; Konishi *et al.*, 1992). Expression of the prM and E proteins in cell culture results in capsid-independent budding and the production of RSPs that resemble the small particles produced during normal flavivirus infection (Allison *et al.*, 1995b; Schalich *et al.*, 1996). Taking advantage of this observation, Ferlenghi and colleagues expressed the prM and E proteins of tick-borne encephalitis virus in a mammalian expression system and were able to collect and purify sufficient amounts of RSPs for cryo-EM analysis (Ferlenghi *et al.*, 2001). Although there was a size distribution that centered on 315 Å, the majority of particles had undergone the maturation cleavage converting prM to M.

The RSP structure showed a dense outer layer nearly covering the relatively thin 30-Å lipid bilayer (Ferlenghi *et al.*, 2001). Examination of the external density features clearly showed the icosahedral threefold axes and pronounced projections around the fivefold positions, although the surface was rather smooth. This smooth surface was predicted by the earlier work of Rey *et al.* on the E protein structure (Rey *et al.*, 1995). They suggested, on the basis of the curved shape of the E dimer when viewed from the side, along with mutational and biochemical data, that the dimer would lie flat on the viral membrane. Thus, it was possible to place the atomic structure of the tick-borne encephalitis E protein dimer onto the icosahedral twofold axis and refine its position using the program EMfit, accounting for 30 E dimers in a $T = 1$ arrangement (Rossmann, 2000). Predicting that the mature native virion would contain a quasi-equivalent arrangement of E, Ferlenghi *et al.* suggested a $T = 3$ organization of E that would be promoted by the presence of the capsid (Ferlenghi *et al.*, 2001). Their predicted flavivirus virion would have a classic arrangement of 90 E dimers somewhat similar to that seen with the E1 protein of alphaviruses. However, the E dimer would lie parallel to the viral membrane, unlike the angled disposition of the alphavirus E1.

The structure of a complete flavivirus particle was determined by cryo-EM of purified dengue virus (Kuhn *et al.*, 2002). The 24-Å resolution reconstruction showed a 500-Å particle with a smooth surface and icosahedral features (Fig. 2A). The particle had a thin, dense outer layer, above a lipid bilayer, and an inner shell of density presumed to be the nucleocapsid core. The density of the inner shell was only about 50% of the height found in the external glycoprotein shell and suggested a disordered capsid shell.

Interpretation of the flavivirus virion cryo-EM density required placement of the E protein atomic structure into the density. This was accomplished with the tick-borne encephalitis E protein structure, as the dengue E structure was not yet available (Rey *et al.*, 1995). Placement of the E protein dimer onto the icosahedral twofold axis yielded an acceptable fit and was followed by the addition of another dimer on a quasi-twofold axis. This gave 3 monomers per asymmetric unit and thus 180 monomers in the virion. The overall fit of the 90 TBEV E dimer structures to the cryo-EM map of dengue virus provided a close packing arrangement with sets of three parallel dimers (Fig. 2B). These sets associate to form a "herringbone" configuration on the viral surface. Although the $T = 3$ organization of the flavivirus virion as predicted by Ferlenghi and colleagues was a reasonable extrapolation from the then available data, it was clearly different from the actual E protein arrangement in dengue virus (Ferlenghi *et al.*, 2001). This arrangement markedly differs from a true $T = 3$ structure because the dimers on the icosahedral twofold axes do not have a quasi-threefold relationship to the dimers on the quasi-twofold axes. The environment of the dimers on the icosahedral twofold axes is substantially different from the environment of the dimers on the quasi-twofold axes. At a resolution of 24 Å, it was not possible to unambiguously identify transmembrane helices and the small M polypeptide, although there was density that could be assigned to those components.

A higher resolution structure of dengue virus was achieved by including more virus particles in the image reconstruction process (W. Zhang *et al.*, 2003). At a resolution of 9.5 Å significantly more features were observed in the dengue virion, especially in the bilayer region. An obvious benefit of the higher resolution structure was the ability to more accurately fit the E protein into the cryo-EM density. Zhang *et al.* did this by creating a homology-modeled structure of the dengue E protein based on the tick-borne E structure (W. Zhang *et al.*, 2003). This was then computationally fitted into the density, using the EMfit program, confirming the previous fitting (Rossmann, 2000). Two glycosylation sites, at residues 67 and 153, were identified by generating a difference map, in which the density corresponding to the fitted E dimers was set to zero. Density peaks close to Asn-67 and Asn-153 appeared for each of the three E glycoproteins in an icosahedral asymmetric unit. These glycosylation sites are also readily observable as small bumps on the virus surface at higher resolution in a surface-shaded view.

The lipid bilayer can be clearly identified at a resolution of 9.5 Å and takes on a polygonal rather than a spherical shape. Densities that can be attributed to protein can be identified as crossing the membrane and this is where constrictions that accentuate the polygonal shape of the bilayer can be seen. The mechanism for this altered appearance of the membrane is not known but the transmembrane proteins must surely be influencing the curvature.

After fitting the homology-modeled dengue E protein into the density, additional well-defined density could be seen in the lipid bilayer and in the space between the outer leaflet of the bilayer and the fitted ectodomain of the E protein. A model proposed by Allison and Heinz to explain the E protein topology suggested that following the C terminus of the E ectodomain, as determined in the X-ray structure, two helices connected by a loop of conserved amino acids would precede the two transmembrane domains (Allison *et al.*, 1999; Heinz and Allison, 2000). This was partially corroborated by the high-resolution dengue structure, in which density could be traced from the C terminus of the fitted E protein at residue 394 through the lipid bilayer (Fig. 2d). The density was consistent with two helical regions (E-H1 and E-H2) connected by a loop (CS, conserved sequence). These helices were found to be positioned in the plane (E-H2) or partially in the plane (E-H1) of the outer leaflet of the membrane. They were connected to two transmembrane $\alpha$ helices (E-T1 and E-T2) that form an antiparallel coiled coil joined by density in the internal lipid leaflet. Thus, the path for the E glycoprotein can be traced from its amino terminus through to its carboxy terminus. After accounting for the three E molecules within the asymmetric unit, additional density could still be seen in the lipid bilayer and in the region between the bilayer and the E ectodomain. This density was assigned to the M protein. Secondary structural predictions of the M protein suggested a weakly amphipathic $\alpha$ helix (M-H) followed by the two transmembrane $\alpha$ helices (M-T1, M-T2). The cryo-EM density verified that prediction and showed that helix M-H is partially buried at its C terminus in the outer lipid leaflet and makes an angle of about 20° with the membrane surface (Fig. 2d). The N-terminal 20 amino acids of M are not visible but they are expected to travel underneath E and terminate close to the position of the "holes" that can be seen between monomers in the E dimer. Exactly what role, if any, M plays in the mature virus is not known. However, the function of the M precursor, prM, is understood and is essential for escorting the immature virus particle through the Golgi and preventing premature fusion.

## V. Virus Assembly

### A. *Role of Glycoproteins in Icosahedral Structure*

As evidence accumulated for the formation of a spherical well-defined nucleocapsid core in the cytoplasm of alphavirus-infected cells, it was assumed that an icosahedral core would drive the formation of a similarly organized glycoprotein network on the virus exterior (Strauss and Strauss, 1994). Unlike the flaviviruses, where capsidless particles could be produced, alphaviruses released only particles that contained capsid. However, it was shown that large deletions in the capsid protein of Semliki Forest virus prevented intracellular core formation and yet icosahedral virus particles were released that contained the crippled capsid protein (Forsell *et al.*, 2000). Forsell and colleagues suggested that the defect in core formation could be rescued by glycoprotein association and budding (Forsell *et al.*, 1996). Attempts to carry out cryo-EM on *in vitro*-assembled alphavirus nucleocapsid cores yielded reconstructions that suggested that the cores were not organized as perfect icosahedrons, or that they were unstable and lost symmetry during purification (Mukhopadhyay *et al.*, 2002). This implies that the exact icosahedral arrangement seen in the nucleocapsid of mature virions is imposed by the glycoproteins after their interaction with the capsid.

The determination of the structure of E1 and the placement of the protein into an ordered icosahedral lattice further argued that alphavirus protein organization might be driven by the outer, not the inner, shell of proteins. This would then be similar to the flaviviruses, in which $T = 1$ particles lacking core are released from infected cells and can be generated by expression of solely the envelope proteins. If flaviviruses can use their glycoprotein scaffold to bud from cells without the help or presence of the capsid then why don't alphaviruses bud particles from infected cells lacking capsid? It seems logical to suggest that the alphavirus E2 protein must interact with the capsid through its 33-amino acid cytoplasmic domain and that such an interaction is lacking in the flaviviruses, where both C-terminal transmembrane domains of M and E remain within the lipid bilayer. However, this difference in budding strategy remains an enigma.

### *B. Assembling the Nucleocapsid Core*

Alphavirus nucleocapsid cores accumulate in the cytoplasm of the cell and associate under the plasma membrane before budding (Söderlund, 1973). Mutations that negate core assembly prevent the accumulation of core particles (Perera *et al.*, 2001). Furthermore, purified capsid protein, in the presence of single-stranded nucleic acid, promotes the *in vitro* assembly of alphavirus capsid cores (Tellinghuisen *et al.*, 1999). The *in vitro*-assembled cores are similar in size and shape to the *in vivo*-assembled cores that are seen within infected cells. Extensive studies have established the requirement for E2–capsid interactions to promote the process of budding (Garoff *et al.*, 1998). Studies have also established that the E1 and E2 transmembrane domains must interact appropriately for successful virus release from the cell plasma membrane (Sjoberg and Garoff, 2003; Strauss *et al.*, 2002). Thus, in alphaviruses, the role of the capsid must be to recruit genome RNA and to promote the budding process, ensuring an infectious particle.

The role of the flavivirus capsid protein is not as well established. After cleavage of its C-terminal signal sequence, the capsid protein probably associates with the ER membrane and perhaps serves as a nucleation site for the recruitment of genome RNA and envelope proteins (Lindenbach and Rice, 2001). The flavivirus capsid differs in several ways from the alphavirus protein. Alphaviruses contain a C-terminal proteinase domain of approximately 150 amino acids, which is lacking in the flavivirus capsid. In contrast, nuclear magnetic resonance (NMR) studies show that the 100-amino acid flavivirus capsid protein has four helices that form a tight core (Jones *et al.*, 2003; Ma *et al.*, 2003). The structure for the N-terminal 105 residues of the alphavirus capsid is not known, although there is predicted to be only a single $\alpha$ helix in that domain (Perera *et al.*, 2001). The alphavirus capsid is monomeric in solution and rapidly oligomerizes into core particles in the presence of single-stranded nucleic acid (Tellinghuisen *et al.*, 1999). In contrast, purified dengue and yellow fever capsid proteins behave as dimers in solution and do not significantly oligomerize in the presence of nucleic acid (Jones *et al.*, 2003). Binding sites on the genomic RNA and the capsid protein have been identified for the Sindbis capsid yet such sites have not been found in the flaviviruses (Weiss *et al.*, 1989, 1994). Finally, and this might be the most important distinction, the alphavirus nucleocapsid core is organized into a $T = 4$ icosahedral structure within the native virus particle. At the present time, a similarly organized flavivirus core is not evident in cryo-EM reconstructions (W. Zhang *et al.*, 2003). Thus, the core of flaviviruses might not have a unique structure, or their icosahedral orientation is not synchronized with the larger and dominant external structure. As neither the E nor M protein appears to extend beyond the limits of the cytoplasmic side of the lipid membrane, the capsid is probably randomly oriented relative to the outer glycoprotein shell, unlike alphaviruses, where a glycoprotein interaction has been firmly established.

### *C. Immature Virus Particles*

In alphaviruses, assembly is a multistep event culminating in the release of an infectious virion at the plasma membrane. In contrast, flaviviruses can bud two forms of immature particles into the ER compartment: those with and without genome RNA and capsid protein. In both alphaviruses and flaviviruses, an immature form of the virion is produced that requires a protease cleavage to render the particle infectious (Allison *et al.*, 1995a; de Curtis and Simons, 1988; Presley *et al.*, 1991; Wengler and Wengler, 1989). In both cases, the fusion protein is paired with a precursor protein that is subsequently cleaved by a furin-like protease in a late secretory compartment. Before cleavage the heterodimer stabilizes the fusion protein, preventing premature release in the slightly acidic Golgi compartments. Immature forms of both an alphavirus and flavivirus have been examined in cryo-EM and image reconstructions (Ferlenghi *et al.*, 1998; Paredes *et al.*, 1998; Y. Zhang *et al.*, 2003). Despite the major differences in the mature virus particles, the immature alphavirus and flavivirus particles appear remarkably similar in structure, suggesting a functional similarity.

Sindbis virus E2 is synthesized in the form of a precursor called PE2. This precursor has an additional 64 amino acids at the N terminus of E2 and contains the signal sequence required for translocation into the ER lumen (Bonatti and Blobel, 1979). Mutations that abrogate the furin cleavage site, and prevent release of the amino-terminal fragment termed E3, are lethal for the virus (Heidner *et al.*, 1994). However, it has been possible to isolate second-site suppressors that restore viability yet retain the PE2 precursor in virus particles, permitting cryo-EM structural studies of the immature alphavirus particles (Paredes *et al.*, 1998). The resulting structure shows that the additional density contributed by E3 partially fills the space between the petals of the trimeric glycoprotein spike (Paredes *et al.*, 1998). Similar observations regarding the location of E3 were also made for a cleavage-defective mutant of Semliki Forest virus (Ferlenghi *et al.*, 1998; Salminen *et al.*, 1992). Unfortunately, high-resolution structures of these immature alphaviruses are not available and hence detailed position assignments of the additional E3 peptide in relation to E1 and its fusion peptide could not be made. However, the data support the hypothesis that the presence of PE2 favors a strong E1–PE2 interaction and prevents premature low pH-activated fusion.

Biochemical experiments on flaviviruses have established that in the immature form of the virus, a prM–E heterodimer is favored (Wengler and Wengler, 1989). On maturation, carried out by a furin-like cleavage of the prM protein, the heterodimer rearranges to form E homodimers (Allison *et al.*, 1995a). Like the alphaviruses, abrogation of the prM cleavage site results in noninfectious virions

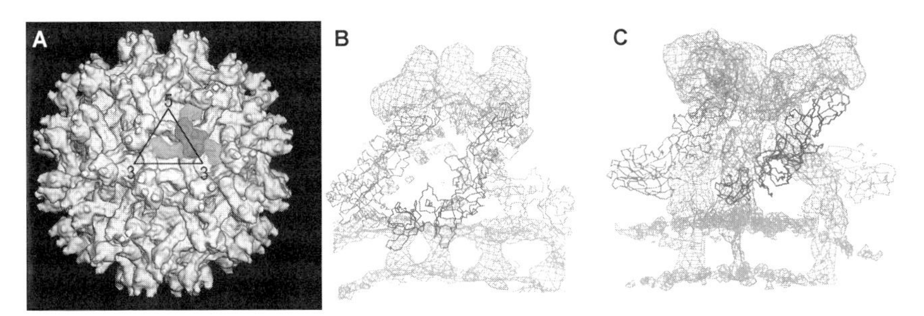


FIG 3. (A) Surface-shaded view of an immature dengue virus at a resolution of 16 Å. An icosahedral asymmetric unit is outlined in black. One of the 60 spikes in each particle is identified in color: prM in gray and E in green. (B and C) Comparison of the spike structure of (B) immature dengue particles and (C) mature Sindbis virus. The $C_\alpha$ backbones of the E and E1 glycoprotein trimers (in blue, red, and green) are shown for the dengue and Sindbis particles, respectively. The corresponding densities have been set to zero, leaving the densities (in gray) for the prM and E2 molecules, respectively. The lipid bilayer is shown in green. The slab used for depicting the lipid bilayer is thinner than that used to define the ectodomain trimer. (See Color Insert.)

(Elshuber *et al.*, 2003). A similar effect can be accomplished by treatment of cells with the acidotropic agent ammonium chloride, which blocks the ability of the furin-like enzyme to cleave the immature particle (Guirakhoo *et al.*, 1991). Using this approach, Zhang *et al.* were able to isolate immature dengue particles and carry out a cryo-EM reconstruction to a resolution of 16 Å (Y. Zhang *et al.*, 2003). The immature virus looks radically different from the mature virus, having projecting trimeric spikes (Fig. 3A). The structure of the E protein was fitted into the surface density and was shown to be tilted with respect to the viral surface, such that the long axis containing domain II makes an angle of $\sim 25^\circ$ to the surface (Fig. 3B). The fusion peptides of the three E monomers were found to be at the tip of each spike. By removing the density contributed by the fitted E proteins, the location of the prM proteins were identified. The prM proteins cover the three E fusion peptides of domain II and density contributed by prM follows the length of each E monomer in a spike down to the membrane. Thus, prM forms a cap over the distal end of the E protein in a manner analogous to E2 covering E1 in mature alphaviruses (Fig. 3B and C). The conversion of the immature flavivirus after cleavage of prM to the mature virus will require a substantial rearrangement. The mechanism of such a conformational change is not known.

## VI. Concluding Remarks

Significant advances have been made in understanding the structure and assembly of icosahedral enveloped viruses. It is now apparent that these viruses undergo significant conformational changes as they mature within the cell. Biochemical results and structural studies have suggested that equally dramatic changes in protein organization occur during the process of fusion. The next goal will be to describe these pathways of assembly and disassembly in molecular detail. Ultimately such an understanding will be necessary to explain the evolutionary and functional relationships of these viruses to those enveloped viruses lacking regular symmetry.

## Acknowledgments

The authors thank Suchetana Mukhopadhyay for help with the figures and many useful discussions, and Felix Rey for providing refined coordinates necessary to generate the E1 structure. This work was supported by a National Institutes of Health Program Project grant to R.J.K. and M.G.R. (AI45976); and by a National Institutes of Health grant to R.J.K. (GM56279).

## References

Aliperti, G., and Schlesinger, M. J. (1978). Evidence for an autoprotease activity of Sindbis virus capsid protein. *Virology* **90:**366–369.

Allison, S. L., Schalich, J., Stiasny, K., Mandl, C. W., Kunz, C., and Heinz, F. X. (1995a). Oligomeric rearrangement of tick-borne encephalitis virus envelope proteins induced by an acidic pH. *J. Virol.* **69:**695–700.

Allison, S. L., Stadler, K., Mandl, C. W., Kunz, C., and Heinz, F. X. (1995b). Synthesis and secretion of recombinant tick-borne encephalitis virus protein E in soluble and particulate form. *J. Virol.* **69:**5816–5820.

Allison, S. L., Stiasny, K., Stadler, K., Mandl, C. W., and Heinz, F. X. (1999). Mapping of functional elements in the stem-anchor region of tick-borne encephalitis virus envelope protein E. *J. Virol.* **73:**5605–5612.

Bonatti, S., and Blobel, G. (1979). Absence of a cleavable signal sequence in Sindbis virus glycoprotein PE2. *J. Biol. Chem.* **254:**12261–12264.

Cheng, R. H., Kuhn, R. J., Olson, N. H., Rossmann, M. G., Choi, H. K., Smith, T. J., and Baker, T. S. (1995). Nucleocapsid and glycoprotein organization in an enveloped virus. *Cell* **80:**621–630.

Choi, H. K., Lu, G., Lee, S., Wengler, G., and Rossmann, M. G. (1997). The structure of Semliki Forest virus core protein. *Proteins* **27:**345–359.

Choi, H. K., Tong, L., Minor, W., Dumas, P., Boege, U., Rossmann, M. G., and Wengler, G. (1991). Structure of Sindbis virus core protein reveals a chymotrypsin-like serine proteinase and the organization of the virion. *Nature* **354:**37–43.

Corver, J., Ortiz, A., Allison, S. L., Schalich, J., Heinz, F. X., and Wilschut, J. (2000). Membrane fusion activity of tick-borne encephalitis virus and recombinant subviral particles in a liposomal model system. *Virology* **269:**37–46.

de Curtis, I., and Simons, K. (1988). Dissection of Semliki Forest virus glycoprotein delivery from the *trans*-Golgi network to the cell surface in permeabilized BHK cells. *Proc. Natl. Acad. Sci. USA* **85:**8052–8056.

Elshuber, S., Allison, S. L., Heinz, F. X., and Mandl, C. W. (2003). Cleavage of protein prM is necessary for infection of BHK-21 cells by tick-borne encephalitis virus. *J. Gen. Virol.* **84:**183–191.

Ferlenghi, I., Clarke, M., Ruttan, T., Allison, S. L., Schalich, J., Heinz, F. X., Harrison, S. C., Rey, F. A., and Fuller, S. D. (2001). Molecular organization of a recombinant subviral particle from tick-borne encephalitis. *Mol. Cell* **7:**593–602.

Ferlenghi, I., Gowen, B., de Haas, F., Mancini, E. J., Garoff, H., Sjöberg, M., and Fuller, S. D. (1998). The first step: Activation of the Semliki Forest virus spike protein precursor causes a localized conformational change in the trimeric spike. *J. Mol. Biol.* **283:**71–81.

Forsell, K., Griffiths, G., and Garoff, H. (1996). Preformed cytoplasmic nucleocapsids are not necessary for alphavirus budding. *EMBO J.* **15:**6495–6505.

Forsell, K., Xing, L., Kozlovska, T., Cheng, R. H., and Garoff, H. (2000). Membrane proteins organize a symmetrical virus. *EMBO J.* **19:**5081–5091.

Fuller, S. D. (1987). The T = 4 envelope of Sindbis virus is organized by interactions with a complementary T = 3 capsid. *Cell* **48:**923–934.

Gaedigk-Nitschko, K., and Schlesinger, M. J. (1990). The Sindbis virus 6K protein can be detected in virions and is acylated with fatty acids. *Virology* **175:**274–281.

Garoff, H., Hewson, R., and Opstelten, D. J. E. (1998). Virus maturation by budding. *Microbiol. Mol. Biol. Rev.* **62:**1171–1190.

Garoff, H., and Simons, K. (1974). Location of the spike glycoproteins in the Semliki Forest virus membrane. *Proc. Natl. Acad. Sci. USA* **71:**3988–3992.

Guirakhoo, F., Heinz, F. X., Mandl, C. W., Holzmann, H., and Kunz, C. (1991). Fusion activity of flaviviruses: Comparison of mature and immature (prM-containing) tick-borne encephalitis virions. *J. Gen. Virol.* **72:**1323–1329.

Hahn, C. S., and Strauss, J. H. (1990). Site-directed mutagenesis of the proposed catalytic amino acids of the Sindbis virus capsid protein autoprotease. *J. Virol.* **64:**3069–3073.

Harrison, S. C. (2001). Principles of virus structure. *In* "Field's Virology" (D. M. Knipe and P. M. Howley, eds.), 4th Ed., pp. 53–85. Lippincott Williams & Wilkins, Philadelphia, PA.

Heidner, H. W., McKnight, K. L., Davis, N. L., and Johnston, R. E. (1994). Lethality of PE2 incorporation into Sindbis virus can be suppressed by second-site mutations in E3 and E2. *J. Virol.* **68:**2683–2692.

Heinz, F. X., and Allison, S. L. (2000). Structures and mechanisms in flavivirus fusion. *Adv. Virus Res.* **55:**231–269.

Heinz, F. X., Mandl, C. W., Holzmann, H., Kunz, C., Harris, B. A., Rey, F., and Harrison, S. C. (1991). The flavivirus envelope protein E: Isolation of a soluble form from tick-borne encephalitis virus and its crystallization. *J. Virol.* **65:**5579–5583.

Jones, C. T., Ma, L., Burgner, J. W., Groesch, T. D., Post, C. B., and Kuhn, R. J. (2003). Flavivirus capsid protein is a dimeric $\alpha$-helical protein. *J. Virol.* **77:**7143–7149.

Jones, K. J., Scupham, R. K., Pfeil, J. A., Wan, K., Sagik, B. P., and Bose, H. R. (1977). Interaction of Sindbis virus glycoproteins during morphogenesis. *J. Virol.* **21:**778–787.

Kielian, M. (1995). Membrane fusion and the alphavirus life cycle. *Adv. Virus Res.* **45:**113–151.

Konishi, E., Pincus, S., Paoletti, E., Shope, R. E., Burrage, T., and Mason, P. W. (1992). Mice immunized with a subviral particle containing the Japanese encephalitis virus prM/M and E proteins are protected from lethal JEV infection. *Virology* **188:**714–720.

Kuhn, R. J., Zhang, W., Rossmann, M. G., Pletnev, S. V., Corver, J., Lenches, E., Jones, C. T., Mukhopadhyay, S., Chipman, P. R., Strauss, E. G., Baker, T. S., and Strauss, J. H. (2002). Structure of dengue virus: Implications for flavivirus organization, maturation, and fusion. *Cell* **108:**717–725.

Lee, S., Owen, K. E., Choi, H. K., Lee, H., Lu, G., Wengler, G., Brown, D. T., Rossmann, M. G., and Kuhn, R. J. (1996). Identification of a protein binding site on the surface of the alphavirus nucleocapsid protein and its implication in virus assembly. *Structure* **4:**531–541.

Lescar, J., Roussel, A., Wein, M. W., Navaza, J., Fuller, S. D., Wengler, G., Wengler, G., and Rey, F. A. (2001). The fusion glycoprotein shell of Semliki Forest virus: An icosahedral assembly primed for fusogenic activation at endosomal pH. *Cell* **105:**137–148.

Levy-Mintz, P., and Kielian, M. (1991). Mutagenesis of the putative fusion domain of the Semliki Forest virus spike protein. *J. Virol.* **65:**4292–4300.

Lindenbach, B. D., and Rice, C. M. (2001). *Flaviviridae*: The viruses and their replication. *In* "Field's Virology" (D. M. Knipe and P. M. Howley, eds.), 4th Ed., pp. 991–1041. Lippincott Williams & Wilkins, Philadelphia, PA.

Ma, L., Jones, C. T., Groesch, T. D., Kuhn, R. J., and Post, C. B. (2003). Solution structure of dengue virus capsid protein reveals a new fold. *Proc. Natl. Acad. Sci. USA* **101:**3414–3419.

Mancini, E. J., Clarke, M., Gowen, B. E., Rutten, T., and Fuller, S. D. (2000). Cryo-electron microscopy reveals the functional organization of an enveloped virus, Semliki Forest virus. *Mol. Cell* **5:**255–266.

Melancon, P., and Garoff, H. (1987). Processing of the Semliki Forest virus structural polyprotein: Role of the capsid protease. *J. Virol.* **61:**1301–1309.

Metsikkö, K., and Garoff, H. (1990). Oligomers of the cytoplasmic domain of the p62/E2 membrane protein of Semliki Forest virus bind to the nucleocapsid *in vitro*. *J. Virol.* **64:**4678–4683.

Modis, Y., Ogata, S., Clements, D., and Harrison, S. C. (2003). A ligand-binding pocket in the dengue virus envelope glycoprotein. *Proc. Natl. Acad. Sci. USA* **100:**6986–6991.

Mukhopadhyay, S., Chipman, P. R., Hong, E. M., Kuhn, R. J., and Rossmann, M. G. (2002). *In vitro*-assembled alphavirus core-like particles maintain a structure similar to that of nucleocapsid cores in mature virus. *J. Virol.* **76:**11128–11132.

Murphy, F. A. (1980). Togavirus morphology and morphogenesis. *In* "The Togaviruses" (R. W. Schlesinger, ed.), pp. 241–316. Academic Press, New York.

Paredes, A., Alwell-Warda, K., Weaver, S. C., Chiu, W. I., and Watowich, S. J. (2001). Venezuelan equine encephalomyelitis virus structure and its divergence from Old World alphaviruses. *J. Virol.* **75:**9532–9537.

Paredes, A. M., Brown, D. T., Rothnagel, R., Chiu, W., Schoepp, R. J., Johnston, R. E., and Prasad, B. V. V. (1993). Three-dimensional structure of a membrane-containing virus. *Proc. Natl. Acad. Sci. USA* **90:**9095–9099.

Paredes, A. M., Heidner, H., Thuman-Commike, P., Prasad, B. V., Johnston, R. E., and Chiu, W. (1998). Structural localization of the E3 glycoprotein in attenuated Sindbis virus mutants. *J. Virol.* **72:**1534–1541.

Perera, R., Owen, K. E., Tellinghuisen, T. L., Gorbalenya, A. E., and Kuhn, R. J. (2001). Alphavirus nucleocapsid protein contains a putative coiled coil $\alpha$-helix important for core assembly. *J. Virol.* **75:**1–10.

Pletnev, S. V., Zhang, W., Mukhopadhyay, S., Fisher, B. R., Hernandez, R., Brown, D. T., Baker, T. S., Rossmann, M. G., and Kuhn, R. J. (2001). Locations of carbohydrate sites on Sindbis virus glycoproteins show that E1 forms an icosahedral scaffold. *Cell* **105:**127–136.

Presley, J. F., Polo, J. M., Johnston, R. E., and Brown, D. T. (1991). Proteolytic processing of the Sindbis virus membrane protein precursor PE2 is nonessential for growth in vertebrate cells but is required for efficient growth in invertebrate cells. *J. Virol.* **65:**1905–1909.

Rey, F. A., Heinz, F. X., Mandl, C., Kunz, C., and Harrison, S. C. (1995). The envelope glycoprotein from tick-borne encephalitis virus at 2 Å resolution. *Nature* **375:**291–298.

Rossmann, M. G. (2000). Fitting atomic models into electron microscopy maps. *Acta Crystallogr. D* **56:**1341–1349.

Rossmann, M. G., Bernal, R., and Pletnev, S. V. (2001). Combining electron microscopic with x-ray crystallographic structures. *J. Struct. Biol.* **136:**190–200.

Salminen, A., Wahlberg, J. M., Lobigs, M., Liljestrom, P., and Garoff, H. (1992). Membrane fusion process of Semliki Forest virus II: Cleavage-dependent reorganization of the spike protein complex controls virus entry. *J. Cell Biol.* **116:**349–357.

Schalich, J., Allison, S. L., Stiasny, K., Mandl, C. W., Kunz, C., and Heinz, F. X. (1996). Recombinant subviral particles from tick-borne encephalitis virus are fusogenic and provide a model system for studying flavivirus envelope glycoprotein functions. *J. Virol.* **70:**4549–4557.

Schlesinger, S., and Schlesinger, M. J. (2001). *Togaviridae*: The viruses and their replication. *In* "Field's Virology" (D. M. Knipe and P. M. Howley, eds.), 4th Ed., pp. 895–916. Lippincott Williams & Wilkins, Philadelphia, PA.

Simmons, D. T., and Strauss, J. H. (1974). Translation of Sindbis virus 26S RNA and 49S RNA in lysates of rabbit reticulocytes. *J. Mol. Biol.* **86:**397–409.

Sjoberg, M., and Garoff, H. (2003). Interactions between the transmembrane segments of the alphavirus E1 and E2 proteins play a role in virus budding and fusion. *J. Virol.* **77:**3441–3450.

Skoging, U., Vihinen, M., Nilsson, L., and Liljeström, P. (1996). Aromatic interactions define the binding of the alphavirus spike to its nucleocapsid. *Structure* **4:**519–529.

Söderlund, H. (1973). Kinetics of formation of Semliki Forest virus nucleocapsid. *Intervirology* **1:**354–361.

Söderlund, H., and Ulmanen, I. (1977). Transient association of Semliki Forest virus capsid protein with ribosomes. *J. Virol.* **24:**907–909.

Strauss, E. G. (1978). Mutants of Sindbis virus III. Host polypeptides present in purified HR and *ts* 103 particles. *J. Virol.* **28:**466–474.

Strauss, E. G., Lenches, E., and Strauss, J. H. (2002). Molecular genetic evidence that the hydrophobic anchors of glycoproteins E2 and E1 interact during assembly of alphaviruses. *J. Virol.* **76:**10188–10194.

Strauss, J. H., and Strauss, E. G. (1994). The alphaviruses: Gene expression, replication, and evolution. *Microbiol. Rev.* **58:**491–562.

Strauss, J. H., and Strauss, E. G. (2001). Virus evolution: How does an enveloped virus make a regular structure? *Cell* **105:**5–8.

Tellinghuisen, T. L., Hamburger, A. E., Fisher, B. R., Ostendorp, R., and Kuhn, R. J. (1999). *In vitro* assembly of alphavirus cores by using nucleocapsid protein expressed in *Escherichia coli*. *J. Virol.* **73:**5309–5319.

Vénien-Bryan, C., and Fuller, S. D. (1994). The organization of the spike complex of Semliki Forest virus. *J. Mol. Biol.* **236:**572–583.

Vogel, R. H., Provencher, S. W., von Bonsdorff, C.-H., Adrian, M., and Dubochet, J. (1986). Envelope structure of Semliki Forest virus reconstructed from cryo-electron micrographs. *Nature* **320:**533–535.

von Bonsdorff, C.-H., and Harrison, S. C. (1975). Sindbis virus glycoproteins form a regular icosahedral surface lattice. *J. Virol.* **16:**141–148.

Weiss, B., Geigenmüller-Gnirke, U., and Schlesinger, S. (1994). Interactions between Sindbis virus RNAs and a 68 amino acid derivative of the viral capsid protein further defines the capsid binding site. *Nucleic Acids Res.* **22:**780–786.

Weiss, B., Nitschko, H., Ghattas, I., Wright, R., and Schlesinger, S. (1989). Evidence for specificity in the encapsidation of Sindbis virus RNAs. *J. Virol.* **63:**5310–5318.

Wengler, G., and Wengler, G. (1989). Cell-associated West Nile flavivirus is covered with E + pre-M protein heterodimers which are destroyed and reorganized by proteolytic cleavage during virus release. *J. Virol.* **63:**2521–2526.

Wengler, G., Wengler, G., and Rey, F. A. (1999). The isolation of the ectodomain of the alphavirus E1 protein as a soluble hemagglutinin and its crystalization. *Virology* **257:**472–482.

Zhang, W., Chipman, P. R., Corver, J., Johnson, P. R., Zhang, Y., Mukhopadhyay, S., Baker, T. S., Strauss, J. H., Rossmann, M. G., and Kuhn, R. J. (2003). Visualization of membrane protein domains by cryo-electron microscopy of dengue virus. *Nat. Struct. Biol.* **10:**907–912.

Zhang, Y., Corver, J., Chipman, P. R., Pletnev, S. V., Sedlak, D., Baker, T. S., Strauss, J. H., Kuhn, R. J., and Rossmann, M. G. (2003). Structures of immature flavivirus particles. *EMBO J.* **22:**2604–2613.

Zhang, W., Fisher, B. R., Olson, N. H., Strauss, J. H., Kuhn, R. J., and Baker, T. S. (2002a). Aura virus structure suggests that the T=4 organization is a fundamental property of viral structural proteins. *J. Virol.* **76:**7239–7246.

Zhang, W., Mukhopadhyay, S., Pletnev, S. V., Baker, T. S., Kuhn, R. J., and Rossmann, M. G. (2002b). Placement of the structural proteins in Sindbis virus. *J. Virol.* **76:**11645–11658.

Ziemiecki, A., Garoff, H., and Simons, K. (1980). Formation of the Semliki Forest virus membrane glycoprotein complexes in the infected cell. *J. Gen. Virol.* **50:**111–123.

# KINETIC AND MASS SPECTROMETRY-BASED INVESTIGATION OF HUMAN IMMUNODEFICIENCY VIRUS TYPE 1 ASSEMBLY AND MATURATION

Jason Lanman* and Peter E. Prevelige, Jr.

Department of Microbiology, University of Alabama at Birmingham
Birmingham, Alabama 35294



## I. Introduction

Advances in mass spectrometry have made it possible to ionize proteins and peptides without fragmentation. Concurrent advances in mass analyzers have made it possible to obtain high-resolution mass spectra. As a result, mass spectrometry is becoming an increasingly important tool in structural biology. It holds particular promise for application to the analysis of complex macromolecular machines, where it serves as a complement to the traditional approaches of X-ray crystallography and electron microscopy. In this article, we describe the application of mass spectrometry to the analysis of the assembly and maturation of human immunodeficiency virus type 1 (HIV-1).

* Present Address: Department of Molecular Biology, The Scripps Research Institute, La Jolla, California 92037.




0065-3527/05 $35.00
DOI: 10.1016/S0065-3527(05)64009-2

### *A. HIV Life Cycle and Maturation*

The assembly of HIV-1 is a two-step process. In the first step, the pr55 Gag polyprotein assembles at the plasma membrane, forming an immature virion. The Gag polyprotein is composed of three structural domains. From the N terminus they are as follows: the matrix (MA) domain, the capsid (CA) domain, and the nucleocapsid (NC) domain. The N terminus of the MA domain is myristoylated and associates with the plasma membrane. In the immature virion, the remainder of the Gag polyprotein is radially arranged with the NC domain pointing toward the center of the spherical virion. The immature virion buds from the host cell and the Gag polyprotein then undergoes a programmed sequence of proteolytic events mediated by the virally encoded protease to liberate the MA, CA, and NC domains. MA remains associated with the viral membrane, whereas CA and NC collapse to form a conical core structure in which the CA surrounds the NC–RNA complex (Fig. 1). Proper formation of the core is essential for viral infectivity. Protease inhibitors block cleavage and hence the structural transformation, thereby inhibiting viral infectivity. Likewise, mutations, or small organic molecules that allow cleavage to proceed but block proper core formation, are invariably deleterious

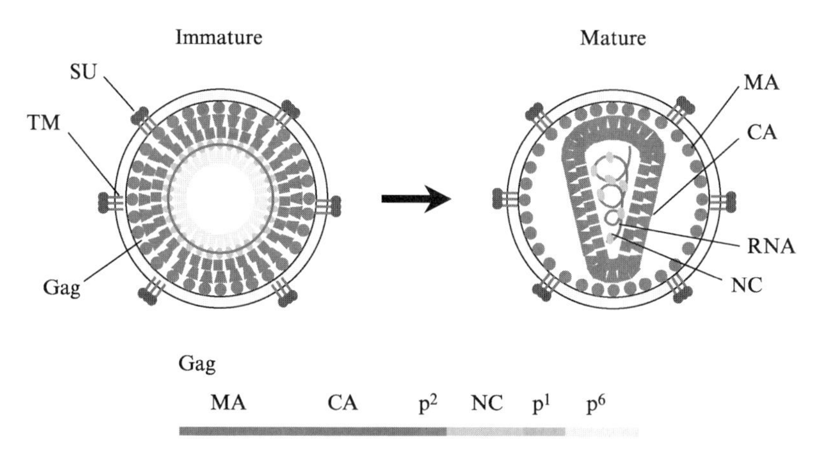


FIG 1. Schematic representation of the maturation stage of the HIV-1 life cycle. The spherical immature virus is composed of the Gag polyprotein. After budding, the virally encoded protease is activated and cleaves Gag at the indicated junctions (*bottom*). The liberated domains are free to rearrange. The MA domain remains membrane-associated while the CA collapses to form the conical core. Contained within the conical core is the diploid viral genome complexed with NC. The locations of p2, p1, and p6 are unknown. (See Color Insert.)

to infectivity. Thus, in addition to posing an intriguing question in form determination, understanding the molecular events leading to the formation of the conical core has important therapeutic implications.

### *B. HIV Structural Biology*

Considerable effort has been expended toward understanding the structural biology of HIV, a goal made particularly challenging by the fact that the virions are nonicosahedral and pleiomorphic, making crystallography and cryoelectron microscopy (cryo-EM) image reconstruction extremely difficult. To circumvent these problems a "divide and conquer" strategy has been employed in which high-resolution crystallographic and nuclear magnetic resonance (NMR)-based structures of the individual domains in isolation have been obtained (Gamble *et al.*, 1996, 1997; Gitti *et al.*, 1996; Hill *et al.*, 1996; Massiah *et al.*, 1994; Worthylake *et al.*, 1999). Now, the challenge is ultimately to put the pieces back together to gain a picture of the whole viral capsid. Of particular interest are the changes in the CA protein accompanying maturation, as this is the protein that forms the conical core of the virion.

The capsid protein is composed of two domains, the N-terminal domain and the C-terminal domain (Fig. 2A). The domains are connected by a "linker" region that appears unstructured in the isolated subunit. The N-terminal domain is composed of seven $\alpha$ helices and a $\beta$ hairpin whereas the C-terminal domain is composed of four $\alpha$ helices. The C-terminal domain of HIV-1 forms dimers in solutions with a $K_d$ of $\sim$20 $\mu$M whereas the N-terminal domain has not been observed to form any oligomeric structures (Gamble *et al.*, 1997; Rose *et al.*, 1992; Yoo *et al.*, 1997).

### *C. Models and Structure of Virus-Like Particles*

A significant advance in understanding the structure of the mature HIV-1 virion was achieved with the attainment of cryo-EM-based reconstructions of a family of tubes of CA protein assembled *in vitro* (Li *et al.*, 2000). CA protein assembles into both tubes and cones *in vitro*, suggesting the two forms are structurally related. Although the cones are regular, falling into discrete families of cone angles, they are not well suited for cryo-EM reconstruction (Ganser *et al.*, 1999). The tubes also fell into helical families and it was possible to reconstruct members of four different families to a resolution of $\sim$30 Å. The different families of tubes displayed structural similarities with regard

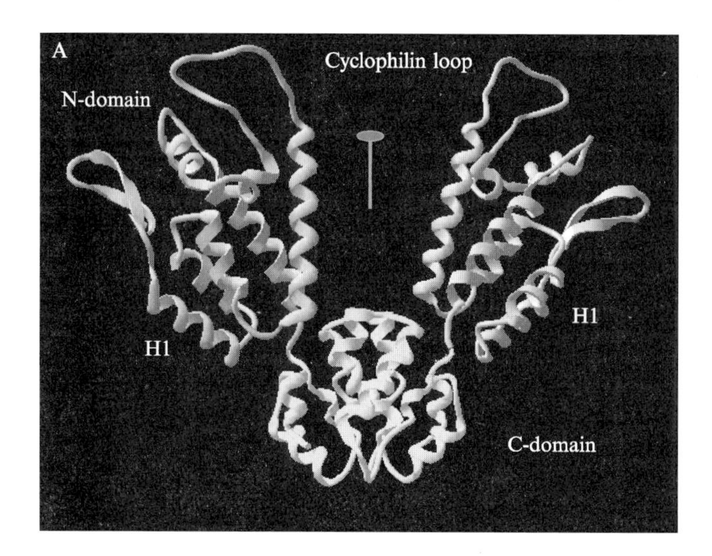


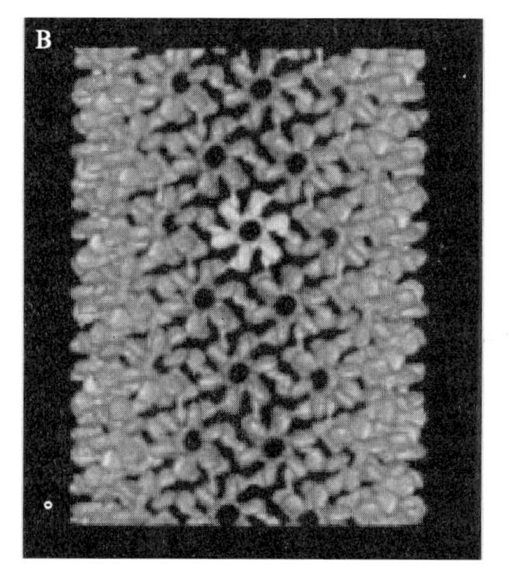


FIG 2. (A) Crystal structure of the N- and C-terminal domains of CA. The crystal structures of the N- and C-terminal domains were solved individually. They are connected by a flexible linker region that is not visualized in the intact protein. The connection is arbitrary and the relative orientation of the two domains is unknown. The C-terminal domain forms a dimer in both the crystal and in solution. (B) Three-dimensional reconstruction of tubes of CA. Three-dimensional reconstructions of tubes of CA polymerized *in vitro* were obtained by cryo-EM. The tubes consist of hexameric protein clusters that are interconnected by dimeric interactions lying slightly below the surface of the tube. (See Color Insert.)

to the overall subunit packing and interaction (Fig. 2B). In the reconstructions, packed hexameric clusters of protein molecules formed the outer surface of the tube; these clusters were interconnected below the tube surface so that each hexamer was tethered to its six nearest neighbors. The crystallographic density of the N-terminal domain of CA was unambiguously merged into the hexameric clusters, with N-terminal domain helices I and II forming the intersubunit interface. The density linking each subunit of a hexamer to its neighbor was assigned to the C-terminal domain, consistent with the observation that the C-terminal domain forms dimers both in solution and in the crystal. A model of the conical cores was built on the basis of the N-terminal domain hexameric clusters and C-terminal domain dimers observed in the tubes, demonstrating that the CA core in the virion could be formed from a similar hexameric lattice.

## II. Hydrogen–Deuterium Exchange Theory and Practice

Hydrogen–deuterium exchange as applied to protein structural studies monitors the rate at which individual amide protons exchange with protons from the solvent. Although the chemical rate of exchange is on the order of $10^1$–$10^3$ $s^{-1}$ under physiological conditions the observed rate of exchange is sensitive to the degree of hydrogen bonding and the extent of solvent accessibility. These structure-based "protection factors" can decrease the observed rate of exchange as much as 10 orders of magnitude, making hydrogen–deuterium exchange an extremely sensitive probe of protein structure and association (Hoofnagle *et al.*, 2003; Lanman and Prevelige, 2004).

Amide proton hydrogen–deuterium exchange as detected by NMR has been widely used in studies of protein folding and ligand binding. The use of NMR as a detection technique limits studies to proteins of high solubility and sufficiently small size to make possible NMR spectroscopy. In contrast, protein mass spectrometry requires little protein and is not size limited. Modern mass analyzers are easily capable of detecting the 1-Da mass increment that accompanies the exchange of a proton for a deuteron and thus can readily monitor the exchange event. The global exchange properties of a protein may be monitored in the intact protein or, alternatively, region-specific information can be obtained by enzymatic digestion of the protein into smaller fragments, typically on the order of 15–30 amino acids.

Experimentally, the protein or complex of interest is transferred into a compatible deuterium-based buffer, generally by dilution. At the desired time points, the exchange reaction is sampled. To obtain a "snapshot" of the extent of exchange the chemical exchange is quenched by simultaneous chilling and acidification to pH 2.5–3. Because exchange is general acid/general base catalyzed, this procedure slows the chemical exchange rate to the tens of minutes time scale, making possible sample handling and analysis. Although exchange is significantly retarded at low pH and temperature, it is not halted. Thus, it is desirable to carry out the analysis as rapidly as possible in order to avoid artifactual in-exchange or back-exchange. For this reason, short steep gradients are generally used for liquid chromatography runs.

If region-specific information is desired, the sample can be rapidly digested to obtain fragments. Because digestion also needs to be done under quench conditions to preserve the exchange information, an excess of pepsin, which is active at acid pH, is employed. For any given peptide, the deuterium content at each time point is determined by calculation of the centroid of the mass distribution, and the shift of the centroid with time corresponds to the exchange profile. The observed exchange profile is the sum of the individual exponential exchange profiles for each amide proton within the peptide. Therefore, it is possible to fit the overall exchange profile as a sum of exponentials and obtain amplitudes and exchange rates. Although in principle it should be possible to extract information about each amide proton in this way, in practice the overall profile is usually fit with three or fewer components representing fast, medium, and slow classes.

Although pepsin cuts reproducibly, it does not cut specifically; all peptide bonds are fair game. Thus any given protein can produce an array of peptides close to one another in mass. With sufficient mass accuracy, such as that obtained by Fourier-transform ion cyclotron resonance (FT-ICR) mass spectrometry, it is sometimes possible to make assignments by exact mass matching but more generally it is necessary to determine the identity of the peptide by MS/MS sequencing (Lanman *et al.*, 2003; Lanman and Prevelige, 2004). Assignment of the peptides allows the exchange information to be mapped back onto the primary structure and, when it is available, onto the tertiary structure.

## III. Studies on CA Polymerized *In Vitro*

### A. *Hydrogen–Deuterium Exchange Studies*

Examination of the structure of the reconstructed tubes of HIV CA presented a conundrum. The hexameric clusters were hypothesized to form via interactions between helices I and II of the N-terminal domain, yet the N-terminal domain shows no evidence of forming hexamers under assembly conditions in solution. To rule out the possibility that the isolated N-terminal domain was conformationally altered and incapable of self-association, point mutations that blocked dimerization were introduced into the C-terminal domain dimer interface and the mutant protein was analyzed by equilibrium ultracentrifugation under assembly conditions. These experiments demonstrated that protein that was incapable of C-domain dimerization was incapable of oligomerization, suggesting the existence of an interaction between the N and C domains (Lanman *et al.*, 2003). Therefore, hydrogen–deuterium exchange experiments were performed on CA in solution and polymerized into tubes to identify the intersubunit interactions stabilizing the tubes.

CA protein in solution and polymerized into tubes was diluted in buffered $D_2O$ solutions and exchanged at room temperature. The exchange reactions were quenched at various time points extending to a total of 68 h of exchange. To uncouple the exchange reactions from the analysis the quenched exchanged samples were flash frozen in liquid nitrogen and stored at −80°C. The exchanged samples were subsequently thawed, digested with pepsin, and analyzed by electrospray FT-ICR mass spectrometry (Lam *et al.*, 2002; Lanman *et al.*, 2003).

While exchange experiments can be analyzed by either matrix-assisted laser desorption ionization (MALDI) or electrospray ionization (ESI) and many mass spectrometers have sufficient resolution, the use of FT-ICR mass spectrometry affords two significant advantages. Pepsin digestion results in a large number of fragments, some of which can be very similar in mass. As exchange occurs, the isotopic envelope broadens, exacerbating this problem. In hydrogen–deuterium exchange experiments, chromatography times are kept short to minimize back-exchange and the experiment relies on the ability of the instrumentation to resolve closely spaced peptides. The superior resolution of FT-ICR makes it possible to resolve ions that differ by as little as 26 milli-Thomson ($m/z$) units (Lam *et al.*, 2002). The second advantage of FT-ICR stems from its extremely high mass accuracy, which makes it possible to assign fragments from a peptide digest by exact mass matching, saving the time and effort of performing a separate MS/MS analysis.

Consideration of the expected exchange behavior of a peptide located at the C-terminal domain dimer interface is illustrative. The C-terminal domain dimer is modulated by hydrophobic packing of two symmetry-related $\alpha$ helices, one from each subunit. In solution CA protein is in a rapid and reversible equilibrium between monomer and dimer, with a $K_d$ of ~20 $\mu$M. While buried in the dimer, in the monomeric form

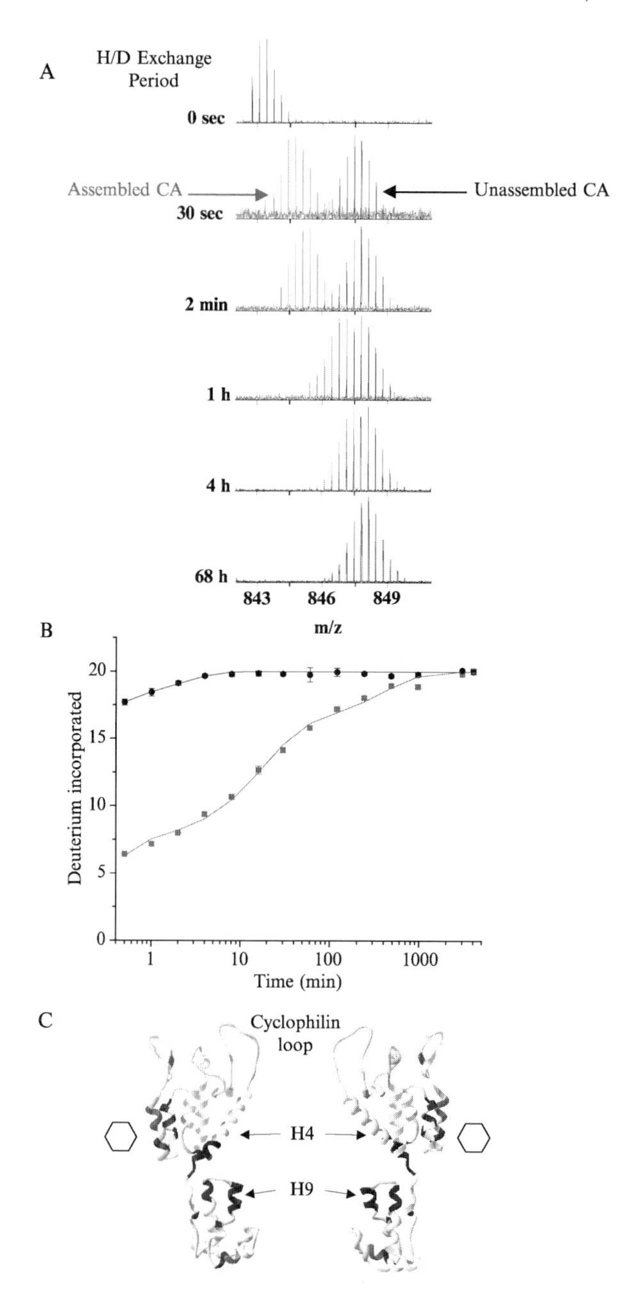


FIG 3. (A) FT-ICR mass spectra of CA peptide spanning residues 169–189 during exchange. The peptide-spanning CA residues 169–189 lie at the C-terminal domain–dimer

the interfacial $\alpha$ helices are surface exposed, which should allow for facile exchange with solvent. When polymerized into tubes, the effective concentration of the C-terminal domain is high and the subunits are unable to diffuse away. Hence, the dimer is expected to show enhanced stability, with a resulting increase in exchange protection. The raw and analyzed exchange profiles of a peptide derived from the C-terminal domain dimer interface are presented in Fig. 3A.

Initially the isotopic distributions for the peptide in both polymerized and unpolymerized CA are identical. The family of related peaks represents the natural isotope distribution. The lowest peak in the isotopic distribution represents the monoisotopic peptide (all carbon atoms are $^{12}C$, all hydrogen atoms are $^{1}H$, all oxygen are $^{16}O$, all nitrogen are $^{14}N$, and all sulfur are $^{32}S$). The peaks to the right represent higher masses due to the random incorporation of naturally occurring isotopes (primarily $^{13}C$ but also $^{15}N$ and $^{34}S$). Over time, as deuterium ($\sim$1 Da higher in mass than hydrogen) replaces the exchangeable protons the profile shifts to higher mass. In the case of the unpolymerized protein, exchange is nearly complete in 30 s, whereas in the case of the polymerized protein complete exchange takes in excess of 4 h. To perform a quantitative analysis of exchange, the centroids of the distribution are determined and plotted versus exchange time (Fig. 3B). After correction for the loss of deuterium due to back exchange, fitting of the exchange profiles with multiple exponentials reveals the amplitudes (number of protons) and exchange rates of each class (fast, medium, or slow).

A single such experiment yields peptides from throughout the primary sequence. In the case of CA, 93% of the sequence was covered, and a total of 83% had sufficient signal to noise for quantitative analysis. The exchange rates and amplitudes in the unassembled and assembled forms were compared and the difference in exchange rate mapped back onto the tertiary structure (Fig. 3C). Considerable

interface. As indicated by the time-dependent increase in mass-to-charge ratio, it exchanges rapidly in the free CA (black) and substantially more slowly in the polymerized form (red). (B) Exchange kinetics of peptide 169–189 in the unassembled and assembled form. The centroid of the distributions in (A) were calculated and plotted versus time. The solid line represents a multiexponential fit to the data. (C) The exchange protection in CA as a result of assembly. The degree of exchange protection was calculated from the exponential fits to the data and mapped onto the structure of CA. Darkest blue is most protected on assembly. Red segments are less protected in the assembled form. Yellow represents no change, and white regions were not analyzed. The two representations are rotated by 180°. (See Color Insert.)

exchange protection was observed in N-terminal domain helices I and II, consistent with the structure suggested by merging the crystal structure into the cryo-EM reconstruction of the tubes. Similarly, considerable protection was also observed in the C-terminal domain dimer interface. Despite the clear importance of the formation of the N-terminal $\beta$ hairpin in maturation (von Schwedler *et al.*, 1998), little to no change in protection was seen to accompany assembly in this region. Unexpectedly, 1000-fold protection was observed in a peptide that spanned the C terminus of helix III, and the N terminus of helix IV. This region had not previously been implicated in assembly.

### *B. Identification of Interacting Interfaces by Cross-Linking*

Although increased exchange protection can identify a region involved in the formation of an interface, it cannot unambiguously identify its binding partner. However, chemical cross-linking when combined with mass spectrometry is a powerful tool with which to obtain distance constraints between atoms in a protein or protein complex. To identify proximal regions in the assembled tube, tubes were sparingly cross-linked with disuccinimidyl tartrate (DST), a 6.4-Å homobifunctional lysine-reactive cross-linker, and then dissociated with sodium dodecyl sulfate (SDS). Size-exclusion chromatography was used to separate monomers, dimers, and higher order oligomers from one another, and the monomeric and dimeric species were digested with the endoproteinase Lys-C, which cuts after lysine groups. Cross-links can be either intramolecular or intermolecular. To identify the cross-links responsible for dimerization, the digests of both the monomeric and dimeric fractions were interrogated to identify species uniquely present in the dimeric fraction.

A single species was identified as being unique to the dimeric fraction, and this species had a neutral average mass of 14,439 Da. The mass of this peptide corresponded to a cross-linked peptide arising from a cross-link between Lys-70 of the N-terminal domain of one subunit to Lys-182 of the C-terminal domain of another. The identification of an intersubunit N-terminal domain:C-terminal domain interaction provides a possible explanation for the conundrum of “no hexamers without dimers.” In this model, C-terminal domain dimerization induces a conformational change in the symmetry-related C-terminal domains that allows them to interact with the N-terminal domains. The ability to form two stabilizing interactions, a homotypic N-terminal domain:N-terminal domain interaction and an N-terminal domain:C-terminal domain interaction, provides sufficient bonding

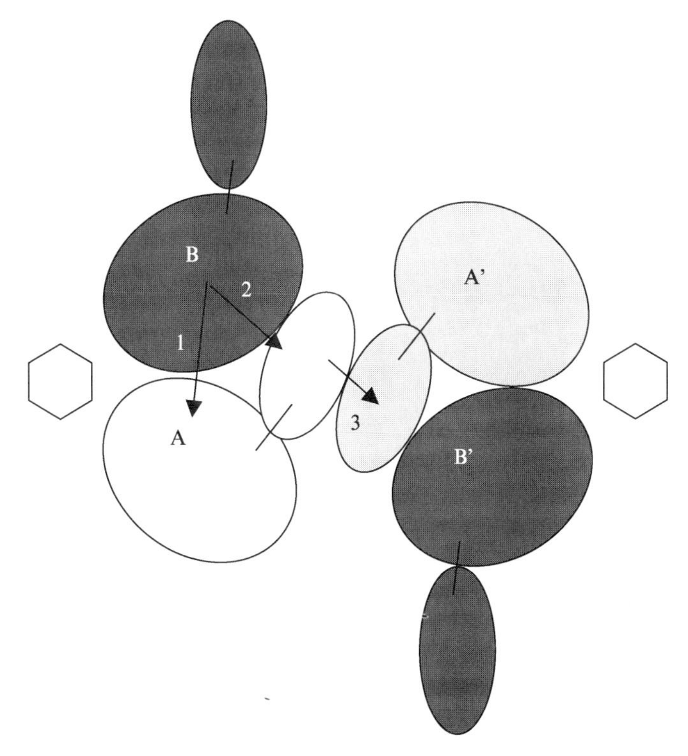


FIG 4. Models for the intersubunit interactions in CA tubes. The proposed interactions between CA protein subunits are indicated by arrows and labeled. The homotypic N-terminal domain interaction (1) is mediated by helices I and II. The homotypic C-terminal domain interaction is mediated by helix IX. The heterotypic N-domain:C-domain interaction involves the base of helix IV. Formation of this interaction on dimerization is proposed to be required for continued subunit addition. (See Color Insert.)

energy for an N-terminal domain to be incorporated into a growing tube (Fig. 4). In the absence of dimerization, the C-terminal domain is not receptive for N-terminal domain binding and as a result the bonding energy is insufficient for N-terminal domain binding. In this model the formation of an N-terminal domain:C-terminal domain interaction is crucial for formation of the mature viral lattice.

## IV. Hydrogen–Deuterium Exchange Studies on Intact Virions

Although the *in vitro* polymerized tubes provide a model for the mature virion, there is no model for the subunit interactions in the immature virion and therefore no model for the changes in interaction accompanying maturation. Cryo-EM studies have demonstrated that the Gag polyprotein is radially arranged, with the matrix domain (MA) embedded in the viral envelope, and cryo-EM analysis of immature virions and *in vitro* assemblies of Gag suggest that Gag is packed with hexameric symmetry (Fuller *et al.*, 1997; Wilk *et al.*, 2001).

To elucidate the conformational changes that occur during capsid maturation we performed comparative hydrogen–deuterium exchange experiments on immature and mature HIV virus-like particles as well as isolated CA protein (Lanman *et al.*, 2004). The immature and mature virus-like particles were produced by transiently transfecting 293T cells with plasmids containing the pNL4-3 viral genome. One plasmid, termed iVIP, encoded in the active site of the protease (D25N) a mutation that blocked the proteolytic processing of Gag necessary for viral maturation whereas the other plasmid, termed mVIP, did not. To remove the complications posed by working with infectious HIV particles the incorporation of gp120 into both types of viral particles was prevented by the creation of a deletion mutation in the transmembrane region. The reverse transcriptase also had a deletion mutation at the active site that made it nonfunctional, thereby blocking viral genome replication.

The immature virus-like particles consisted primarily of the Gag polyprotein whereas the mature virus-like particles contained primarily MA, CA, and, NC, which result from the proteolytic processing of Gag (Fig. 5). The mature virus-like particles were estimated to be approximately 70% cleaved. The mass of the proteins in mature viral particles was analyzed by liquid chromatography-mass spectrometry (LC-MS) before hydrogen–deuterium exchange. The observed mass of MA (14,922 Da) was in agreement with the predicted mass (14,922 Da), assuming that the N-terminal methionine was removed and the resultant N-terminal glycine was myristoylated. The observed mass of CA (25,601 Da) was in agreement with the predicted mass of 25,602 or 25,600 Da, assuming the cysteines are reduced or oxidized, respectively. No posttranslational modifications were detected, suggesting the proteins are not extensively modified. NC was not observed, possibly because it interacted with RNA, making oligomers that were too large to detect.

A unique feature of hydrogen–deuterium exchange when analyzed by mass spectrometry is its capacity to detect whether a protein exists in multiple conformations (Miranker *et al.*, 1993). For example, if half the MA molecules were folded and the other half were unfolded, the amide hydrogen exchange rate for the two states would be different. The unfolded protein molecules would rapidly exchange, thereby

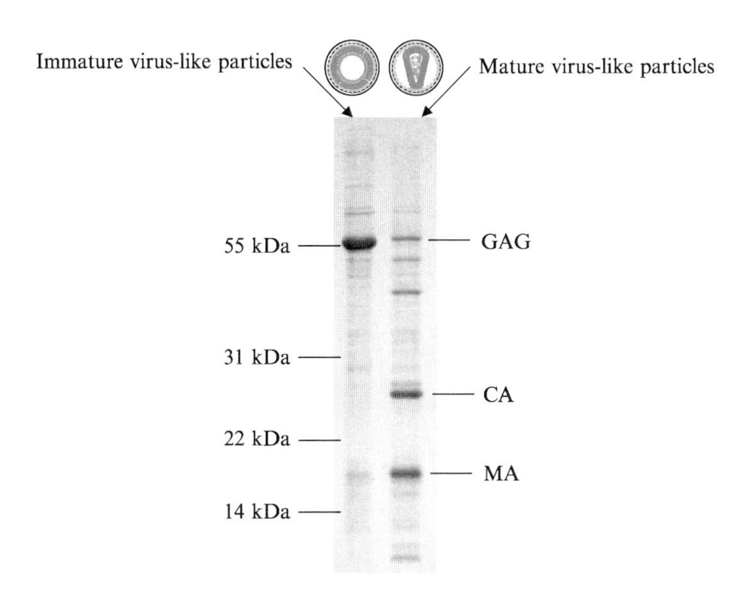


FIG 5. SDS–PAGE analysis of immature and mature virus-like particles. Immature and mature virus-like particles used for hydrogen–deuterium exchange experiments were produced by transient transfection, purified, and analyzed by SDS–PAGE. (See Color Insert.)

incorporating more deuterium than the folded protein at early times. The two different masses for MA, produced by the two populations, could readily be detected by mass spectrometry. At later times, the profiles would be expected to converge as the folded protein slowly became fully exchanged. To detect conformational heterogeneity within mature virus-like particles the amide hydrogen exchange pattern of MA and CA was analyzed.

### *A. MA in Mature Virus-Like Particles Is in One State*

Electrospray ionization (ESI) of large molecules such as proteins produces multiply protonated positive ions of various charge states, and mass analyzers detect these molecules as a mass-to-charge ratio ($m/z$). The $m/z$ value and the charge state can then be used to calculate the number-average mass for the protein. At the 0-min time point, the MA protein in mature virus-like particles, as represented by the peak at 933.6 $m/z$ (+16 charge state), has a mass of 14,922 Da, similar to the predicted mass for MA (Fig. 6A). The MA peak shifts from approximately 933.6 to 935.8 $m/z$ within 30 s of exchange, indicating a mass increase of ~35 Da due to the incorporation of deuterium. This rapid

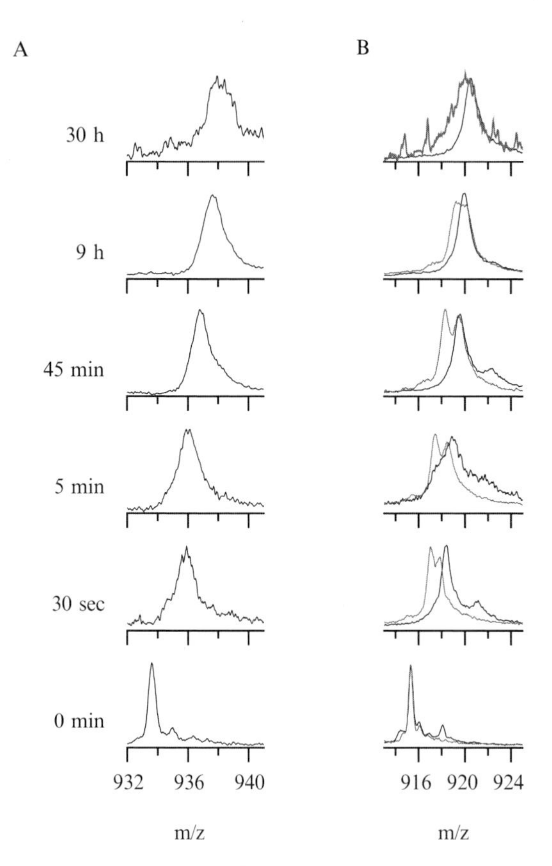


FIG 6. Mass-to-charge ratio spectra for structural proteins in mature HIV capsids during exchange. Mature virus-like particles were exchanged in deuterated phosphate-buffered saline (PBS) and then analyzed by mass spectrometry without further proteolytic digestion. (A) The +16 charge state of MA as a function of exchange time. (B) The +28 charge state of CA in mature virions (red) and free in solution (black) as a function of exchange time. (See Color Insert.)

deuterium incorporation suggests that transfer of $D_2O$ across the viral lipid membrane is not limiting. At all time points the MA mass is a single peak, consistent with the MA being present as only one conformation in the mature virus-like particles.

### *B. CA in Mature Virus-Like Particles Is in Two States*

To determine the extent of CA protection in the conical core of mature virus-like particles relative to CA in solution, samples of each were analyzed by hydrogen–deuterium exchange. As expected, both forms of CA initially have the same mass (25,601 Da) as represented by the peak at ~915.4 *m/z* (+28 charge state) (Fig. 6B). With the passage of time, the mass of both free CA and CA in the mature virus-like particles shifts to higher *m/z* values, reflecting deuterium incorporation. However, the profiles are distinct. After 30 s of exchange, the peak for CA in solution is unimodal and has shifted from 915.4 to 918.3 *m/z*, an increase corresponding to an 84-Da mass increase. In contrast, the peak for CA in mature virus-like particles displays a bimodal distribution composed of fast ($CA_{fast}$) and slow ($CA_{slow}$) exchanging peaks. The faster exchanging peak ($CA_{fast}$) has shifted to approximately 917.5, an increase corresponding to ~59 Da, whereas the slower exchanging CA peak ($CA_{slow}$) has shifted to 917.2, corresponding to an increase of approximately ~50 Da. The bimodal distribution becomes even more pronounced after 5 and 45 min of exchange, with the difference between $CA_{fast}$ and $CA_{slow}$ corresponding to 25 and 34 Da, respectively. Thus, CA contained within mature virus-like particles exists in two states with different exchange behavior.

The exchange profile for $CA_{fast}$ is similar to that of free CA in solution (Fig. 6B), suggesting that $CA_{fast}$ is not substantially protected. However, at early times $CA_{fast}$ incorporated less deuterium than free CA, indicating a small degree of protection. The protection seen at early times is probably due to dimerization of the C-terminal domains. The published estimate of the concentration of CA within a virion is ~7.6 mM. If 50% of the total CA is in the form of $CA_{fast}$, as suggested by the area under the respective peaks, the effective concentration of $CA_{fast}$ is 3.8 mM, a concentration at which, given a $K_d$ of 18 $\mu$M, it would be predominantly dimeric. In CA assembled *in vitro*, 30 residues in the C domain, 20 of which were located at the dimer interface, became protected. This degree of protection is consistent with the maximum protection (17 Da) observed for $CA_{fast}$ when compared with free CA in solution, suggesting that the observed protection is probably due to dimerization of the C-terminal domain.

The $CA_{slow}$ deuterium incorporation profile is different from $CA_{fast}$ and CA in solution, indicating that it is a different conformation. If CA in the conical core forms the same interactions as does *in vitro*-assembled CA, the N domain should display a similar degree of protection (~40 residues). The 34-Da protection of $CA_{slow}$ compared with $CA_{fast}$ is consistent with the $CA_{slow}$ forming stable N-domain interactions similar to those observed in *in vitro*-assembled CA. This analysis suggests that approximately half the CA molecules form both N-domain and C-domain interactions, presumably in the conical core, whereas the other half of the CA molecules form only the dimeric interaction.

A)

```
1
GARASVLSGG ELDKWEKIRL RPGGKKQYKL KHIVWASREL ERFAVNPGLL ETSEGCRQIL GQLQPSLQTG

71
SEELRSLYNT IAVLYCVHQR IDVKDTKEAL DKIEEEQNKS KKKAQQAAAD TGNNSQVSQN YPIVQNLQGQ

141
MVHQAISPRT LNAWVKVVEE KAFSPEVIPM FSALSEGATP QDLNTMLNTV GGHQAAMQML KETINEEAAE

211
WDRLHPVHAG PIAPGQMREP RGSDIAGTTS TLQEQIGWMT HNPPIPVGEI YKRWIILGLN KIVRMYSPTS

281
ILDIRQGPKE PFRDYVDRFY KTLRAEQASQ EVKNWMTETL LVQNANPDCK TILKALGPGA TLEEMMTACQ

351
GVGGPGHKAR VLAEAMSQVT NPATIMIQKG NFRNQRKTVK CFNCGKEGHI AKNCRAPRKK GCWKCGKEGH

421
QMKDCTERQA NFLGKIWPSH KGRPGNFLQS RPEPTAPPEE SFRFGEETTT PSQKQEPIDK ELYPLASLRS

491
LFGSDPSSQ
```

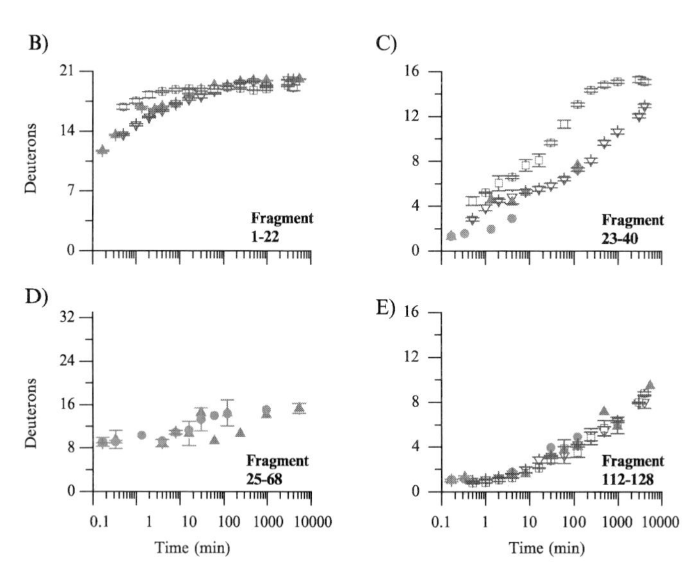


FIG. 7. Peptide fragments produced by digestion of immature and mature virus-like particles. (A) Observed peptide fragments for immature virus-like particles (red dashes), mature virus-like particles (green dots), and both (black solid lines). (B–E)

The bimodal exchange distribution in mature virus-like particles indicates that CA exists in two different conformations. One possible mechanism for the formation of the bimodal distribution is an interconversion between the slow- and fast-exchanging conformations, with complete exchange occurring rapidly in the fast-exchanging conformation (EX1 mechanism). In this case every time $CA_{slow}$ switched to $CA_{fast}$ the intensity of $CA_{fast}$ would increase and that of $CA_{slow}$ would decrease, but the relative *m/z* difference between the peaks would remain constant. Another possible mechanism is that $CA_{slow}$ and $CA_{fast}$ are not in dynamic equilibrium: $CA_{slow}$ remains a slower hydrogen–deuterium exchanging conformation and $CA_{fast}$ remains a faster exchanging conformation throughout the experiment. Over time, $CA_{slow}$ incorporates more and more deuterium, eventually merging with $CA_{fast}$ by traversing intermediate exchange states. This model predicts that the intensities of $CA_{slow}$ and $CA_{fast}$ relative to each other should remain constant as $CA_{slow}$ shifts toward higher mass. The $CA_{fast}$ and $CA_{slow}$ peaks that are formed from mature virus-like particles are approximately equal in intensity at all time points, indicating that approximately 50% of the CA in mature virus-like particles is in the slow-exchanging form and the other 50% of CA is in the fast-exchanging form, and that the two forms do not interconvert.

### *C. Exchange Patterns of Individual Peptides*

To determine the hydrogen–deuterium exchange profile for localized regions of the protein, viral particles were digested with pepsin and the extent of deuterium incorporation was analyzed by mass spectrometry. Thirty peptic fragments, each approximately 10 to 30 residues in length, were observed in the peptic digest and identified by exact mass matching. Six of the peptides identified were previously observed in *in vitro* CA hydrogen–deuterium exchange experiments, making possible the direct comparison of CA in viral particles with CA assembled *in vitro*. Peptides covering 44 and 52% of the Gag primary sequence in the immature and mature virus-like particles, respectively, had sufficient signal to noise to enable hydrogen–deuterium exchange analysis (Fig. 7A). Of these, peptides covering 42% of the Gag sequence were

Number-average mass versus hydrogen–deuterium exchange period for immature viral particles (red diamonds), and mature virus-like particles (green triangles) superimposed on the deuterium incorporation profiles for *in vitro*–unassembled (open squares) and *in vitro*–assembled (open triangles) CA. (See Color Insert.)

common to both immature and mature virus-like particles, allowing a direct comparison of the two forms.

A peptide (805.774 *m/z*) arising from residues 1–22 of CA was observed in both mature virus-like particles and *in vitro*-assembled CA. This peptide became protected on *in vitro* assembly of CA because of the formation of an N-domain:N-domain interaction. This peptide displayed a similar degree of protection in mature virus-like particles (Fig. 7B), *suggesting* that the involvement of this peptide in the N-domain:N-domain interaction among mature virus-like particles is similar to that seen in *in vitro*-assembled CA.

The N-domain:N-domain interaction mediated through helices I and II is necessary for *in vitro* assembly of CA, suggesting that it should also be present in mature virus-like particles, and perhaps in immature virions as well. A peptide spanning residues 23 to 40 (helices I and II) was identified in digests of both immature and mature virus-like particles and CA studied *in vitro* (Fig. 7C). Although the inherent difficulty of examining virus-like particles made it impossible to determine the number-average mass of this peptide at all exchange times in the virus-like particles, a sufficient number of time points were obtained to gain an understanding of the interactions in immature and mature viral particles. The greatest difference in deuterium incorporation between unassembled and assembled CA *in vitro* is observed at 2 h. At 2 h of exchange, deuterium incorporation by immature and mature virus-like particles coincides exactly with that of assembled CA, suggesting this peptide is involved in the N-domain: N-domain interactions in both immature and mature forms of the virus-like particles.

Another peptide in the N domain spans residues 25–68 (helix I to the middle of the loop region between helix III and IV) (Fig. 7D). There is no difference in the deuterium incorporation profile of this peptide between immature and mature virus-like particles. These data are again consistent with the presence of an N-domain:N-domain interaction in both immature and mature virus-like particles. The last eight residues of this peptide overlap with a peptide spanning residues 55–68, which was observed to become substantially protected on assembly *in vitro*.

A peptide spanning CA residues 112–128 is observed in both immature virus-like particles and mature virus-like particles (Fig. 7E). The deuterium incorporation profile for this peptide in immature and mature virus-like particles, as well as in unassembled and assembled CA *in vitro*, are superimposable, indicating that CA has the same conformation in all forms.

The deuterium incorporation profile for a peptide spanning MA residues 44–75 also displays little difference between immature and mature virus-like particles. A peptide spanning P6 residues 455–496 also displays little difference in deuterium incorporation profile for both immature and mature virus-like particles. This region contains both fast-exchanging and slow-exchanging residues and has not completely exchanged even after 270 h, indicating that a portion of this peptide probably forms a stable folded core of the P6 peptide. The deuterium incorporation profile for the peptides from virus-like particles undergoes little change, indicating that small conformational changes accompany maturation.

### *D. N-Domain:C-Domain Interface Is Bimodal*

The significant protection observed in $CA_{slow}$ suggests that many of the interactions necessary for CA assembly *in vitro* are also present in the conical core of mature viral particles. Although most peptides also have basically the same deuterium incorporation profile in immature and mature virus-like particles, the peptide spanning CA residues 55–68 (744.84 *m/z*) displays significant differences (Fig. 8). This peptide has previously been identified as contributing to an intersubunit N–C domain interaction resulting from core formation. In immature virus-like particles this peptide exchanged rapidly, becoming nearly completely exchanged within approximately 16 min. The exchange profile of this peptide in the immature virion is similar to that observed for unassembled CA in solution, suggesting that the CA in immature virus-like particles does not form the previously identified N-domain: C-domain interaction.

The spectra for this peptide in mature virus-like particles are bimodal. Whereas after 20 s of exchange the isotopic envelope has shifted on average the same as in immature virions, it is broader, indicating that even at these early time points it has begun separating into two distributions (bimodal distributions). At 4 min of exchange the bimodal distribution is apparent. The separation between the two distributions increases at exchange times up to 1 h and then, as expected, the two distributions begin to converge after 4 h of exchange and continue converging at the later time points. The hydrogen–deuterium exchange kinetics of the faster component of the distribution is similar to the hydrogen–deuterium exchange kinetics of immature virus-like particles, and to unassembled CA. This indicates that the CA molecules contributing to the faster distribution are probably in a conformation similar to the unassembled subunits. The fraction comprising the

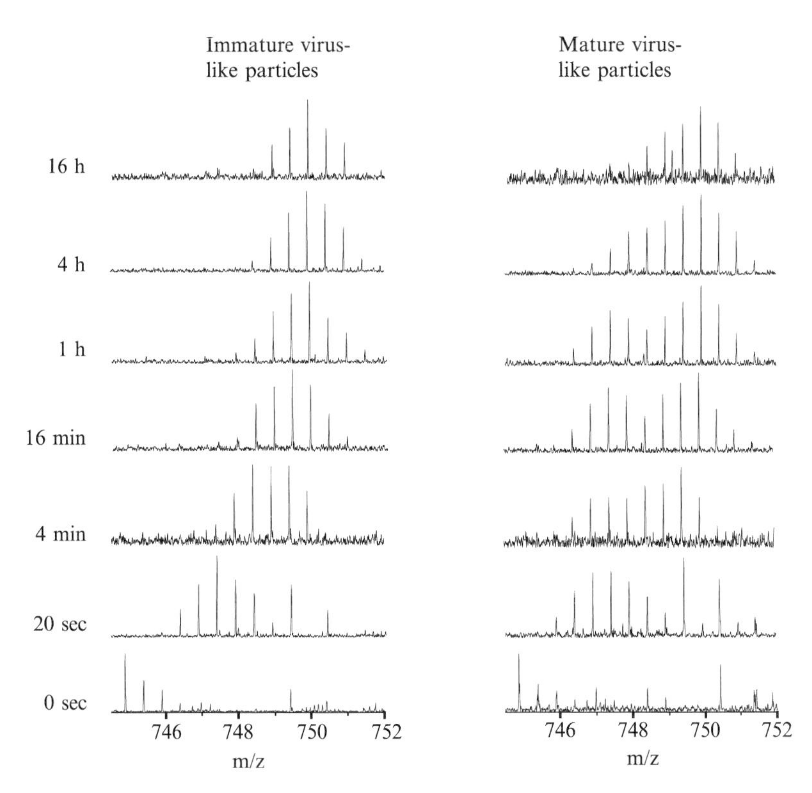


FIG. 8. Hydrogen–deuterium exchange profile for peptide-spanning CA residues 55–68. Shown are the mass-to-charge ratio spectra for doubly charged peptide fragments spanning CA residues 55–68 for immature and mature virus-like particles after zero time, 20 s, 4 min, 16 min, 1 h, 4 h, and 16 h of hydrogen–deuterium exchange.

slower exchanging distribution is approximately 50%, indicating that approximately half the CA molecules in the mature virus-like particles have become protected in this region. The hydrogen–deuterium exchange kinetics for the slower exchanging distribution are similar to those observed in the *in vitro*-assembled CA, suggesting that approximately half the CA forms an N-domain:C-domain interaction in the mature virus-like particles whereas the other half does not.

## *E. Model for Maturation*

How, then, does the formation of the N-terminal domain:C-terminal domain interaction cause maturation? In most of CA there is little difference in deuterium incorporation profiles between immature and mature virus-like particles, suggesting the CA protein structure stays the same. The conical core formation occurs by the concerted movements of folded domains to form new protein–protein interactions. The homotypic N-terminal and C-terminal domain interactions are present in both immature and mature particles. After proteolytic processing of Gag, CA forms the N-terminal domain:C-terminal domain interaction, which causes the CA protein to collapse into the conical core.

The formation of this is somehow prevented in the immature virions, presumably because the CA domain is fused to the MA and NC domains. Fusion to the MA and NC domains may prevent CA from adopting the proper position or conformation for formation of the N-terminal domain:C-terminal domain interaction. Cleavage between MA and CA allows the formation of the CA N-terminal $\beta$ hairpin and the salt bridge linking the N-terminal proline and Asp-51 (Tang *et al.*, 2002; von Schwedler *et al.*, 1998). This conformational change could promote the formation of the conical core by stabilizing the N-domain: C-domain interaction. Addition of four residues of the MA protein to the N terminus of the CA protein blocks the formation of the N-terminal $\beta$ hairpin and causes the *in vitro* assembly of spheres instead of tubes. Disruption of $\beta$ hairpin formation through fusion with MA and/or the C-domain:C-domain interaction through fusion with NC may provide complementary mechanisms for maintaining a spherical immature morphology.

## V. Kinetics of Tube Assembly and Inhibition

The demonstration that CA tubes polymerized *in vitro* recapitulate the interactions found in the mature virus suggests that studies of CA polymerization *in vitro* can shed light on the mechanism of assembly and perhaps provide a route to the discovery of inhibitors of assembly and/or maturation, and a number of investigators have embraced this strategy. The *in vitro* polymerization of CA requires molar concentrations of NaCl, suggesting the need for charge/charge shielding during assembly. Polymerization can be achieved either by dialysis against appropriate salt solutions, or by dilution of a stock solution of CA protein into buffered salt solutions. A strategy of alanine scanning mutagenesis and dialysis-based polymerization has been employed to identify the interfaces involved in core assembly (Ganser-Pornillos *et al.*, 2004). This series of experiments identified N-terminal domain helices I and II and the C-terminal dimer interface as being important in assembly, consistent with the structural model.

The kinetic strategy affords the advantage of allowing quantitative analysis of the effect of mutation on assembly rate, making it possible to evaluate more precisely the role of specific amino acid changes and to obtain information about the pathway of assembly. The rate of assembly displays a high-order dependence on CA protein concentration (Lanman *et al.*, 2002). A doubling of the protein concentration results in a 16-fold increase in the rate of assembly, indicating the involvement of multiple subunits in the rate-determining step. Kinetic analysis can also be used to determine the importance of particular residues in assembly. For example, the replacement of Met-185 with alanine results in a significant decrease in the rate of assembly. This particular residue is located at the C-domain dimer interface. Replacement with an alanine results in an approximately 10-fold weaker association (J. Lanman and P. E. Prevelige, Jr., unpublished data). The N-terminal domain has a patch of charged residues located proximal to the intersubunit-binding site. Replacement of the glutamate residue at position 45 with alanine resulted in the formation of a more stable viral core, and decreased infectivity. To evaluate the charge network in this region a series of mutations were made that substituted neutral or oppositely charged residues for the original charge (Douglas *et al.*, 2004). The effect of these substitutions on assembly was evaluated *in vitro*.

The mutations did not alter either the stability of the subunits or their ability to dimerize, indicating that the effects were localized. The E45A mutation resulted in an increased rate of assembly, and a decreased critical concentration consistent with the observation that it resulted in a more stable core *in vivo*. Analysis of the family of charge change mutants suggested that a delicate balance is maintained between charge/charge repulsion, which disfavors polymerization, and attractive forces, which favor assembly.

The structural model of the core and tube suggests that assembly should be inhibited by inhibiting C-terminal domain dimerization because it is the C-terminal domain dimerization that ties the hexamers together. The isolated C-terminal domain has been shown to dimerize with an association constant identical to that of the intact CA protein. Thus, the introduction of C-terminal domain into an assembly reaction should inhibit assembly through the formation of heterodimers. Indeed, when C-terminal domain was introduced into *in vitro* assembly reactions it resulted in a concentration-dependent decrease in the rate of assembly (Fig. 9), providing proof of concept for direct inhibition of assembly. Screening for small molecule inhibitors of *in vitro* CA assembly is currently underway in the authors' laboratory,

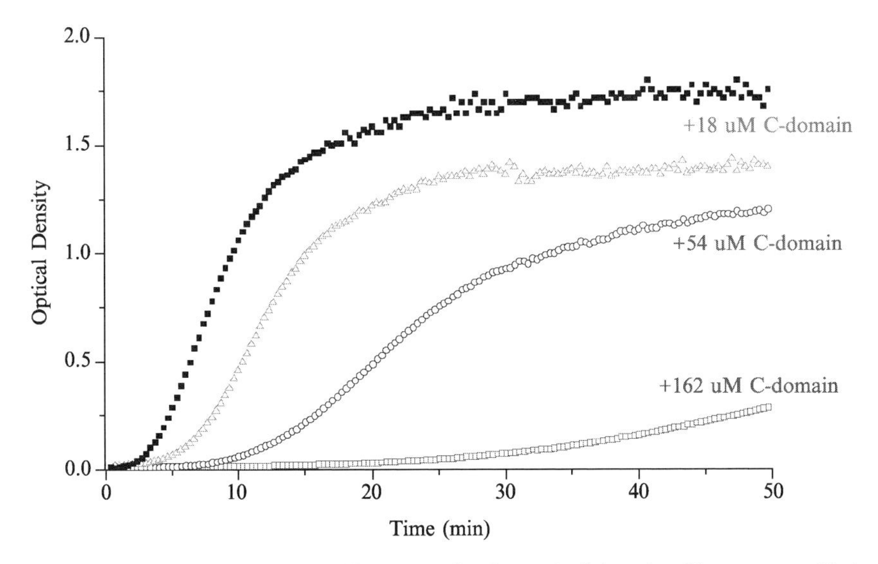


FIG. 9. Inhibition of CA assembly *in vitro* by C-terminal domain. CA was assembled *in vitro* in the absence (black) and presence of increasing concentrations of C-terminal domain. The rate of assembly decreases with increasing concentration of added C-terminal domain.

and one report of a small molecule inhibitor of maturation has been published.

## VI. CONCLUSION

Viral capsids, because of their high degree of symmetry, have long been a favored platform for the development of analytical methods of structural biology, and these methods have found broad utility. A current focus of structural biology is to develop a picture of complex cellular machines. The most expedient route to doing so appears to be one of obtaining high-resolution structures of the individual components and then assembling these pieces into the whole.

When the whole is symmetric, such as icosahedral or helical viral capsids, reconstruction by electron microscopy is capable of providing appropriate-resolution structures into which the crystal density may be unambiguously merged. When the particles are asymmetric the challenge is greater but may be met through the use of single-particle reconstruction techniques. However, for specimens that are polymorphous, such as the virions of HIV, alternative approaches are required.

Electron tomography has the potential to deliver reconstructions of such specimens but the beam damage associated with repeatedly imaging biological specimens currently limits the resolution.

The use of a combination of hydrogen–deuterium exchange mass spectrometry and chemical cross-linking provides a pathway to obtain intermediate-resolution information that can be used to identify the inter subunit interfaces and physical arrangement of subunits within complex biological structures. A particular advantage of mass spectrometry is the ability to resolve individual subpopulations as well as to obtain information about dynamic processes. The techniques and instrumentation available to the investigator are rapidly evolving and mass spectrometry will play an increasingly important role in structural biology.

## References

Douglas, C. C., Lanman, J., Thomas, D., and Prevelige, P. E. (2004). Investigation of N-terminal domain charged residues on the assembly and stability of HIV-1 CA. *Biochemistry* **43:**10435–10441.

Fuller, S. D., Wilk, T., Gowen, B. E., Krausslich, H. G., and Vogt, V. M. (1997). Cryo-electron microscopy reveals ordered domains in the immature HIV-1 particle. *Curr. Biol.* **7:**729–738.

Gamble, T. R., Vajdos, F. F., Yoo, S., Worthylake, D. K., Houseweart, M., Sundquist, W. I., and Hill, C. P. (1996). Crystal structure of human cyclophilin A bound to the amino-terminal domain of HIV-1 capsid. *Cell* **87:**1285–1294.

Gamble, T. R., Yoo, S. H., Vajdos, F. F., von Schwedler, U. K., Worthylake, D. K., Wang, H., McCutcheon, J. P., Sundquist, W. I., and Hill, C. P. (1997). Structure of the carboxyl-terminal dimerization domain of the HIV-1 capsid protein. *Science* **278:**849–853.

Ganser, B. K., Li, S., Klishko, V. Y., Finch, J. T., and Sundquist, W. I. (1999). Assembly and analysis of conical models for the HIV-1 core. *Science* **283:**80–83.

Ganser-Pornillos, B. K., von Schwedler, U. K., Stray, K. M., Aiken, C., and Sundquist, W. I. (2004). Assembly properties of the human immunodeficiency virus type 1 CA protein. *J. Virol.* **78:**2545–2552.

Gitti, R. K., Lee, B. M., Walker, J., Summers, M. F., Yoo, S., and Sundquist, W. I. (1996). Structure of the amino-terminal core domain of the HIV-1 capsid protein. *Science* **273:**231–235.

Hill, C. P., Worthylake, D., Bancroft, D. P., Christensen, A. M., and Sundquist, W. I. (1996). Crystal structures of the trimeric human immunodeficiency virus type 1 matrix protein: Implications for membrane association and assembly. *Proc. Natl. Acad. Sci. USA* **93:**3099–3104.

Hoofnagle, A. N., Resing, K. A., and Ahn, N. G. (2003). Protein analysis by hydrogen exchange mass spectrometry. *Annu. Rev. Biophys. Biomol. Struct.* **32:**1–25.

Lam, T. T., Lanman, J. K., Emmett, M. R., Hendrickson, C. L., Marshall, A. G., and Prevelige, P. E. (2002). Mapping of protein:protein contact surfaces by hydrogen/deuterium exchange, followed by on-line high-performance liquid chromatography-electrospray ionization Fourier-transform ion-cyclotron-resonance mass analysis. *J. Chromatogr. A* **982:**85–95.

Lanman, J., Lam, T. T., Barnes, S., Sakalian, M., Emmett, M. R., Marshall, A. G., and Prevelige, P. E. (2003). Identification of novel interactions in HIV-1 capsid protein assembly by high-resolution mass spectrometry. *J. Mol. Biol.* **325:**759–772.

Lanman, J., Lam, T. T., Emmett, M. R., Sakalian, M., Marshall, A. G., and Prevelige, P. E. (2004). Key interactions in HIV-1 maturation identified by mass spectrometry based H/D exchange. *Nat. Struct. Mol. Biol.* **11:**676–677.

Lanman, J., and Prevelige, P. E., Jr. (2004). High-sensitivity mass spectrometry for imaging subunit interactions: Hydrogen/deuterium exchange. *Curr. Opin. Struct. Biol.* **14:**181–188.

Lanman, J., Sexton, J., Sakalian, M., and Prevelige, P. E., Jr. (2002). Kinetic analysis of the role of intersubunit interactions in human immunodeficiency virus type 1 capsid protein assembly *in vitro*. *J. Virol.* **76:**6900–6908.

Li, S., Hill, C. P., Sundquist, W. I., and Finch, J. T. (2000). Image reconstructions of helical assemblies of the HIV-1 CA protein. *Nature* **407:**409–413.

Massiah, M. A., Starich, M. R., Paschall, C., Summers, M. F., Christensen, A. M., and Sundquist, W. I. (1994). Three-dimensional structure of the human immunodeficiency virus type 1 matrix protein. *J. Mol. Biol.* **244:**198–223.

Miranker, A., Robinson, C. V., Radford, S. E., Aplin, R. T., and Dobson, C. M. (1993). Detection of transient protein folding populations by mass spectrometry. *Science* **262:**896–900.

Rose, S., Hensley, P., O'Shannessy, D. J., Culp, J., Debouck, C., and Chaiken, I. (1992). Characterization of HIV-1 p24 self-association using analytical affinity chromatography. *Proteins* **13:**112–119.

Tang, C., Ndassa, Y., and Summers, M. F. (2002). Structure of the N-terminal 283-residue fragment of the immature HIV-1 Gag polyprotein. *Nat. Struct. Biol.* **9:**537–543.

von Schwedler, U. K., Stemmler, T. L., Klishko, V. Y., Li, S., Albertine, K. H., Davis, D. R., and Sundquist, W. I. (1998). Proteolytic refolding of the HIV-1 capsid protein amino-terminus facilitates viral core assembly. *EMBO J.* **17:**1555–1568.

Wilk, T., Gross, I., Gowen, B. E., Rutten, T., de Haas, F., Welker, R., Krausslich, H. G., Boulanger, P., and Fuller, S. D. (2001). Organization of immature human immunodeficiency virus type 1. *J. Virol.* **75:**759–771.

Worthylake, D. K., Wang, H., Yoo, S., Sundquist, W. I., and Hill, C. P. (1999). Structures of the HIV-1 capsid protein dimerization domain at 2.6 Å resolution. *Acta Crystallogr. D Biol. Crystallogr.* **55:**85–92.

Yoo, S., Myszka, D. G., Yeh, C., McMurray, M., Hill, C. P., and Sundquist, W. I. (1997). Molecular recognition in the HIV-1 capsid/cyclophilin A complex. *J. Mol. Biol.* **269:**780–795.

ADVANCES IN VIRUS RESEARCH, VOL 64

# ROLE OF LIPID RAFTS IN VIRUS REPLICATION*

Akira Ono and Eric O. Freed

Virus–Cell Interaction Section, HIV Drug Resistance Program
National Cancer Institute at Frederick, National Institutes of Health
Frederick, Maryland 21702



## I. Introduction

Viruses need to both enter and exit their host cell to complete a cycle of replication. Because the plasma membrane acts as a barrier between cells and their surroundings, viruses have had to evolve strategies to cross this lipid bilayer. In the case of many enveloped viruses, entry into the host cell early in the replication cycle, and egress during budding, involve membrane fusion/fission reactions between the viral envelope and the host cell plasma membrane. Advances in cell biology have given rise to the concept that the plasma membrane is not a simple homogeneous lipid bilayer with transmembrane and peripheral membrane proteins distributed uniformly, but rather consists of various microdomains with specific lipid and protein compositions. One type of microdomain that has been the subject of much excitement and controversy is the lipid raft. A number of animal viruses have been reported to associate with rafts upon their entry into, or exit from, their host cells. In this review, we outline the current knowledge on lipid rafts, describe the most recent findings on the relationship between lipid rafts and virus replication, and discuss possible models to explain the evolution of virus–raft association.

* This chapter was written by Akira Ono and Eric O. Freed in their personal capacity. The views expressed in this chapter do not necessarily reflect the views of the NIH, DHHS, or the United States.




0065-3527/05 $35.00
DOI: 10.1016/S0065-3527(05)64010-9

## II. General Properties of Lipid Rafts

### *A. Structure and Components*

Lipid rafts are microdomains present on the plasma membrane and the membranes of various intracellular organelles such as the Golgi and endosomes. Rafts are highly enriched in cholesterol and sphingolipids with saturated acyl chains (Fig. 1) (Brown and London, 2000; Simons and Toomre, 2000). These lipids reportedly associate laterally to promote the formation of liquid-ordered membrane domains in the surrounding highly fluid (liquid-disordered) glycerolipid-rich bilayer, thereby causing phase separation (Brown and London, 1998; Rietveld and Simons, 1998). Early work (Brown and Rose, 1992) established that lipid rafts can be isolated biochemically as detergent-resistant membrane (DRM) by nonionic detergent treatment (typically Triton

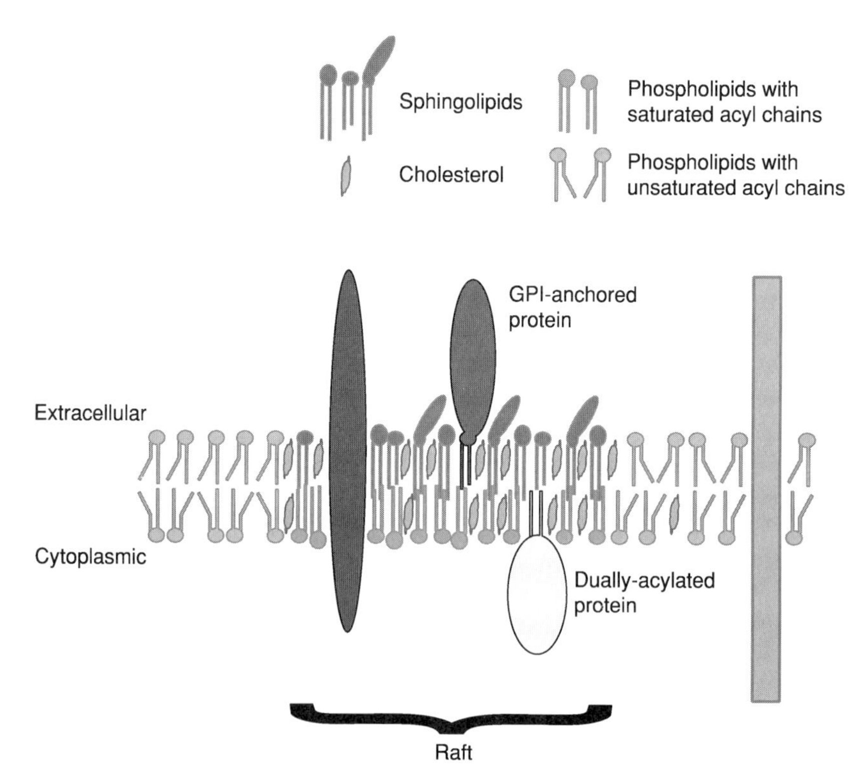


Fig 1. A simplified model of lipid raft structure. Nonraft lipid and protein molecules are shown in gray. (See Color Insert.)

X-100) at low temperature coupled with equilibrium flotation centrifugation (Brown and London, 2000; Simons and Toomre, 2000). By using this approach, several classes of proteins, including glycosylphosphatidylinositol (GPI)-anchored proteins and doubly acylated Src family kinases, were identified as being raft associated (Brown and London, 2000; Simons and Toomre, 2000). Interactions of these proteins with lipid rafts are mediated by their saturated lipid linkers (e.g., palmitic acid), which have higher affinity for the ordered lipid environment of the raft relative to nonraft membrane. In contrast, proteins modified by prenylation, which involves the attachment of unsaturated lipids (e.g., farnesyl and geranylgeranyl moieties), are generally excluded from rafts (Melkonian *et al.*, 1999; Zacharias *et al.*, 2002). Several transmembrane proteins have also been classified as being either raft [e.g., CD4 and influenza hemagglutinin (HA)] or nonraft (e.g., transferrin receptor and CD45) associated on the basis of their isolation with DRM. These proteins are now commonly used as marker proteins in biochemical analyses to distinguish raft from nonraft fractions. In addition to proteins, the GM1 glycosphingolipid, which can be detected by specific binding to the cholera toxin B subunit, is widely used as a raft marker in DRM analyses. With an increasing number of molecules identified as being raft or nonraft associated, immunofluorescence microscopy has become a powerful technique to evaluate the association of proteins with rafts, especially with copatching methods (Fig. 2) (Simons and Toomre, 2000).

### *B. Evidence Supporting the Presence of Lipid Rafts on Living Cell Membranes*

DRM association and copatching approaches are widely used for raft analyses. However, because these techniques modify intact membrane organization, one must keep in mind that positive data show only a preference for rafts by proteins or lipids of interest, rather than proving that raft association precedes experimental manipulation. Furthermore, the legitimate concern has been raised that lipid rafts might represent artifacts of detergent treatment (Mayor and Maxfield, 1995; Munro, 2003). However, much evidence supports the existence of lipid microdomains in living cell membranes: (1) *in vitro* experiments with liposomes or planar-supported lipid monolayers containing sphingolipids and cholesterol show strong correlations between the formation of an ordered phase, detergent insolubility, cholesterol dependence, and the behavior of GPI-anchored proteins (Dietrich *et al.*, 2001a,b; Schroeder *et al.*, 1994, 1998); (2) the GPI-anchored

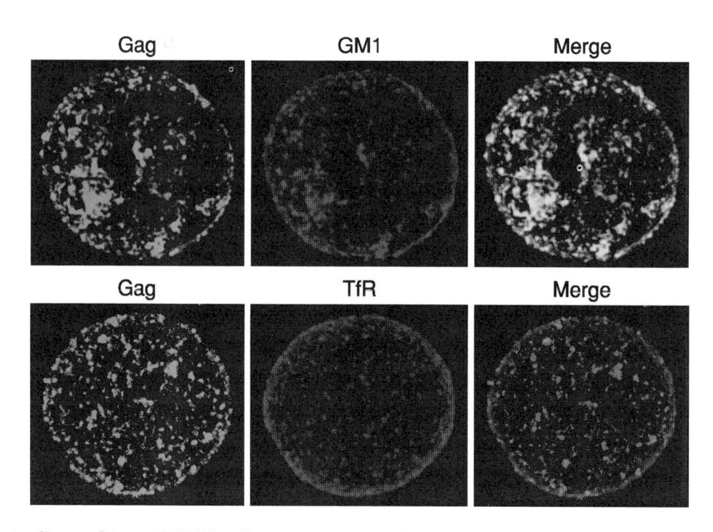


FIG 2. Copatching of HIV-1 Gag proteins with a raft marker but not with a nonraft marker. Jurkat cells expressing HIV-1 Gag were labeled with cholera toxin B subunit (GM1) or antitransferrin receptor (TfR) before fixation. After fixation and permeabilization, Gag proteins were detected with an anti-p17MA antibody (Gag). Images were collected in *z* series with a confocal microscope and projected in 3D volume with Imaris software. Merged images of Gag and GM1 or TfR are also shown (Merge). (See Color Insert.)

protein CD59, and the Src-family kinase Fyn, exhibit a delay in DRM association after membrane binding (van den Berg *et al.*, 1995; van't Hof and Resh, 1997), arguing against the idea that these proteins partition randomly in DRM and detergent-soluble membrane; (3) chemical cross-linking and fluorescence resonance energy transfer (FRET) between GPI-anchored or dually acylated proteins indicate that these proteins are in close proximity within a small region of the living cell plasma membrane (Friedrichson and Kurzchalia, 1998; Varma and Mayor, 1998; Zacharias *et al.*, 2002). Although similar experiments previously yielded conflicting results (Kenworthy and Edidin, 1998; Kenworthy *et al.*, 2000), the most recent FRET data (Sharma *et al.*, 2004) seem to resolve the discrepancy by demonstrating the small size of lipid rafts; (4) two different raft-associated molecules can copatch when cross-linked independently, whereas nonraft molecules form segregated patches (Harder *et al.*, 1998; Janes *et al.*, 1999); (5) microscopic methods that monitor the lateral mobility of single or small groups of membrane molecules reveal the presence of confined zones in which the mobility of raft-associated molecules is restricted

(Dietrich *et al.*, 2002; Pralle *et al.*, 2000; Schutz *et al.*, 2000; Sheets *et al.*, 1997); (6) fluorescence recovery after photobleaching (FRAP) analysis reveals cholesterol-sensitive diffusion of green fluorescent protein (GFP)-tagged, raft-associated proteins on the plasma membrane (Niv *et al.*, 2002; Shvartsman *et al.*, 2003); (7) two-photon microscopy utilizing a membrane probe whose fluorescence is sensitive to phase transition of its surroundings shows the existence of cholesterol-dependent membrane domains (Gaus *et al.*, 2003); and (8) immuno electron microscopy (EM) of sheets derived from the plasma membrane reveals a clustering of raft-associated proteins (Prior *et al.*, 2003; Wilson *et al.*, 2000, 2001). These studies not only support the existence of rafts, but also provide rough estimates on the size of rafts *in vivo*. These estimates vary widely, however, ranging from approximately 50–70 nm (Pralle *et al.*, 2000; Varma and Mayor, 1998) to several hundred nanometers (Dietrich *et al.*, 2002; Schutz *et al.*, 2000) depending on the methods used. The most advanced FRET and single-molecule techniques suggest that rafts can be even smaller (~10 nm) and very short-lived (Kusumi *et al.*, 2004; Mayor and Rao, 2004; Sharma *et al.*, 2004; Subczynski and Kusumi, 2003). On the basis of these studies, lipid rafts are now viewed as submicroscopic, highly dynamic entities that can coalesce upon oligomerization or cross-linking of their components to form larger stabilized platforms (Brown and London, 2000; Jacobson and Dietrich, 1999; Kusumi *et al.*, 2004; Mayor and Rao, 2004; Simons and Toomre, 2000).

The various methods used in raft studies, including those described previously, have advantages and disadvantages and sometimes produce conflicting results. For example, some proteins display high sensitivity to detergent treatment in DRM isolation assays, but copatch with known raft-associated proteins (Harder *et al.*, 1998). Because viral structural proteins have an inherent propensity to form high molecular weight complexes, simple separation of cell homogenates into supernatant and pellet fractions cannot distinguish between membrane-bound material and high molecular weight aggregates. For this reason, flotation centrifugation is the method of choice for the biochemical analysis of DRM association. Even in flotation assays, it has been suggested that detergent resistance may in some instances represent an artifact of detergent-induced protein aggregation (Ding *et al.*, 2003). Thus, although rafts are often operationally equated with DRM, it is frequently necessary to employ multiple methodologies to definitively establish raft association of a particular molecule or a cellular function.

### *C. Lipid Raft Subsets*

In the widely used raft isolation procedure based on detergent resistance, various subsets of detergent-resistant microdomains are recovered as a mixture in the DRM fraction. Caveolae, originally observed by EM as flask-shaped invaginations in the plasma membrane, constitute one such subset. These structures, which appear to be formed from rafts by multimerization of the protein caveolin (Harris *et al.*, 2002; Kurzchalia and Parton, 1999), play important roles in signal transduction, non-clathrin-mediated endocytosis, and cholesterol transport (Harris *et al.*, 2002; Kurzchalia and Parton, 1999). As discussed later, caveola-mediated endocytosis has attracted considerable interest as a route for pathogen uptake (Harris *et al.*, 2002).

Increasing evidence suggests that in addition to caveolae, other distinct types of rafts exist. Two different GPI-anchored proteins reportedly display differential detergent resistance (Madore *et al.*, 1999). Another study reports that one subset of proteins is resistant to the detergent Lubrol WX but not to Triton X-100, whereas another subset is resistant to the latter but not to the former (Roper *et al.*, 2000). These observations could result from a differential affinity of different proteins for the same microdomain. Indeed, some nonionic detergents are less selective than Triton X-100 in their ability to solubilize raft versus nonraft proteins (Schuck *et al.*, 2003). However, proteins that show distinct behavior in the detergent resistance analysis also localize to distinct regions of the plasma membrane (Madore *et al.*, 1999; Roper *et al.*, 2000). In addition, two glycosphingolipids, GM1 and GM3, are also distributed to the opposite ends of polarized T cells (Gomez-Mouton *et al.*, 2001). Finally, immuno-EM reportedly reveals that two different raft proteins form separate clusters at distinct sites on the plasma membrane (Wilson *et al.*, 2000, 2001).

Because in theory the concentration of raft-associated molecules to a limited area of the plasma membrane is a primary mechanism by which rafts function, it is important to consider how much of the plasma membrane possesses raft-like properties. On the basis of detergent resistance analyses of lipids and GPI-anchored proteins in various cell types, fluorescence anisotropy measurement of lipid order in the mast cell plasma membrane, and immuno-EM-based analyses of plasma membrane sheets, the area occupied by rafts on the cell surface is estimated to be relatively large, ranging from 35 to 70% of the total plasma membrane surface (Brown and London, 1998; Gidwani *et al.*, 2001; Hao *et al.*, 2001; Mayor and Maxfield, 1995; Mescher and Apgar, 1985; Pierini and Maxfield, 2001; Prior *et al.*, 2003). Thus, the plasma membrane may be dominated by multiple types of rafts that differ from each other in their lipid and protein compositions and therefore possess distinct functions.

### *D. Raft Functions*

Lipid rafts can be disrupted by depleting cellular cholesterol with a variety of drugs. Popular compounds include methyl-$\beta$-cyclodextrin (M$\beta$CD), a cyclic sugar molecule that binds cholesterol in its hydrophobic cavity, thereby acutely extracting cholesterol from the plasma membrane, and the statins, which inhibit the activity of HMG-CoA reductase, a key enzyme in cholesterol biosynthesis. Reagents such as filipin and nystatin, which sequester cholesterol in the plasma membrane, are also widely used to disrupt rafts. By examining the effects of these drugs on protein function and DRM association, it has been proposed that lipid rafts are involved in a wide variety of cellular processes including signal transduction, polarized protein sorting, endocytosis, cell–cell attachment, front–rear polarity, and the binding of toxins and pathogens (reviewed in Brown and London, 1998, 2000; Simons and Ikonen, 1997; Simons and Toomre, 2000). In many cases, lipid rafts may serve as concentrating platforms for molecules involved in these processes. However, caution should be used when interpreting results obtained by cholesterol depletion as these treatments may exert widespread effects on membrane properties (Kwik *et al.*, 2003; Rodal *et al.*, 1999; Subtil *et al.*, 1999). The relationship between putative raft functions and virus replication are discussed in more detail later.

## III. Membrane Microdomains and Virus Replication

### *A. Nonenveloped Viruses*

Although this review focuses primarily on the role of rafts in the replication of enveloped viruses, findings on caveola-mediated entry of nonenveloped viruses deserve mention. Replication of many nonenveloped viruses is initiated by internalization of virus after receptor binding on the host cell plasma membrane (for a review, see Sieczkarski and Whittaker, 2002). Clathrin-mediated endocytosis is well described as an entry route for some enveloped viruses as well as many nonenveloped viruses (e.g., adenoviruses and parvoviruses). However, studies reveal that some viruses enter host cells through clathrin-independent endocytosis mediated by caveolae. For example,

Simian virus 40 (SV40), a nonenveloped DNA virus in the *Polyomaviridae* family, is reportedly internalized through caveolae (Anderson *et al.*, 1998; Stang *et al.*, 1997). In support of this model, SV40 particles are observed to be associated with caveolae by EM analysis (Stang *et al.*, 1997) and viral proteins fractionate with DRM (Stang *et al.*, 1997). In addition, SV40 infection is inhibited by disruption of caveolar function by cholesterol depletion or expression of a dominant-negative caveolin mutant (Anderson *et al.*, 1996; Pelkmans *et al.*, 2001; Roy *et al.*, 1999). It was shown that caveolae containing single virus particles fuse with larger preexisting organelles termed caveosomes (Pelkmans *et al.*, 2001), from which viruses are transported to the endoplasmic reticulum (ER) by a brefeldin A-sensitive pathway (Norkin *et al.*, 2002; Richards *et al.*, 2002). Interestingly, caveola-mediated endocytosis is generally inefficient and requires SV40-induced signaling events that lead to recruitment of molecules necessary for the formation of caveola-derived endocytic vesicles (Chen and Norkin, 1999; Pelkmans *et al.*, 2002). Internalization of several other nonenveloped viruses, including mouse polyomavirus, echovirus 1 and 11, and coxsackievirus A9, has also been reported to involve caveolae or rafts and requires signaling from these locations (Marjomaki *et al.*, 2002; Richterova *et al.*, 2001; Stuart *et al.*, 2002; Triantafilou and Triantafilou, 2003; Upla *et al.*, 2004; but see Gilbert *et al.*, 2003). As discussed later, several enveloped viruses also appear to enter host cells through caveolae or rafts. Thus, the caveola/raft-mediated endocytic pathway may be widely used for virus entry.

In contrast to the function of rafts and caveolae in the early phase of nonenveloped virus replication, a role for these microdomains in the late phase of the replication cycle is just beginning to be explored. The structural proteins of polioviruses and rotaviruses are found in the DRM fractions of virus-producing cells (Cuadras and Greenberg, 2003; Martin-Belmonte *et al.*, 2000; Sapin *et al.*, 2002). In addition, rotavirus release is inhibited by cholesterol depletion (Cuadras and Greenberg, 2003). How rafts might promote the assembly and/or release of nonenveloped viruses remains to be defined.

### *B. Enveloped Viruses*

#### *1. Enveloped Virus Structure*

Advances in our understanding of the relationship between enveloped virus replication and plasma membrane structure derive from the study of a broad range of viruses, including the orthomyxoviruses,

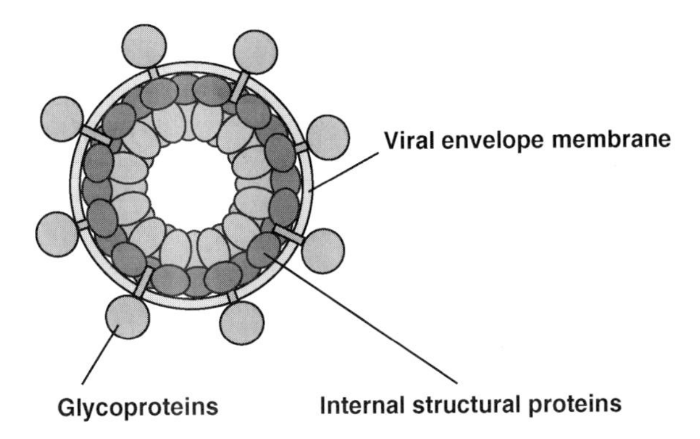


| | Glycoproteins | Internal structural proteins |
|---|---|---|
| Orthomyxoviridae | HA, NA, M2 | M1 |
| Paramyxoviridae | | |
| Sendai virus | F, HN | M |
| Measles virus | F, H | M |
| Respiratory syncytial virus | F, G | M |
| Retroviridae | Env (SU + TM) | Gag |
| Filoviridae | GP | VP40, VP24 |
| Rhabdoviridae | G | M |

FIG 3. *Top*: Enveloped virus structure. *Bottom*: Viral membrane-associated structural proteins mentioned in text.

paramyxoviruses, retroviruses, filoviruses, rhabdoviruses, and alphaviruses. These enveloped viruses share several structural similarities. They are surrounded by a lipid bilayer originating from the host plasma membrane. This envelope membrane contains glycoproteins (Fig. 3) that mediate receptor binding and fusion between viral and host cell membranes. Immediately underneath the lipid bilayer of the viral envelope is located another group of viral structural proteins; these form capsid structures that encase the viral nucleic acid (Fig. 3).

For many enveloped viruses, including those in the six families listed previously, particle assembly and release occur primarily at the host cell plasma membrane. Although the lipid bilayer of the viral envelope is derived from the producer cell plasma membrane, virologists have long been intrigued by the finding in a number of viral systems that the lipid composition of viral and plasma membranes differs significantly (Aloia *et al.*, 1993; McSharry and Wagner, 1971; Pessin and Glaser, 1980; Quigley *et al.*, 1971, 1972; Renkonen *et al.*, 1971; Slosberg and Montelaro, 1982; reviewed in Aloia *et al.*, 1992). This observation led to the proposal that viruses bud from specific microdomains at the cell surface (Pessin and Glaser, 1980; Scheiffele *et al.*, 1999). As detailed later, a large number of enveloped virus structural proteins, not only transmembrane glycoproteins but also internal structural proteins, have been identified as DRM associated. Cholesterol and sphingomyelin (Aloia *et al.*, 1992), as well as some raft-associated cellular proteins, are often enriched in viral lipid bilayers. Some viruses display properties consistent with the presence of liquid-ordered domains in their envelopes (Scheiffele *et al.*, 1999), suggesting that at least some viruses select rafts for their assembly at the plasma membrane. Although reports suggest that herpesvirus and flavivirus proteins associate with lipid microdomains in intracellular organelles (Lee *et al.*, 2003; Shi *et al.*, 2003), this review focuses on the role of plasma membrane rafts in virus replication.

### 2. *Role of Membrane Microdomains in Enveloped Virus Assembly*

A large number of cellular factors are reportedly either transiently or constitutively associated with rafts. Localization of these molecules to a confined area in the plasma membrane has been predicted to facilitate the biological processes in which they are involved (e.g., signal transduction) by increasing the opportunity for intermolecular interactions (Brown and London, 2000; Simons and Toomre, 2000). This "concentration platform" function of rafts may also play a role in virus assembly, as multiple distinct viral components must colocalize to form a virus particle. Alternatively, other cellular functions, physical properties, or components associated with rafts may facilitate multiple steps in virus particle production; for example, the transport, targeting, membrane association, or multimerization of internal structural and transmembrane proteins, the interaction or incorporation of transmembrane envelope glycoproteins, or the budding and release of virus particles (Fig. 5). In the following sections we summarize what has been reported regarding the relationship between virus assembly and membrane microdomains, and discuss the roles that microdomains might play in the late stages of the virus replication cycle.

*a. Orthomyxoviruses* A substantial body of evidence supports the hypothesis that assembly of the orthomyxoviruses occurs in raft-rich domains (Nayak and Barman, 2002; Suomalainen, 2002). The influenza A virus encodes three transmembrane proteins: hemagglutinin (HA), neuraminidase (NA), and the ion channel-forming protein M2, and one peripherally membrane-associated protein, M1 (matrix). These proteins are all structural components of the virus particle. HA, a type I transmembrane glycoprotein, was the first viral protein identified as being DRM associated (Skibbens *et al.*, 1989). DRM association of HA occurs only after it arrives at the Golgi complex (Skibbens *et al.*, 1989), as is also observed for GPI-anchored proteins (Brown and Rose, 1992). The detergent resistance of HA is sensitive to cellular cholesterol depletion (Scheiffele *et al.*, 1997). The transmembrane domain, cytoplasmic tail, and palmitoylation sites located within these domains have all been reported to be involved in DRM association of HA (Lin *et al.*, 1998; Melkonian *et al.*, 1999; Scheiffele *et al.*, 1997; Zhang *et al.*, 2000).

The NA glycoprotein has also been observed to be DRM associated; chimeras between NA and the nonraft protein transferrin receptor revealed that two regions of the NA transmembrane domain are involved in its association with DRM (Barman and Nayak, 2000; Kundu *et al.*, 1996). In addition, the cytoplasmic tail of NA was also reported to play a role in DRM association (Zhang *et al.*, 2000). Importantly, virus particles with tailless HA and NA were found to contain less cholesterol and sphingomyelin relative to wild-type virus (Zhang *et al.*, 2000), suggesting that HA and NA may determine the location at which virus particle assembly and release occur.

The M1 protein, which is located at the inner leaflet of the viral membrane, was also found to be DRM associated in virus-infected cells (Ali *et al.*, 2000; Zhang *et al.*, 2000). The fraction of membrane-bound M1 that was DRM associated increased over time in pulse–chase analyses, supporting the notion that association of M1 with rafts is due to specific, not random, targeting (Ali *et al.*, 2000). Interestingly, this association was diminished if the viral glycoproteins were not present or if their cytoplasmic tails were truncated (Ali *et al.*, 2000; Zhang *et al.*, 2000). Similar observations were also made with virus-associated M1 (Zhang *et al.*, 2000). Expression of either HA or NA was sufficient for recruitment of M1 to DRM (Ali *et al.*, 2000), providing evidence of a direct interaction between the glycoproteins and M1.

In contrast to HA, NA, and M1, M2 reportedly is not associated with DRM (Zhang *et al.*, 2000). Consistent with the hypothesis that influenza A virus assembly occurs in rafts, M2 incorporation into virus particles is normally minimal (Zebedee and Lamb, 1988). Interestingly, however, removal of the cytoplasmic tails of HA and NA, which prevents DRM association of these glycoproteins and reduces incorporation of raft lipids into virus particles, increases M2 incorporation (Zhang *et al.*, 2000).

These observations collectively support the model that, through interactions with raft-associated HA and NA, M1 is targeted to rafts or raft-rich regions of the plasma membrane; M1 in turn recruits other virion components such as the ribonucleoproteins (RNPs). Consistent with a functional significance of rafts in influenza virus assembly, removal of the cytoplasmic tails of both HA and NA reduces particle production 10-fold and results in the formation of virions with aberrant morphology (Jin *et al.*, 1997). It should be noted, however, that mutations in these viral proteins may have multiple effects on virus assembly in addition to disrupting raft association. Indeed, three amino acid substitutions in the HA transmembrane domain, which abrogate HA–raft association, reduce virus particle production but have no impact on the morphology of progeny virions (Takeda *et al.*, 2003).

Finally, the hypothesis that orthomyxoviruses select ordered lipid domains for budding is supported by the observation that cholesterol and sphingomyelin are less detergent extractable when present in the envelope of the fowl plague virus than they are when packed into the envelope of a rhabdovirus [vesicular stomatitis virus (VSV)] or an alphavirus [Semliki Forest virus (SFV)] (Scheiffele *et al.*, 1999). Similarly, the envelope of fowl plague virus displays lower fluidity than those of VSV and SFV, and HA is more resistant to detergent solubilization when incorporated into fowl plague virus particles than when incorporated into VSV or SFV (Scheiffele *et al.*, 1999).

*b. Paramyxoviruses* Paramyxoviruses encode at least two envelope glycoproteins, one of which (the F protein) mediates fusion of viral and host plasma membranes and the other of which functions as an attachment protein. The latter is designated HN, H, or G, depending on the particular paramyxovirus genus (Lamb and Kolakofsky, 2001). These viruses also encode a peripherally membrane-associated matrix (M) protein that underlies the viral lipid bilayer. In the case of Sendai virus, both F and HN are detected in DRM fractions of flotation gradients (Sanderson *et al.*, 1995) and this DRM association is observed only after these proteins reach the Golgi or *trans*-Golgi network (Ali and Nayak, 2000). The membrane association of F and HN is detergent resistant even when these proteins are expressed singly (Sanderson *et al.*, 1995). Similar to the influenza A M1 protein described previously (Ali *et al.*, 2000; Zhang *et al.*, 2000), Sendai virus M proteins are associated with DRM only when coexpressed with either F or HN (Ali and Nayak, 2000). Analysis of influenza HA–Sendai viral F protein chimeras suggests that specific interactions between the M protein and either the cytoplasmic or transmembrane domain of F are important for recruiting the M protein to DRM (Ali and Nayak, 2000).

In pulse–chase analyses of virus-infected cells, a significant portion of newly synthesized M protein is membrane bound, but, as seen for the viral glycoproteins, DRM association of M is not observed immediately after pulse labeling (Ali and Nayak, 2000). The M protein becomes associated with DRM only after a chase period during which F and HN also acquire detergent resistance. These observations support the hypothesis that raft association of M requires that it interact with the viral glycoproteins. Interestingly, the nucleoprotein (NP), which is a component of the viral nucleocapsid, is also recruited to DRM with kinetics similar to those of the M protein, suggesting that nucleocapsids also associate with DRM through interaction with M (Ali and Nayak, 2000). Together, these results support a model for the involvement of rafts in Sendai virus assembly similar to that described previously for the orthomyxoviruses.

Measles virus, like Sendai, belongs to the family *Paramyxoviridae* and, like Sendai, all the measles membrane-associated proteins (F, H, and M), as well as its nucleocapsid (N) protein, are DRM associated in infected cells (Manie *et al.*, 2000). However, the determinants of DRM association of measles proteins are reportedly different in several respects from those of the corresponding Sendai proteins. First, unlike the Sendai virus attachment protein HN, DRM association of the measles H protein depends on coexpression of the F protein, which has an intrinsic ability to associate with DRM (Vincent *et al.*, 2000). Second, distribution of measles virus M and N proteins to DRM does not require specific interaction with its viral glycoproteins; cells infected with measles virus encoding the non-DRM-associated VSV G protein (MGV), instead of its own H and F proteins, show a similar level of M and N proteins associated with DRM (Manie *et al.*, 2000; Vincent *et al.*, 2000). Single expression of either M or N by recombinant vaccinia virus reveals that M but not N has an intrinsic ability to interact with DRM (Vincent *et al.*, 2000). However, the level of DRM-associated M protein expressed by vaccinia virus infection is markedly lower than that in cells infected with measles virus or MGV, and is not increased by coinfection with recombinant vaccinia viruses that express N proteins or other nucleocapsid proteins (Vincent *et al.*, 2000). These observations support the notion that distribution of viral internal proteins to rafts may require the measles virus genome. On the basis of these results, the following model for measles virus assembly was proposed (Vincent *et al.*, 2000). F associates with rafts and recruits H into these microdomains. Independently, the RNP, whose components include the genome and the N protein, forms in the cytosol and interacts with M. This interaction leads to efficient raft association of the M–RNP complex. In rafts, the M–RNP complex recruits viral glycoproteins and virus particles are released.

In addition to Sendai and measles viruses, the structural proteins of another paramyxovirus, respiratory syncytial virus (RSV), are detected in DRM fractions (Henderson *et al.*, 2002; Marty *et al.*, 2004; McCurdy and Graham, 2003). When the RSV M and F proteins are singly expressed or coexpressed in a vaccinia system, M appears to require the presence of F to associate with Triton X-100-resistant membrane (Henderson *et al.*, 2002). Confocal and EM data obtained from cells infected with RSV suggest that assembly of this virus occurs in a region of the cell surface enriched in caveolin-1 and the glycosphingolipid GM1 (Brown *et al.*, 2002a,b; Jeffree *et al.*, 2003). Thus these studies, which use a variety of techniques and several different viruses, support the model that at least to some extent paramyxovirus assembly takes place within rafts or raft-rich areas of the plasma membrane. However, because functional studies that examine the effect of raft disruption on paramyxovirus assembly have not yet been performed, whether rafts are required for the assembly of these viruses remains an open question.

*c. Retroviruses* Retroviruses typically encode two membrane-associated structural proteins: the Env glycoprotein and the peripherally membrane-associated Gag protein. Virus-like particle (VLP) formation is mediated solely by Gag; for most retroviruses, Env is dispensable for this process. Sequential steps in virus assembly and release are generally promoted by three functional domains within Gag: the membrane binding (M), Gag–Gag interaction (I), and late (L) domains (Fig. 4).

The lipid bilayer of retroviral envelopes has long been known to be enriched, relative to the host cell plasma membrane, in the raft components sphingomyelin and cholesterol (Aloia *et al.*, 1993; Pessin

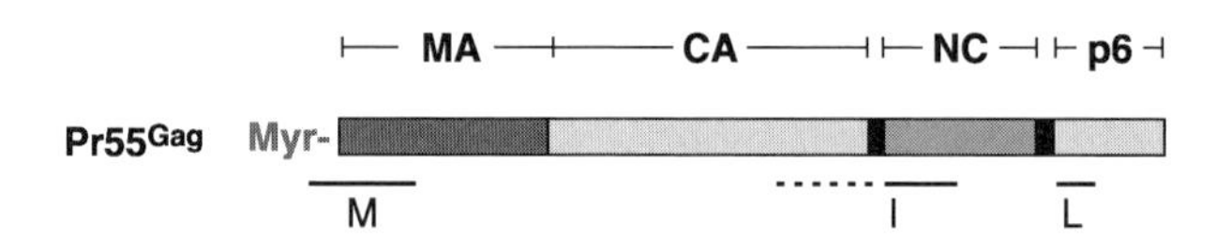


FIG 4. Schematic representation of the HIV-1 Gag protein. The domain structure of the precursor Pr55Gag is shown at the top. Positions of the N-terminal myristate (Myr-) and three functional domains (membrane binding [M], Gag–Gag interaction [I], and late [L] domains) are shown. While the I domain in NC promotes strong Gag–Gag interaction, a C-terminal region of CA (dashed line) also promotes Gag multimerization.

and Glaser, 1980; Quigley *et al.*, 1971, 1972; Slosberg and Montelaro, 1982; for a review, see Aloia *et al.*, 1992). In addition, some GPI-anchored proteins have been detected in human immunodeficiency virus type 1 (HIV-1) virions (Nguyen and Hildreth, 2000; Ott, 1997; Saifuddin *et al.*, 1995), whereas the non-raft-associated protein CD45 is largely excluded (Nguyen and Hildreth, 2000). Although the data on virion composition are consistent with the concept that retrovirus particles assemble in raft microdomains, several caveats are worth noting: (1) HIV-1 virions incorporate not only raft-associated molecules, but also nonraft markers, for example, the transferrin receptor (CD71) (Orentas and Hildreth, 1993); and (2) exclusion of CD45 from virions is likely due to steric interference induced by its long cytoplasmic domain. In addition, it has been reported that murine leukemia virus (MLV) particles display minimal exclusion of host plasma membrane proteins (Hammarstedt *et al.*, 2000).

The hypothesis that retroviruses assemble in rafts gained support from biochemical analyses demonstrating that the Gag proteins of several retroviral genera, including the lentiviruses (e.g., HIV-1), gammaretroviruses (e.g., MLV), and deltaretroviruses (e.g., human T cell leukemia virus type 1 [HTLV-1]), are associated with DRM (Ding *et al.*, 2003; Feng *et al.*, 2003; Halwani *et al.*, 2003; Lindwasser and Resh, 2001; Nguyen and Hildreth, 2000; Ono and Freed, 2001; Pickl *et al.*, 2001; Yang and Ratner, 2002). Microscopic analyses have also shown extensive copatching of HIV-1 Gag with the glycosphingolipid GM1 (Fig. 2) (Holm *et al.*, 2003), supporting the data obtained in biochemical assays. DRM association of HIV-1 Gag relative to total Gag-membrane binding increases with time postsynthesis (Halwani *et al.*, 2003; Ono and Freed, 2001), suggesting that Gag–DRM association is not the result of random targeting of Gag to raft and nonraft domains in the plasma membrane. Analysis of Gag truncation mutants indicates that the MA domain by itself or MA–CA (Fig. 4) is sufficient to confer DRM association (Lindwasser and Resh, 2001; Ono and Freed, 2001), but that the kinetics and extent of this association are increased by sequences downstream of MA, particularly within the I domain of NC (Lindwasser and Resh, 2001; Ono and Freed, 2001). These results suggest that Gag–Gag interaction enhances or stabilizes Gag association with lipid rafts, or that conformational changes or interactions with cellular factors mediated by sequences encompassing the I domain may increase DRM binding. It is worth noting, however, that a full-length Gag mutant containing substitutions in NC residues that constitute the I domain still efficiently associates with DRM (A. Ono, A. A. Waheed, and E. O. Freed, unpublished data), indicating that although Gag–Gag interactions may promote DRM association they are not essential for raft binding. Gag-containing DRM has been reported to possess a higher density than DRM containing classical raft markers, suggesting that the presence of multimeric Gag in rafts increases the density of the lipid microdomain (Lindwasser and Resh, 2001; but see Ding *et al.*, 2003). It has also been reported that HIV-1 Gag associates with lipid rafts that are disrupted by cold Triton X-100 treatment but are resistant to another detergent, Brij 98 (Holm *et al.*, 2003). Together with the data on viral membrane composition, these observations support the hypothesis that retrovirus assembly takes place in lipid rafts or in raft-rich areas of the plasma membrane. However, the exact biochemical nature of the microdomain (s) with which Gag associates, and the manner in which Gag itself alters the properties of such microdomain(s), remain to be fully defined.

Studies using pharmacological reagents have demonstrated a functional relevance for the association of Gag with lipid rafts. Raft disruption by cholesterol depletion with methyl-$\beta$-cyclodextrin or HMG-CoA reductase inhibitors (Ono and Freed, 2001), or cholesterol sequestration with nystatin (Pickl *et al.*, 2001), significantly decreases virus release efficiency from HIV-1- or MLV-producing cells. The addition of a polyunsaturated fatty acid to HIV-1-expressing cells, which replaces the myristate at the N terminus of Gag and thereby decreases the affinity of Gag for a liquid-ordered environment, also inhibits HIV-1 particle production (Lindwasser and Resh, 2002). Interferon $\alpha$, which reduces the association of HTLV-1 Gag with DRM by an unknown mechanism, also blocks HTLV-1 release (Feng *et al.*, 2003). These results strongly suggest that the interaction of Gag with lipid rafts is beneficial to efficient retrovirus assembly. As discussed later, multiple steps in virus assembly could be facilitated by Gag–raft binding.

HIV-1 and MLV Env glycoproteins have also been detected in DRM fractions (Li *et al.*, 2002; Nguyen and Hildreth, 2000; Pickl *et al.*, 2001; Rousso *et al.*, 2000). Because, as mentioned previously, a major portion of the plasma membrane may possess raft-like properties, whether the Env glycoproteins are specifically targeted to DRM remains to be determined. However, immunostaining and/or antibody-copatching data provide supporting evidence for specific raft association of both HIV-1 (Nguyen and Hildreth, 2000; Pickl *et al.*, 2001) and MLV (Pickl *et al.*, 2001) Env glycoproteins. Several retroviral Env glycoproteins, including those of avian sarcoma/leukosis virus (ASLV), MLV, and HIV, are known to be palmitoylated in their transmembrane (TM) subunits (Ochsenbauer *et al.*, 2000; Yang and Compans, 1996; Yang *et al.*, 1995).

Although palmitoylation is often a driving force for protein association with rafts (Melkonian *et al.*, 1999), ASLV Env glycoproteins are not detected in DRM fractions (Ochsenbauer *et al.*, 2000). For MLV Env, which does show raft association in biochemical and microscopic assays (Li *et al.*, 2002; Pickl *et al.*, 2001), a mutation that abolishes TM palmitoylation diminishes recovery of Env in DRM fractions (Li *et al.*, 2002). It is tempting to speculate that the concentration of Env glycoproteins in rafts would increase the efficiency of Env incorporation into retroviral particles assembled in rafts. Indeed, palmitoylation-defective MLV Env is incorporated into virions with reduced efficiency (Yang and Compans, 1996). However, this incorporation defect was attributed to reduced cell surface Env expression (Li *et al.*, 2002) as observed for ASLV (Ochsenbauer-Jambor *et al.*, 2001). Furthermore, the MLV palmitoylation site mutation has no significant impact on virus replication or syncytium formation (Li *et al.*, 2002; Yang and Compans, 1996). Thus, the functional significance of MLV Env palmitoylation and DRM association remains to be determined.

The role of HIV-1 Env glycoprotein palmitoylation has also been investigated. Amino acid substitutions at palmitoylation sites in the cytoplasmic domain of the gp41 TM glycoprotein increased the detergent solubility of the Env glycoprotein precursor gp 160, reduced incorporation of gp160 into virus particles produced in 293T cells, and impaired the infectivity of progeny virions (Rousso *et al.*, 2000). These observations are consistent with a role for TM palmitoylation in gp160 DRM association and Env incorporation. However, in this study, raft association was not measured by flotation centrifugation. It is also unclear whether mutation of cysteine residues in the gp41 cytoplasmic domain affects only DRM association of Env or additional, unrelated aspects of Env glycoprotein folding and transport. In this regard, it has been demonstrated that the entire cytoplasmic domain of Env (which includes the sites of gp41 palmitoylation) can be removed with little effect on Env incorporation and virus infectivity when virions are produced in 293T cells (Murakami and Freed, 2000). Clearly, more detailed biochemical analysis is needed to define the role of HIV-1 Env palmitoylation in raft association, Env incorporation into virus particles, and virus infectivity. In addition, alternative approaches should be used to block Env–raft association, as mutations may have pleiotropic effects.

*d. Filoviruses* The filoviruses encode three membrane-associated proteins: the transmembrane glycoprotein GP, the peripherally membrane-associated matrix protein VP40, and VP24, a highly hydrophobic protein of uncertain function (Sanchez *et al.*, 2001). Filovirus VLPs can be produced by coexpression of GP and VP40 (Bavari *et al.*, 2002; Noda *et al.*, 2002). These VLPs, as well as authentic filovirus virions, contain readily detectable levels of the GM1 sphingolipid, and both GP and VP40 are detected in DRM (Bavari *et al.*, 2002). GP association with DRM is abolished either by cholesterol depletion of virus-producing cells or by mutation of putative palmitoylation sites within GP (Bavari *et al.*, 2002). In addition, confocal microscopy demonstrates that GP displays significant copatching with GM1 on the cell surface (Bavari *et al.*, 2002). These results collectively suggest that filovirus assembly and/or release is associated with rafts (Bavari *et al.*, 2002; Freed, 2002b). Consistent with this notion, DRM-associated VP40 is more oligomerized than is VP40 in detergent-soluble membrane (Panchal *et al.*, 2003). Tsg101, a protein involved in the budding/release of Ebola and several retroviruses (Freed, 2002a; Pornillos *et al.*, 2002), was recruited to rafts on coexpression with VP40 (Panchal *et al.*, 2003). Furthermore, the efficiency of VLP release driven by a series of C-terminally truncated VP40 mutants appears to correlate with the extent of VP40–DRM association (Panchal *et al.*, 2003), although both release and DRM association may simply reflect plasma membrane-binding efficiency (Panchal *et al.*, 2003). Future studies will define the mechanism by which VP40 associates with lipid rafts, and the significance of this association for filovirus replication.

*e. Rhabdoviruses and Alphaviruses* It is currently unclear whether rhabdovirus or alphavirus assembly takes place in rafts. Rhabdovirus particles are enriched in cholesterol and sphingomyelin (Blough *et al.*, 1977; McSharry and Wagner, 1971; Patzer *et al.*, 1978; Pessin and Glaser, 1980; Welti and Glaser, 1994) and incorporate GPI- anchored proteins (Calafat *et al.*, 1983). The G glycoprotein of the rhabdovirus vesicular stomatitis virus (VSV) is palmitoylated (Schmidt and Schlesinger, 1979) and exhibits partial copatching with cellular raft-associated proteins on the surface of expressing cells (Harder *et al.*, 1998). Furthermore, the presence of VSV G and matrix proteins in unilamellar vesicles induces the lateral organization of lipid domains enriched in sphingomyelin and cholesterol (Luan *et al.*, 1995). VSV G reportedly organizes into membrane microdomains, 100–150 nm in diameter, in the absence of other viral proteins or an intact cytoplasmic domain (Brown and Lyles, 2003a,b). Virions of the alphavirus Semliki Forest virus (SFV) are also enriched in cholesterol (Renkonen *et al.*, 1971), and studies analyzing cholesterol-depleted cells demonstrate that plasma membrane cholesterol is essential for alphavirus assembly and release (Lu and Kielian, 2000; Lu *et al.*, 1999; Marquardt *et al.*, 1993; Vashishtha *et al.*, 1998) and virus entry (see later). These observations suggest the possibility that rhabdovirus and alphavirus particle production is associated with raft-like microdomains.

However, in apparent contradiction, the VSV and SFV glycoproteins are not associated with DRM in biochemical assays (Harder *et al.*, 1998; Lu and Kielian, 2000; Scheiffele *et al.*, 1999; but see Pickl *et al.*, 2001); indeed, VSV G is often used as a negative control for raft association in such assays. The influenza virus glycoprotein HA, which is normally associated with DRM, displays a high detergent sensitivity when incorporated into VSV virions (Scheiffele *et al.*, 1999). A comparative study of VSV, SFV, and influenza A showed that cholesterol and sphingomyelin in SFV and VSV lipid bilayers are markedly more sensitive to extraction by detergent or methyl-$\beta$-cyclodextrin than are these lipids from the influenza A virus envelope (Scheiffele *et al.*, 1999). The lipid bilayer of VSV and SFV also displayed higher fluidity than that of influenza A virus, although the fluidity of VSV and SFV envelopes is still lower than that of an artificial membrane in liquid crystalline phase (Scheiffele *et al.*, 1999). These results suggest that lipids derived from SFV and VSV membranes are less ordered relative to those from influenza envelopes, arguing that SFV and VSV bud from non-DRM domains of the plasma membrane. Further study is needed to characterize the nature of microdomains from which rhabdovirus and alphavirus particle production takes place.

### 3. *Possible Benefits of Raft Association for Virus Assembly*

The association of viral proteins with lipid rafts could promote virus assembly in a variety of ways (Fig. 5). Experimental evidence in support of several nonmutually exclusive models has been obtained.

*a. Concentration of Viral Structural Proteins* As has been suggested for the association of signal transduction molecules with rafts (Brown and London, 2000; Simons and Toomre, 2000), one of the possible benefits for viruses of raft binding during assembly is that various viral components can be efficiently concentrated to a confined area of the plasma membrane, thereby facilitating multimerization. Although a relatively large area of the plasma membrane in some cell types may display detergent-resistant properties, a concentration effect can still be imposed if the plasma membrane is composed of different subsets of rafts and if viral proteins interact specifically with one subset (Pierini and Maxfield, 2001; Simons and Toomre,

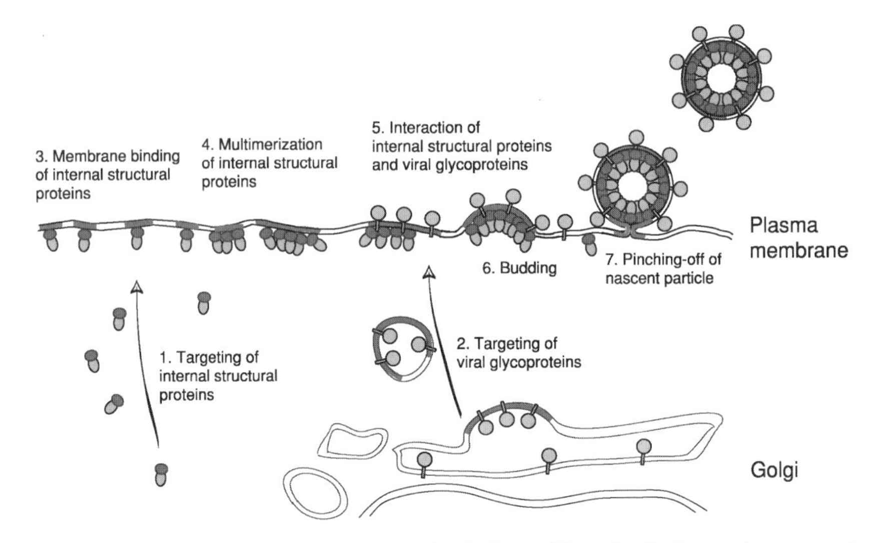


FIG 5. Steps in virus assembly that may be facilitated by rafts. Rafts are shown in red. (See Color Insert.)

2000). Rafts may provide a common destination for multiple envelope glycoproteins or for envelope glycoproteins and inner structural proteins and thereby promote the incorporation of glycoproteins into virus particles (Briggs *et al.*, 2003; Pickl *et al.*, 2001; Rousso *et al.*, 2000; Takeda *et al.*, 2003). This may apply to the incorporation of heterologous viral glycoproteins (i.e., pseudotype formation), as copatching experiments suggest that a wide variety of viral glycoproteins are at least weakly associated with rafts (Harder *et al.*, 1998; Pickl *et al.*, 2001). It is noteworthy that incorporation of VSV G into MLV virions appears to be more sensitive to cholesterol depletion of virus-producing cells than is the incorporation of the homologous MLV Env glycoprotein (Pickl *et al.*, 2001).

As proposed for the oligomerization of bacterial toxins in DRM (Abrami and van der Goot, 1999, but see Nelson and Buckley, 2000; Miyata *et al.*, 2002; Zhuang *et al.*, 2002), rafts may also serve as concentration platforms that increase the multimerization efficiency of raft-associated internal structural proteins. Consistent with this model, cholesterol depletion of HIV-1-producing cells appears to reduce the efficiency of HIV-1 Gag multimerization on the plasma membrane (A. Ono and E. O. Freed, unpublished data). Interactions between raft-associated viral proteins would stabilize protein–raft association and, as observed with antibody cross-linking of cellular raft proteins (e.g., Harder *et al.*, 1998), may promote the coalescence of small, highly dynamic rafts into large, stable rafts that serve as centers of virus assembly.

*b. Efficient Targeting of Viral Components and Virus Assembly* In polarized epithelial cells, many DRM-associated proteins, such as those that contain GPI anchors, are targeted to the apical surface. The DRM association of these proteins begins after their sorting from the *trans*-Golgi network (e.g., Brown and Rose, 1992). Similarly, viral proteins such as influenza HA and NA are also DRM associated and are apically targeted (see previous discussion). Moreover, apical targeting of HA in polarized epithelial cells is abolished by cholesterol depletion (Keller and Simons, 1998). These results support the idea that rafts mediate sorting by serving as transport platforms to the apical surface (for a review, see Simons and Ikonen, 1997). However, there are a number of instances in which this model does not apply: (1) apical sorting and DRM association of certain HA and NA mutants is not coupled (Barman and Nayak, 2000; Lin *et al.*, 1998; Takeda *et al.*, 2003; Zhang *et al.*, 2000); (2) the influenza A M2 protein is apically targeted yet is not raft associated (Zhang *et al.*, 2000); (3) the measles virus F protein and the HIV-1 Env glycoprotein are DRM associated yet are sorted to the basolateral surface in polarized epithelial cells (Lodge *et al.*, 1994, 1997; Naim *et al.*, 2000); (4) in some circumstances, DRM association and apical sorting of GPI-anchored proteins do not correlate (Arreaza and Brown, 1995; Benting *et al.*, 1999; Lipardi *et al.*, 2000); and (5) caveolae, which constitute a distinct class of raft-associated structures (see previous discussion), are localized to the basolateral surface (Scheiffele *et al.*, 1998; Vogel *et al.*, 1998). Thus, although association with rafts can promote apical transport, there appear to be various factors that modulate this process. In other types of polarized cells, such as stimulated lymphocytes, rafts are involved in mobilization of membrane-associated proteins to specific sites, for example, pseudopodia or points of cell–cell contact (e.g., Gomez-Mouton *et al.*, 2001; Viola *et al.*, 1999).

Because rafts can be concentrated to specific regions of the plasma membrane, raft association of viral structural proteins may determine the site on the cell surface to which virus assembly is targeted. Coexpression of measles M, F, and H proteins in the context of infectious virus production enhances levels of DRM association of these proteins compared with levels observed on single expression (Vincent *et al.*, 2000). Interestingly, in polarized cells, coexpression redirects F and H proteins to the apical surface and induces virus budding from that surface (Naim *et al.*, 2000). This targeted budding would in theory contribute to airway shedding of infectious virus particles. Raft association may promote the targeting of viral proteins to other cellular structures such as the cytoskeleton. A link between the actin cytoskeleton and DRM may contribute to the formation of filamentous influenza virions (Simpson-Holley *et al.*, 2002). HIV-1 budding in activated lymphocytes and monocytes appears to take place predominantly from pseudopodia, which protrude in a polarized manner (Pearce-Pratt *et al.*, 1994; Perotti *et al.*, 1996), and are raft rich (Gomez-Mouton *et al.*, 2001). Because pseudopodia are enriched in molecules that mediate cell–cell adhesion (Sanchez-Madrid and del Pozo, 1999), it is tempting to speculate that the targeting of virus budding to raft-rich sites could have major implications for cell-to-cell virus transmission. Consistent with this possibility, HTLV-1 and HIV-1 are observed to concentrate at cell–cell contact sites in a cytoskeleton-dependent manner (Igakura *et al.*, 2003; Jolly *et al.*, 2004).

As the plasma membrane is highly enriched in rafts relative to intracellular membranes, rafts may assist in directing virus assembly to the cell surface. Substitution of the myristic acid at the N terminus of HIV-1 Gag with an unsaturated myristate analog impairs Gag–DRM association and redirects Gag from the plasma membrane to the ER (Lindwasser and Resh, 2002). In contrast, cholesterol depletion at the plasma membrane does not cause redirection of Gag to intracellular membranes (A. Ono and E. O. Freed, unpublished data). It therefore remains unclear what role DRM association plays in the targeting of Gag to the cell surface.

*c. Additional Steps in Virus Particle Formation* For viral structural proteins peripherally associated with the lipid bilayer, ensuring stable binding to the cytoplasmic leaflet of the plasma membrane is essential for efficient virus assembly. Rafts may facilitate this process directly by providing essential cellular components or a suitable lipid bilayer microenvironment for membrane binding. Alternatively, rafts could enhance membrane binding indirectly by promoting multimerization of viral proteins thereby increasing the valency of membrane binding. Consistent with a role for rafts in the interaction between viral proteins and the lipid bilayer, cholesterol depletion reduces the efficiency of HIV-1 Gag membrane binding (A. Ono and E. O. Freed, unpublished data). It is also possible that the final steps in virus assembly and release, for example, the budding or pinching off of virus particles from the cell surface, may be facilitated by rafts or their components. In this regard, it has been proposed that membrane microdomains modulate a number of cellular processes that involve budding or fission of membrane vesicles (reviewed by Huttner and Zimmerberg, 2001).

In addition to these theoretical and apparent benefits of raft association for virus assembly and release, budding from rafts or raft-rich membrane may also enhance particle infectivity in the subsequent round of replication. This concept is supported by the observation that truncation of the cytoplasmic tail of the influenza A HA glycoprotein blocks raft association of HA (Ali *et al.*, 2000; Zhang *et al.*, 2000), impairs virus production, disrupts proper virion morphogenesis, and reduces the infectivity of released particles (Jin *et al.*, 1997). Cholesterol depletion of HIV-1-producing cells also reduces both virus production and infectivity of released particles (Ono and Freed, 2001; Zheng *et al.*, 2001). This reduced infectivity could be due to structural changes in the virion, inefficient incorporation of viral or cellular factors, or impaired function of incorporated proteins. One study found that reduced infectivity of HIV-1 occurs only when Nef, a viral accessory protein involved in several aspects of HIV replication, is expressed in the cholesterol-depleted producer cells (Zheng *et al.*, 2001). Interestingly, Nef is itself a DRM-associated protein (Walk *et al.*, 2001; Wang *et al.*, 2000). The mechanism by which cholesterol depletion might disrupt Nef function remains to be determined. The role of rafts in the early events of virus replication is discussed in the next section.

### 4. *Role of Membrane Microdomains in Enveloped Virus Entry*

The entry of enveloped viruses is initiated upon binding of viral spike or Env glycoproteins on the virion to receptors on the surface of the target cell. After virion binding, the viral genome, generally in a complex with inner structural proteins, gains entry into the cell by one of two distinct pathways: (1) direct fusion between the lipid bilayer of the virion and the target cell plasma membrane, or (2) low pH-triggered fusion between the viral lipid bilayer and cellular endosomal membrane after endocytosis of the virus particle. Lipid rafts may play important roles in virus entry through both pathways. A number of virus receptors are reportedly associated with rafts, and virus interaction with such receptors may lead to raft-mediated endocytosis or virus–cell fusion in rafts at the plasma membrane. In this section, we describe the information currently available regarding viruses that use either GPI-anchored or transmembrane proteins as their receptors and discuss the possible significance of raft involvement in the early phase of virus replication.

*a. Viruses Using GPI-Anchored Receptors* Although the involvement of clathrin-mediated endocytosis in the internalization of raft proteins has been reported (Abrami *et al.*, 2003; Stoddart *et al.*, 2002; Wilson *et al.*, 2000), most raft-associated molecules studied thus far are internalized via clathrin-independent pathways (for review, see Nichols and Lippincott-Schwartz, 2001; Sharma *et al.*, 2002). GPI-anchored proteins (Nichols *et al.*, 2001; Sabharanjak *et al.*, 2002), as well as raft-associated transmembrane proteins (Lamaze *et al.*, 2001) and lipids (e.g., Puri *et al.*, 2001), have been used to study clathrin-independent endocytosis, which is constitutive but may be enhanced by cross-linking of raft-associated molecules (Nichols and Lippincott-Schwartz, 2001; Sharma *et al.*, 2002). Cross-linking of GPI-anchored proteins may also drive them to caveolae (Mayor *et al.*, 1994), from which some clathrin-independent endocytosis may occur. However, internalization of raft-associated molecules certainly occurs by non-caveola-mediated mechanisms because this process has been observed in cells lacking caveolae (Orlandi and Fishman, 1998). The itinerary of raft-associated molecules internalized by nonclathrin pathways is currently under investigation (Fivaz *et al.*, 2002; Nichols *et al.*, 2001; Sabharanjak *et al.*, 2002).

Raft-associated endocytosis may provide an ideal portal for virus entry because it can mediate virus internalization constitutively and can be enhanced by cross-linking mediated by virus particle binding. Indeed, two filoviruses and two retroviruses appear to use GPI-anchored proteins as entry receptors. Expression of the GPI-anchored folate receptor $\alpha$ confers susceptibility to Marburg and Ebola pseudotypes in otherwise nonpermissive Jurkat cells (Chan *et al.*, 2001; but see Simmons *et al.*, 2003), although it should be noted that other receptors likely promote filovirus entry in some cell types (for review, see Aman *et al.*, 2003). The receptor for subgroup A ASLV, Tva, is expressed either as a transmembrane or a GPI-anchored protein due to alternative splicing; both isoforms can function as virus receptors (Bates *et al.*, 1993). Jaagsiekte sheep retrovirus also uses a GPI-anchored receptor, HYAL2 (Rai *et al.*, 2001).

As filovirus entry is pH dependent (Takada *et al.*, 1997; Wool-Lewis and Bates, 1998), it likely occurs through endocytosis. It is tempting to hypothesize that internalization of filoviruses after binding to folate receptor $\alpha$ is mediated by either caveolae or rafts. Consistent with this hypothesis, filovirus-GP pseudotypes colocalize with caveolin, whereas VSV-G pseudotypes, which use the clathrin-mediated pathway, do not (Empig and Goldsmith, 2002). Furthermore, infection by both filovirus pseudotypes and bona fide Ebola virus is inhibited by cholesterol-perturbing reagents (Bavari *et al.*, 2002; Empig and Goldsmith, 2002), although whether these reagents inhibit caveola/raft-mediated endocytosis or other steps in virus entry is unknown. Retrovirus entry typically occurs by direct fusion between the viral envelope and the target cell plasma membrane. ASLV entry, however, has been reported to be dependent on low-pH triggering (Diaz-Griffero *et al.*, 2002; Mothes *et al.*, 2000), suggesting that ASLV entry may proceed via endocytosis [although data that oppose this model have also been presented (Earp *et al.*, 2003)]. Further analysis of ASLV stability after entry suggested that the GPI-linked Tva mediates a different pathway of entry from that mediated by the non-raft-associated transmembrane isoform of Tva, and that raft disruption forces virus bound to the GPI-linked Tva to follow the pathway usually mediated by the non-raft Tva (Narayan *et al.*, 2003). Although more study will be required to fully characterize ASLV entry, this virus may provide a useful tool with which to analyze raft-mediated endocytosis.

*b. Viruses Using Transmembrane Receptors* The role of rafts in virus entry mediated by direct fusion between the viral envelope and the target cell plasma membrane has been studied most for the retroviruses, in particular for HIV-1 (Campbell *et al.*, 2001). HIV-1 entry is initiated by interaction between the surface (SU) Env glycoprotein gp120 and two receptors: CD4, and a member of the seven-transmembrane G protein-coupled chemokine receptor family (usually CCR5 or CXCR4). Because gp120 first binds CD4 and then subsequently interacts with the chemokine receptor, CD4 is viewed as the primary receptor whereas CCR5 and CXCR4 are often referred to as "coreceptors." Sequential interaction of gp120 with CD4 and coreceptor induces a series of conformational changes in gp120 that ultimately triggers the formation of a six-helix bundle in the ectodomain of gp41 and exposure of the "fusion peptide" at the gp41 N terminus (reviewed in Berger *et al.*, 1999; Doms, 2000). The ultimate outcome of these conformational changes in gp120 and gp41 is fusion between the lipid bilayer of the virion and the host cell plasma membrane. Contact between HIV Env-expressing cells and those positive for CD4 and coreceptor can also lead to cell–cell fusion, or syncytium formation. It has also been reported that HIV-1 can be internalized in a raft-mediated manner when virus particles are transcytosed across epithelial cell layers (Alfsen and Bomsel, 2002; Alfsen *et al.*, 2001; Liu *et al.*, 2002; but see Argyris *et al.*, 2003).

The involvement of rafts in HIV-1 entry is supported by several reports, although this topic is not free from controversy. CD4 is widely accepted as being raft associated (Parolini *et al.*, 1996, 1999). In some studies, the coreceptors CCR5 and CXCR4 have also been detected in DRM fractions (Manes *et al.*, 2000; Popik *et al.*, 2002; Viard *et al.*, 2002), and binding of gp120 to these coreceptors appears to enhance their DRM association (Manes *et al.*, 2000; Sorice *et al.*, 2001; Viard *et al.*, 2002). In contrast, other reports observed that CXCR4 and CCR5 are weakly DRM associated or are not present in raft fractions (Kozak *et al.*, 2002; Percherancier *et al.*, 2003). Copatching methods have also been used to assess CD4 and coreceptor association with rafts. CD4 copatching with the GM1 sphingolipid has been observed, but again, divergent results were obtained for coreceptors: they copatch with GM1 but not with CD4 unless gp120 is present (Manes *et al.*, 2000), they copatch with both GM1 and CD4 (Popik *et al.*, 2002), or they copatch with neither (Kozak *et al.*, 2002; Percherancier *et al.*, 2003). These divergent findings may have arisen from differences in experimental design, cell types, or condition of the cells, as the interaction of raft-associated transmembrane proteins with other cellular structures such as the actin cytoskeleton can change the phenotype of raft association (Oliferenko *et al.*, 1999). Interestingly, in immuno-EM analysis, CD4 and coreceptors form microclusters that are distinct but close to each other (Singer *et al.*, 2001). These microclusters are predominantly found on microvilli, actin-based structures at the cell surface (Singer *et al.*, 2001). This point is noteworthy as actin disruption inhibits HIV Env-mediated cell–cell fusion (Viard *et al.*, 2002). Further analysis of the relationship between rafts and other cellular structures may provide additional insight into the connection between HIV entry and the association of CD4 and coreceptor with lipid rafts. A link between rafts and HIV-1 entry is further supported by the finding that CD4 binding induces gp120 to bind raft-associated glycosphingolipids (Hammache *et al.*, 1999).

The functional relevance of lipid rafts in HIV-1 entry has been addressed by a variety of methods. Mixed results have been obtained by genetic approaches analyzing CD4 mutants with varying degrees of raft association. Among a panel of chimeric proteins in which the extracellular domain of CD4 was fused to various transmembrane and cytoplasmic domains or a GPI anchor, only those that associated with rafts could support HIV entry (Del Real *et al.*, 2002). In contrast, two other studies reported that a CD4 mutant defective in raft association fully supported Env-mediated cell–cell fusion and virus entry (Percherancier *et al.*, 2003; Popik and Alce, 2004). This discrepancy may again be due to differences in experimental conditions; for example, the density of receptors on the target cell may affect the extent to which virus entry or fusion is dependent on raft integrity (Viard *et al.*, 2002). It is also important to note that changing CD4 structure by generating protein chimeras may affect an array of CD4 properties in addition to its raft association.

Pharmacological approaches generally support the involvement of cholesterol and glycosphingolipids in HIV-1 fusion and entry. Consistent with the observed reduction in infectivity of viruses derived from cholesterol-depleted cells (Liao *et al.*, 2003; Ono and Freed, 2001; Zheng *et al.*, 2001), direct treatment of virions with cholesterol-depleting agents also inhibits virus infectivity (Campbell *et al.*, 2002; Guyader *et al.*, 2002). Extraction of the majority of HIV-1 virion cholesterol creates openings in the viral lipid bilayer, resulting in the loss of a portion of viral proteins (Graham *et al.*, 2003). However, reductions in infectivity are evident even after milder cholesterol depletion treatments that do not cause substantial loss of viral components (Campbell *et al.*, 2002; Graham *et al.*, 2003; Guyader *et al.*, 2002; Liao *et al.*, 2003). Significantly, infectivity can be restored by replenishing cholesterol after depletion (Campbell *et al.*, 2002; Guyader *et al.*, 2002; Liao *et al.*, 2003), suggesting that at least under relatively gentle conditions cholesterol depletion does not globally disrupt virion structure. A shift in virion density was observed upon mild cholesterol depletion, but this effect was reversed by cholesterol replenishment (Campbell *et al.*, 2002). Cholestenone, a cholesterol analog that inhibits domain formation in artificial membranes (Xu and London, 2000), restored to normal the density of cholesterol-depleted virus particles but reduced virus infectivity even further (Campbell *et al.*, 2002). These observations suggest an important role for cholesterol in the HIV-1 lipid bilayer in virus infectivity. Interestingly, the entry of HIV-1 virions pseudotyped with VSV G was less susceptible to cholesterol depletion than was the internalization of virions bearing HIV-1 Env glycoproteins (Guyader *et al.*, 2002). Inhibition of virus–cell fusion but not virus–cell binding was observed after mild cholesterol depletion (Liao *et al.*, 2003). These results, as well as data on the role of cholesterol in the influenza virus lipid bilayer (Sun and Whittaker, 2003; but see Takeda *et al.*, 2003), suggest that glycoprotein functions are highly dependent on cholesterol content and, possibly, the microdomain structure of viral membranes (see later discussion).

A number of studies have also examined the role of target cell cholesterol and glycosphingolipids in fusion and entry. Cholesterol depletion of target cells reduces virus entry and fusion efficiency, apparently without affecting the surface expression level of CD4 and coreceptors (Liao *et al.*, 2001; Manes *et al.*, 2000; Nguyen and Taub,

2002b; Percherancier *et al.*, 2003; Popik *et al.*, 2002; Viard *et al.*, 2002). This inhibition could reflect an inherent role for cholesterol in HIV-induced membrane fusion or could result from a disruption of raft-mediated gp120/CD4/coreceptor interactions (Manes *et al.*, 2000; Popik *et al.*, 2002). Interestingly, the effect of cholesterol depletion on membrane fusion appears to be mitigated by high CD4/coreceptor densities (Viard *et al.*, 2002). Inhibition of glycosphingolipid (GSL) synthesis in target cells also reduces cell–cell fusion mediated by Env (Puri *et al.*, 1998b) and disrupts HIV-1 infectivity (Hug *et al.*, 2000). This effect was reversed when the target membrane was supplemented with a particular GSL species, Gb3 (Hug *et al.*, 2000; Puri *et al.*, 1998a). Antibodies against the glycosphingolipid galactosyl ceramide were reported to inhibit HIV-1 infection in neural cell lines (Harouse *et al.*, 1991).

In addition to HIV-1, other retroviruses, including ecotropic MLV (Lu and Silver, 2000; Lu *et al.*, 2002), may also utilize rafts to gain entry into target cells. The receptor for ecotropic MLV is MCAT-1, a 14-transmembrane protein that functions as a cationic amino acid transporter and colocalizes with caveolin (McDonald *et al.*, 1997). GFP-tagged MCAT-1, which supports MLV entry (Lee *et al.*, 1999; Masuda *et al.*, 1999), is readily detected in DRM fractions (Lu and Silver, 2000). MLV entry is impaired by depleting cholesterol in target cells expressing endogenous MCAT-1 or the MCAT-1–GFP chimera (Lu and Silver, 2000). Although it is generally thought that amphotropic MLV fuses directly at the plasma membrane, the route of ecotropic MLV entry appears to be largely cell type dependent (McClure *et al.*, 1990). Whereas cell–cell fusion mediated by ecotropic MLV Env can occur at neutral pH (Klement *et al.*, 1969; Zarling and Keshet, 1979), entry of ecotropic MLV into a number of cell types is sensitive to lysosomotropic agents (McClure *et al.*, 1990). Interestingly, internalization of ecotropic MLV by a non-clathrin pathway was reported (Lee *et al.*, 1999), suggesting that disruption of ecotropic MLV entry by cholesterol depletion may result from inhibition of endocytosis. Alternatively, the observed block to MLV entry by cholesterol depletion may be explained by a variety of models (see later discussion). Further studies are needed to elucidate the role of cholesterol and rafts in MLV entry.

Some evidence suggests that HTLV-1 fusion and entry are also raft dependent. It was reported that several monoclonal antibodies that inhibit HTLV-1-induced syncytium formation bind proteins that are GPI linked and raft associated (Niyogi and Hildreth, 2001). Furthermore, depletion of plasma membrane cholesterol with cyclodextrin inhibited syncytium formation (Niyogi and Hildreth, 2001) and binding of virus particles to target cells (Hague *et al.*, 2003). These results support the hypothesis that HTLV-1 uses raft-associated molecules as entry receptors and that receptor binding, fusion, and entry occur in raft microdomains. Whether the putative HTLV-1 receptor GLUT-1 (Manel *et al.*, 2003) is raft associated remains to be defined.

*c. Viruses Using Raft-Associated Lipids as Receptors* Several viruses in addition to HIV-1 have been reported to utilize specific lipid molecules in the entry process (Suzuki *et al.*, 1985a,b). For example, the requirement for cholesterol and sphingolipids on the target membrane for efficient entry of SFV is well established (Nieva *et al.*, 1994; Phalen and Kielian, 1991; for review, see Kielian *et al.*, 2000). SFV enters target cells by clathrin-mediated endocytosis. The subsequent fusion reaction between viral and endosomal membranes is catalyzed by the E1 transmembrane protein, which undergoes conformational changes upon exposure to low pH. This fusion event requires both cholesterol (Kielian and Helenius, 1984) and small amounts of sphingolipids (Nieva *et al.*, 1994). As discussed previously, these lipid molecules are major components of rafts, and the fusion peptide in the E1 ectodomain associates with DRM (Ahn *et al.*, 2002). However, using liposomes as the target membrane, cholesterol analogs that do not partition into rafts (Xu and London, 2000) can still support the interaction between the E1 ectodomain and the target membrane and can promote SFV fusion (Kielian and Helenius, 1984). SFV and another alphavirus, Sindbis virus, fuse with large unilamellar vesicles composed of various combinations of synthetic sphingolipids and sterols irrespective of detergent-resistant microdomain formation (Waarts *et al.*, 2002). Conversely, interaction of the E1 ectodomain with vesicles is not observed in the absence of sphingolipids even when DRM can still form in the target liposomes (Ahn *et al.*, 2002). Together, these studies suggest that although alphaviruses can interact and perhaps fuse with rafts, at least in *in vitro* systems these microdomains are not essential for the fusion process. Indeed, it appears that the requirement for cholesterol in SFV-mediated fusion is due to an interaction between a specific residue in E1 and the $3\beta$-hydroxyl group on cholesterol, rather than the partitioning of cholesterol into DRM (Kielian and Helenius, 1984). As mentioned previously, cholesterol is also required for efficient alphavirus egress; intriguingly, an SFV E1 mutant that displays a reduced requirement for cholesterol in virus entry is also released from cells in a cholesterol-independent manner (Vashishtha *et al.*, 1998). The connection between the role of cholesterol in alphavirus egress and entry, and what function rafts might play in alphavirus replication *in vivo*, remain to be determined.

*d. Models for Raft Involvement in Virus Entry* Cholesterol, sphingolipids, and the raft microdomains in which they are concentrated could play a variety of roles in virus entry (Fig. 6). (1) Rafts may increase the local concentration of entry receptors, thereby increasing the efficiency of binding and fusion events. Consistent with this model, the effect of cholesterol depletion on HIV-1 entry appears to be counteracted by high receptor levels (Viard *et al.*, 2002). Rafts may also enhance binding of virus to the target cell through interaction with nonreceptor molecules localized to rafts. For example, the ability of the adhesion molecule LFA-1 to interact with its ligand on the viral membrane could be compromised by cholesterol depletion, thereby reducing virus binding to target cells (Liao *et al.*, 2001); (2) cholesterol or sphingolipids could serve as binding sites for viral spike or Env proteins at the target cell surface and could affect their conformation after binding. Two GSL species, Gb3 and GM3, can reportedly interact with gp120 in a CD4-dependent manner *in vitro* (Hammache *et al.*, 1999), and GM3 and CXCR4 can form a complex with gp120 on the T cell surface (Sorice *et al.*, 2001). Gb3 can promote Env-mediated

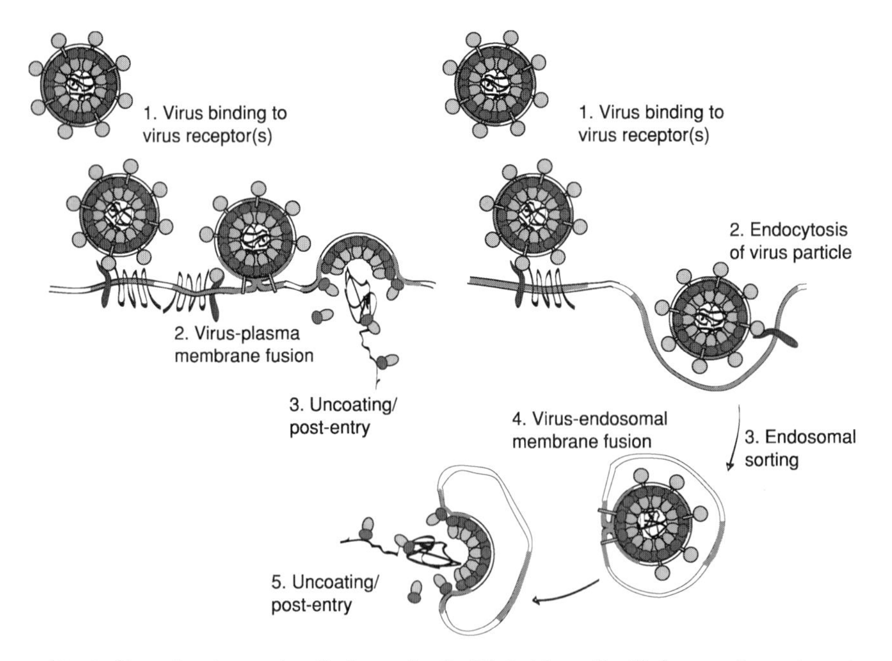


FIG 6. Steps in virus entry that may be facilitated by rafts. Rafts are shown in red. Two modes of virus entry (direct fusion and low pH-triggered fusion) are shown. (See Color Insert.)

cell–cell fusion in cells treated with GSL synthesis inhibitors (Hug *et al.*, 2000). Depletion of GSLs in target cells reportedly impedes conformational changes that lead to HIV-1 gp41 fusion peptide exposure (Hug *et al.*, 2000). It is noteworthy that although GSLs are major components of rafts, the finding that only particular GSL species enhance HIV-1 fusion (Hug *et al.*, 2000; Puri *et al.*, 1998a) suggests that effects of these molecules are independent of raft structure itself. As discussed previously, an interaction between a specific domain of the alphavirus E2 protein and cholesterol appears to play a critical role in triggering membrane fusion; again, it is not clear whether raft structure is important in this process; (3) rafts may modulate receptor conformation. Cell surface binding of the CXCR4 ligand SDF-1 and the CCR5 ligand MIP-1$\beta$ is reportedly reduced on cholesterol depletion or oxidation, apparently by altering the conformation of these chemokine receptors (Nguyen and Taub, 2002a,b, 2003). Such changes in coreceptor conformation could contribute to the inhibitory effect of cholesterol depletion on HIV-1 fusion and entry; (4) cholesterol, sphingolipids, other raft components, or raft structure, may provide a suitable environment for membrane fusion (Rawat *et al.*, 2003). Data to support this model include the observation that the presence of cholesterol in artificial target membranes strongly affects fusion catalyzed by fusion peptides derived from HIV-1 gp41 (Pereira *et al.*, 1997) or by influenza HA expressed on the donor cell surface (Razinkov and Cohen, 2000). The importance of viral membrane cholesterol in the fusion process was also demonstrated for HIV and influenza (Liao *et al.*, 2003; Sun and Whittaker, 2003; but see Takeda *et al.*, 2003). In contrast to these observations on virus–cell fusion, depletion of cholesterol from cells bearing the MLV Env glycoprotein or the Newcastle disease virus F protein did not disrupt cell–cell fusion (Dolganiuc *et al.*, 2003; Lu *et al.*, 2002). The relative contribution of cholesterol in the donor cell, target cell, or viral membrane to viral glycoprotein-induced membrane fusion remains to be defined; finally, (5) rafts may facilitate signal transduction pathways initiated by the binding of viral proteins to raft-localized receptors, thereby enhancing entry or postentry events. Indeed, signal transduction induced by binding of HIV-1 gp120 to surface receptors is reported to be inhibited by raft disruption (Kinet *et al.*, 2002). Any or all of these models might be operative in different viral and target cell systems. The addition of two herpesviruses (herpes simplex and Epstein-Barr) to the growing list of viruses whose entry is susceptible to cholesterol depletion of target cell membranes (Bender *et al.*, 2003; Katzman and Longnecker, 2003) supports the possibility that cholesterol-dependent microdomain structures in cellular membranes are widely used by a variety of enveloped viruses not only to promote virus particle production but also to enhance virus entry.

## IV. Perspective

It is now clear that a large number of viruses interact with membrane microdomains during entry into and egress from their host cells. The mechanism by which lipid microdomains promote the steps that constitute the virus replication cycle will no doubt be the focus of numerous future studies. Most studies to date have been conducted with a limited number of relatively crude methods (e.g., DRM isolation, cholesterol depletion, and copatching); application of new techniques such as video microscopy, chromophore-assisted laser inactivation, FRET, and photobleaching will add depth and detail to our mechanistic insights.

It is likely that lipid rafts play significant roles in steps of the virus replication cycle other than entry and assembly. Lipid raft association of several viral and cellular components has been implicated in virus-induced pathogenesis (Alexander *et al.*, 2004; Brown *et al.*, 2001; Cambi *et al.*, 2004; Clausse *et al.*, 1997; Dykstra *et al.*, 2001; Gummuluru *et al.*, 2003; Hamilton *et al.*, 2003; Higuchi *et al.*, 2001; Kaykas *et al.*, 2001; Koshizuka *et al.*, 2002; Park *et al.*, 2003; Rothenberger *et al.*, 2002; Walk *et al.*, 2001; Wang *et al.*, 2000; Zheng *et al.*, 2001, 2003). Rafts are involved not only in signal transduction and protein transport but also in the generation or regulation of cellular structures; for example, endosomes, cytoskeleton, pseudopodia, and cell–cell contact points. How viruses use these structures, in particular points of cell–cell contact, to facilitate their replication is of particular interest.

The existence of microdomains in a variety of cellular membranes is now well documented, but a variety of key aspects of raft behavior and function remain to be elucidated. For example, what is the nature of the cytoplasmic leaflet of the lipid bilayer in raft microdomains, and how do raft-associated molecules communicate between the inner and outer leaflets? What distinctions can be made between "different" types of rafts within the plasma membrane and within intracellular membranes? What modulates raft coalescence/dispersal and regulates the size of rafts under various conditions? Answers to these and many additional questions will be particularly important to our understanding of the relationships between rafts and virus replication. Conversely, further studies on the interaction between viruses and rafts will provide key insights into the biological functions of these microdomains. Ultimately, one would hope that a more in-depth understanding of the relationship between viruses and membrane microdomains will contribute to our ability to combat viral diseases.

## References

Abrami, L., and van der Goot, F. G. (1999). Plasma membrane microdomains act as concentration platforms to facilitate intoxication by aerolysin. *J. Cell Biol.* **147:**175–184.

Abrami, L., Liu, S., Cosson, P., Leppla, S. H., and van der Goot, F. G. (2003). Anthrax toxin triggers endocytosis of its receptor via a lipid raft-mediated clathrin-dependent process. *J. Cell Biol.* **160:**321–328.

Ahn, A., Gibbons, D. L., and Kielian, M. (2002). The fusion peptide of Semliki Forest virus associates with sterol-rich membrane domains. *J. Virol.* **76:**3267–3275.

Alexander, M., Bor, Y. C., Ravichandran, K. S., Hammarskjold, M. L., and Rekosh, D. (2004). Human immunodeficiency virus type 1 Nef associates with lipid rafts to downmodulate cell surface CD4 and class I major histocompatibility complex expression and to increase viral infectivity. *J. Virol.* **78:**1685–1696.

Alfsen, A., and Bomsel, M. (2002). HIV-1 gp41 envelope residues 650–685 exposed on native virus act as a lectin to bind epithelial cell galactosyl ceramide. *J. Biol. Chem.* **277:**25649–25659.

Alfsen, A., Iniguez, P., Bouguyon, E., and Bomsel, M. (2001). Secretory IgA specific for a conserved epitope on gp41 envelope glycoprotein inhibits epithelial transcytosis of HIV-1. *J. Immunol.* **166:**6257–6265.

Ali, A., Avalos, R. T., Ponimaskin, E., and Nayak, D. P. (2000). Influenza virus assembly: Effect of influenza virus glycoproteins on the membrane association of M1 protein. *J. Virol.* **74:**8709–8719.

Ali, A., and Nayak, D. P. (2000). Assembly of Sendai virus: M protein interacts with F and HN proteins and with the cytoplasmic tail and transmembrane domain of F protein. *Virology* **276:**289–303.

Aloia, R. C., Curtain, C. C., and Jensen, F. C. (1992). Membrane cholesterol and human immunodeficiency virus infectivity. *In* "Advances in Membrane Fluidity" (R. C. Aloia and C. C. Curtain, eds.), Vol. 6, pp. 283–304. Wiley-Liss, New York.

Aloia, R. C., Tian, H., and Jensen, F. C. (1993). Lipid composition and fluidity of the human immunodeficiency virus envelope and host cell plasma membranes. *Proc. Natl. Acad. Sci. USA* **90:**5181–5185.

Aman, M. J., Bosio, C. M., Panchal, R. G., Burnett, J. C., Schmaljohn, A., and Bavari, S. (2003). Molecular mechanisms of filovirus cellular trafficking. *Microbes Infect.* **5:**639–649.

Anderson, H. A., Chen, Y., and Norkin, L. C. (1996). Bound simian virus 40 translocates to caveolin-enriched membrane domains, and its entry is inhibited by drugs that selectively disrupt caveolae. *Mol. Biol. Cell* **7:**1825–1834.

Anderson, H. A., Chen, Y., and Norkin, L. C. (1998). MHC class I molecules are enriched in caveolae but do not enter with simian virus 40. *J. Gen. Virol.* **79:**1469–1477.

Argyris, E. G., Acheampong, E., Nunnari, G., Mukhtar, M., Williams, K. J., and Pomerantz, R. J. (2003). Human immunodeficiency virus type 1 enters primary human brain microvascular endothelial cells by a mechanism involving cell surface proteoglycans independent of lipid rafts. *J. Virol.* **77:**12140–12151.

Arreaza, G., and Brown, D. A. (1995). Sorting and intracellular trafficking of a glycosyl-phosphatidylinositol-anchored protein and two hybrid transmembrane proteins with the same ectodomain in Madin-Darby canine kidney epithelial cells. *J. Biol. Chem.* **270:**23641–23647.

Barman, S., and Nayak, D. P. (2000). Analysis of the transmembrane domain of influenza virus neuraminidase, a type II transmembrane glycoprotein, for apical sorting and raft association. *J. Virol.* **74:**6538–6545.

Bates, P., Young, J. A., and Varmus, H. E. (1993). A receptor for subgroup A Rous sarcoma virus is related to the low density lipoprotein receptor. *Cell* **74:**1043–1051.

Bavari, S., Bosio, C. M., Wiegand, E., Ruthel, G., Will, A. B., Geisbert, T. W., Hevey, M., Schmaljohn, C., Schmaljohn, A., and Aman, M. J. (2002). Lipid raft microdomains: A gateway for compartmentalized trafficking of Ebola and Marburg viruses. *J. Exp. Med.* **195:**593–602.

Bender, F. C., Whitbeck, J. C., Ponce de Leon, M., Lou, H., Eisenberg, R. J., and Cohen, G. H. (2003). Specific association of glycoprotein B with lipid rafts during herpes simplex virus entry. *J. Virol.* **77:**9542–9552.

Benting, J. H., Rietveld, A. G., and Simons, K. (1999). *N*-Glycans mediate the apical sorting of a GPI-anchored, raft-associated protein in Madin-Darby canine kidney cells. *J. Cell Biol.* **146:**313–320.

Berger, E. A., Murphy, P. M., and Farber, J. M. (1999). Chemokine receptors as HIV-1 coreceptors: Roles in viral entry, tropism, and disease. *Annu. Rev. Immunol.* **17:**657–700.

Blough, H. A., Tiffany, J. M., and Aaslestad, H. G. (1977). Lipids of rabies virus and BHK-21 cell membranes. *J. Virol.* **21:**950–955.

Briggs, J. A., Wilk, T., and Fuller, S. D. (2003). Do lipid rafts mediate virus assembly and pseudotyping? *J. Gen. Virol.* **84:**757–768.

Brown, D. A., and London, E. (1998). Structure and origin of ordered lipid domains in biological membranes. *J. Membr. Biol.* **164:**103–114.

Brown, D. A., and London, E. (2000). Structure and function of sphingolipid- and cholesterol-rich membrane rafts. *J. Biol. Chem.* **275:**17221–17224.

Brown, D. A., and Rose, J. K. (1992). Sorting of GPI-anchored proteins to glycolipid-enriched membrane subdomains during transport to the apical cell surface. *Cell* **68:**533–544.

Brown, E. L., and Lyles, D. S. (2003a). A novel method for analysis of membrane microdomains: Vesicular stomatitis virus glycoprotein microdomains change in size during infection, and those outside of budding sites resemble sites of virus budding. *Virology* **310:**343–358.

Brown, E. L., and Lyles, D. S. (2003b). Organization of the vesicular stomatitis virus glycoprotein into membrane microdomains occurs independently of intracellular viral components. *J. Virol.* **77:**3985–3992.

Brown, G., Aitken, J., Rixon, H. W., and Sugrue, R. J. (2002a). Caveolin-1 is incorporated into mature respiratory syncytial virus particles during virus assembly on the surface of virus-infected cells. *J. Gen. Virol.* **83:**611–621.

Brown, G., Rixon, H. W., and Sugrue, R. J. (2002b). Respiratory syncytial virus assembly occurs in GM1-rich regions of the host-cell membrane and alters the cellular distribution of tyrosine phosphorylated caveolin-1. *J. Gen. Virol.* **83:**1841–1850.

Brown, K. D., Hostager, B. S., and Bishop, G. A. (2001). Differential signaling and tumor necrosis factor receptor-associated factor (TRAF) degradation mediated by CD40 and the Epstein-Barr virus oncoprotein latent membrane protein 1 (LMP1). *J. Exp. Med.* **193:**943–954.

Calafat, J., Janssen, H., Demant, P., Hilgers, J., and Zavada, J. (1983). Specific selection of host cell glycoproteins during assembly of murine leukaemia virus and vesicular stomatitis virus: Presence of Thy-1 glycoprotein and absence of H-2, Pgp-1 and T-200 glycoproteins on the envelopes of these virus particles. *J. Gen. Virol.* **64:**1241–1253.

Cambi, A., De Lange, F., Van Maarseveen, N. M., Nijhuis, M., Joosten, B., Van Dijk, E. M., De Bakker, B. I., Fransen, J. A., Bovee-Geurts, P. H., Van Leeuwen, F. N., Van Hulst, N. F., and Figdor, C. G. (2004). Microdomains of the C-type lectin DC-SIGN are portals for virus entry into dendritic cells. *J. Cell Biol.* **164:**145–155.

Campbell, S. M., Crowe, S. M., and Mak, J. (2001). Lipid rafts and HIV-1: From viral entry to assembly of progeny virions. *J. Clin. Virol.* **22:**217–227.

Campbell, S. M., Crowe, S. M., and Mak, J. (2002). Virion-associated cholesterol is critical for the maintenance of HIV-1 structure and infectivity. *AIDS* **16:**2253–2261.

Chan, S. Y., Empig, C. J., Welte, F. J., Speck, R. F., Schmaljohn, A., Kreisberg, J. F., and Goldsmith, M. A. (2001). Folate receptor-$\alpha$ is a cofactor for cellular entry by Marburg and Ebola viruses. *Cell* **106:**117–126.

Chen, Y., and Norkin, L. C. (1999). Extracellular simian virus 40 transmits a signal that promotes virus enclosure within caveolae. *Exp. Cell Res.* **246:**83–90.

Clausse, B., Fizazi, K., Walczak, V., Tetaud, C., Wiels, J., Tursz, T., and Busson, P. (1997). High concentration of the EBV latent membrane protein 1 in glycosphingolipid-rich complexes from both epithelial and lymphoid cells. *Virology* **228:**285–293.

Cuadras, M. A., and Greenberg, H. B. (2003). Rotavirus infectious particles use lipid rafts during replication for transport to the cell surface *in vitro* and *in vivo*. *Virology* **313:**308–321.

Del Real, G., Jimenez-Baranda, S., Lacalle, R. A., Mira, E., Lucas, P., Gomez-Mouton, C., Carrera, A. C., Martinez, A. C., and Manes, S. (2002). Blocking of HIV-1 infection by targeting CD4 to nonraft membrane domains. *J. Exp. Med.* **196:**293–301.

Diaz-Griffero, F., Hoschander, S. A., and Brojatsch, J. (2002). Endocytosis is a critical step in entry of subgroup B avian leukosis viruses. *J. Virol.* **76**(24)**:**12866–12876.

Dietrich, C., Bagatolli, L. A., Volovyk, Z. N., Thompson, N. L., Levi, M., Jacobson, K., and Gratton, E. (2001a). Lipid rafts reconstituted in model membranes. *Biophys. J.* **80:**1417–1428.

Dietrich, C., Volovyk, Z. N., Levi, M., Thompson, N. L., and Jacobson, K. (2001b). Partitioning of Thy-1, GM1, and cross-linked phospholipid analogs into lipid rafts reconstituted in supported model membrane monolayers. *Proc. Natl. Acad. Sci. USA* **98**(19)**:**10642–10647.

Dietrich, C., Yang, B., Fujiwara, T., Kusumi, A., and Jacobson, K. (2002). Relationship of lipid rafts to transient confinement zones detected by single particle tracking. *Biophys. J.* **82:**274–284.

Ding, L., Derdowski, A., Wang, J. J., and Spearman, P. (2003). Independent segregation of human immunodeficiency virus type 1 Gag protein complexes and lipid rafts. *J. Virol.* **77:**1916–1926.

Dolganiuc, V., McGinnes, L., Luna, E. J., and Morrison, T. G. (2003). Role of the cytoplasmic domain of the Newcastle disease virus fusion protein in association with lipid rafts. *J. Virol.* **77:**12968–12979.

Doms, R. W. (2000). Beyond receptor expression: The influence of receptor conformation, density, and affinity in HIV-1 infection. *Virology* **276:**229–237.

Dykstra, M. L., Longnecker, R., and Pierce, S. K. (2001). Epstein-Barr virus coopts lipid rafts to block the signaling and antigen transport functions of the BCR. *Immunity* **14:**57–67.

Earp, L. J., Delos, S. E., Netter, R. C., Bates, P., and White, J. M. (2003). The avian retrovirus avian sarcoma/leukosis virus subtype A reaches the lipid mixing stage of fusion at neutral pH. *J. Virol.* **77:**3058–3066.

Empig, C. J., and Goldsmith, M. A. (2002). Association of the caveola vesicular system with cellular entry by filoviruses. *J. Virol.* **76:**5266–5270.

Feng, X., Heyden, N. V., and Ratner, L. (2003). Alpha interferon inhibits human T-cell leukemia virus type 1 assembly by preventing Gag interaction with rafts. *J. Virol.* **77:**13389–13395.

Fivaz, M., Vilbois, F., Thurnheer, S., Pasquali, C., Abrami, L., Bickel, P. E., Parton, R. G., and van der Goot, F. G. (2002). Differential sorting and fate of endocytosed GPI-anchored proteins. *EMBO J.* **21:**3989–4000.

Freed, E. O. (2002a). Viral late domains. *J. Virol.* **76:**4679–4687.

Freed, E. O. (2002b). Virology: Rafting with Ebola. *Science* **296:**279.

Friedrichson, T., and Kurzchalia, T. V. (1998). Microdomains of GPI-anchored proteins in living cells revealed by crosslinking. *Nature* **394:**802–805.

Gaus, K., Gratton, E., Kable, E. P., Jones, A. S., Gelissen, I., Kritharides, L., and Jessup, W. (2003). Visualizing lipid structure and raft domains in living cells with two-photon microscopy. *Proc. Natl. Acad. Sci. USA* **100:**15554–15559.

Gidwani, A., Holowka, D., and Baird, B. (2001). Fluorescence anisotropy measurements of lipid order in plasma membranes and lipid rafts from RBL-2H3 mast cells. *Biochemistry* **40:**12422–12429.

Gilbert, J. M., Goldberg, I. G., and Benjamin, T. L. (2003). Cell penetration and trafficking of polyomavirus. *J. Virol.* **77:**2615–2622.

Gomez-Mouton, C., Abad, J. L., Mira, E., Lacalle, R. A., Gallardo, E., Jimenez-Baranda, S., Illa, I., Bernad, A., Manes, S., and Martinez, A. C. (2001). Segregation of leading-edge and uropod components into specific lipid rafts during T cell polarization. *Proc. Natl. Acad. Sci. USA* **98:**9642–9647.

Graham, D. R., Chertova, E., Hilburn, J. M., Arthur, L. O., and Hildreth, J. E. (2003). Cholesterol depletion of human immunodeficiency virus type 1 and simian immunodeficiency virus with $\beta$-cyclodextrin inactivates and permeabilizes the virions: Evidence for virion-associated lipid rafts. *J. Virol.* **77:**8237–8248.

Gummuluru, S., Rogel, M., Stamatatos, L., and Emerman, M. (2003). Binding of human immunodeficiency virus type 1 to immature dendritic cells can occur independently of DC-SIGN and mannose binding C-type lectin receptors via a cholesterol-dependent pathway. *J. Virol.* **77:**12865–12874.

Guyader, M., Kiyokawa, E., Abrami, L., Turelli, P., and Trono, D. (2002). Role for human immunodeficiency virus type 1 membrane cholesterol in viral internalization. *J. Virol.* **76:**10356–10364.

Hague, B. F., Zhao, T. M., and Kindt, T. J. (2003). Binding of HTLV-1 virions to T cells occurs by a temperature and calcium-dependent process and is blocked by certain type 2 adenosine receptor antagonists. *Virus Res.* **93:**31–39.

Halwani, R., Khorchid, A., Cen, S., and Kleiman, L. (2003). Rapid localization of Gag/GagPol complexes to detergent-resistant membrane during the assembly of human immunodeficiency virus type 1. *J. Virol.* **77:**3973–3984.

Hamilton, V. T., Stone, D. M., and Cantor, G. H. (2003). Translocation of the B cell receptor to lipid rafts is inhibited in B cells from BLV-infected, persistent lymphocytosis cattle. *Virology* **315:**135–147.

Hammache, D., Yahi, N., Maresca, M., Pieroni, G., and Fantini, J. (1999). Human erythrocyte glycosphingolipids as alternative cofactors for human immunodeficiency virus type 1 (HIV-1) entry: Evidence for CD4-induced interactions between HIV-1 gp120 and reconstituted membrane microdomains of glycosphingolipids (Gb3 and GM3). *J. Virol.* **73:**5244–5248.

Hammarstedt, M., Wallengren, K., Pedersen, K. W., Roos, N., and Garoff, H. (2000). Minimal exclusion of plasma membrane proteins during retroviral envelope formation. *Proc. Natl. Acad. Sci. USA* **97:**7527–7532.

Hao, M., Mukherjee, S., and Maxfield, F. R. (2001). Cholesterol depletion induces large scale domain segregation in living cell membranes. *Proc. Natl. Acad. Sci. USA* **98:**13072–13077.

Harder, T., Scheiffele, P., Verkade, P., and Simons, K. (1998). Lipid domain structure of the plasma membrane revealed by patching of membrane components. *J. Cell Biol.* **141:**929–942.

Harouse, J. M., Bhat, S., Spitalnik, S. L., Laughlin, M., Stefano, K., Silberberg, D. H., and Gonzalez-Scarano, F. (1991). Inhibition of entry of HIV-1 in neural cell lines by antibodies against galactosyl ceramide. *Science* **253:**320–323.

Harris, J., Werling, D., Hope, J. C., Taylor, G., and Howard, C. J. (2002). Caveolae and caveolin in immune cells: Distribution and functions. *Trends Immunol.* **23:**158–164.

Henderson, G., Murray, J., and Yeo, R. P. (2002). Sorting of the respiratory syncytial virus matrix protein into detergent-resistant structures is dependent on cell-surface expression of the glycoproteins. *Virology* **300:**244–254.

Higuchi, M., Izumi, K. M., and Kieff, E. (2001). Epstein-Barr virus latent-infection membrane proteins are palmitoylated and raft-associated: Protein 1 binds to the cytoskeleton through TNF receptor cytoplasmic factors. *Proc. Natl. Acad. Sci. USA* **98:**4675–4680.

Holm, K., Weclewicz, K., Hewson, R., and Suomalainen, M. (2003). Human immunodeficiency virus type 1 assembly and lipid rafts: Pr55(gag) associates with membrane domains that are largely resistant to Brij98 but sensitive to Triton X-100. *J. Virol.* **77:**4805–4817.

Hug, P., Lin, H. M., Korte, T., Xiao, X., Dimitrov, D. S., Wang, J. M., Puri, A., and Blumenthal, R. (2000). Glycosphingolipids promote entry of a broad range of human immunodeficiency virus type 1 isolates into cell lines expressing CD4, CXCR4, and/or CCR5. *J. Virol.* **74:**6377–6385.

Huttner, W. B., and Zimmerberg, J. (2001). Implications of lipid microdomains for membrane curvature, budding and fission. *Curr. Opin. Cell Biol.* **13:**478–484.

Igakura, T., Stinchcombe, J. C., Goon, P. K., Taylor, G. P., Weber, J. N., Griffiths, G. M., Tanaka, Y., Osame, M., and Bangham, C. R. (2003). Spread of HTLV-I between lymphocytes by virus-induced polarization of the cytoskeleton. *Science* **299:**1713–1716.

Jacobson, K., and Dietrich, C. (1999). Looking at lipid rafts? *Trends Cell Biol.* **9:**87–91.

Janes, P. W., Ley, S. C., and Magee, A. I. (1999). Aggregation of lipid rafts accompanies signaling via the T cell antigen receptor. *J. Cell Biol.* **147:**447–461.

Jeffree, C. E., Rixon, H. W., Brown, G., Aitken, J., and Sugrue, R. J. (2003). Distribution of the attachment (G) glycoprotein and GM1 within the envelope of mature respiratory syncytial virus filaments revealed using field emission scanning electron microscopy. *Virology* **306:**254–267.

Jin, H., Leser, G. P., Zhang, J., and Lamb, R. A. (1997). Influenza virus hemagglutinin and neuraminidase cytoplasmic tails control particle shape. *EMBO J.* **16:**1236–1247.

Jolly, C., Kashefi, K., Hollinshead, M., and Sattentau, Q. J. (2004). HIV-1 cell to cell transfer across an Env-induced, actin-dependent synapse. *J. Exp. Med.* **199:**283–293.

Katzman, R. B., and Longnecker, R. (2003). Cholesterol-dependent infection of Burkitt's lymphoma cell lines by Epstein-Barr virus. *J. Gen. Virol.* **84:**2987–2992.

Kaykas, A., Worringer, K., and Sugden, B. (2001). CD40 and LMP-1 both signal from lipid rafts but LMP-1 assembles a distinct, more efficient signaling complex. *EMBO J.* **20:**2641–2654.

Keller, P., and Simons, K. (1998). Cholesterol is required for surface transport of influenza virus hemagglutinin. *J. Cell Biol.* **140:**1357–1367.

Kenworthy, A. K., and Edidin, M. (1998). Distribution of a glycosylphosphatidylinositol-anchored protein at the apical surface of MDCK cells examined at a resolution of <100 Å using imaging fluorescence resonance energy transfer. *J. Cell Biol.* **142:**69–84.

Kenworthy, A. K., Petranova, N., and Edidin, M. (2000). High-resolution FRET microscopy of cholera toxin B-subunit and GPI-anchored proteins in cell plasma membranes. *Mol. Biol. Cell* **11:**1645–1655.

Kielian, M., Chatterjee, P. K., Gibbons, D. L., and Lu, Y. E. (2000). Specific roles for lipids in virus fusion and exit: Examples from the alphaviruses. *Subcell. Biochem.* **34:**409–455.

Kielian, M. C., and Helenius, A. (1984). Role of cholesterol in fusion of Semliki Forest virus with membranes. *J. Virol.* **52:**281–283.

Kinet, S., Bernard, F., Mongellaz, C., Perreau, M., Goldman, F. D., and Taylor, N. (2002). gp120-mediated induction of the MAPK cascade is dependent on the activation state of $CD4^+$ lymphocytes. *Blood* **100:**2546–2553.

Klement, V., Rowe, W. P., Hartley, J. W., and Pugh, W. E. (1969). Mixed culture cytopathogenicity: A new test for growth of murine leukemia viruses in tissue culture. *Proc. Natl. Acad. Sci. USA* **63:**753–758.

Koshizuka, T., Goshima, F., Takakuwa, H., Nozawa, N., Daikoku, T., Koiwai, O., and Nishiyama, Y. (2002). Identification and characterization of the UL56 gene product of herpes simplex virus type 2. *J. Virol.* **76:**6718–6728.

Kozak, S. L., Heard, J. M., and Kabat, D. (2002). Segregation of CD4 and CXCR4 into distinct lipid microdomains in T lymphocytes suggests a mechanism for membrane destabilization by human immunodeficiency virus. *J. Virol.* **76:**1802–1815.

Kundu, A., Avalos, R. T., Sanderson, C. M., and Nayak, D. P. (1996). Transmembrane domain of influenza virus neuraminidase, a type II protein, possesses an apical sorting signal in polarized MDCK cells. *J. Virol.* **70:**6508–6515.

Kurzchalia, T. V., and Parton, R. G. (1999). Membrane microdomains and caveolae. *Curr. Opin. Cell Biol.* **11:**424–431.

Kusumi, A., Koyama-Honda, I., and Suzuki, K. (2004). Molecular dynamics and interactions for creation of stimulation-induced stabilized rafts from small unstable steady-state rafts. *Traffic* **5:**213–230.

Kwik, J., Boyle, S., Fooksman, D., Margolis, L., Sheetz, M. P., and Edidin, M. (2003). Membrane cholesterol, lateral mobility, and the phosphatidylinositol 4,5-bisphosphate-dependent organization of cell actin. *Proc. Natl. Acad. Sci. USA* **100:**13964–13969.

Lamaze, C., Dujeancourt, A., Baba, T., Lo, C. G., Benmerah, A., and Dautry-Varsat, A. (2001). Interleukin 2 receptors and detergent-resistant membrane domains define a clathrin-independent endocytic pathway. *Mol. Cell* **7:**661–671.

Lamb, R. A., and Kolakofsky, D. (2001). *Paramyxoviridae*: The viruses and their replication. *In* "Field's Virology" (D. M. Knipe and P. M. Howley, eds.), 4th Ed., Vol. 1, pp. 1305–1340. Lippincott Williams & Wilkins, Philadelphia.

Lee, G. E., Church, G. A., and Wilson, D. W. (2003). A subpopulation of tegument protein Vhs localizes to detergent-insoluble lipid rafts in herpes simplex virus-infected cells. *J. Virol.* **77:**2038–2045.

Lee, S., Zhao, Y., and Anderson, W. F. (1999). Receptor-mediated Moloney murine leukemia virus entry can occur independently of the clathrin-coated-pit-mediated endocytic pathway. *J. Virol.* **73:**5994–6005.

Li, M., Yang, C., Tong, S., Weidmann, A., and Compans, R. W. (2002). Palmitoylation of the murine leukemia virus envelope protein is critical for lipid raft association and surface expression. *J. Virol.* **76:**11845–11852.

Liao, Z., Cimakasky, L. M., Hampton, R., Nguyen, D. H., and Hildreth, J. E. (2001). Lipid rafts and HIV pathogenesis: Host membrane cholesterol is required for infection by HIV type 1. *AIDS Res. Hum. Retroviruses* **17:**1009–1019.

Liao, Z., Graham, D. R., and Hildreth, J. E. (2003). Lipid rafts and HIV pathogenesis: Virion-associated cholesterol is required for fusion and infection of susceptible cells. *AIDS Res. Hum. Retroviruses* **19:**675–687.

Lin, S., Naim, H. Y., Rodriguez, A. C., and Roth, M. G. (1998). Mutations in the middle of the transmembrane domain reverse the polarity of transport of the influenza virus hemagglutinin in MDCK epithelial cells. *J. Cell. Biol.* **142:**51–57.

Lindwasser, O. W., and Resh, M. D. (2001). Multimerization of human immunodeficiency virus type 1 Gag promotes its localization to barges, raft-like membrane microdomains. *J. Virol.* **75:**7913–7924.

Lindwasser, O. W., and Resh, M. D. (2002). Myristoylation as a target for inhibiting HIV assembly: Unsaturated fatty acids block viral budding. *Proc. Natl. Acad. Sci. USA* **99:**13037–13042.

Lipardi, C., Nitsch, L., and Zurzolo, C. (2000). Detergent-insoluble GPI-anchored proteins are apically sorted in Fischer rat thyroid cells, but interference with cholesterol or sphingolipids differentially affects detergent insolubility and apical sorting. *Mol. Biol. Cell* **11:**531–542.

Liu, N. Q., Lossinsky, A. S., Popik, W., Li, X., Gujuluva, C., Kriederman, B., Roberts, J., Pushkarsky, T., Bukrinsky, M., Witte, M., Weinand, M., and Fiala, M. (2002). Human immunodeficiency virus type 1 enters brain microvascular endothelia by macropinocytosis dependent on lipid rafts and the mitogen-activated protein kinase signaling pathway. *J. Virol.* **76:**6689–6700.

Lodge, R., Gottlinger, H., Gabuzda, D., Cohen, E. A., and Lemay, G. (1994). The intracytoplasmic domain of gp41 mediates polarized budding of human immunodeficiency virus type 1 in MDCK cells. *J. Virol.* **68:**4857–4861.

Lodge, R., Lalonde, J. P., Lemay, G., and Cohen, E. A. (1997). The membrane-proximal intracytoplasmic tyrosine residue of HIV-1 envelope glycoprotein is critical for basolateral targeting of viral budding in MDCK cells. *EMBO J.* **16:**695–705.

Lu, X., and Silver, J. (2000). Ecotropic murine leukemia virus receptor is physically associated with caveolin and membrane rafts. *Virology* **276:**251–258.

Lu, X., Xiong, Y., and Silver, J. (2002). Asymmetric requirement for cholesterol in receptor-bearing but not envelope-bearing membranes for fusion mediated by ecotropic murine leukemia virus. *J. Virol.* **76:**6701–6709.

Lu, Y. E., Cassese, T., and Kielian, M. (1999). The cholesterol requirement for Sindbis virus entry and exit and characterization of a spike protein region involved in cholesterol dependence. *J. Virol.* **73:**4272–4278.

Lu, Y. E., and Kielian, M. (2000). Semliki Forest virus budding: Assay, mechanisms, and cholesterol requirement. *J. Virol.* **74:**7708–7719.

Luan, P., Yang, L., and Glaser, M. (1995). Formation of membrane domains created during the budding of vesicular stomatitis virus: A model for selective lipid and protein sorting in biological membranes. *Biochemistry* **34:**9874–9883.
Madore, N., Smith, K. L., Graham, C. H., Jen, A., Brady, K., Hall, S., and Morris, R. (1999). Functionally different GPI proteins are organized in different domains on the neuronal surface. *EMBO J.* **18:**6917–6926.
Manel, N., Kim, F. J., Kinet, S., Taylor, N., Sitbon, M., and Battini, J. L. (2003). The ubiquitous glucose transporter GLUT-1 is a receptor for HTLV. *Cell* **115:**449–459.
Manes, S., del Real, G., Lacalle, R. A., Lucas, P., Gomez-Mouton, C., Sanchez-Palomino, S., Delgado, R., Alcami, J., Mira, E., and Martinez, A. C. (2000). Membrane raft microdomains mediate lateral assemblies required for HIV-1 infection. *EMBO Rep.* **1:**190–196.
Manie, S. N., Debreyne, S., Vincent, S., and Gerlier, D. (2000). Measles virus structural components are enriched into lipid raft microdomains: A potential cellular location for virus assembly. *J. Virol.* **74:**305–311.
Marjomaki, V., Pietiainen, V., Matilainen, H., Upla, P., Ivaska, J., Nissinen, L., Reunanen, H., Huttunen, P., Hyypia, T., and Heino, J. (2002). Internalization of echovirus 1 in caveolae. *J. Virol.* **76:**1856–1865.
Marquardt, M. T., Phalen, T., and Kielian, M. (1993). Cholesterol is required in the exit pathway of Semliki Forest virus. *J. Cell Biol.* **123:**57–65.
Martin-Belmonte, F., Lopez-Guerrero, J. A., Carrasco, L., and Alonso, M. A. (2000). The amino-terminal nine amino acid sequence of poliovirus capsid VP4 protein is sufficient to confer N-myristoylation and targeting to detergent-insoluble membranes. *Biochemistry* **39:**1083–1090.
Marty, A., Meanger, J., Mills, J., Shields, B., and Ghildyal, R. (2004). Association of matrix protein of respiratory syncytial virus with the host cell membrane of infected cells. *Arch. Virol.* **149:**199–210.
Masuda, M., Kakushima, N., Wilt, S. G., Ruscetti, S. K., Hoffman, P. M., and Iwamoto, A. (1999). Analysis of receptor usage by ecotropic murine retroviruses, using green fluorescent protein-tagged cationic amino acid transporters. *J. Virol.* **73:**8623–8629.
Mayor, S., and Maxfield, F. R. (1995). Insolubility and redistribution of GPI-anchored proteins at the cell surface after detergent treatment. *Mol. Biol. Cell* **6**(7):929–944.
Mayor, S., and Rao, M. (2004). Rafts: Scale-dependent, active lipid organization at the cell surface. *Traffic* **5:**231–240.
Mayor, S., Rothberg, K. G., and Maxfield, F. R. (1994). Sequestration of GPI-anchored proteins in caveolae triggered by cross-linking. *Science* **264:**1948–1951.
McClure, M. O., Sommerfelt, M. A., Marsh, M., and Weiss, R. A. (1990). The pH independence of mammalian retrovirus infection. *J. Gen. Virol.* **71:**767–773.
McCurdy, L. H., and Graham, B. S. (2003). Role of plasma membrane lipid microdomains in respiratory syncytial virus filament formation. *J. Virol.* **77:**1747–1756.
McDonald, K. K., Zharikov, S., Block, E. R., and Kilberg, M. S. (1997). A caveolar complex between the cationic amino acid transporter 1 and endothelial nitric-oxide synthase may explain the "arginine paradox." *J. Biol. Chem.* **272:**31213–31216.
McSharry, J. J., and Wagner, R. R. (1971). Lipid composition of purified vesicular stomatitis viruses. *J. Virol.* **7:**59–70.
Melkonian, K. A., Ostermeyer, A. G., Chen, J. Z., Roth, M. G., and Brown, D. A. (1999). Role of lipid modifications in targeting proteins to detergent-resistant membrane rafts: Many raft proteins are acylated, while few are prenylated. *J. Biol. Chem.* **274:**3910–3917.

Mescher, M. F., and Apgar, J. R. (1985). The plasma membrane "skeleton" of tumor and lymphoid cells: A role in cell lysis? *Adv. Exp. Med. Biol.* **184:**387–400.
Miyata, S., Minami, J., Tamai, E., Matsushita, O., Shimamoto, S., and Okabe, A. (2002). *Clostridium perfringens* ε-toxin forms a heptameric pore within the detergent-insoluble microdomains of Madin-Darby canine kidney cells and rat synaptosomes. *J. Biol. Chem.* **277:**39463–39468.
Mothes, W., Boerger, A. L., Narayan, S., Cunningham, J. M., and Young, J. A. (2000). Retroviral entry mediated by receptor priming and low pH triggering of an envelope glycoprotein. *Cell* **103:**679–689.
Munro, S. (2003). Lipid rafts: Elusive or illusive? *Cell* **115:**377–388.
Murakami, T., and Freed, E. O. (2000). The long cytoplasmic tail of gp41 is required in a cell type-dependent manner for HIV-1 envelope glycoprotein incorporation into virions. *Proc. Natl. Acad. Sci. USA* **97:**343–348.
Naim, H. Y., Ehler, E., and Billeter, M. A. (2000). Measles virus matrix protein specifies apical virus release and glycoprotein sorting in epithelial cells. *EMBO J.* **19:**3576–3585.
Narayan, S., Barnard, R. J., and Young, J. A. (2003). Two retroviral entry pathways distinguished by lipid raft association of the viral receptor and differences in viral infectivity. *J. Virol.* **77:**1977–1983.
Nayak, D. P., and Barman, S. (2002). Role of lipid rafts in virus assembly and budding. *Adv. Virus Res.* **58:**1–28.
Nelson, K. L., and Buckley, J. T. (2000). Channel formation by the glycosylphosphatidylinositol-anchored protein binding toxin aerolysin is not promoted by lipid rafts. *J. Biol. Chem.* **275:**19839–19843.
Nguyen, D. H., and Hildreth, J. E. (2000). Evidence for budding of human immunodeficiency virus type 1 selectively from glycolipid-enriched membrane lipid rafts. *J. Virol.* **74:**3264–3272.
Nguyen, D. H., and Taub, D. (2002a). Cholesterol is essential for macrophage inflammatory protein 1 β binding and conformational integrity of CC chemokine receptor 5. *Blood* **99:**4298–4306.
Nguyen, D. H., and Taub, D. (2002b). CXCR4 function requires membrane cholesterol: Implications for HIV infection. *J. Immunol.* **168:**4121–4126.
Nguyen, D. H., and Taub, D. D. (2003). Inhibition of chemokine receptor function by membrane cholesterol oxidation. *Exp. Cell Res.* **291:**36–45.
Nichols, B. J., Kenworthy, A. K., Polishchuk, R. S., Lodge, R., Roberts, T. H., Hirschberg, K., Phair, R. D., and Lippincott-Schwartz, J. (2001). Rapid cycling of lipid raft markers between the cell surface and Golgi complex. *J. Cell Biol.* **153:**529–541.
Nichols, B. J., and Lippincott-Schwartz, J. (2001). Endocytosis without clathrin coats. *Trends Cell Biol.* **11:**406–412.
Nieva, J. L., Bron, R., Corver, J., and Wilschut, J. (1994). Membrane fusion of Semliki Forest virus requires sphingolipids in the target membrane. *EMBO J.* **13:**2797–2804.
Niv, H., Gutman, O., Kloog, Y., and Henis, Y. I. (2002). Activated K-Ras and H-Ras display different interactions with saturable nonraft sites at the surface of live cells. *J. Cell Biol.* **157:**865–872.
Niyogi, K., and Hildreth, J. E. (2001). Characterization of new syncytium-inhibiting monoclonal antibodies implicates lipid rafts in human T-cell leukemia virus type 1 syncytium formation. *J. Virol.* **75:**7351–7361.
Noda, T., Sagara, H., Suzuki, E., Takada, A., Kida, H., and Kawaoka, Y. (2002). Ebola virus VP40 drives the formation of virus-like filamentous particles along with GP. *J. Virol.* **76:**4855–4865.

Norkin, L. C., Anderson, H. A., Wolfrom, S. A., and Oppenheim, A. (2002). Caveolar endocytosis of simian virus 40 is followed by brefeldin A-sensitive transport to the endoplasmic reticulum, where the virus disassembles. *J. Virol.* **76:**5156–5166.

Ochsenbauer, C., Dubay, S. R., and Hunter, E. (2000). The Rous sarcoma virus Env glycoprotein contains a highly conserved motif homologous to tyrosine-based endocytosis signals and displays an unusual internalization phenotype. *Mol. Cell Biol.* **20:**249–260.

Ochsenbauer-Jambor, C., Miller, D. C., Roberts, C. R., Rhee, S. S., and Hunter, E. (2001). Palmitoylation of the Rous sarcoma virus transmembrane glycoprotein is required for protein stability and virus infectivity. *J. Virol.* **75:**11544–11554.

Oliferenko, S., Paiha, K., Harder, T., Gerke, V., Schwarzler, C., Schwarz, H., Beug, H., Gunthert, U., and Huber, L. A. (1999). Analysis of CD44-containing lipid rafts: Recruitment of annexin II and stabilization by the actin cytoskeleton. *J. Cell Biol.* **146:**843–854.

Ono, A., and Freed, E. O. (2001). Plasma membrane rafts play a critical role in HIV-1 assembly and release. *Proc. Natl. Acad. Sci. USA* **98:**13925–13930.

Orentas, R. J., and Hildreth, J. E. (1993). Association of host cell surface adhesion receptors and other membrane proteins with HIV and SIV. *AIDS Res. Hum. Retroviruses* **9:**1157–1165.

Orlandi, P. A., and Fishman, P. H. (1998). Filipin-dependent inhibition of cholera toxin: Evidence for toxin internalization and activation through caveolae-like domains. *J. Cell Biol.* **141:**905–915.

Ott, D. E. (1997). Cellular proteins in HIV virions. *Rev. Med. Virol.* **7:**167–180.

Panchal, R. G., Ruthel, G., Kenny, T. A., Kallstrom, G. H., Lane, D., Badie, S. S., Li, L., Bavari, S., and Aman, M. J. (2003). *In vivo* oligomerization and raft localization of Ebola virus protein VP40 during vesicular budding. *Proc. Natl. Acad. Sci. USA* **100:**15936–15941.

Park, J., Cho, N. H., Choi, J. K., Feng, P., Choe, J., and Jung, J. U. (2003). Distinct roles of cellular Lck and p80 proteins in herpesvirus saimiri Tip function on lipid rafts. *J. Virol.* **77:**9041–9051.

Parolini, I., Sargiacomo, M., Lisanti, M. P., and Peschle, C. (1996). Signal transduction and glycophosphatidylinositol-linked proteins (Lyn, Lck, CD4, CD45, G proteins, and CD55) selectively localize in Triton-insoluble plasma membrane domains of human leukemic cell lines and normal granulocytes. *Blood* **87:**3783–3794.

Parolini, I., Topa, S., Sorice, M., Pace, A., Ceddia, P., Montesoro, E., Pavan, A., Lisanti, M. P., Peschle, C., and Sargiacomo, M. (1999). Phorbol ester-induced disruption of the CD4–Lck complex occurs within a detergent-resistant microdomain of the plasma membrane: Involvement of the translocation of activated protein kinase C isoforms. *J. Biol. Chem.* **274:**14176–14187.

Patzer, E. J., Moore, N. F., Barenholz, Y., Shaw, J. M., and Wagner, R. R. (1978). Lipid organization of the membrane of vesicular stomatitis virus. *J. Biol. Chem.* **253:**4544–4550.

Pearce-Pratt, R., Malamud, D., and Phillips, D. M. (1994). Role of the cytoskeleton in cell-to-cell transmission of human immunodeficiency virus. *J. Virol.* **68:**2898–2905.

Pelkmans, L., Kartenbeck, J., and Helenius, A. (2001). Caveolar endocytosis of simian virus 40 reveals a new two-step vesicular-transport pathway to the ER. *Nat. Cell Biol.* **3:**473–483.

Pelkmans, L., Puntener, D., and Helenius, A. (2002). Local actin polymerization and dynamin recruitment in SV40-induced internalization of caveolae. *Science* **296:**535–539.

Percherancier, Y., Lagane, B., Planchenault, T., Staropoli, I., Altmeyer, R., Virelizier, J. L., Arenzana-Seisdedos, F., Hoessli, D. C., and Bachelerie, F. (2003). HIV-1 entry into T-cells is not dependent on CD4 and CCR5 localization to sphingolipid-enriched, detergent-resistant, raft membrane domains. *J. Biol. Chem.* **278:**3153–3161.

Pereira, F. B., Goni, F. M., and Nieva, J. L. (1997). Membrane fusion induced by the HIV type 1 fusion peptide: Modulation by factors affecting glycoprotein 41 activity and potential anti-HIV compounds. *AIDS Res. Hum. Retroviruses* **13:**1203–1211.

Perotti, M. E., Tan, X., and Phillips, D. M. (1996). Directional budding of human immunodeficiency virus from monocytes. *J. Virol.* **70:**5916–5921.

Pessin, J. E., and Glaser, M. (1980). Budding of Rous sarcoma virus and vesicular stomatitis virus from localized lipid regions in the plasma membrane of chicken embryo fibroblasts. *J. Biol. Chem.* **255:**9044–9050.

Phalen, T., and Kielian, M. (1991). Cholesterol is required for infection by Semliki Forest virus. *J. Cell Biol.* **112:**615–623.

Pickl, W. F., Pimentel-Muinos, F. X., and Seed, B. (2001). Lipid rafts and pseudotyping. *J. Virol.* **75:**7175–7183.

Pierini, L. M., and Maxfield, F. R. (2001). Flotillas of lipid rafts fore and aft. *Proc. Natl. Acad. Sci. USA* **98:**9471–9473.

Popik, W., and Alce, T. M. (2004). CD4 receptor localized to non-raft membrane microdomains supports HIV-1 entry: Identification of a novel raft localization marker in CD4. *J. Biol. Chem.* **279:**704–712.

Popik, W., Alce, T. M., and Au, W. C. (2002). Human immunodeficiency virus type 1 uses lipid raft-colocalized CD4 and chemokine receptors for productive entry into $CD4^+$ T cells. *J. Virol.* **76:**4709–4722.

Pornillos, O., Garrus, J. E., and Sundquist, W. I. (2002). Mechanisms of enveloped RNA virus budding. *Trends Cell Biol.* **12:**569–579.

Pralle, A., Keller, P., Florin, E. L., Simons, K., and Horber, J. K. (2000). Sphingolipid–cholesterol rafts diffuse as small entities in the plasma membrane of mammalian cells. *J. Cell Biol.* **148:**997–1008.

Prior, I. A., Muncke, C., Parton, R. G., and Hancock, J. F. (2003). Direct visualization of Ras proteins in spatially distinct cell surface microdomains. *J. Cell Biol.* **160:**165–170.

Puri, A., Hug, P., Jernigan, K., Barchi, J., Kim, H. Y., Hamilton, J., Wiels, J., Murray, G. J., Brady, R. O., and Blumenthal, R. (1998a). The neutral glycosphingolipid globotriaosylceramide promotes fusion mediated by a CD4-dependent CXCR4-utilizing HIV type 1 envelope glycoprotein. *Proc. Natl. Acad. Sci. USA* **95:**14435–14440.

Puri, A., Hug, P., Munoz-Barroso, I., and Blumenthal, R. (1998b). Human erythrocyte glycolipids promote HIV-1 envelope glycoprotein-mediated fusion of $CD4^+$ cells. *Biochem. Biophys. Res. Commun.* **242:**219–225.

Puri, V., Watanabe, R., Singh, R. D., Dominguez, M., Brown, J. C., Wheatley, C. L., Marks, D. L., and Pagano, R. E. (2001). Clathrin-dependent and -independent internalization of plasma membrane sphingolipids initiates two Golgi targeting pathways. *J. Cell Biol.* **154:**535–547.

Quigley, J. P., Rifkin, D. B., and Reich, E. (1971). Phospholipid composition of Rous sarcoma virus, host cell membranes and other enveloped RNA viruses. *Virology* **46:**106–116.

Quigley, J. P., Rifkin, D. B., and Reich, E. (1972). Lipid studies of Rous sarcoma virus and host cell membranes. *Virology* **50:**550–557.

Rai, S. K., Duh, F. M., Vigdorovich, V., Danilkovitch-Miagkova, A., Lerman, M. I., and Miller, A. D. (2001). Candidate tumor suppressor HYAL2 is a glycosylphosphatidylinositol (GPI)-anchored cell-surface receptor for Jaagsiekte sheep retrovirus, the envelope protein of which mediates oncogenic transformation. *Proc. Natl. Acad. Sci. USA* **98:**4443–4448.

Rawat, S. S., Viard, M., Gallo, S. A., Rein, A., Blumenthal, R., and Puri, A. (2003). Modulation of entry of enveloped viruses by cholesterol and sphingolipids [review]. *Mol. Membr. Biol.* **20:**243–254.

Razinkov, V. I., and Cohen, F. S. (2000). Sterols and sphingolipids strongly affect the growth of fusion pores induced by the hemagglutinin of influenza virus. *Biochemistry* **39:**13462–13468.

Renkonen, O., Kaarainen, L., Simons, K., and Gahmberg, C. G. (1971). The lipid class composition of Semliki Forest virus and plasma membranes of the host cells. *Virology* **46:**318–326.

Richards, A. A., Stang, E., Pepperkok, R., and Parton, R. G. (2002). Inhibitors of COP-mediated transport and cholera toxin action inhibit simian virus 40 infection. *Mol. Biol. Cell.* **13:**1750–1764.

Richterova, Z., Liebl, D., Horak, M., Palkova, Z., Stokrova, J., Hozak, P., Korb, J., and Forstova, J. (2001). Caveolae are involved in the trafficking of mouse polyomavirus virions and artificial VP1 pseudocapsids toward cell nuclei. *J. Virol.* **75:**10880–10891.

Rietveld, A., and Simons, K. (1998). The differential miscibility of lipids as the basis for the formation of functional membrane rafts. *Biochim. Biophys. Acta* **1376:**467–479.

Rodal, S. K., Skretting, G., Garred, O., Vilhardt, F., van Deurs, B., and Sandvig, K. (1999). Extraction of cholesterol with methyl-$\beta$-cyclodextrin perturbs formation of clathrin-coated endocytic vesicles. *Mol. Biol. Cell* **10:**961–974.

Roper, K., Corbeil, D., and Huttner, W. B. (2000). Retention of prominin in microvilli reveals distinct cholesterol-based lipid micro-domains in the apical plasma membrane. *Nat. Cell Biol.* **2:**582–592.

Rothenberger, S., Rousseaux, M., Knecht, H., Bender, F. C., Legler, D. F., and Bron, C. (2002). Association of the Epstein-Barr virus latent membrane protein 1 with lipid rafts is mediated through its N-terminal region. *Cell. Mol. Life Sci.* **59:**171–180.

Rousso, I., Mixon, M. B., Chen, B. K., and Kim, P. S. (2000). Palmitoylation of the HIV-1 envelope glycoprotein is critical for viral infectivity. *Proc. Natl. Acad. Sci. USA* **97:**13523–13525.

Roy, S., Luetterforst, R., Harding, A., Apolloni, A., Etheridge, M., Stang, E., Rolls, B., Hancock, J. F., and Parton, R. G. (1999). Dominant-negative caveolin inhibits H-Ras function by disrupting cholesterol-rich plasma membrane domains. *Nat. Cell Biol.* **1:**98–105.

Sabharanjak, S., Sharma, P., Parton, R. G., and Mayor, S. (2002). GPI-anchored proteins are delivered to recycling endosomes via a distinct cdc42-regulated, clathrin-independent pinocytic pathway. *Dev. Cell* **2:**411–423.

Saifuddin, M., Parker, C. J., Peeples, M. E., Gorny, M. K., Zolla-Pazner, S., Ghassemi, M., Rooney, I. A., Atkinson, J. P., and Spear, G. T. (1995). Role of virion-associated glycosylphosphatidylinositol-linked proteins CD55 and CD59 in complement resistance of cell line-derived and primary isolates of HIV-1. *J. Exp. Med.* **182:**501–509.

Sanchez, A., Khan, A. S., Zaki, S. R., Nabel, G. J., Ksiazek, T. G., and Peters, C. J. (2001). *Filoviridae*: Marburg and Ebola viruses. *In* "Field's Virology" (D. M. Knipe and P. M. Howley, eds.), 4th Ed., Vol. 1, pp. 1279–1304. Lippincott Williams & Wilkins, Philadelphia.

Sanchez-Madrid, F., and del Pozo, M. A. (1999). Leukocyte polarization in cell migration and immune interactions. *EMBO J.* **18:**501–511.

Sanderson, C. M., Avalos, R., Kundu, A., and Nayak, D. P. (1995). Interaction of Sendai viral F, HN, and M proteins with host cytoskeletal and lipid components in Sendai virus-infected BHK cells. *Virology* **209:**701–707.

Sapin, C., Colard, O., Delmas, O., Tessier, C., Breton, M., Enouf, V., Chwetzoff, S., Ouanich, J., Cohen, J., Wolf, C., and Trugnan, G. (2002). Rafts promote assembly and atypical targeting of a nonenveloped virus, rotavirus, in Caco-2 cells. *J. Virol.* **76:**4591–4602.

Scheiffele, P., Rietveld, A., Wilk, T., and Simons, K. (1999). Influenza viruses select ordered lipid domains during budding from the plasma membrane. *J. Biol. Chem.* **274:**2038–2044.

Scheiffele, P., Roth, M. G., and Simons, K. (1997). Interaction of influenza virus haemagglutinin with sphingolipid–cholesterol membrane domains via its transmembrane domain. *EMBO J.* **16:**5501–5508.

Scheiffele, P., Verkade, P., Fra, A. M., Virta, H., Simons, K., and Ikonen, E. (1998). Caveolin-1 and -2 in the exocytic pathway of MDCK cells. *J. Cell Biol.* **140:**795–806.

Schmidt, M. F., and Schlesinger, M. J. (1979). Fatty acid binding to vesicular stomatitis virus glycoprotein: A new type of post-translational modification of the viral glycoprotein. *Cell* **17:**813–819.

Schroeder, R., London, E., and Brown, D. (1994). Interactions between saturated acyl chains confer detergent resistance on lipids and glycosylphosphatidylinositol (GPI)-anchored proteins: GPI-anchored proteins in liposomes and cells show similar behavior. *Proc. Natl. Acad. Sci. USA* **91:**12130–12134.

Schroeder, R. J., Ahmed, S. N., Zhu, Y., London, E., and Brown, D. A. (1998). Cholesterol and sphingolipid enhance the Triton X-100 insolubility of glycosylphosphatidylinositol-anchored proteins by promoting the formation of detergent-insoluble ordered membrane domains. *J. Biol. Chem.* **273:**1150–1157.

Schuck, S., Honsho, M., Ekroos, K., Shevchenko, A., and Simons, K. (2003). Resistance of cell membranes to different detergents. *Proc. Natl. Acad. Sci. USA* **100:**5795–5800.

Schutz, G. J., Kada, G., Pastushenko, V. P., and Schindler, H. (2000). Properties of lipid microdomains in a muscle cell membrane visualized by single molecule microscopy. *EMBO J.* **19:**892–901.

Sharma, P., Sabharanjak, S., and Mayor, S. (2002). Endocytosis of lipid rafts: An identity crisis. *Semin. Cell Dev. Biol.* **13:**205–214.

Sharma, P., Varma, R., Sarasij, R. C., Ira, Gousset, K., Krishnamoorthy, G., Rao, M., and Mayor, S. (2004). Nanoscale organization of multiple GPI-anchored proteins in living cell membranes. *Cell* **116:**577–589.

Sheets, E. D., Lee, G. M., Simson, R., and Jacobson, K. (1997). Transient confinement of a glycosylphosphatidylinositol-anchored protein in the plasma membrane. *Biochemistry* **36:**12449–12458.

Shi, S. T., Lee, K. J., Aizaki, H., Hwang, S. B., and Lai, M. M. (2003). Hepatitis C virus RNA replication occurs on a detergent-resistant membrane that cofractionates with caveolin-2. *J. Virol.* **77:**4160–4168.

Shvartsman, D. E., Kotler, M., Tall, R. D., Roth, M. G., and Henis, Y. I. (2003). Differently anchored influenza hemagglutinin mutants display distinct interaction dynamics with mutual rafts. *J. Cell Biol.* **163:**879–888.

Sieczkarski, S. B., and Whittaker, G. R. (2002). Dissecting virus entry via endocytosis. *J. Gen. Virol.* **83:**1535–1545.

Simmons, G., Rennekamp, A. J., Chai, N., Vandenberghe, L. H., Riley, J. L., and Bates, P. (2003). Folate receptor $\alpha$ and caveolae are not required for Ebola virus glycoprotein-mediated viral infection. *J. Virol.* **77:**13433–13438.

Simons, K., and Ikonen, E. (1997). Functional rafts in cell membranes. *Nature* **387:**569–572.
Simons, K., and Toomre, D. (2000). Lipid rafts and signal transduction. *Nat. Rev.* **1:**31–39.
Simpson-Holley, M., Ellis, D., Fisher, D., Elton, D., McCauley, J., and Digard, P. (2002). A functional link between the actin cytoskeleton and lipid rafts during budding of filamentous influenza virions. *Virology* **301:**212–225.
Singer, II, Scott, S., Kawka, D. W., Chin, J., Daugherty, B. L., DeMartino, J. A., DiSalvo, J., Gould, S. L., Lineberger, J. E., Malkowitz, L., Miller, M. D., Mitnaul, L., Siciliano, S. J., Staruch, M. J., Williams, H. R., Zweerink, H. J., and Springer, M. S. (2001). CCR5, CXCR4, and CD4 are clustered and closely apposed on microvilli of human macrophages and T cells. *J. Virol.* **75:**3779–3790.
Skibbens, J. E., Roth, M. G., and Matlin, K. S. (1989). Differential extractability of influenza virus hemagglutinin during intracellular transport in polarized epithelial cells and nonpolar fibroblasts. *J. Cell Biol.* **108:**821–832.
Slosberg, B. N., and Montelaro, R. C. (1982). A comparison of the mobilities and thermal transitions of retrovirus lipid envelopes and host cell plasma membranes by electron spin resonance spectroscopy. *Biochim. Biophys. Acta* **689:**393–402.
Sorice, M., Garofalo, T., Misasi, R., Longo, A., Mattei, V., Sale, P., Dolo, V., Gradini, R., and Pavan, A. (2001). Evidence for cell surface association between CXCR4 and ganglioside GM3 after gp120 binding in SupT1 lymphoblastoid cells. *FEBS Lett.* **506:**55–60.
Stang, E., Kartenbeck, J., and Parton, R. G. (1997). Major histocompatibility complex class I molecules mediate association of SV40 with caveolae. *Mol. Biol. Cell* **8:**47–57.
Stoddart, A., Dykstra, M. L., Brown, B. K., Song, W., Pierce, S. K., and Brodsky, F. M. (2002). Lipid rafts unite signaling cascades with clathrin to regulate BCR internalization. *Immunity* **17:**451–462.
Stuart, A. D., Eustace, H. E., McKee, T. A., and Brown, T. D. (2002). A novel cell entry pathway for a DAF-using human enterovirus is dependent on lipid rafts. *J. Virol.* **76:**9307–9322.
Subczynski, W. K., and Kusumi, A. (2003). Dynamics of raft molecules in the cell and artificial membranes: Approaches by pulse EPR spin labeling and single molecule optical microscopy. *Biochim. Biophys. Acta* **1610:**231–243.
Subtil, A., Gaidarov, I., Kobylarz, K., Lampson, M. A., Keen, J. H., and McGraw, T. E. (1999). Acute cholesterol depletion inhibits clathrin-coated pit budding. *Proc. Natl. Acad. Sci. USA* **96:**6775–6780.
Sun, X., and Whittaker, G. R. (2003). Role for influenza virus envelope cholesterol in virus entry and infection. *J. Virol.* **77:**12543–12551.
Suomalainen, M. (2002). Lipid rafts and assembly of enveloped viruses. *Traffic* **3:**705–709.
Suzuki, Y., Matsunaga, M., and Matsumoto, M. (1985a). *N*-Acetylneuraminyllactosylceramide, GM3-NeuAc, a new influenza A virus receptor which mediates the adsorption-fusion process of viral infection: Binding specificity of influenza virus A/Aichi/2/68 (H3N2) to membrane-associated GM3 with different molecular species of sialic acid. *J. Biol. Chem.* **260:**1362–1365.
Suzuki, Y., Suzuki, T., Matsunaga, M., and Matsumoto, M. (1985b). Gangliosides as paramyxovirus receptor. Structural requirement of sialo-oligosaccharides in receptors for hemagglutinating virus of Japan (Sendai virus) and Newcastle disease virus. *J. Biochem. (Tokyo)* **97:**1189–1199.

Takada, A., Robison, C., Goto, H., Sanchez, A., Murti, K. G., Whitt, M. A., and Kawaoka, Y. (1997). A system for functional analysis of Ebola virus glycoprotein. *Proc. Natl. Acad. Sci. USA* **94:**14764–14769.

Takeda, M., Leser, G. P., Russell, C. J., and Lamb, R. A. (2003). Influenza virus hemagglutinin concentrates in lipid raft microdomains for efficient viral fusion. *Proc. Natl. Acad. Sci. USA* **100:**14610–14617.

Triantafilou, K., and Triantafilou, M. (2003). Lipid raft microdomains: Key sites for coxsackievirus A9 infectious cycle. *Virology* **317:**128–135.

Upla, P., Marjomaki, V., Kankaanpaa, P., Ivaska, J., Hyypia, T., Van Der Goot, F. G., and Heino, J. (2004). Clustering induces a lateral redistribution of $\alpha_2\beta_1$ integrin from membrane rafts to caveolae and subsequent protein kinase C-dependent internalization. *Mol. Biol. Cell* **15:**625–636.

van den Berg, C. W., Cinek, T., Hallett, M. B., Horejsi, V., and Morgan, B. P. (1995). Exogenous glycosyl phosphatidylinositol-anchored CD59 associates with kinases in membrane clusters on U937 cells and becomes $Ca^{2+}$-signaling competent. *J. Cell Biol.* **131:**669–677.

van't Hof, W., and Resh, M. D. (1997). Rapid plasma membrane anchoring of newly synthesized $p59^{fyn}$: Selective requirement for $NH_2$-terminal myristoylation and palmitoylation at cysteine-3. *J. Cell Biol.* **136:**1023–1035.

Varma, R., and Mayor, S. (1998). GPI-anchored proteins are organized in submicron domains at the cell surface. *Nature* **394:**798–801.

Vashishtha, M., Phalen, T., Marquardt, M. T., Ryu, J. S., Ng, A. C., and Kielian, M. (1998). A single point mutation controls the cholesterol dependence of Semliki Forest virus entry and exit. *J. Cell Biol.* **140:**91–99.

Viard, M., Parolini, I., Sargiacomo, M., Fecchi, K., Ramoni, C., Ablan, S., Ruscetti, F. W., Wang, J. M., and Blumenthal, R. (2002). Role of cholesterol in human immunodeficiency virus type 1 envelope protein-mediated fusion with host cells. *J. Virol.* **76:**11584–11595.

Vincent, S., Gerlier, D., and Manie, S. N. (2000). Measles virus assembly within membrane rafts. *J. Virol.* **74:**9911–9915.

Viola, A., Schroeder, S., Sakakibara, Y., and Lanzavecchia, A. (1999). T lymphocyte costimulation mediated by reorganization of membrane microdomains. *Science* **283:**680–682.

Vogel, U., Sandvig, K., and van Deurs, B. (1998). Expression of caveolin-1 and polarized formation of invaginated caveolae in Caco-2 and MDCK II cells. *J. Cell Sci.* **111:**825–832.

Waarts, B. L., Bittman, R., and Wilschut, J. (2002). Sphingolipid and cholesterol dependence of alphavirus membrane fusion: Lack of correlation with lipid raft formation in target liposomes. *J. Biol. Chem.* **277:**38141–38147.

Walk, S. F., Alexander, M., Maier, B., Hammarskjold, M. L., Rekosh, D. M., and Ravichandran, K. S. (2001). Design and use of an inducibly activated human immunodeficiency virus type 1 Nef to study immune modulation. *J. Virol.* **75:**834–843.

Wang, J. K., Kiyokawa, E., Verdin, E., and Trono, D. (2000). The Nef protein of HIV-1 associates with rafts and primes T cells for activation. *Proc. Natl. Acad. Sci. USA* **97:**394–399.

Welti, R., and Glaser, M. (1994). Lipid domains in model and biological membranes. *Chem. Phys. Lipids* **73:**121–137.

Wilson, B. S., Pfeiffer, J. R., and Oliver, J. M. (2000). Observing $Fc_\varepsilon RI$ signaling from the inside of the mast cell membrane. *J. Cell Biol.* **149:**1131–1142.

Wilson, B. S., Pfeiffer, J. R., Surviladze, Z., Gaudet, E. A., and Oliver, J. M. (2001). High resolution mapping of mast cell membranes reveals primary and secondary domains of $Fc_{\varepsilon}RI$ and LAT. *J. Cell Biol.* **154:**645–658.

Wool-Lewis, R. J., and Bates, P. (1998). Characterization of Ebola virus entry by using pseudotyped viruses: Identification of receptor-deficient cell lines. *J. Virol.* **72:**3155–3160.

Xu, X., and London, E. (2000). The effect of sterol structure on membrane lipid domains reveals how cholesterol can induce lipid domain formation. *Biochemistry* **39:**843–849.

Yang, C., and Compans, R. W. (1996). Palmitoylation of the murine leukemia virus envelope glycoprotein transmembrane subunits. *Virology* **221:**87–97.

Yang, C., Spies, C. P., and Compans, R. W. (1995). The human and simian immunodeficiency virus envelope glycoprotein transmembrane subunits are palmitoylated. *Proc. Natl. Acad. Sci. USA* **92:**9871–9875.

Yang, L., and Ratner, L. (2002). Interaction of HIV-1 Gag and membranes in a cell-free system. *Virology* **302:**164–173.

Zacharias, D. A., Violin, J. D., Newton, A. C., and Tsien, R. Y. (2002). Partitioning of lipid-modified monomeric GFPs into membrane microdomains of live cells. *Science* **296:**913–916.

Zarling, D. A., and Keshet, I. (1979). Fusion activity of virions of murine leukemia virus. *Virology* **95:**185–196.

Zebedee, S. L., and Lamb, R. A. (1988). Influenza A virus M2 protein: Monoclonal antibody restriction of virus growth and detection of M2 in virions. *J. Virol.* **62:**2762–2772.

Zhang, J., Pekosz, A., and Lamb, R. A. (2000). Influenza virus assembly and lipid raft microdomains: A role for the cytoplasmic tails of the spike glycoproteins. *J. Virol.* **74:**4634–4644.

Zheng, Y. H., Plemenitas, A., Fielding, C. J., and Peterlin, B. M. (2003). Nef increases the synthesis of and transports cholesterol to lipid rafts and HIV-1 progeny virions. *Proc. Natl. Acad. Sci. USA* **100:**8460–8465.

Zheng, Y.-H., Plemenitas, A., Linneman, T., Fackler, O. T., and Peterlin, B. M. (2001). Nef increases infectivity of HIV via lipid rafts. *Curr. Biol.* **11:**875–879.

Zhuang, M., Oltean, D. I., Gomez, I., Pullikuth, A. K., Soberon, M., Bravo, A., and Gill, S. S. (2002). *Heliothis virescens* and *Manduca sexta* lipid rafts are involved in CrylA toxin binding to the midgut epithelium and subsequent pore formation. *J. Biol. Chem.* **277:**13863–13872.

# POLYMORPHISM OF FILOVIRUS GLYCOPROTEINS

Viktor E. Volchkov,* Valentina A. Volchkova,* Olga Dolnik,* Heinz Feldmann,† and Hans-Dieter Klenk‡

*Biologie des Filovirus, Claude Bernard University Lyon
INSERM U412 69365 Lyon, France
†Canadian Science Centre for Human and Animal Health, Winnipeg
Manitoba R3E 3R2, Canada
‡Institut für Virologie Philipps-Universität Marburg, 35037 Marburg, Germany



## I. Expression Strategies of the Glycoprotein Gene

Filoviral glycoproteins are encoded by gene 4 (GP gene) of the nonsegmented negative-strand RNA genome (Fig. 1A). Gene 4 of Marburg virus (MBGV) encodes a single open reading frame of 2043 nucleotides that translates into the membrane glycoprotein (GP) (Bukreyev *et al.*, 1995; Feldmann *et al.*, 1992; Sanchez *et al.*, 1993, 1998a; Will *et al.*, 1993). In contrast, gene 4 of Ebola virus (EBOV) has two overlapping open reading frames, and expression involves transcriptional editing (Sanchez *et al.*, 1996; Volchkov *et al.*, 1995, 1998b). As shown in Fig. 1B, the primary structure of the editing site is a run of seven uridine residues in the genomic sequence. Transcriptional editing is performed by the viral RNA-dependent RNA polymerase (L protein). Unedited mRNA (~80% of GP-specific mRNAs) encodes the primary product of gene 4, which is a secreted nonstructural glycoprotein (sGP).




0065-3527/05 $35.00
DOI: 10.1016/S0065-3527(05)64011-0

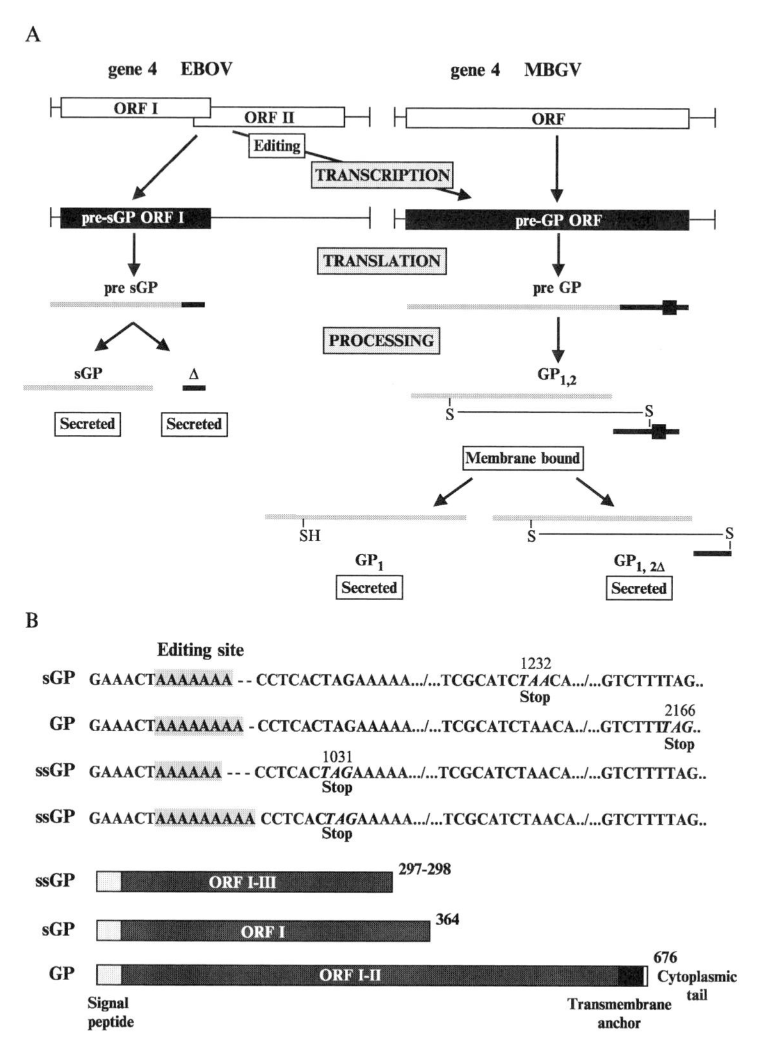


FIG 1. The glycoproteins of filoviruses. (A) Expression strategies. Unlike MBGV, EBOV requires transcriptional editing for expression of the membrane glycoprotein. The primary product of the EBOV glycoprotein gene (gene 4) is the precursor of the secreted glycoprotein (pre-sGP) that is expressed from unedited transcripts and post-translationally cleaved into the soluble products sGP and Δ-peptide. The precursor of the membrane glycoprotein pre-GP is cleaved into the disulfide-linked (S–S) subunits $GP_1$ and $GP_2$. The ectodomain of the membrane glycoprotein ($GP_{1,2\Delta}$) is released in soluble form after removal of the membrane anchor. Free sGP is generated by release of the disulfide bond with $GP_2$. (B) Transcriptional editing of EBOV gene 4. mRNAs with editing site and in-frame stop codons and the respective translation products are shown. The primary transcript of gene 4 with seven A's at the editing site encodes sGP, a soluble

GP is translated from edited mRNA, which is the result of the addition of a single adenosine residue at the editing site. This event, which shifts the open reading frame into −1, seems to occur in about 20% of the GP gene-specific transcripts (Sanchez *et al.*, 1996; Volchkov *et al.*, 1995). In addition, deletion of one or insertion of two adenosine residues has also been observed, and these changes allow a switch into a third open reading frame (−2). This open reading frame terminates two amino acids downstream of the editing site and generates another nonstructural small secreted glycoprotein (ssGP) (Volchkov *et al.*, 2000a; Volchkova *et al.*, 1998).

EBOV Zaire variants have been selected in guinea pigs (Volchkov *et al.*, 2000a) and tissue culture (Sanchez *et al.*, 1993) that have incorporated an additional uridine residue at the editing site. The insertion leads to an inversion of the sGP:GP ratio, with about 80% of the messengers encoding the membrane glycoprotein GP, 10% encoding sGP, and about 10% ssGP (Volchkov *et al.*, 2000a). Adaptation of EBOV to guinea pigs resulted in increased pathogenicity for these animals. It should be pointed out that not all the virus variants adapted to guinea pigs showed insertions at the editing site of GP. Thus, it appears that the GP mutations were, if at all, only partially responsible for the change in virulence. The only markers that could be linked to virulence in this study were specific point mutations in VP24 that all resulted in an increased electrophoretic mobility of this protein (Volchkov *et al.*, 2000a), and evidence suggests that interferon antagonism may be the mechanism by which VP24 determines host range and pathogenicity (Basler and Palese, 2004; Bray, 2004).

sGP as well as EBOV and MBGV GP undergo several posttranslational processing steps giving rise to a whole series of different glycoproteins (Fig. 1A). The GP precursor pre-GP is cleaved into the amino-terminal fragment $GP_1$ and the carboxy-terminal fragment $GP_2$, which are linked by a disulfide bond forming the $GP_{1,2}$ complex. $GP_{1,2}$ is converted into $GP_{1,2\Delta}$ by proteolytic removal of its membrane anchor. Release of the disulfide bond between $GP_1$ and $GP_2$ gives rise to soluble $GP_1$. sGP is derived from pre-sGP by proteolytic cleavage of the carboxy-terminal $\Delta$ peptide. $GP_{1,2}$ is a membrane protein, whereas all other cleavage products are secretory glycoproteins.

glycoprotein 364 amino acids in length. Insertion of two or deletion of one A residue results in expression of ssGP, a soluble glycoprotein 297 or 298 amino acids long. Insertion of one A residue fuses both open reading frames, resulting in the expression of the 676-amino acid-long membrane glycoprotein.

## II. Biosynthesis, Processing, and Maturation of Glycoproteins

### A. *Transmembrane Glycoprotein*

GP is a type I membrane protein that matures during export to the cell surface through the exocytotic transport route (Fig. 2). Co- and posttranslational modifications include removal of the signal peptide (Sanchez *et al.*, 1998b; Will *et al.*, 1993), N-glycosylation (Becker *et al.*, 1996; Feldmann *et al.*, 1991, 1994; Sanchez *et al.*, 1998b; Volchkov *et al.*, 1995), and oligomerization (Feldmann *et al.*, 1991; Sanchez *et al.*, 1998b) in the endoplasmic reticulum (ER). ER processing is followed by acylation in a pre-Golgi compartment (Funke *et al.*, 1995; Ito *et al.*, 2001), and by O-glycosylation and maturation of *N*-glycans in the Golgi apparatus (Becker *et al.*, 1996; Feldmann *et al.*, 1991, 1994; Geyer *et al.*, 1992; Volchkov *et al.*, 1995; Will *et al.*, 1993). Pre-GP is proteolytically cleaved in the *trans*-Golgi network into $GP_{1,2}$ by the proprotein convertase furin (Volchkov *et al.*, 1998a, 2000b). $GP_{1,2}$ is then transported to the plasma membrane, where it appears to accumulate in cholesterol-rich microdomains (Bavari *et al.*, 2002). Transport of viral proteins into these microdomains may be mediated by multivesicular bodies (Kolesnikova *et al.*, 2004), but GP has so far not been identified yet in these organelles. GP is released from the cell in virus particles or in membrane vesicles. Alternatively, the ectodomain ($GP_{1,2\Delta}$) may be cleaved from the cell surface by the metalloprotease TACE (tumor necrosis factor $\alpha$-converting enzyme) in a shedding process (Dolnik *et al.*, 2004).

GP of MBGV, strains Musoke and Popp, and of EBOV, strain Mayinga, has a total length of 681 and 676 amino acids, respectively. Comparison of the GP sequences of MBGV and EBOV shows conservation at the amino-terminal and carboxy-terminal ends, whereas the middle region is variable. GP can be subdivided into a large ectodomain, a lipid membrane-spanning domain of approximately 30 amino acids, and a short cytoplasmic tail of 4 (EBOV) and 8 (MBGV) amino acids, respectively (Fig. 3A). There are three proteolytic cleavage sites: one for the signal peptidase, the polybasic recognition motif for furin that cleaves $GP_1$ from $GP_2$, and the cleavage site of TACE that removes the ectodomain from the membrane anchor. The ectodomain consists therefore of $GP_1$ and most of $GP_2$. $GP_1$ is believed to contain the receptor-binding site, which, however, has not been identified yet. $GP_1$ carries also the bulk of *N*- and *O*-glycans that account for more than one-third of the molecular weight of mature $GP_{1,2}$ (Becker *et al.*, 1996; Feldmann *et al.*, 1991; Geyer *et al.*, 1992; Volchkov *et al.*, 1995; Will *et al.*, 1993).

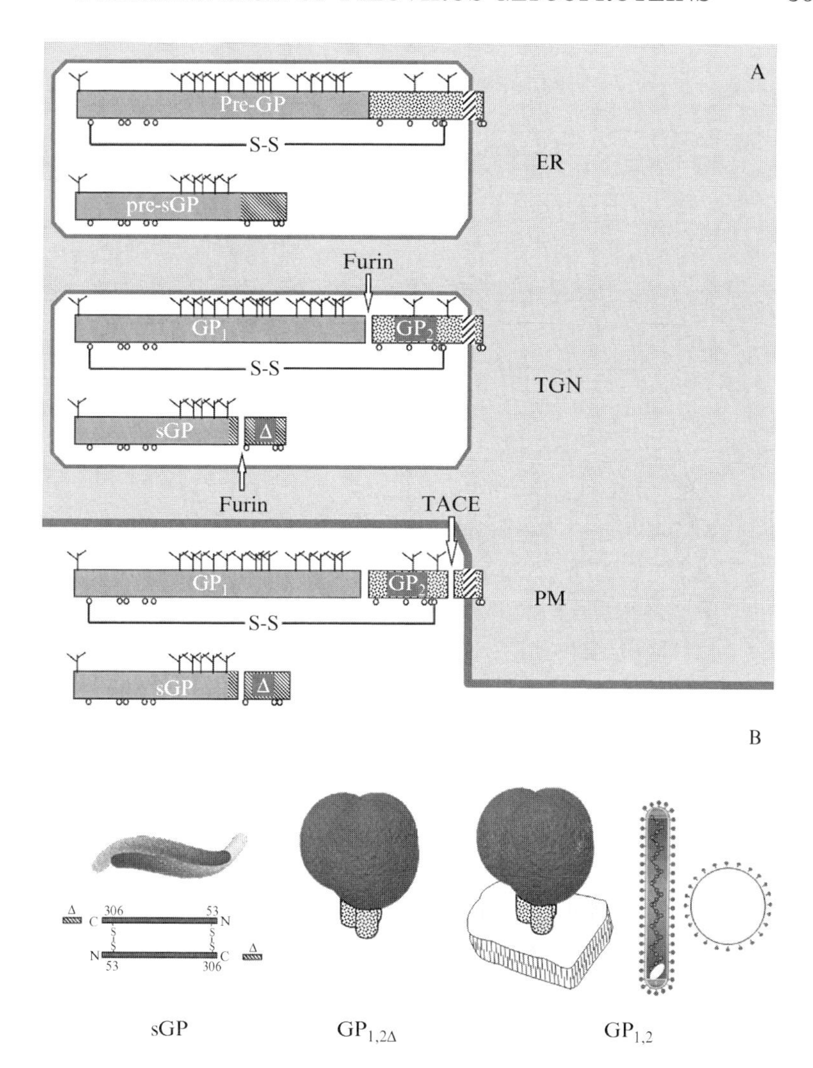


FIG 2. Maturation of EBOV glycoproteins. (A) Compartmentalization of glycoprotein processing. Pre-GP is synthesized as a type I membrane protein in the endoplasmic reticulum (ER). Pre-sGP is secreted into the lumen of the ER. Furin cleavage of pre-GP and pre-sGP occurs in the *trans*-Golgi network (TGN). The ectodomain of GP is released by TACE at the cell surface (peripheral membrane, PM). ▩, $GP_2$; ▨, membrane anchor; ▧, carboxy-terminal sequence of sGP not identical with $GP_1$; Y, *N*-glycans; o, cysteine residues. (B) The mature forms of the glycoproteins. sGP is a dimer with antiparallel orientation. $GP_{1,2}$ is a trimeric spike inserted into virus particles and empty membrane vesicles. $GP_{1,2\Delta}$ is the ectodomain released by shedding.

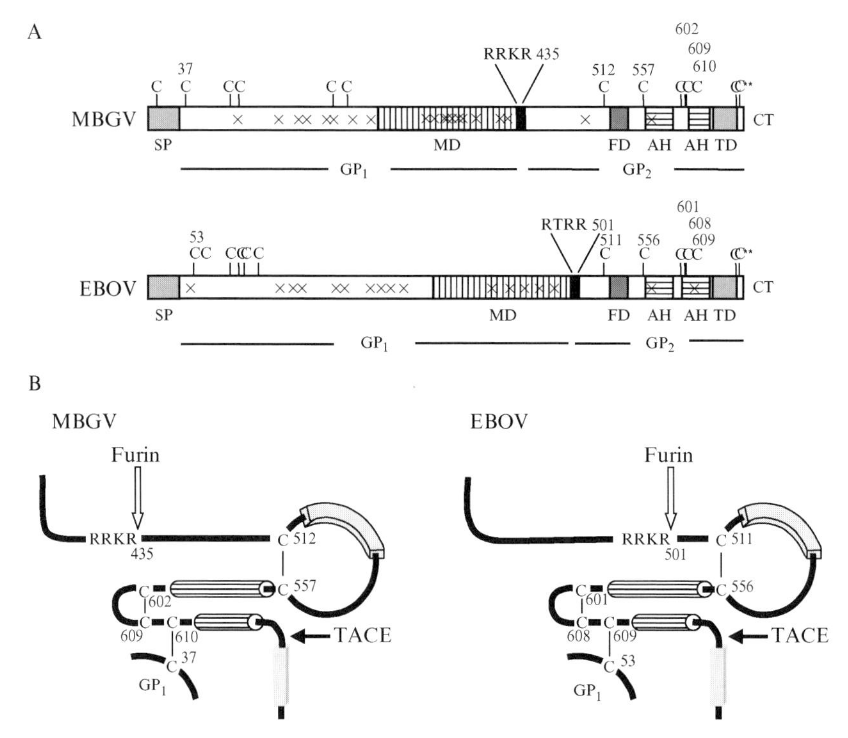


FIG 3. The glycoprotein GP. (A) Primary structure. $GP_{1,2}$ has three hydrophobic domains (gray boxes): a signal peptide (SP) at the amino-terminal end, a fusion domain (FD), and a transmembrane domain (TD) at the carboxy-terminal end. The mucin-like domain (MD) of $GP_1$, the amphipathic helices (AH) of $GP_2$, the furin cleavage site of MBGV ($RRKR_{495}$) and EBOV ($RTRR_{501}$), cysteine residues (C), and N-glycosylation sites (x) are also indicated. Acylated cysteines are marked with asterisks (*). Cysteine residues forming disulfide bonds shown in (B) are numbered. (B) Folding of $GP_2$. The fusion peptide is present in a loop flanked by the disulfide-linked cysteine residues 512 and 557 (MBGV) and 511 and 556 (EBOV). This loop is followed by a hairpin formed by the amphipathic helices and an intervening loop flanked by cysteine residues 602 and 609 (MBGV) and 601 and 608 (EBOV). The carboxy-terminal amphipathic helix is followed by the membrane anchor. Cysteine residues 610 and 609 form disulfide bridges with cysteine residues 37 and 53 on $GP_1$ of MBGV and EBOV, respectively. The cleavage sites of furin and TACE are also indicated.

Detailed structural analyses of filoviral carbohydrates are available for MBGV only. The *N*-glycans comprise oligomannosidic and hybrid types as well as bi-, tri-, and tetraantennary complex species. In addition, large amounts of neutral mucin-type *O*-glycans are present (Geyer *et al.*, 1992). MBGV GP contains about 30 side chains of the latter type, which are likely to be attached to clusters of hydroxyamino acids and proline residues on $GP_1$ (Will *et al.*, 1993). Such a "mucin-like" domain is also present on EBOV $GP_1$. Oligosaccharide side chains differ in their terminal sialylation patterns, which seem to be virus isolate as well as cell line dependent. Interestingly, MBGV GP has been found to be completely sialic acid free when the virus was grown in Vero cells (Feldmann *et al.*, 1994). Because filoviruses do not contain a neuraminidase, there is no explanation for this phenomenon. MBGV $GP_1$ carries also phosphate residues, but the exact phosphorylation sites are unknown (Sänger *et al.*, 2002).

Sequence comparison has already revealed several striking similarities between $GP_2$ of filoviruses and the transmembrane glycoprotein of retroviruses (Gallaher, 1996; Volchkov *et al.*, 1992). These structural features were subsequently definitely proved by biochemical (Weissenhorn *et al.*, 1998a) and X-ray crystallographic studies (Malashkevich *et al.*, 1999; Weissenhorn *et al.*, 1998b). According to these data, $GP_2$ contains a hydrophobic fusion peptide at a distance of 22 (EBOV) or 91 (MBGV) amino acids from the amino terminus. Between the fusion domain, which is flanked by two cysteines, and the transmembrane domain there are two amphipathic helices separated by a $CX_6CC$ motif (Fig. 3A). The amphipathic helices have a hairpin orientation. The linker region, which reverses the chain direction of the helices, contains a short disulfide-linked loop formed by the $CX_6CC$ motif. The cysteine residues flanking the fusion domain (C512 and C557 for MBGV, C511 and C556 for EBOV) have also been shown to form a disulfide bridge (Jeffers *et al.*, 2002), which would place the fusion peptide in a loop at the tip of the hairpin (Fig. 3B). As indicated by the crystallographic data, three hairpins form a six-helix bundle as the central structural feature of $GP_2$ (Weissenhorn *et al.*, 1998a,b), confirming the concept that the filovirus spike is a homotrimer of disulfide-bonded $GP_{1,2}$ molecules (Feldmann *et al.*, 1991; Sanchez *et al.*, 1998b). Besides the cysteine residues already mentioned, at the carboxy-terminal tail of $GP_2$ there are two cysteines that are acylated (Funke *et al.*, 1995; Ito *et al.*, 2001). Four cysteine residues of $GP_1$ form two intermolecular disulfide bridges, whereas the fifth, an amino-terminal cysteine (C37 for MBGV and C53 for EBOV), binds to C610 (MBGV) or C609 (EBOV) on $GP_2$ (Jeffers *et al.*, 2002; Volchkova *et al.*, 1998). This indicates that $GP_{1,2}$ is basically a loop where the amino-terminal end of $GP_1$ is bound to the membrane-adjacent region of $GP_2$.

$GP_{1,2}$ trimers form the spikes of mature virions. In addition, empty membrane vesicles studded with spikes are released from infected cells. Such vesicles have also been observed when GP was expressed from a vaccinia virus vector. This indicates that maturation and release of GP in membrane-bound form does not depend on other viral proteins (Volchkov *et al.*, 1998b). Furthermore, GP-containing membrane vesicles released from transfected cells may prove suitable for vaccination purposes.

### *B. Soluble GP Ectodomain*

Soluble $GP_1$ is released into the culture medium from HeLa cells (Volchkov *et al.*, 1998a). It has been hypothesized that the disulfide bond between $GP_1$ and $GP_2$ (cysteine residues 53 and 609) may not always be completed during early processing of GP in the ER. Thus, proteolytic processing in the Golgi would lead to the release of $GP_1$ molecules. Studies have shown that $GP_1$ is also released as a complex with $GP_2$ ($GP_{1,2\Delta}$) after removal of the membrane anchor by proteolytic cleavage between Asp-637 and Gln-638. $GP_{1,2\Delta}$ is present in trimeric form, which, however, is more labile than $GP_{1,2}$ trimers, indicating that the membrane anchor has a stabilizing function. The enzyme responsible for shedding of the ectodomain has been identified as the metalloprotease TACE (Dolnik *et al.*, 2004).

### *C. Secreted Glycoprotein*

The secreted glycoprotein precursor (pre-sGP) of EBOV, species Zaire, has a length of 364 amino acids and shares the amino-terminal 295 amino acids with the membrane glycoprotein (GP) (Fig. 2). The different carboxy terminus (69 amino acids) contains several charged residues as well as cysteine residues. As with pre-GP, pre-sGP undergoes several co- and posttranslational processing events, such as signal peptide cleavage, glycosylation, oligomerization, and proteolytic cleavage (Volchkova *et al.*, 1998, 1999). The limiting step during maturation and transport seems to be oligomerization in the ER (Volchkova *et al.*, 1998). After oligomerization pre-sGP is transported into the Golgi compartments, where glycosylation is completed and posttranslational cleavage into sGP and a small peptide, designated Δ peptide, occurs (Volchkova *et al.*, 1999). Cleavage is mediated by furin, which is also responsible for cleavage of pre-GP (Volchkov *et al.*, 1998a) (Figs. 2 and 3). sGP is efficiently secreted from infected cells because of the lack of a transmembrane anchor. Secretion also occurs if cleavage is abolished (Volchkova *et al.*, 1999). sGP and uncleaved pre-sGP appear as a disulfide-linked homodimer that shows an antiparallel orientation of monomers (Fig. 2B) (Sanchez *et al.*, 1998b; Volchkova *et al.*, 1998). Dimerization is due to an intermolecular disulfide linkage between the amino- and carboxy-terminal cysteine residues at positions 53 and 306, respectively. The remaining four highly conserved cysteine residues at the amino terminus seem to be involved in intramolecular folding of monomers (Volchkova *et al.*, 1998).

### *D. Δ Peptide*

Δ Peptide, the small cleavage product of pre-sGP, varies in length between 40 and 48 amino acids with the different EBOVs (Volchkova *et al.*, 1999) (Fig. 2). Its molecular mass of approximately 10 to 14 kDa is significantly larger than that predicted from the amino acid sequence (~5 kDa). The difference is due to the attachment of several *O*-glycans that carry terminal sialic acids. In this respect it differs from sGP, which seems to carry mainly N-linked carbohydrates. Δ Peptide is secreted from cells, but this process seems not to be as efficient as the secretion of sGP (Volchkova *et al.*, 1999).

### *E. Small Secreted Glycoprotein*

The small secreted glycoprotein (ssGP) of EBOV resembles a natural carboxy terminus-truncated variant of sGP. Because of the lack of three carboxy-terminal cysteine residues including the one at position 306, which is involved in dimerization of sGP, ssGP is secreted in a monomeric form (Volchkov *et al.*, 1995; Volchkova *et al.*, 1998).

### *F. Proteases Involved in Processing of Filovirus Glycoproteins*

#### *1. Furin*

Cleavage of pre-GP into $GP_1$ and $GP_2$ is done by furin. The enzyme belongs to the proprotein convertases, a family of subtilisin-like eukaryotic serine endoproteases that also includes PC1/PC3, PC2, PC4, PACE4, PC5/PC6, and LPC/PC7 (Seidah *et al.*, 1996). Proprotein convertases are expressed differentially in cells and tissues and display similar, but not identical, specificity for basic motifs, such as R-X-K/R-R, at the cleavage sites of their substrates. Besides furin, which is thought to be ubiquitous, PACE4, PC5/PC6, and LPC/PC7 also exhibit widespread tissue distribution. They are processing enzymes present within the constitutive secretory pathway of mammalian cells (Seidah *et al.*, 1996). The expression of PC1/PC3 and PC2 is restricted to endocrine cells displaying the regulated secretory pathway. PC4 mRNA is found exclusively in germ cells (Nakayama *et al.*, 1992). In contrast to most other subtilisins, LPC/PC7 and furin are synthesized as membrane-anchored endoproteases (van den Ouweland *et al.*, 1990). Furin is localized predominantly in the *trans*-Golgi network (Molloy *et al.*, 1994; Schäfer *et al.*, 1995) but is also secreted from cells (Vey *et al.*, 1995, Vidricaire *et al.*, 1993). Furin cleaves preferentially at the consensus sequence RXR/KR, but also at RXXR and more rarely at RXXXXR sites, and it appears to be the key activating enzyme of proproteins containing such sites. Enzymatic activity depends on calcium ions and neutral pH. Furin is a type I transmembrane glycoprotein containing a signal peptide, an autocatalytically cleavable prodomain, a catalytic domain with histidine, aspartic acid, serine, and asparagines arranged at the catalytic site as in bacterial subtilisins, a membrane anchor, and a cytoplasmic domain (Nakayama, 1997; Van de Ven *et al.*, 1990).

Proprotein convertases are thought to be involved in the proteolytic activation of a large variety of different cellular substrates, including precursor molecules of hormones, receptors, cell adhesion molecules, and neurotransmitters (Barr *et al.*, 1991). Besides their role in activation of cellular substrates, subtilisin-like proteases play a key role in the activation of many viral envelope proteins (Klenk and Garten, 1994).

### 2. *TACE*

TACE (ADAM 17) has been shown to be involved in shedding of the soluble ectodomain of EBOV GP. The ADAM (a disintegrin and metalloprotease) proteases comprise a large group of zinc-dependent cell surface proteases that were first identified in guinea pig sperm, but have since been found in most mammalian tissues as well as lower eukaryotes, such as *Caenorhabditis elegans, Drosophila*, and *Xenopus*. ADAMs are multidomain proteins consisting of a signal sequence, a prodomain, a metalloprotease domain, a disintegrin domain, a cysteine-rich domain, an epidermal growth factor (EGF) repeat, and a transmembrane domain, followed by a cytoplasmic tail (Blobel *et al.*, 1992; Killar *et al.*, 1999). ADAMs are produced as zymogens that are formed by an intramolecular complex between a single cysteine residue in the prodomain and the essential zinc ion in the catalytic domain, a complex that blocks the active site. The catalytic site of the metalloprotease domain has the consensus sequence HEXGHXXGXXHD that is present in most ADAMs, including ADAM 17 (Primakoff and Myles, 2000). The disintegrin domain is believed to be a ligand for integrins or other receptors (Zhang *et al.*, 1998). The cysteine-rich domain is characterized by the presence of numerous cysteinyl residues displayed in a distinct sequence pattern. The function of this domain remains unclear, but it has been suggested to play a role in protein–protein interaction (Iba *et al.*, 2000). Although the transmembrane domain functions as a membrane anchor, the presence of an EGF repeat and a cytoplasmic tail suggests that these proteins may serve as signal transducers between the extracellular and the intracellular space.

The presence of these multiple domains in a single polypeptide suggests that the ADAM proteins are capable of several functions, including proteolysis, adhesion, and signaling. Accordingly, these proteins have been implicated in diverse developmental events, such as fertilization, remodeling of the extracellular matrix (ECM), growth factor ectodomain shedding, and neurogenesis. As proteases, ADAMs have a wide variety of substrates. They can degrade ECM components, shed cell-bound ectodomains to free growth factors and ligands from the cell surface, and cleave other integral membrane proteins (Blobel, 2000; Kheramand and Werb, 2002; Moss *et al.*, 2001). EBOV GP is the first viral glycoprotein shown to be processed by such an enzyme.

TACE is responsible for shedding of tumor necrosis factor $\alpha$ (TNF-$\alpha$) from the cell surface (Black *et al.*, 1997; Moss *et al.*, 1997). In addition, it has been found to mediate release of the ectodomains of several other membrane proteins. These include interleukin (IL)-6 receptor, TNF receptors p55 and p75, transforming growth factor $\alpha$ (TGF-$\alpha$), L-selectin, and IL-1 receptor II (Althoff *et al.*, 2000; Peschon *et al.*, 1998, Reddy *et al.*, 2000). Furthermore, TACE has been shown to have $\alpha$-secretase activity that removes the soluble ectodomain (sAPP$\alpha$) from the amyloidal precursor protein (APP) (Buxbaum *et al.*, 1998; Lammich *et al.*, 1999). Because of the great diversity of these substrates it has been proposed that ADAM 17 has the function of a common sheddase (Althoff *et al.*, 2001). Most, but not all, substrates are cleaved between two hydrophobic residues. The available evidence indicates, however, that TACE requires neither a specific recognition sequence nor a specific secondary structure at the cleavage site (Althoff *et al.*, 2000, 2001; Arribas *et al.*, 1997; Ehlers *et al.*, 1996).

## III. Roles of Glycoproteins in Viral Life Cycle and in Pathogenesis

### A. *Cell Entry*

When analyzed by electron microscopy filovirus particles were found to be associated with coated pits along the plasma membrane, suggesting endocytosis as a possible mechanism for entry (Geisbert and

Jahrling, 1995). This concept is supported by studies employing lysosomotropic agents (Chan *et al.*, 2000b; Mariyankova *et al.*, 1993). Using vesicular stomatitis virus and retrovirus pseudotypes, several groups showed independently that virus entry is mediated by $GP_{1,2}$ (Chan *et al.*, 2000b; Takada *et al.*, 1997; Wool-Lewis and Bates, 1998; Yang *et al.*, 1998).

A number of different cellular proteins have been proposed to act as filovirus receptors. The first study presented evidence that MBGV uses the asialoglycoprotein receptor to infect hepatocytes (Becker *et al.*, 1995). For EBOV, it was suggested that integrins, especially the $\beta_1$ group, might interact with the glycoprotein and perhaps be involved in virus entry into cells (Takada *et al.*, 2000). Other studies indicate that the folate receptor $\alpha$ serves as a cofactor for cellular entry by MBGV and EBOV (Chan *et al.*, 2001). It is therefore reasonable to assume that filoviruses use different receptors on different cells. Alvarez and co-authors demonstrated that on binding of EBOV GP by DC-SIGN [dendritic cell-specific intercellular adhesion molecule 3 (ICAM-3)-grabbing nonintegrin], infectivity of GP-specific pseudotype virus particles increased *in trans*, most likely by a similar manner previously described for human (HIV) and simian (SIV) immunodeficiency virus infections (Alvarez *et al.*, 2002; Bashirova *et al.*, 2001; Pohlmann *et al.*, 2001).

Although direct experimental evidence of the fusion activity of filoviruses is still lacking, it is clear that the $GP_2$ subunit of both EBOV and MBGV glycoproteins shows the typical structural features of a type I fusion protein. These features include a fusion peptide and two antiparallel helices forming a six-helix bundle in the trimer (Fig. 3). The EBOV fusion peptide was found to induce fusion with liposomes (Ruiz-Argüello *et al.*, 1998). This observation together with mutational analysis of the putative fusion domain (Ito *et al.*, 1999) offers compelling support for a fusion peptide role for these conserved regions in the filovirus membrane glycoprotein. The central structural features of the EBOV $GP_2$ ectodomain suggest that the fusion peptide and membrane anchor are located at one end of the rod-like six-helix bundle (Weissenhorn *et al.*, 1998a,b). Such structures have been observed with the transmembrane subunits $HA_2$ of the influenza virus hemagglutinin (Bullough *et al.*, 1994), gp41 of the HIV Env protein (Chan *et al.*, 1997; Weissenhorn *et al.*, 1997), and $F_1$ of the paramyxovirus fusion protein (Joshi *et al.*, 1998). The rod-like structures are believed to be the fusion-active form of these proteins, and studies have shown that the underlying coiled-coil motif of $GP_2$ plays an important role in facilitating EBOV entry (Watanabe *et al.*, 2000).

An important control mechanism of the fusion activity of viral surface glycoproteins is the processing by proprotein convertases (Klenk and Garten, 1994). Proteolytic cleavage, which occurs often next to the fusion peptide, is the first step in the activation of these proteins and is followed by extensive refolding of the molecule, resulting in formation of the six-helix bundle and exposure of the fusion domain. Cleavage and conformational change, which may be triggered by low pH, such as in endosomes (Skehel *et al.*, 1982), or by the interaction with a secondary receptor protein at the cell surface (Feng *et al.*, 1996), are indispensable for the infectivity of influenza virus, HIV, and paramyxoviruses.

The structural similarities between the fusion proteins of these viruses and the filovirus membrane glycoprotein, GP cleavage by furin as a conserved trait of filoviruses, and folding differences between uncleaved and cleaved GP as revealed by mobility on sodium dodecyl sulfate (SDS) gels (Volchkov *et al.*, 1998b) would strongly suggest that fusion activity and infectivity of filoviruses depend on GP cleavage. However, studies using pseudotyped viruses have demonstrated that proteolytic cleavage of the transmembrane glycoprotein is dispensable for replication of EBOV in cell culture (Ito *et al.*, 2001; Wool-Lewis and Bates, 1999). Similar results have been obtained when recombinant EBOV with a cleavage site mutation in GP has been analyzed (Neumann *et al.*, 2002). To date, there is no explanation yet for these apparently conflicting observations. It is interesting to see that filovirus $GP_2$ has an internal fusion peptide, unlike the fusion proteins of influenza virus, HIV, and paramyxoviruses, which have an amino-terminal fusion peptide (Fig. 3). This difference may account for cleavage requirement in some, but not in other, cells. The data available so far would also be compatible with the concept that cleavage activation does not necessarily have to occur during virus maturation, but may also take place after entry into a new host cell, as has been observed with influenza virus (Boycott *et al.*, 1994). On the other hand, about 5% of GP have been found in cleaved form even when all basic amino acids at the furin cleavage site were exchanged (RRTRR → SGTGG) (V. E. Volchkov, unpublished data). This residual cleavage may explain infectivity of recombinant EBOV mutants containing "uncleavable" surface GP. Further work is necessary to clarify these problems. It also remains to be seen whether inhibition of furin cleavage, which can be achieved with peptidyl chloromethylketones or other components (Anderson *et al.*, 1993; Stieneke-Gröber *et al.*, 1992), is a valuable concept for treatment of acute infections with filoviruses.

Variation in the cleavage site of the glycoprotein GP may account for differences in the pathogenicity of EBOV (Volchkov *et al.*, 1998a). MBGV and all highly pathogenic EBOV strains display the canonical furin motif R-X-K/R-R at the cleavage site and are highly susceptible to cleavage. Only the glycoprotein of the EBOV Reston strain, which appears to be less pathogenic for humans and only moderately pathogenic for at least some monkey species (Fisher-Hoch *et al.*, 1992), shows reduced cleavability because of the suboptimal cleavage site sequence K-Q-K-R (Volchkov *et al.*, 1998a). Expression of the Reston virus glycoprotein in transfected mammalian cells demonstrated a lower cleavability of this protein that could be increased by a single amino acid change (Volchkov *et al.*, 1998a). Thus, highly pathogenic variants may emerge from Reston-like strains by mutations restricted to the cleavage site (Klenk *et al.*, 1998).

### *B. Target Cell Destruction*

Filovirus infections lead to a moderate cytopathogenic effect in target cells. However, the mechanism causing cell destruction is unknown. It is possible that either massive production and accumulation of viral proteins or maturation of viral particles at the plasma membrane is involved in this process. Alternatively, a viral protein may have specific cytotoxic potential. Studies demonstrated cell destruction on expression of EBOV transmembrane glycoprotein (Chan *et al.*, 2000a; Takada *et al.*, 2000; Yang *et al.*, 2000). In one study it was reported that the serine–threonine-rich mucin-like domain located on $GP_1$ mediates cytotoxicity in 293T and endothelial cells. This was confirmed in vessel explants by infections with recombinant adenovirus vectors expressing EBOV GP (Yang *et al.*, 2000). A second study demonstrated cell detachment of 293 T cells after expression of EBOV, but not MBGV, GP. Cell detachment in this case occurred without cell death. It was largely attributed to a domain in the extracellular region of $GP_2$ and seemed to involve a phosphorylation-dependent signal cascade (Chan *et al.*, 2000a). The ectodomain of the glycoprotein and its anchorage to the membrane are required for GP-induced morphological changes (Takada *et al.*, 2000). Using a reverse genetics system it was demonstrated that the cytotoxicity of EBOV depends on the level of GP expression. Overexpression of GP leads to early detachment and cytotoxicity of infected cells (Volchkov *et al.*, 2001). These data show that GP expression is controlled by RNA editing, which allows expression from approximately 20% of the GP-specific transcripts and therefore downregulates its synthesis.

It appears that editing of the GP gene of EBOV, although not required for virus replication, is linked with the need to control cytotoxicity.

### *C. Immunogenicity*

Neutralizing anti-GP antibodies have been generated from several species, including humans, that were immunized or infected with Ebola virus. These neutralizing antibodies showed protective and therapeutic properties in animal models (Maruyama *et al.*, 1999; Wilson *et al.*, 2000). Protective properties, most likely due to neutralizing antibodies, were also associated with convalescent sera (Mupapa *et al.*, 1999). The successful use of the transmembrane glycoprotein in different immunization approaches has clearly demonstrated the immunogenic and protective properties of this protein in small animal models and nonhuman primates (Garbutt *et al.*, 2004; Hevey *et al.*, 1998; Pushko *et al.*, 2000; Sullivan *et al.*, 2000; Vanderzanden *et al.*, 1998; Xu *et al.*, 1998).

### *D. Inhibition of Immune Defense*

Immunosuppression seems to be an important factor in the pathogenesis of filovirus hemorrhagic fever. However, the mechanisms leading to immunosuppressed status are only partially understood. For EBOV, it has been reported that sGP interacts with the host immune response by binding to neutrophils through CD16b, the neutrophil-specific form of $Fc_{\gamma}$ receptor III. Subsequently, sGP binding appears to inhibit early activation of these cells (Kindzelskii *et al.*, 2000; Yang *et al.*, 1998). This concept, however, has been challenged by a report from Maruyama and colleagues (1998). Relatively high amounts of EBOV glycoproteins are released into the medium as sGP and $GP_{1,2\Delta}$. sGP may effectively bind anti-EBOV antibodies that might otherwise be neutralizing or protective (Sanchez *et al.*, 1996; Volchkov *et al.*, 1998b). $GP_{1,2\Delta}$ may also have a decoy function, because it is able to block neutralizing antibodies *in vivo* (Dolnik *et al.*, 2004). In addition, filovirus transmembrane glycoprotein molecules possess a sequence close to the carboxy terminus, resembling a presumptive immunosuppressive domain found in retrovirus glycoproteins (Bukreyev *et al.*, 1995; Volchkov *et al.*, 1992; Will *et al.* 1993). Peptides synthesized according to this 26-amino acid-long region inhibited the blastogenesis of lymphocytes in response to mitogens, inhibited production of cytokines, and decreased proliferation of mononuclear cells *in vitro* (Ignatyev, 1999). However, it is not yet known whether the immunosuppressive domain on the GP is functional in mature molecules.

### *E. Endothelial Damage*

Disturbance of the blood–tissue barrier, which is controlled primarily by endothelial cells, is another important factor in pathogenesis. The endothelium seems to be affected in two ways: directly by virus infection, leading to activation and eventual cytopathogenic replication, and indirectly by a mediator-induced inflammatory response. These mediators originate from virus-activated cells of the mononuclear phagocytic system (MPS), especially macrophages, which are the primary target cells (Feldmann *et al.*, 1996; Schnittler and Feldmann, 1999). Current data indicate that the activation of mononuclear phagocytotic cells is triggered at an early stage of virus infection (Ströher *et al.*, 2001).

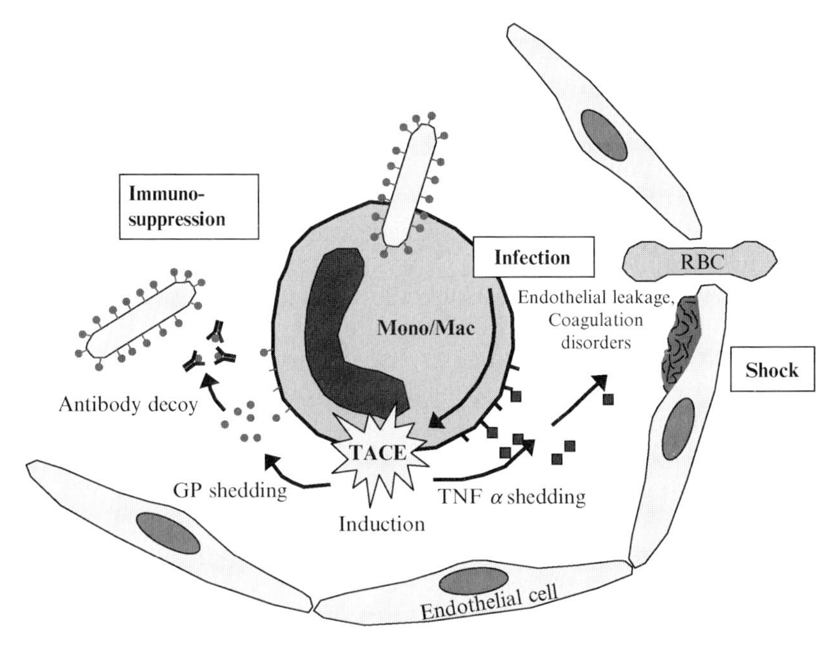


FIG 4. The role of TACE in pathogenesis. Infection of macrophages, the primary target cell of filoviruses, upregulates expression of TACE. TACE causes endothelial leakage and coagulation disorders (Feldmann *et al.*, 1996) and is thus involved in shock development. TACE is also responsible for $GP_{1,2\Delta}$ shedding, which may be an important mechanism in immunosuppression (Dolnik *et al.*, 2004).

### *F. Central Role for TACE in Pathogenesis*

One of the mediators responsible for endothelial leakage is TNF-$\alpha$, which is released from the surface of macrophages by TACE cleavage. On the other hand, TACE is involved in $GP_{1,2\Delta}$ shedding and therefore is also a factor in immunosuppression. Finally, there is evidence that TACE expression is dramatically upregulated in monkeys infected with Ebola virus (D. Relman, T. Geisbert, and P. Jahrling, personal communication). These observations taken together indicate that TACE may play a central role in the pathogenesis of filovirus infections (Fig. 4).

## Acknowledgments

V.E.V., H.F., and H.D.K. hold several grants on filoviruses provided by the Deutsche Forschungsgemeinschaft (SFB 286 and 593, Fe 286/4-1), the Kempkes-Stiftung (21/95), INSERM, the NIH, the Fondation pour la Recherche Medicale, the DGA (P01008SH), the Canadian Institutes of Health Research (MOP-43921), and the European Community (INCO-grant ERBIC 18 CT9803832). O.D. is the recipient of a Feodor Lynen Fellowship from the Alexander von Humboldt Foundation.

## References

Althoff, K., Müllberg, J., Aasland, D., Voltz, N., Kallen, K.-J., Grötzinger, J., and Rose-Hohn, S. (2001). Recognition sequences and structural elements contribute to shedding susceptibility of membrane proteins. *Biochem. J.* **35:**663–672.

Althoff, K., Reddy, P., Voltz, N., Rose-John, S., and Müllberg, J. (2000). Shedding of interleukin-6 receptor and tumor necrosis factor $\alpha$: Contribution of the stalk sequence to the cleavage pattern of transmembrane proteins. *Eur. J. Biochem.* **267:**2624–2631.

Alvarez, C. P., Lasala, F., Carrillo, J., Muniz, O., Corbi, A. L., and Delgado, R. (2002). C-type lectins DC-SIGN and L-SIGN mediate cellular entry by Ebola virus *in cis* and *in trans*. *J. Virol.* **76:**6841–6844.

Anderson, E. D., Thomas, L., Hayflick, J. S., and Thomas, G. (1993). Inhibition of HIV-1 gp160-dependent membrane fusion by a furin-directed 1-antitrypsin variant. *J. Biol. Chem.* **268:**24887–24891.

Arribas, J., López-Casillas, F., and Massagué, J. (1997). Role of the juxtamembrane domain of the transforming growth factor-$\alpha$ precursor and the $\beta$-amyloid precursor protein in regulated ectodomain shedding. *J. Biol. Chem.* **272:**17160–17165.

Barr, P. J., Mason, O. B., Landsberg, K. E., Wong, P. A., Kiefer, M. C., and Brake, A. J. (1991). cDNA and gene structure for a human subtilisin-like protease with cleavage specificity for paired basic amino acid residues. DNA *Cell Biol.* **10:**319–328.

Bashirova, A. A., Geijtenbeek, T. B., van Duijnhoven, G. C., van Vliet, S. J., Eilering, J. B., Martin, M. P., Wu, L., Martin, T. D., Viebig, N., Knolle, P. A., KewalRamani, V. N., van Kooyk, Y., and Carrington, M. (2001). A dendritic cell-specific intercellular adhesion molecule 3-grabbing nonintegrin (DC-SIGN)-related protein is highly expressed on human liver sinusoidal endothelial cells and promotes HIV-1 infection. *J. Exp. Med.* **193:**671–678.

Basler, C., and Palese, P. (2004). Modulation of innate immunity by filoviruses. *In* "Ebola and Marburg Viruses: Molecular and Cellular Biology" (H. D. Klenk and H. Feldmann, eds.), pp. 305–349. Horizon Bioscience, Wymondham, UK.

Bavari, S., Bosio, C. M., Wiegand, E., Ruthel, G., Will, A. B., Geisbert, T. W., Hevey, M., Schmaljohn, C., Schmaljohn, A., and Aman, M. J. (2002). Lipid raft microdomains: A gateway for compartmentalized trafficking of Ebola and Marburg viruses. *J. Exp. Med.* **195:**593–602.

Becker, S., Klenk, H.-D., and Mühlberger, E. (1996). Intracellular transport and processing of the Marburg virus surface protein in vertebrate and insect cells. *Virology* **225:**145–155.

Becker, S., Spiess, M., and Klenk, H.-D. (1995). The asialoglycoprotein receptor is a potential liver-specific receptor for Marburg virus. *J. Gen. Virol.* **76:**393–399.

Black, R. A., Rauch, C. T., Kozlosky, C. J., Peschon, J. J., Slack, J. L., Wolfson, M. F., Castner, B. J., Stocking, K. L., Reddy, P., Srinivasan, S., Nelson, N., Boiani, N., Schooley, K. A., Gerhart, M., Davis, R., Fitzner, J. N., Johnson, R. S., Paxton, R. J., March, C. J., and Cerretti, D. P. (1997). A metalloproteinase desintegrin that release tumor-necrosis factor-$\alpha$ from cells. *Nature* **385:**729–733.

Blobel, C. P. (2000). Remarkable roles of proteolysis on and beyond the cell surface. *Curr. Opin. Cell Biol.* **12:**606–612.

Blobel, C. P., Wolfsberg, T. G., Turck, C. W., Myles, D. G., Primakoff, P., and White, J. M. (1992). A potential fusion peptide and an integrin ligand domain in a protein active in sperm–egg fusion. *Nature* **356:**248–252.

Boycott, R., Klenk, H.-D., and Ohuchi, M. (1994). Cell tropism of influenza A virus mediated by hemagglutinin activation at the stage of virus entry. *Virology* **203:**313–319.

Bray, M. (2004). Pathogenesis of filovirus infection in mice. *In* "Ebola and Marburg Viruses: Molecular and Cellular Biology" (H. D. Klenk and H. Feldmann, eds.), pp. 255–277. Horizon Bioscience, Wymondham, UK.

Bukreyev, A. A., Volchkov, V. E., Blinov, V. M., Dryga, S. A., and Netesov, S. V. (1995). The complete nucleotide sequence of the Popp (1967) strain of Marburg virus: A comparison with the Musoke 1980 strain. *Arch. Virol.* **140:**1589–1600.

Bullough, P. A., Hughson, F. M., Skehel, J. J., and Wiley, D. C. (1994). Structure of influenza haemagglutinin at the pH of membrane fusion. *Nature* **371:**37–43.

Buxbaum, J. D., Liu, K.-N., Luo, Y., Slack, J. L., Stocking, K. L., Peschon, J. J., Johnson, R. S., Castner, B. J., Cerretti, D. P., and Black, R. A. (1998). Evidence that tumor necrosis factor $\alpha$ converting enzyme is involved in regulated $\alpha$-secretase cleavage of the Alzheimer amyloid protein precursor. *J. Biol. Chem.* **273:**27765–27767.

Chan, D. C., Fass, D., Berger, J. M., and Kim, P. S. (1997). Core structure of gp41 from the HIV envelope glycoprotein. *Cell* **89:**263–273.

Chan, S. Y., Empig, C. J., Welte, F. J., Speck, R. F., Schmaljohn, A., Kreisberg, J. F., and Goldsmith, M. A. (2001). Folate receptor-$\alpha$ is a cofactor for cellular entry by Marburg and Ebola viruses. *Cell* **106:**117–126.

Chan, S. Y., Ma, M. C., and Goldsmith, M. A. (2000a). Differential induction of cellular detachment by envelope glycoproteins of Marburg and Ebola (Zaire) viruses. *J. Gen. Virol.* **81:**2155–2159.

Chan, S. Y., Speck, R. F., Ma, M. C., and Goldsmith, M. A. (2000b). Distinct mechanisms of entry by envelope glycoproteins of Marburg and Ebola (Zaire) viruses. *J. Virol.* **74:**4933–4937.

Dolnik, O., Volchkova, V. A., Garten, W., Carbonnelle, C., Becker, S., Kahnt, J., Ströher, U., Klenk, H.-D., and Volchkov, V. E. (2004). Ectodomain shedding of the glycoprotein GP of Ebola virus. *EMBO J.* **23:**2175–2184.

Ehlers, M. R. W., Schwager, S. L. U., Scholle, R. R., Manji, G. A., Brandt, W. F., and Riordan, J. F. (1996). Proteolytic release of membrane-bound angiotensin-converting enzyme: Role of the juxtamembrane stalk sequence. *Biochemistry* **35:**9549–9559.

Feldmann, H., Bugany, H., Mahner, F., Klenk, H.-D., Drenckhahn, D., and Schnittler, H. J. (1996). Filovirus-induced endothelial leakage triggered by infected monocytes/macrophages. *J. Virol.* **70:**2208–2214.

Feldmann, H., Mühlberger, E., Randolf, A., Will, C., Kiley, M. P., Sanchez, A., and Klenk, H.-D. (1992). Marburg virus, a filovirus: Messenger RNAs, gene order, and regulatory elements of the replication cycle. *Virus Res.* **24:**1–19.

Feldmann, H., Nichol, S. T., Klenk, H.-D., Peters, C. J., and Sanchez, A. (1994). Characterization of filoviruses based on differences in structure and antigenicity of the virion glycoprotein. *Virology* **199:**469–473.

Feldmann, H., Will, C., Schikore, M., Slenczka, W., and Klenk, H.-D. (1991). Glycosylation and oligomerization of the spike protein of Marburg virus. *Virology* **182:**353–356.

Feng, Y., Broder, C. C., Kennedy, P. E., and Berger, E. A. (1996). HIV-1 entry cofactor: Functional cDNA cloning of a seven-transmembrane, G protein-coupled receptor. *Science* **272:**872–877.

Fisher-Hoch, S. P., Brammer, T. L., Trappier, S. G., Hutwagner, L. C., Farrar, B. B., Ruo, S. L., Brown, B. G., Hermann, L. M., Perez-Oronoz, G. I., Goldsmith, C. S. *et al.* (1992). Pathogenic potential of filoviruses: Role of geographic origin of primate host and virus strain. *J. Infect. Dis.* **166:**753–763.

Funke, C., Becker, B., Dartsch, H., Klenk, H.-D., and Mühlberger, E. (1995). Acylation of the Marburg virus glycoprotein. *Virology* **208:**289–297.

Gallaher, W. R. (1996). Similar structural models of the transmembrane proteins of Ebola and avian sarcoma viruses [letter]. *Cell* **85:**477–478.

Garbutt, M., Liebscher, R., Wahl-Jensen, V., Jones, S., Möller, P., Wagner, R., Volchkov, V., Klenk, H.-D., Feldmann, H., and Ströher, U. (2004). Properties of replication-competent vesicular stomatitis virus vectors expressing glycoproteins of filoviruses and arenaviruses. *J. Virol.* **78:**5458–5465.

Geisbert, T. W., and Jahrling, P. B. (1995). Differentiation of filoviruses by electron microscopy. *Virus Res.* **39:**129–150.

Geyer, H., Will, C., Feldmann, H., Klenk, H.-D., and Geyer, R. (1992). Carbohydrate structure of Marburg virus glycoprotein. *Glycobiology* **2:**299–312.

Hevey, M., Negley, D., Geisbert, J., Jahrling, P., and Schmaljohn, A. (1998). Antigenicity and vaccine potential of Marburg virus glycoprotein expressed by baculovirus recombinants. *Virology* **239:**206–216.

Iba, K., Albrechtsen, R., Gilpin, B., Frohlich, C., Loechel, F., Zolkiewska, A., Ishiguro, K., Kojima, T., Liu, W., Langford, J. K. *et al.* (2000). The cysteine-rich domain of human ADAM 12 supports cell adhesion through syndecans and triggers signalling events that lead to $\beta_1$ integrin-dependent cell spreading. *J. Cell Biol.* **149:**1143–1156.

Ignatyev, G. M. (1999). Immune response to filovirus infections. *Curr. Top. Microbiol. Immunol.* **235:**205–217.

Ito, H., Watanabe, S., Sanchez, A., Whitt, M. A., and Kawaoka, Y. (1999). Mutational analysis of the putative fusion domain of Ebola virus glycoprotein. *J. Virol.* **73:**8907–8912.

Ito, H., Watanabe, S., Takada, A., and Kawaoka, Y. (2001). Ebola virus glycoprotein: Proteolytic processing, acylation, cell tropism, and detection of neutralizing antibodies. *J. Virol.* **75:**1576–1580.

Jeffers, S. A., Sanders, A. D., and Sanchez, A. (2002). Covalent modifications of the Ebola virus glycoprotein. *J. Virol.* **76:**12463–12472.

Joshi, S. B., Dutch, R. E., and Lamb, R. A. (1998). A core trimer of the paramyxovirus fusion protein: Parallels to influenza virus hemagglutinin and HIV-1 gp41. *Virology* **248:**20–34.

Kheramand, F., and Werb, Z. (2002). Shedding light on sheddases: Role in growth and development. *Bioessays* **24:**8–12.

Killar, L., White, J., Black, R., and Peschon, J. (1999). Adamalysins. A family of metzincins including TNF-$\alpha$ converting enzyme (TACE). *Ann. N.Y. Acad. Sci.* **878:**442–452.

Kindzelskii, A. L., Yang, Z., Nabel, G. J., Todd, R. F., III, and Petty, H. R. (2000). Ebola virus secretory glycoprotein (sGP) diminishes Fc$\chi$RIIIB-to-CR3 proximity on neutrophils. *J. Immunol.* **164:**953–958.

Klenk, H.-D., and Garten, W. (1994). Activation cleavage of viral spike proteins by host proteases. *In* "Cellular Receptors for Animal Viruses" (E. Wimmer, ed.), pp. 241–280. Cold Spring Harbor Laboratory Press, Cold Spring Harbor, NY.

Klenk, H.-D., Volchkov, V. E., and Feldmann, H. (1998). Two strings to the bow of Ebola virus. *Nat. Med.* **4:**388–389.

Kolesnikova, L., Berghöfer, B., Bamberg, S., and Becker, S. (2004). Multivesicular bodies as a platform for the formation of Marburgvirus envelope. *J. Virol.* **78:**12277–12287.

Lammich, S., Kojro, E., Postina, R., Gilbert, S., Pfeiffer, R., Jasionowski, M., Haass, C., and Fahrenholz, F. (1999). Constitutive and regulated $\alpha$-secretase cleavage of Alzheimer's amyloid precursor protein by a disintegrin metalloprotease. *Proc. Natl. Acad. Sci. USA* **96:**3922–3927.

Malashkevich, V. N., Schneider, B. J., McNally, M. L., Milhollen, M. A., Pang, J. X., and Kim, P. S. (1999). Core structure of the envelope glycoprotein $GP_2$ from Ebola virus at 1.9-Å resolution. *Proc. Natl. Acad. Sci. USA* **96:**2662–2667.

Mariyankova, R. F., Giushakowa, S. E., Pyzhik, E. V., and Lukashevich, I. S. (1993). Marburg virus penetration into eukaryotic cells. *Vopr. Virusol.* **2:**74–76.

Maruyama, T., Buchmeier, M. J., Parren, P. W., and Burton, D. R. (1998). Ebola virus, neutrophils, and antibody specificity. *Science* **282:**845a.

Maruyama, T., Rodriguez, L. L., Jahrling, P. B., Sanchez, A., Khan, A. S., Nichol, S. T., Petres, C. J., Parren, P. W., and Burton, D. R. (1999). Ebola virus can be effectively neutralized by antibody produced in natural human infection. *J. Virol.* **73:**6024–6030.

Molloy, S. S., Thomas, L., van Slyke, J. K., Stenberg, P. E., and Thomas, G. (1994). Intracellular trafficking and activation of the furin proprotein convertase: Localization to the TGN and recycling from the cell surface. *EMBO J.* **13:**18–33.

Moss, M. L., Jin, S. L., Milla, M. E., Bickett, D. M., Burkhart, W., Cartner, H. L., Chen, W. J., Clay, W. C., Didsbury, J. R., Hassler, D., Hoffman, C. R., Kost, T. A., Lambert, M. H., Leesnitzer, M. A., McCauley, P., McGeehan, G., Mitchell, J., Moyer, M., Pahel., G., Rocque, W., Overton, L. K., Schoenen, F., Seaton, T., Su, J. L., Warner, J., Willars, D., and Becherer, J. D. (1997). Cloning of a disintegrin metalloproteinase that processes precursor tumor-necrosis factor-$\alpha$. *Nature* **385:**733–736.

Moss, M. L., White, J. M., Lambert, M. H., and Andrews, R. C. (2001). TACE and other ADAM proteases as targets for drug discovery. *Drug Discov. Today* **6:**417–426.

Mupapa, K. D., Massamba, M., Kibadi, K., Kuvula, K., Bwaka, A., Kipasa, M., Colebunders, R., Muyembe-Tamfum, J. J., and on behalf of the International Scientific and Technical

Committee (1999). Treatment of Ebola hemorrhagic fever with blood transfusions from convalescent patients. *J. Infect. Dis.* **179**(Suppl. 1)**:**S18–S23.

Nakayama, K. (1997). Furin: A mammalian subtilisin/Kex2p-like endoprotease involved in processing of a wide variety of precursor proteins. *Biochem. J.* **327:**625–635.

Nakayama, K., Kim, W.-S., Torii, S., Hosaka, M., Nakagawa, T., Ikemizu, J., Baba, T., and Murakami, K. (1992). Identification of the fourth member of the mammalian endoprotease family homologous to the yeast Kex2 protease. *J. Biol. Chem.* **267:**5897–5900.

Neumann, G., Feldmann, H., Watanabe, S., Lukahevich, I., and Kawaoka, Y. (2002). Reverse genetics demonstrates that proteolytic processing of the Ebola virus glycoprotein is not essential for replication in cell culture. *J. Virol.* **76:**406–410.

Peschon, J. J., Slack, J. L., Reddy, P., Stocking, K. L., Sunnarborg, S. W., Lee, D. C., Russell, W. E., Castner, B. J., Johnson, R. S., Fitzner, J. N., Boyce, R. W., Nelson, N., Kozlosky, C. J., Wolfson, M. F., Rauch, C. T., Cerretti, D. P., Paxton, R. J., March, C. J., and Black, R. A. (1998). An essential role for ectodomain shedding in mammalian development. *Science* **282:**1281–1284.

Pohlmann, S., Baribaud, F., Lee, B., Leslie, G. J., Sanchez, M. D., Hiebenthal-Millow, K., Munch, J., Kirchhoff, F., and Doms, R. W. (2001). DC-SIGN interactions with human immunodeficiency virus type 1 and 2 and simian immunodeficiency virus. *J. Virol.* **75:**4664–4672.

Primakoff, P., and Myles, D. G. (2000). The ADAM gene family: Surface proteins with adhesion and protease activity. *Trends Genet.* **16:**63–87.

Pushko, P., Bray, M., Ludwig, G. V., Parker, M., Schmaljohn, A., Sanchez, A., Jahrling, P. B., and Smith, J. F. (2000). Recombinant RNA replicons derived from attenuated Venezuelan equine encephalitis virus protect guinea pigs and mice from Ebola hemorrhagic fever virus. *Vaccine* **19:**142–153.

Reddy, P., Slack, J. L., Davis, R., Cerretti, D. P., Kozlosky, C. J., Blanton, R. A., Shows, D., Peschon, J. J., and Black, R. A. (2000). Functional analysis of the domain structure of tumor necrosis factor-$\alpha$ converting enzyme. *J. Biol. Chem.* **275:**14608–14614.

Ruiz-Argüello, M. B., Goni, F. M., Pereira, F. B., and Nieva, J. L. (1998). Phosphatidylinositol-dependent membrane fusion induced by a putative fusogenic sequence of Ebola virus. *J. Virol.* **72:**1775–1781.

Sanchez, A., Kiley, M. P., Holloway, B. P., and Auperin, D. D. (1993). Sequence analysis of the Ebola virus genome: Organization, genetic elements, and comparison with the genome of Marburg virus. *Virus Res.* **29:**215–240.

Sanchez, A., Trappier, S. G., Mahy, B. W. J., Peters, C. J., and Nichol, S. T. (1996). The virion glycoprotein of Ebola viruses are encoded in two reading frames and are expressed through transcriptional editing. *Proc. Natl. Acad. Sci. USA* **93:**3602–3607.

Sanchez, A., Trappier, S. G., Ströher, U., Nichol, S. T., Bowen, M. D., and Feldmann, H. (1998a). Variation in the glycoprotein and VP35 genes of Marburg virus strains. *Virology* **240:**138–146.

Sanchez, A., Yang, Z. Y., Xu, L., Nabel, G. J., Crews, T., and Peters, C. J. (1998b). Biochemical analysis of the secreted and virion glycoproteins of Ebola virus. *J. Virol.* **72:**6442–6447.

Sänger, C., Mühlberger, E., Lötfering, B., Klenk, H.-D., and Becker, S. (2002). The Marburg virus surface protein GP is phosphorylated at its ectodomain. *Virology* **295:**20–29.

Schäfer, W., Stroh, A., Berghöfer, S., Seiler, J., Vey, M., Kruse, M. L., Kern, H.-F., Klenk, H.-D., and Garten, W. (1995). Two independent targeting signals in the cytoplasmic domain determine *trans*-Golgi network localization and endosomal trafficking of the proprotein convertase furin. *EMBO J.* **14:**2424–2435.

Schnittler, H. J., and Feldmann, H. (1999). Molecular pathogenesis of filovirus infections: Role of macrophages and endothelial cells. *Curr. Top. Microbiol. Immunol.* **235:**175–204.

Seidah, N. G., Hamelin, J., Mamarbachi, M., Dong, W., Tadro, H., Mbikay, M., Chretien, M., and Day, R. (1996). cDNA structure, tissue distribution, and chromosomal localization of rat PC7, a novel mammalian proprotein convertase closest to yeast kexin-like proteinases. *Proc. Natl. Acad. Sci. USA* **93:**3388–3393.

Skehel, J. J., Baylay, P. M., Brown, E. B., Martin, S. R., Waterfield, M. D., White, J. M., Wilson, I. A., and Wiley, D. C. (1982). Changes in the conformation of the influenza hemagglutinin at the pH optimum of virus mediated membrane fusion. *Proc. Natl. Acad. Sci. USA* **79:**968–972.

Stieneke-Gröber, A., Vey, M., Angliker, H., Shaw, E., Thomas, G., Roberts, C., Klenk, H.-D., and Garten, W. (1992). Influenza virus hemagglutinin with multibasic cleavage site is activated by furin, a subtilisin-like endoprotease. *EMBO J.* **11:**2407–2414.

Ströher, U., West, E., Bugany, H., Klenk, H. D., Schnittler, H. J., and Feldmann, H. (2001). Infection and activation monocytes by Marburg and Ebola viruses. *J. Virol.* **75:**11025–11033.

Sullivan, N. J., Sanchez, A., Rollin, P. E., Yang, Z. Y., and Nabel, G. J. (2000). Development of a preventive vaccine for Ebola virus infection in primates. *Nature* **408:**605–609.

Takada, A., Robison, C., Goto, H., Sanchez, A., Murti, K. G., Whitt, M. A., and Kawaoka, Y. (1997). A system for functional analysis of Ebola virus glycoprotein. *Proc. Natl. Acad. Sci. USA* **94:**14764–14769.

Takada, A., Watanabe, S., Ito, H., Okazaki, K., Kida, H., and Kawaoka, Y. (2000). Down-regulation of $\beta_1$ integrins by Ebola virus glycoprotein: Implication for virus entry. *Virology* **278:**20–26.

Van den Ouweland, A. M. W., Duijnhoven, H. P., Kreizer, G. D., Dossers, L. C. J., and Van de Ven, W. J. M. (1990). Structural homology between the human fur gene product and the subtilisin-like protease encoded by yeast Kex2. *Nucleic Acids Res.* **18:**644.

Vanderzanden, L., Bray, M., Fuller, D., Roberts, T., Custer, D., Spik, K., Jahrling, P., Huggins, J., Schmaljohn, A., and Schmaljohn, C. (1998). DNA vaccines expressing either the GP or NP genes of Ebola virus protect mice from lethal challenge. *Virology* **246:**134–144.

Van de Ven, W. J., Voorberg, J., Fontijn, R., Pannekoek, H., van den Ouweland, A. M., van Duijnhoven, H. L., Roebroek, A. J., and Siezen, R. J. (1990). Furin is a subtilisin-like proprotein processing enzyme in higher eukaryotes. *Mol. Biol. Rep.* **14:**265–275.

Vey, M., Schäfer, W., Reis, B., Ohuchi, R., Britt, W., Garten, W., Klenk, H.-D., and Radsak, K. (1995). Proteolytic processing of human cytomegalovirus glycoprotein B (gpUL55) is mediated by the human endoprotease furin. *Virology* **206:**746–749.

Vidricaire, G., Denault, J. B., and Leduc, R. (1993). Characterization of a secreted form of human furin endoprotease. *Biochem. Biophys. Res. Commun.* **195:**1011–1018.

Volchkov, V. E., Becker, S., Volchkova, V. A., Ternovoj, V. A., Kotov, A. N., Netesov, S. V., and Klenk, H.-D. (1995). GP mRNA of Ebola virus is edited by the Ebola virus polymerase and by T7 and vaccinia virus polymerases. *Virology* **214:**421–430.

Volchkov, V. E., Blinov, V. M., and Netesov, S. V. (1992). The envelope glycoprotein of Ebola virus contains an immunosuppressive-like domain similar to oncogenic retroviruses. *FEBS Lett.* **305:**181–184.

Volchkov, V. E., Chepurnov, A. A., Volchkove, V. A., Ternovoj, V. A., and Klenk, H.-D. (2000a). Molecular characterization of guinea-pig-adapted variants of Ebola virus. *Virology* **277:**147–155.

Volchkov, V. E., Feldmann, H., Volchkova, V. A., and Klenk, H.-D. (1998a). Processing of the Ebola virus glycoprotein by the proprotein convertase furin. *Proc. Natl. Acad. Sci. USA* **95:**5762–5767.

Volchkov, V. E., Volchkova, V. A., Mühlberger, E., Kolesnikova, L. V., Weik, M., Dolnik, O., and Klenk, H.-D. (2001). Recovery of infectious Ebola virus from complementary DAN: RNA editing of the GP gene and viral cytotoxicity. *Science* **291:**1965–1969.

Volchkov, V. E., Volchkova, V. A., Slenczka, W., Klenk, H.-D., and Feldmann, H. (1998b). Release of viral glycoproteins during Ebola virus infection. *Virology* **245:**110–119.

Volchkov, V. E., Volchkova, V. A., Ströher, U., Cieplik, M., Becker, S., Dolnik, O., Garten, W., Klenk, H.-D., and Feldmann, H. (2000b). Proteolytic processing of Marburg virus glycoprotein. *Virology* **268:**1–6.

Volchkova, V. A., Feldmann, H., Klenk, H.-D., and Volchkov, V. E. (1998). The nonstructural small glycoprotein of Ebola virus is secreted as an antiparallel-orientated homodimer. *Virology* **250:**408–414.

Volchkova, V., Klenk, H.-D., and Volchkov, V. (1999). Δ-Peptide is the carboxy-terminal cleavage fragment of the nonstructural small glycoprotein sGP of Ebola virus. *Virology* **265:**64–171.

Watanabe, S., Takada, A., Watanabe, T., Ito, H., Kida, H., and Kawaoka, Y. (2000). Functional importance of the coiled-coil of the Ebola virus glycoprotein. *J. Virol.* **74:**10194–10201.

Weissenhorn, W., Calder, L. J., Wharton, S. A., Skehel, J. J., and Wiley, D. C. (1998a). The central structural feature of the membrane fusion protein subunit from the Ebola virus glycoprotein is a long triple-stranded coiled coil. *Proc. Natl. Acad. Sci. USA* **95:**6032–6036.

Weissenhorn, W., Carfi, A., Lee, K. H., Skehel, J. J., and Wiley, D. C. (1998b). Crystal structure of the Ebola virus membrane fusion subunit, $GP_2$, from the envelope glycoprotein ectodomain. *Mol. Cell* **2:**605–616.

Weissenhorn, W., Dessen, A., Harrison, S. C., Skehel, J. J., and Wiley, D. C. (1997). Atomic structure of the ectodomain from HIV-1 gp41. *Nature* **387:**426–430.

Will, C., Mühlberger, E., Linder, D., Slenczka, W., Klenk, H.-D., and Feldmann, H. (1993). Marburg virus gene 4 encodes the virion membrane protein, a type I transmembrane glycoprotein. *J. Virol.* **67:**1203–1210.

Wilson, J. A., Hevey, M., Bakken, R., Guest, S., Bray, M., Schmaljohn, A. L., and Hart, M. K. (2000). Epitopes involved in antibody-mediated protection from Ebola virus. *Science* **287:**1664–1666.

Wool-Lewis, R. J., and Bates, P. (1998). Characterization of Ebola virus entry by using pseudotyped viruses: Identification of receptor-deficient cell lines. *J. Virol.* **72:**3155–3160.

Wool-Lewis, R. J., and Bates, P. (1999). Endoproteolytic processing of the Ebola virus envelope glycoprotein: Cleavage is not required for function. *J. Virol.* **73:**1419–1426.

Xu, L., Sanchez, A., Yang, Z. Y., Zaki, S. R., Nabel, E. G., Nichol, S. T., and Nabel, G. J. (1998). Immunization for Ebola virus infection. *Nat. Med.* **4:**37–42.

Yang, Z., Delgado, R., Xu, L., Todd, R. F., Nabek, E. G., Sanchez, A., and Nabel, G. J. (1998). Distinct cellular interactions of secreted and transmembrane Ebola virus glycoproteins. *Science* **279:**1034–1036.

Yang, Z.-Y., Duckers, H. J., Sullivan, N. J., Sanchez, A., Nabel, E., and Nabel, G. J. (2000). Identification of the Ebola virus glycoprotein as the main viral determinant of vascular cell cytotoxicity and injury. *Nat. Med.* **8:**886–889.

Zhang, X. P., Kamata, T., Kokoyama, K., Puzon-McLaughlin, W., and Takada, Y. (1998). Specific interaction of the recombinant disintegrin-like domain of MDC-15 (metargidin, ADAM-15) with integrin $\alpha_v\beta_3$. *J. Biol. Chem.* **273:**7345–7350.

# INFLUENZA VIRUS ASSEMBLY AND BUDDING AT THE VIRAL BUDOZONE

Anthony P. Schmitt* and Robert A. Lamb*,†

*Northwestern University, Evanston, Illinois 60208
†Howard Hughes Medical Institute, Northwestern University, Evanston, Illinois 60208



## I. Introduction

The family *Orthomyxoviridae* contains four genera: *Influenzavirus A*, *Influenzavirus B*, *Influenzavirus C*, and *Thogotovirus*. Influenza A virus is the virus that infects humans and a variety of avian species, horses, and pigs. Influenza A virus caused the major pandemics of respiratory illness during the twentieth century. Influenza B virus strains appear to infect naturally only humans and caused epidemics every few years. Influenza C virus causes more limited outbreaks in humans and also infects pigs. The natural transmission of influenza A, B, and C viruses is by aerosol for humans and nonaquatic hosts whereas transmission is waterborne for aquatic species. Thogoto viruses are transmitted by ticks and replicate in both ticks and in mammalian species and are not discussed further here.

Influenza viruses are enveloped viruses, often spherical (80 to 120 nm in diameter) but also filamentous (several micrometers in length). The lipid envelope is derived from the plasma membrane of the cell in which the virus replicates and is acquired by a budding process




0065-3527/05 $35.00
DOI: 10.1016/S0065-3527(05)64012-2

from the cell plasma membrane as one of the last steps of virus assembly. Inserted into the virion envelope are viral glycoprotein spikes (10 to 14 nm in length). For influenza A and B viruses, the surface spike glycoproteins are hemagglutinin (HA) and neuraminidase (NA). The envelope also contains small amounts of a proton-selective ion channel (M2 for influenza A virus and BM2 for influenza B virus). Influenza C virus contains only one spike glycoprotein, hemagglutinin-esterase-fusion (HEF). Influenza A, B, and C viruses contain a matrix (M1) protein and it is now virtually dogma that the matrix protein underlies the lipid envelope and provides rigidity to the membrane. In addition, it is widely believed that the M1 protein interacts with the cytoplasmic tails of the HA, NA, and M2 (or BM2) proteins and also interacts with the ribonucleoprotein (RNP) structures, thereby organizing the process of virus assembly. The viruses contain a segmented single-stranded RNA genome (influenza A and B viruses, eight RNA segments; and influenza C viruses, seven RNA segments), and each RNA segment forms its own RNP structure. The RNP strands usually exhibit loops at one end and a periodicity of alternating major and minor grooves, suggesting that the structure is formed by a strand that is folded back on itself and then coiled on itself to form a type of twin-stranded helix.

The RNPs contain four proteins and RNA. The major protein is the scaffold protein, nucleocapsid (NP), and each subunit associates with approximately 20 bases of RNA. Associated with the RNPs is the RNA-dependent RNA polymerase (RDRP) that transcribes the genome RNA segments into messenger RNAs. The RDRP complex consists of 3 polymerase proteins, PB1, PB2, and PA, which are present at 30 to 60 copies per virion. The RDRP complex carries out a complex series of reactions including cap binding, endonucleolytic cleavage, RNA synthesis, and polyadenylation (reviewed in Lamb and Krug, 2001).

Influenza virus infection is spread from cell to cell and from host to host in the form of infectious particles that are assembled and released from infected cells. A series of events must occur for production of an infectious influenza virus particle, including the organization and concentration of viral proteins at selected sites on the cell plasma membrane, recruitment of a full complement of eight RNP segments to the assembly sites, and the budding and release of particles by membrane fission. Here, we review progress that has been made toward understanding these steps in influenza virus particle formation. Throughout this review, we use the term influenza virus to mean influenza A virus unless specifically stated. This is because most research on virus assembly has been done studying influenza A virus.

## II. Organizing Assembly

### A. *M1 Protein as the Key Driving Force of Virus Assembly and Determinant of Virion Morphology*

The key tenet of virus assembly and subsequent uncoating in a newly infected cell is that virus assembly and disassembly (uncoating) is a reversible process. A control mechanism must exist to shift the equilibrium of semistable interactions that lead to assembly and disassembly in a newly infected cell. The influenza virus matrix protein M1 has long been thought to have a critical role in organizing virus assembly and disassembly. For assembly, morphological characterizations of influenza virus particles by electron microscopy suggest that M1 protein forms an electron-dense layer underlying the lipid envelope and enclosing the RNP segments (Murti *et al.*, 1992; Nermut, 1972; Ruigrok *et al.*, 1989, 2000; Schulze, 1972) (Fig. 1). These studies indicate that, similar to matrix proteins of other negative-strand RNA viruses (see Schmitt and Lamb, 2004, for a review), the influenza virus M1 protein is positioned in virions so that it has the potential to contact both the RNP segments making up the viral core and the envelope glycoprotein cytoplasmic tails. Thus, the M1 protein is likely to be the key organizer of virus assembly that directs these different viral components to concentrate together at the plasma membranes of infected cells in preparation for virus budding.

Additional evidence supporting the view that the M1 protein is important for directing influenza virus assembly comes from studies of abortively infected cells as well as from studies using temperature-sensitive virus mutants. Abortive replication of avian influenza virus has been studied in mammalian cell lines, and the resulting defect in virus particle formation appears to be due to lack of M1 protein accumulation (Lohmeyer *et al.*, 1979). In addition, an influenza virus mutant has been characterized that has a temperature-sensitive defect in its M1 protein (Li *et al.*, 1995; Rey and Nayak, 1992), and for this mutant virus particle formation is reduced in infected cells at the nonpermissive temperature. These results suggest that assembly and release of influenza virus particles from virus-infected cells require accumulation of functional M1 protein.

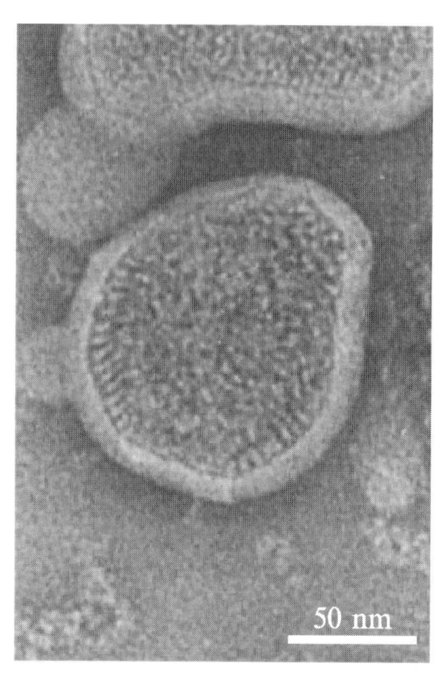


FIG 1. Electron micrograph of a purified influenza A/PR/8/34 virion that has been treated with bromelain to remove the glycoprotein spikes. The negative stain has penetrated the particle to reveal outlines of the presumptive submembranal layer of M1 protein, which is seen as a series of rods that are approximately 60 Å in length. Taken from Ruigrok *et al.* (2000).

Another approach has been employed more recently to study M1 protein function. Influenza virus proteins were expressed from cDNAs in various combinations in cells, and assembly and release of virus-like particles (VLPs) were monitored. For example, expression of all or nearly all the influenza virus proteins from plasmid DNAs results in the release of influenza VLPs, but on omission of M1 protein expression VLP release no longer occurred (Gomez-Puertas *et al.*, 1999, 2000; Mena *et al.*, 1996; Neumann *et al.*, 2000b; Watanabe *et al.*, 2002). Furthermore, expression of M1 protein alone either from plasmid DNA (Gomez-Puertas *et al.*, 2000) or from a recombinant baculovirus (Latham and Galarza, 2001) resulted in the release of spikeless particles, indicating that the M1 protein contains all of the information necessary for assembly and budding of particles from cell membranes. In some cases, replication-incompetent influenza VLPs (containing glycoprotein spikes) have been tested as possible vaccine agents. Not unexpectedly, these VLPs have proved capable of providing

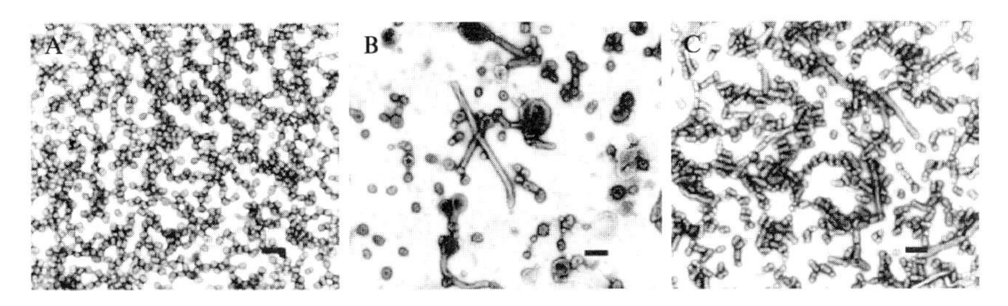


FIG 2. Electron micrographs of purified influenza virus particles. (A) WSN strain, having spherical morphology; (B) Udorn strain, having filamentous morphology; (C) WSN-Mud, in which the M segment of Udorn virus has been transferred into WSN virus, resulting in filamentous particle production. Scale bars: 300 nm. Taken from Bourmakina and Garcia-Sastre (2003).

protective immunity to lethal influenza virus challenge in mice (Watanabe *et al.*, 2002).

A considerable body of evidence indicates the influenza virus M1 protein is an important determinant of virion morphology. Whereas influenza virions are classically described as roughly spherical particles having a diameter of 80–120 nm, this description accurately reflects only a subset of laboratory-adapted influenza virus strains. Fresh clinical isolates of influenza virus contain particles having a wide range of morphologies, with a significant proportion of long, filamentous virions (Chu *et al.*, 1949). In some cases (e.g., A/WSN/33 and A/PR8/34 strains), extensive passaging of these viruses in eggs or in tissue culture has resulted in a morphological shift such that virus preparations now contain mostly spherical particles (Choppin *et al.*, 1960; Kilbourne and Murphy, 1960). In other cases (e.g., A/Udorn/72 and A/Victoria/3/75 strains), virus preparations still contain mostly filamentous particles despite extensive passaging (Elleman and Barclay, 2004) (Fig. 2). A number of studies have used reassortant viruses, or viruses obtained by reverse genetics technology, in an attempt to determine the genetic features of these filamentous viruses that account for the morphological differences. These studies indicate that the M1 gene determines virion morphology (Bourmakina and Garcia-Sastre, 2003; Elleman and Barclay, 2004; Roberts *et al.*, 1998; Smirnov *et al.*, 1991). For example, replacing the A/WSN/33 virus M gene with that of the A/Udorn/72 virus results in acquisition of the filamentous phenotype (Bourmakina and Garcia-Sastre, 2003). Additional support for the notion that spherical versus filamentous virion morphology is specified by the M1 protein was obtained in studies in which the A/Victoria/3/75 M1 protein was expressed alone from cDNA and the released VLPs had a filamentous morphology, similar in morphology to A/Victoria/3/75 virions (Gomez-Puertas *et al.*, 2000).

It is not known how M1 protein contributes in the determination of virion morphology. It has been proposed that filamentous versus spherical morphology could be related to the ratio of M1 protein to NP protein in virions, as filamentous particles have a higher M1:NP protein ratio than do spherical particles (Liu *et al.*, 2002; Roberts *et al.*, 1998). It is possible that virus strains that form filamentous particles have weaker M1–RNP binding (Liu *et al.*, 2002). The actin cytoskeleton has also been proposed as important for the formation of filamentous virions, as both inhibition of actin polymerization by cytochalasin D (Roberts and Compans, 1998) and perturbation of the actin treadmilling cycle by jasplakinolide (Simpson-Holley *et al.*, 2002) disrupt formation of filamentous (but not spherical) virus particles. A possible role for actin filaments in influenza virus assembly has also been suggested on the basis of colocalization studies (Bucher *et al.*, 1989) and biochemical studies showing association of viral proteins with microfilaments (Avalos *et al.*, 1997). Another factor that has been shown to influence influenza virion morphology is host cell type. Production of filamentous viral particles is favored in polarized epithelial cells whereas spherical particle production is favored in nonpolarized cells (Roberts and Compans, 1998).

### *B. Interactions Between Viral Components That Organize Virus Assembly*

Influenza virus assembly requires coordinated localization of different viral components at sites of virus budding, including the viral membrane proteins HA, NA, and M2, which are transported to the plasma membrane of the infected cell by the exocytic pathway, and soluble viral components such as the RNP segments. Coordination is thought to occur as a result of a series of protein–protein and protein–lipid interactions involving M1 protein, as this protein can potentially interact with the viral membrane proteins via their cytoplasmic tails and M1 protein can also potentially interact with the RNP segments that make up the viral core. M1 is a two-domain protein and the atomic structure of the N-terminal domain has been determined by X-ray crystallographic methods (Arzt *et al.*, 2001, 2004; Harris *et al.*, 2001; Sha and Luo, 1997). This domain is made up of two four-helix bundles connected by a helical linker, with a positively charged surface that overlaps a nuclear localization signal

(Fig. 3). A number of investigations of influenza virus assembly have focused on the characterization of interactions involving the M1 protein, including interactions between M1 protein and membranes, interactions between M1 protein and glycoprotein cytoplasmic tails, self-association of M1 protein, and interactions between M1 protein and RNPs.

### *1. Membrane-Binding Properties of M1 Protein*

M1 is a membrane-binding protein that underlies the virion lipid envelope (Nermut, 1972; Ruigrok *et al.*, 2000; Schulze, 1972). When expressed in living cells, M1 protein binds to cellular membranes (Kretzschmar *et al.*, 1996; Zhang and Lamb, 1996), and *in vitro*, the purified M1 protein also binds to membranes (Baudin *et al.*, 2001; Bucher *et al.*, 1980; Gregoriades, 1980; Gregoriades and Frangione, 1981; Ruigrok *et al.*, 2000). Membrane binding by M1 protein is quite strong, and M1 protein was initially thought to interact with membranes as a result of hydrophobic interactions leading to a partial insertion of the protein into the viral lipid bilayer (Bucher *et al.*, 1980; Gregoriades, 1980; Gregoriades and Frangione, 1981). However, more recent experiments based on electron microscopy examination of virions, as well as photolabeling of virions, suggest that interaction of M1 protein with membranes is largely electrostatic (Ruigrok *et al.*, 2000). In addition, purified M1 protein has been found to bind to negatively charged liposomes *in vitro* but not to neutral liposomes, and this binding could be prevented by treatments with salt or high pH, further indicating an important electrostatic component to membrane binding by M1 protein (Ruigrok *et al.*, 2000). Liposome binding *in vitro* was observed for the N-terminal domain of M1 protein, but not the C-terminal domain (Baudin *et al.*, 2001), and it has been proposed that the positively charged patch seen in the atomic structure at the surface of M1 protein could interact with phosphate head groups of membranes (Arzt *et al.*, 2001; Baudin *et al.*, 2001; Sha and Luo, 1997). Studies with M1 protein expressed in living cells from cDNA have also found the protein to be associated with cellular membranes (Kretzschmar *et al.*, 1996; Zhang and Lamb, 1996). In this case, M1 protein could not be extracted from membranes with treatments such as high salt or alkaline pH that characteristically remove peripheral membrane proteins (Kretzschmar *et al.*, 1996; Zhang and Lamb, 1996). However, alterations to multiple hydrophobic regions of M1 protein also failed to reduce membrane association, raising the possibility that no one region of M1 protein is responsible for membrane association and that perhaps both electrostatic and hydrophobic

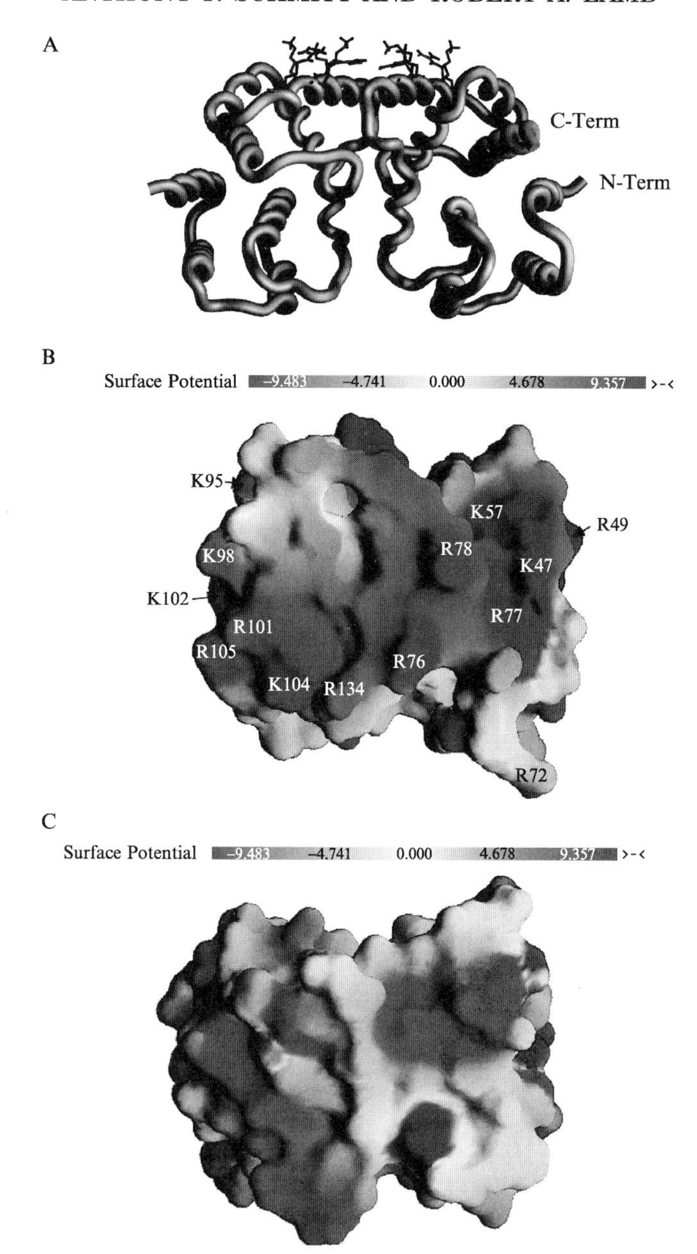


FIG 3. Atomic structure of the influenza virus M1 protein. (A) Structure of a dimer of M1 residues 1–158. Residues that make up the basic surface of the protein are shown. Taken from Lamb and Krug (2001). (B and C) Surface potential of an M1 monomer. Basic residues are indicated in (A), and acidic residues on the opposite side (180° rotation) are shown in (C). Taken from Arzt *et al.* (2001). (See Color Insert.)

components might contribute to membrane binding (Kretzschmar *et al.*, 1996; Zhang and Lamb, 1996).

### 2. *Interactions Between M1 Protein and Glycoprotein Cytoplasmic Tails*

Biochemical evidence to support the notion that M1 protein binds to the influenza virus glycoprotein cytoplasmic tails during virus assembly has been difficult to obtain, most likely because the interactions are weak. One approach that has been used to detect these interactions indirectly has been to monitor incorporation of chimeric glycoproteins having foreign cytoplasmic tail sequences into influenza virions. Wild-type HA glycoprotein when expressed *in trans* in influenza virus infected cells is incorporated efficiently into virions, but a chimeric HA protein having a foreign cytoplasmic tail sequence is incorporated poorly into virions (Naim and Roth, 1993). Furthermore, addition of the HA protein transmembrane (TM) domain and cytoplasmic tail (CT) to a foreign glycoprotein is sufficient to direct efficient incorporation of that protein into influenza virions (Garcia-Sastre *et al.*, 1994) or into VLPs (Latham and Galarza, 2001), suggesting an interaction involving the cytoplasmic tails during virus assembly.

Another approach that has been used to study potential M1 protein–cytoplasmic tail interactions has been to monitor the fraction of M1 protein that is membrane bound in transfected cells, with an increase in membrane binding on coexpression of influenza virus glycoproteins being taken as evidence of M1 protein–glycoprotein interaction. Two independent studies have found that a minority of M1 protein is membrane bound in cells that express M1 protein alone, and that coexpression of either HA glycoprotein or NA glycoprotein results in a significant increase in the membrane association of M1 protein (Enami and Enami, 1996; Gomez-Puertas *et al.*, 2000). However, similar experiments led to conflicting results in two other laboratories, which found a large fraction of M1 protein bound to membrane even when this protein was expressed alone in cells, and found no enhancement of membrane binding on coexpression of the HA or NA glycoprotein (Kretzschmar *et al.*, 1996; Zhang and Lamb, 1996).

The importance of glycoprotein cytoplasmic tails in influenza virus assembly has been tested directly by a genetic approach in which viruses whose glycoproteins lack these cytoplasmic tail sequences were generated. A virus has been isolated that lacks both the HA and NA protein cytoplasmic tails, so that neither protein is expected to be capable of making contacts with M1 protein. This virus was found to have severe assembly defects (Fig. 4). A 10-fold reduction in particle production

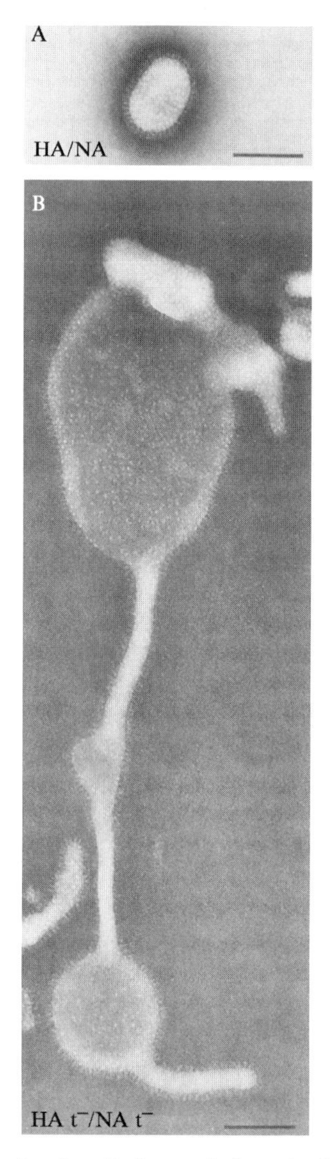


FIG 4. Electron micrographs of purified, negatively stained influenza virions that lack glycoprotein cytoplasmic tails and have distorted morphologies. (A) HA/NA, wild-type virus; (B) HA t-/NA t-, virus lacking both the HA and NA protein cytoplasmic tails. Scale bar: 80 nm. Electron microscopy courtesy of George Leser (Northwestern University).

compared with wild-type virus was observed, and particles that were formed exhibited striking morphological defects (Jin *et al.*, 1997). These distended and irregularly shaped particles were much less infectious than wild-type virions and were defective in genome packaging such that they contained a much broader distribution in the number of packaged RNA segments (Zhang *et al.*, 2000a). Interestingly, these severe defects were observed only when both the HA and NA glycoprotein cytoplasmic tails were deleted. Deletion of only the HA protein cytoplasmic tail led to virus that assembled efficiently and formed particles with the normal spherical morphology (Jin *et al.*, 1994), and deletion of only the NA protein cytoplasmic tail led to defects in virus assembly and particle morphology that were much less severe than those observed in the double mutant virus (Garcia-Sastre and Palese, 1995; Mitnaul *et al.*, 1996). These results suggest that the cytoplasmic domains of influenza virus HA and NA glycoproteins have redundant functions that allow efficient budding of spherically shaped virions.

Additional studies have measured the association of M1 protein specifically with glycolipid-enriched raft membrane microdomains (see Section III) that can be biochemically separated from nonraft membranes on the basis of resistance of the raft membranes to solubilization with nonionic detergents such as Triton X-100 (TX-100) at low temperatures. Influenza virus assembles and buds from raft membranes (Scheiffele *et al.*, 1999; Zhang *et al.*, 2000b) and the influenza viral glycoproteins are sorted intrinsically to raft membranes when expressed in cells (Kundu *et al.*, 1996; Simons and Ikonen, 1997; Skibbens *et al.*, 1989). The influenza virus M1 protein when expressed alone in cells does not bind TX-100-resistant membranes (Ali *et al.*, 2000). Thus, it has been proposed that interactions between viral glycoproteins and M1 protein could induce M1 protein to assemble together with the glycoproteins at raft membranes during viral infection. Consistent with this view, coexpression of either the HA or NA glycoprotein together with M1 protein in transfected cells resulted in redistribution of M1 protein such that a substantial amount of this protein was now bound to raft membrane that resists detergent solubilization (Ali *et al.*, 2000). Similar finding have been made in studies with recombinant influenza virus. M1 protein is found associated mainly with raft membranes in wild-type influenza virus-infected cells. However, on deletion of both the HA and NA glycoprotein cytoplasmic tails, so that these glycoproteins were no longer able to associate with raft membranes efficiently, M1 protein in virus-infected cells was found associated mainly with nonraft membranes (Zhang *et al.*, 2000b), data that support a model in which M1 protein is recruited to raft membranes for virus budding as a result of interactions between M1 protein and the glycoprotein cytoplasmic tails.

### 3. *Self-Association of M1 Protein*

Self-assembly of M1 protein into organized structures has been observed in virus-infected cells, in cells expressing M1 protein alone, and *in vitro* with purified protein. In influenza virus-infected cells and in virions, electron microscopy reveals a layer of protein attached to the inner surface of the membrane that likely consists of M1 protein (Nermut, 1972; Ruigrok *et al.*, 1989, 2000; Schulze, 1972) (see Fig. 1). This M1 protein layer is made up of radially arranged rods about 60 Å in length, touching the membrane with one of its ends (Ruigrok *et al.*, 2000). In cells transfected to produce M1 protein alone, formation of long M1-containing tubular structures in the cell nucleus and cytoplasm has been observed, and it was proposed that these structures are made up exclusively of self-assembled M1 protein (Gomez-Puertas *et al.*, 2000). Purified M1 protein has a tendency to polymerize into flexible ribbon structures *in vitro*, and examination of these ribbons by electron microscopy suggests that they consist of rods about 57 Å long in a side-by-side orientation (Ruigrok *et al.*, 2000).

At present it is not clear what forces drive M1 protein polymerization during virus assembly. M1 protein in which the positively charged surface around the nuclear localization sequence has been altered by mutagenesis was found to be much more soluble than wild-type M1 protein (Baudin *et al.*, 2001), suggesting that this surface could interact with a negatively charged surface in self-polymerization, as was observed in the packing of one of the crystal forms of M1 protein (Arzt *et al.*, 2001). Other studies based on the M1 protein atomic structure suggest that the protein could associate via M1–M1 stacking interactions to form an elongated ribbon of either single or double M1 molecules, with the positively charged areas oriented together on the same side of the oligomer (Harris *et al.*, 2001).

### 4. *Interactions Between M1 Protein and RNPs*

M1–RNP interactions are presumed to be important for packaging of the viral genome segments into budding virions and, in fact, when nucleocapsid segments are isolated from influenza virions, they are found bound with M1 protein (Murti *et al.*, 1992). M1–RNP interactions have also been inferred by experiments showing inhibition of RNP transcription *in vitro* by M1 protein (Elster *et al.*, 1997; Watanabe *et al.*, 1996; Ye *et al.*, 1987, 1989). In addition, reconstitution of M1–RNP complexes has been achieved by incubating virion-derived RNPs that have been stripped of M1 protein by high-salt treatment with recombinant purified M1 protein at physiological salt concentrations (Baudin *et al.*, 2001; Watanabe *et al.*, 1996; Ye *et al.*, 1999). Unfortunately, this reconstitution approach is limited in that it is difficult to know for certain that all the virion-derived M1 protein has been removed by the salt treatment; M1 reassociation with RNPs might be due in part to self-association with residually bound M1 protein. Reconstitution of M1–RNP complexes could be achieved by using the C-terminal domain of M1 protein, but not the N-terminal domain (Baudin *et al.*, 2001), suggesting that the 60-Å-long rods of M1 protein visualized by electron microscopy (Ruigrok *et al.*, 2000) are oriented with the N-terminal domain touching the membrane and the C-terminal domain interacting with RNPs.

M1–RNP interactions are pH dependent. M1 protein can be removed from virion-derived nucleocapsids not only by high salt concentrations but also by low-pH treatment (Zhirnov, 1992) and reconstitution of M1 protein with nucleocapsids was efficient at neutral pH but binding was essentially abolished at low pH (Ye *et al.*, 1999). During viral entry, dissociation of M1 protein from RNP is required so that the RNP can enter the nucleus of the cell (Bui *et al.*, 1996; Helenius, 1992; Martin and Helenius, 1991). The dissociation of M1 protein from the RNP is triggered by the acidification of virions with protons that flow via the M2 proton-selective ion channel from the endosomal lumen to the interior of the internalized virion. When ion channel activity is blocked by the antiviral drug amantadine, viral replication is inhibited because of a block in viral uncoating (Hay, 1992; Helenius, 1992; Lamb *et al.*, 1994; Martin and Helenius, 1991; Sugrue and Hay, 1991; Takeda *et al.*, 2002). These observations underscore the importance of having control mechanisms to shift the equilibrium of interactions that drive virus assembly versus disassembly. Viruses must be assembled in such a way that this assembly can be reversed later on in subsequent infections, and for influenza virus this reversibility is accomplished at least in part through the pH dependence of RNP–M1 protein interactions.

In addition to binding to assembled RNPs, M1 protein can also bind to naked RNA. M1 protein contains a basic stretch of amino acids, 101-RKLKR-105, that can act as a nuclear localization signal (Ye *et al.*, 1995), and this basic stretch is required for M1 protein binding to naked RNA (Elster *et al.*, 1997; Wakefield and Brownlee, 1989; Watanabe *et al.*, 1996; Ye *et al.*, 1989). It is unclear whether M1 protein inside virions is bound to RNP in part through direct binding to the viral RNA, and it has been argued that in fact the negatively charged backbone of the viral RNA is instead more likely bound to the positively charged viral nucleocapsid protein, leaving the positively charged surface of M1 protein free to interact with lipid or to participate in M1–M1 protein interactions (Arzt *et al.*, 2001; Ruigrok *et al.*, 2000).

## III. Selecting Assembly Sites

### *A. Polarized Budding of Influenza Virus from Epithelial Cell Surfaces*

Many viruses including influenza virus infect host organisms by initially infecting epithelial body surfaces such as the epithelial cells of the respiratory tract. These cells are polarized, having distinct apical (outward facing) and basolateral (inward facing) surfaces, and virus budding often occurs asymmetrically in these cells, so that progeny virions are released predominantly from either the apical side (as has been shown to be the case for influenza virus and several paramyxoviruses) or the basolateral side [e.g., vesicular stomatitis virus (VSV) and Marburg virus] (Compans, 1995; Rodriguez-Boulan and Sabatini, 1978; Sanger *et al.*, 2001). Polarized budding is thought to have an important role in modulating viral pathogenicity, as apical budding could favor restriction of the infection to the epithelial cell layer, whereas basolateral budding could allow the infection to spread more readily to underlying tissues and favor systemic infection. This view is supported by results that have been obtained with a mutant Sendai virus, in which virus budding occurs in a nonpolarized fashion rather than apically, and this mutant causes a systemic infection that is more virulent than wild-type Sendai virus infection (Tashiro *et al.*, 1990).

Each of the influenza virus membrane proteins HA (Roth *et al.*, 1983), NA (Jones *et al.*, 1985), and M2 (Hughey *et al.*, 1992) is intrinsically sorted to the apical surface of polarized cells when expressed alone. Unlike basolateral sorting of membrane proteins such as VSV G, which typically rely on the presence of tyrosine-containing signals in the cytoplasmic tail regions to direct polarized sorting (Thomas and Roth, 1994), apical sorting of influenza virus integral membrane proteins occurs efficiently even in the absence of the cytoplasmic tails (Zhang *et al.*, 2000b). Instead, apical targeting has been found to be directed by signals in the TM domains of the glycoproteins HA and NA (Barman *et al.*, 2001; Barman and Nayak, 2000; Kundu *et al.*, 1996;

Lin *et al.*, 1998). These signals are not localized to discrete motifs in the TM domains, but rather seem to be dispersed throughout the lengths of the HA and NA protein TM regions (Barman *et al.*, 2001; Barman and Nayak, 2000; Lin *et al.*, 1998). Amino acids that would reside in the outer leaflet of the membrane have been found to be particularly important for targeting HA protein to the apical membrane (Tall *et al.*, 2003).

To test whether altered sorting of viral glycoproteins could cause a redistribution of virus budding, recombinant viruses were generated. One virus was generated containing HA protein that is distributed in a nonpolarized fashion (Mora *et al.*, 2002), and another virus was generated containing HA protein that is predominantly sorted to the basolateral surface (Barman *et al.*, 2003; Brewer and Roth, 1991). In both cases, budding of virus was still restricted to the apical surface, suggesting that HA protein does not determine the site of influenza virus assembly. Similar observations have been made with measles virus, in which it was found that polarized virus budding is not directed by the viral glycoproteins, but rather is likely directed by the viral matrix protein (Maisner *et al.*, 1998; Moll *et al.*, 2001; Naim *et al.*, 2000).

### *B. Influenza Virus Assembly on Membrane Rafts*

Many cellular membranes contain dynamic liquid-ordered phase microdomains, often called lipid rafts, that are enriched in sphingolipids and cholesterol and that can preferentially incorporate or exclude certain proteins (Brown and London, 2000; Simons and Ikonen, 1997). These lipid microdomains are thought to act as platforms to allow concentration of selected proteins together on membranes, and in many cases rafts appear to be used by enveloped viruses as nucleation points to facilitate the efficient concentration and assembly of viral proteins in preparation for budding (for reviews, see Chazal and Gerlier, 2003; Nayak and Barman, 2002; Suomalainen, 2002).

The precise size of the lipid raft is a topic of much discussion (Munro, 2003; Pralle *et al.*, 2000; Varma and Mayor, 1998). However, from the point of view of influenza virus budding, the direct relationship of the sphingomyelin/cholesterol-enriched budding patch (the viral budozone) to a raft is a semantic argument. It is quite clear that influenza virus assembles at the plasma membrane at discrete patches that are visible by electron microscopy (Fig. 5). Viruses containing mutated HA proteins that fail to concentrate at these sites of budding, assemble and bud inefficiently (Takeda *et al.*, 2003). The initial evidence that

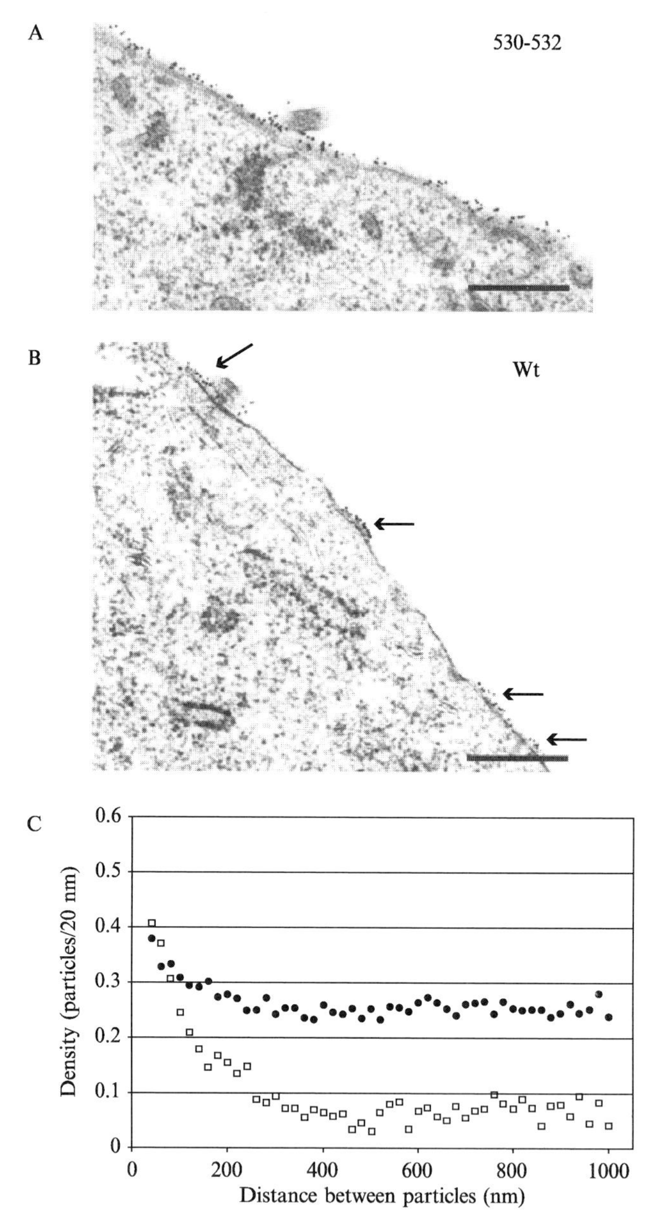


FIG 5. Clustering of raft-associated HA proteins at the surface of infected cells. (A) Uniform distribution of HA protein at the surface of a cell infected with influenza virus harboring HA protein that fails to associate with rafts because of mutations in its transmembrane domain. (B) Clusters of HA protein at the surface of a cell infected with

influenza virus budding occurs from lipid raft structures was obtained through biochemical analysis of purified influenza virions, in which it was found that the HA and NA proteins were bound to membrane that had solubility properties characteristic of membrane rafts. In addition, the lipid composition of purified influenza virus particles is consistent with that of membrane rafts (Scheiffele *et al.*, 1999; Zhang *et al.*, 2000b), suggesting that the virion envelope is composed largely of raft-like membrane and that the virus therefore buds through rafts. In cells infected with influenza virus, both of the viral glycoproteins HA and NA as well as the matrix protein M1 are found associated predominantly with raft membranes (Skibbens *et al.*, 1989; Zhang *et al.*, 2000b), and raft incorporation occurs both in virus strains that bud spherical virions as well as in those that bud filamentous particles (Simpson-Holley *et al.*, 2002).

Influenza virus glycoproteins have intrinsic signals that specify sorting to raft membranes. Expression of either HA protein (Simons and Ikonen, 1997; Skibbens *et al.*, 1989) or NA protein (Kundu *et al.*, 1996) alone in transfected cells results in its association with raft membranes. The signals directing these proteins to associate with lipid rafts have been mapped to the TM domains, and in particular to amino acid residues predicted to span the outer leaflets of the lipid bilayers (Barman *et al.*, 2001; Barman and Nayak, 2000; Kundu *et al.*, 1996; Lin *et al.*, 1998; Scheiffele *et al.*, 1997; Takeda *et al.*, 2003; Tall *et al.*, 2003). Although apical targeting signals also reside in the TM domains of these proteins, it has been possible to construct both influenza virus HA and NA proteins with altered TM domains in which raft association was disrupted but apical targeting was retained (Barman *et al.*, 2001; Barman and Nayak, 2000; Lin *et al.*, 1998; Takeda *et al.*, 2003). It is unclear at this point whether apical sorting and raft association signals are in fact distinct, or alternatively if sorting signal affinity that is required for biochemical recovery of protein attached to detergent-resistant membrane is greater than the affinity that is needed for apical sorting in the cell (Tall *et al.*, 2003). Glycoprotein cytoplasmic tails have also been found to have a role in raft targeting of influenza virus glycoproteins, as raft association (but not apical targeting) was disrupted on removal of the glycoprotein cytoplasmic

wild-type influenza virus. The HA protein associates with lipid rafts. HA protein is visualized with immunogold labeling. Scale bar: 500 nm. (C) Distribution of gold particles measured from approximately 30 random cell profiles as described (Brown and Lyles, 2003). Solid circles represent measurements with nonraft HA virus; open squares represent measurements with wild-type virus. Taken from Takeda *et al.* (2003).

tails both in transfection experiments and in a recombinant virus (Zhang *et al.*, 2000b). It is possible that the changes to the cytoplasmic tails directly alter the conformation of the TM domain, hence mimicking the effect of TM domain mutations. However, for influenza virus HA protein it has also been shown that mutation of its cytoplasmic tail cysteine residues that are normally palmitoylated, leads to decreased raft association of HA protein (B. Chen and R. A. Lamb, unpublished observations). Thus, it seems probable that the signal for raft association involves both the TM domain and palmitoylation. Although the viral matrix protein M1 is found associated with raft membranes in virus-infected cells, this association does not appear to be due to any intrinsic raft sorting signal in the M1 protein, as M1 protein is associated only with detergent-sensitive nonraft membrane when expressed alone in transfected cells (Ali *et al.*, 2000). Rather, M1 protein appears to be recruited to sites of virus assembly on membrane rafts by interactions involving the glycoprotein cytoplasmic tails (see Section II).

Budding from membrane rafts may provide a mechanism for controlling which viral proteins are incorporated efficiently into virions. Influenza virions normally are densely packed with glycoproteins HA (approximately 500 trimers per virion) and NA, but contain a relatively small amount of the M2 membrane protein (5 to 15 tetramers per virion) despite abundant cell surface expression of M2 protein in infected cells (Zebedee and Lamb, 1988). Whereas HA and NA proteins associate with membrane rafts, M2 protein is largely excluded from raft microdomains, hence providing an explanation for its considerable underrepresentation in virus as compared with HA and NA proteins (Zhang *et al.*, 2000b). Alteration of the HA protein TM domain in recombinant virus so that HA is no longer targeted to raft membranes results in a dramatic (3 logs) decrease in maximum virus titer, a reduction in budding efficiency, and altered polypeptide composition of virus particles such that the amount of HA protein is reduced and the amount of NA protein is increased, leading to a defect in the step of virus to cell membrane fusion (Takeda *et al.*, 2003). Recombinant viruses having TM domain-altered NA proteins defective for raft targeting also exhibit severe replication defects and altered polypeptide composition with less NA protein incorporation, causing aggregation of virions at the cell surface, a defect that could be partially reversed on addition of exogenous bacterial neuraminidase enzyme (Barman *et al.*, 2004). Disruption of raft targeting for both the HA and NA proteins in recombinant virus as a result of deletion of both glycoprotein cytoplasmic tails results in a substantial reduction in the preferential incorporation of these glycoproteins into virions compared with incorporation of the M2 membrane protein, providing further data in support of the suggestion that raft targeting may provide a mechanism for the selective inclusion or exclusion of viral proteins into budding virions (Zhang *et al.*, 2000b).

## IV. Packaging Cargo

The influenza virus genome is divided into eight viral RNA segments, and each of these segments, in the form of a viral RNP, must be represented in a virus particle to allow normal infectivity. Viral RNPs are assembled in the nuclei of infected cells, and consist of repeating subunits of NP protein in a helical configuration with the RNA wrapped around the NP subunits such that the RNA is susceptible to RNase digestion (Heggeness *et al.*, 1982; Lamb and Choppin, 1983; Murti *et al.*, 1988; Pons *et al.*, 1969). Electron micrographs indicate the RNP forms a superhelical complex, twisted on itself with the 5′ and 3′ ends of the RNA coming together (Compans *et al.*, 1972; Pons *et al.*, 1969). Associated with each RNP segment is the viral RNA polymerase complex. Immunogold staining of the RNPs suggests that the polymerase complex is located at one end of the RNA (Murti *et al.*, 1988), and three-dimensional reconstruction of recombinant RNPs containing short model vRNAs suggests the viral polymerase complex forms a compact structure that interacts with template-associated nucleoprotein monomers in the RNP (Area *et al.*, 2004; Martin-Benito *et al.*, 2001). M1 protein is imported to the nucleus, where it assembles together with RNPs, and reconstitution experiments *in vitro* suggest that assembly of viral RNA with NP protein to produce the helical RNP structure is facilitated by the presence of M1 protein (Huang *et al.*, 2001). Nuclear import of M1 protein has been proposed to be mediated by a nuclear localization signal that comprises a basic surface on the protein (Ye *et al.*, 1989, 1995), the same surface that has also been found to be important for electrostatic interaction of M1 protein with membranes (Ruigrok *et al.*, 2000). However, it should be noted that M1 protein is of a size such that it could diffuse into the nucleus. Studies using reverse genetics have confirmed the importance of this basic region of M1 protein in virus replication (Hui *et al.*, 2003; Liu and Ye, 2002), but it is not yet clear whether these replication defects were caused by loss of nuclear localization function or by disruption of other proposed functions of this domain, such as membrane binding, RNA–RNP binding, and pinching off of virus particles. Once synthesized, RNP complexes must be transported from the nucleus to the cytoplasm so that they can participate in virus assembly. M1 protein binding has been found to promote RNP nuclear export (Bui *et al.*, 1996, 2000; Whittaker *et al.*, 1996) and also prevents RNPs from reentering the nucleus (Martin and Helenius, 1991). Other factors in addition to M1 protein have also been found to play a role in nuclear export of viral RNPs, including the viral NS2 (NEP) protein that may act as an adapter, linking M1–RNP complexes with the nuclear pore complex (O'Neill *et al.*, 1998), as well as cellular chromosome region maintenance 1 (CRM1) protein that may contribute to RNP export by binding to viral NP protein directly (Elton *et al.*, 2001), or by binding to NS2 protein (Neumann *et al.*, 2000a).

One model for packaging of influenza virus segments into virions predicts random incorporation of the different segments into particles. The feasibility of such a mechanism is supported by observations that influenza virions can contain more than eight viral RNPs (Enami *et al.*, 1991). Aberrant virus assembly as a result of deletion of the glycoprotein cytoplasmic tails leads to even greater variation in the number of RNA segments incorporated into virions (Zhang *et al.*, 2000a). However, newer observations favor a different model for the packaging of viral genome segments in which specific structures in each viral RNA lead to the individual incorporation of segments into virus particles (Fujii *et al.*, 2003). Observations in support of this view were made in the course of studying influenza viruses containing large deletions in their NA segments, so that no functional NA proteins could be produced (Hughes *et al.*, 2000; Liu and Air, 1993; Liu *et al.*, 1995; Yang *et al.*, 1997). These internally deleted NA viral RNA segments were found to be stably maintained by the viruses even after multiple passages, suggesting that the presence of these altered viral RNA segments could be important to the virus for reasons other than expression of functional NA protein. Further investigation revealed that both ends of the NA viral RNA coding region were critically important for the incorporation of this segment into virions and that a foreign gene could be stably maintained as an influenza viral RNA segment so long as it was flanked at both ends by short sequences taken from the coding region of the NA viral RNA (Fujii *et al.*, 2003). Similar signals have also been identified in the coding region of the HA viral RNA, and a virus has been produced that stably expresses both VSV G, whose gene is flanked by sequences normally used for incorporation of the HA segment, and green fluorescent protein (GFP), whose gene is flanked by sequences normally used for incorporation of the NA segment (Watanabe *et al.*, 2003). In earlier studies in which foreign genes were flanked only by the 3′ and 5′ noncoding regions of influenza viral RNA segments, loss of the new segments was observed after several passages (Luytjes *et al.*, 1989). These new results suggest that coding sequences at the ends of influenza viral RNA segments have an important function as packaging signals to allow the selective incorporation of these segments into budding virions, and it has been suggested that this selection could occur through a base-pairing mechanism in which the eight viral RNA segments are joined to form a larger unit that is then recognized at the plasma membrane during virus assembly (Fujii *et al.*, 2003). Characterization of these packaging signals for influenza virus has facilitated studies on the use of this virus as a vaccine vector for expression of foreign genes (Shinya *et al.*, 2004), and also has allowed the generation of chimeric viruses containing segments derived from both type A and type B influenza viruses (Horimoto *et al.*, 2003).

## V. Releasing Virions

### *A. Membrane Fission*

Among the least understood aspects of influenza virus replication are the events that occur subsequent to assembly of viral proteins and genome segments together at cellular membranes; events that allow the formation of virus particles and the pinching off of these particles by membrane fission. Much of our understanding of these late events in virus budding has come from advances in the study of retrovirus budding. In many retroviruses, protein–protein interaction domains called late domains have been identified that manipulate host machinery to allow release of virus particles (reviewed in Freed, 2002; Pornillos *et al.*, 2002). Cellular factors recruited by late domains in many cases are part of the vacuolar protein-sorting (VPS) pathway of the cell, and once recruited, this interconnected network of proteins appears to assist in the release of viral particles in a way that is similar to its normal function of sorting selected proteins into cellular vesicles that bud into endosomes to form multivesicular bodies (Martin-Serrano *et al.*, 2003; Pornillos *et al.*, 2002; von Schwedler *et al.*, 2003).

Multiple types of late domains having distinct amino acid sequences have been characterized. One type contains the consensus amino acid sequence P(T/S)AP and was first identified in the p6 region of the human immunodeficiency virus type 1 (HIV-1) Gag polyprotein (Gottlinger *et al.*, 1991; Huang *et al.*, 1995). A second type of late domain containing the consensus sequence PPxY was identified in the p2b region of the Rous sarcoma virus (RSV) Gag polyprotein (Wills *et al.*, 1994; Xiang *et al.*, 1996), and a third late domain having the sequence YPDL was identified in the p9 region of equine infectious anemia virus Gag polyprotein (Puffer *et al.*, 1997). Each of these late domains functions to recruit a different host factor to facilitate virus budding. The cellular partner protein for P(T/S)AP late domains has been identified as Tsg101 (Garrus *et al.*, 2001; Martin-Serrano *et al.*, 2001; VerPlank *et al.*, 2001), and the partner protein for the YPDL late domain is now thought to be AIP1 (Martin-Serrano *et al.*, 2003; Strack *et al.*, 2003; Vincent *et al.*, 2003; von Schwedler *et al.*, 2003). Both Tsg101 and AIP1 are part of the cellular VPS pathway that allows formation of multivesicular bodies. Although the biologically relevant cellular partner protein for PPxY late domains has not yet been defined unequivocally, this domain has been shown to interact with WW domains from a variety of cellular proteins, including the cellular Nedd4-related E3 ubiquitin ligases (Harty *et al.*, 1999; Kikonyogo *et al.*, 2001; Timmins *et al.*, 2003).

Matrix proteins of negative-strand RNA viruses VSV (Craven *et al.*, 1999; Harty *et al.*, 1999; Jayakar *et al.*, 2000) and Ebola virus (Harty *et al.*, 2000; Licata *et al.*, 2003; Martin-Serrano *et al.*, 2001, 2004) contain the same P(T/S)AP and PPxY late domain sequences that were originally defined in retroviruses, and these late domains have been found to be important for the budding of these viruses, indicating that the strategy of host factor recruitment via late domains to assist in particle release is conserved even among distantly related viruses. Interestingly, the influenza virus M1 protein lacks any of the late domain sequences that have been characterized so far, suggesting that if late steps for influenza virus budding are in fact similar to the late steps of retrovirus, rhabdovirus, and filovirus budding, the late domain sequence used to recruit cellular budding machinery must be different. Ubiquitin does not appear to be involved in influenza virus budding, as depletion of cellular ubiquitin levels using the proteasome inhibitors MG-132 and lactacystin fails to block influenza virus budding (Hui and Nayak, 2001; Khor *et al.*, 2003), in contrast to the situation with other viruses such as HIV-1, RSV, and VSV that use P(T/S)AP and PPxY-type late domains and whose budding is sensitive to ubiquitin depletion (Harty *et al.*, 2001; Patnaik *et al.*, 2000; Schubert *et al.*, 2000). It has been suggested that in influenza virus replication, ubiquitin may instead play a role during low pH-mediated virus entry by allowing sorting to occur within the multivesicular body/late endosome (Khor *et al.*, 2003). Evidence of involvement of late domains in influenza virus budding has been obtained in studies that were initially meant to examine the role of the basic surface/nuclear localization signal of M1 protein in the influenza virus life cycle (Hui *et al.*, 2003). Some alterations to this basic stretch of amino acids led to arrested virus budding and production of filamentous particles that appeared to contain multiple spherical particles joined together. A putative late domain sequence overlapping this basic stretch was identified having amino acid sequence YRKL. Lack of the YRKL sequence could be compensated for by addition of known late domain sequences PTAP or YPDL, but not by addition of PPPY (Hui *et al.*, 2003). These results suggest influenza virus budding may in fact rely on recruitment of host factors to facilitate membrane fission, similar to other enveloped viruses, but this recruitment may be mediated by a YRKL late domain sequence that is different from other late domain sequences that have been characterized so far.

### *B. Preventing Reattachment and Aggregation of Particles*

Once formed and released from cells by membrane fission, influenza virus particles are designed to mediate the safe transit of viral genomes into new target cells. Attachment to target cells is directed by the HA glycoprotein, which binds to terminal sialic acid residues on cell surface glycoproteins and glycolipids. Virus particles are then internalized by endocytosis and subsequent entry via membrane fusion is triggered by the low-pH environment of cellular endosomes (Skehel and Wiley, 2000). The viral NA glycoprotein possesses receptor-destroying sialidase activity. This sialidase activity is seemingly counterproductive but in fact it is quite important to prevent newly formed virus particles from reattaching to infected cells and failing to be released, and to prevent particles from aggregating with each other. Hence, temperature-sensitive defects in NA protein function cause defects at the restrictive temperature in the ability of virus particles to separate from infected cells and from each other (Palese *et al.*, 1974; Shibata *et al.*, 1993). Similarly, introduction of a large internal deletion within the NA gene, induced by growth of virus in the presence of exogenous bacterial sialidase together with an antibody to NA protein, results in aggregation of virions and failure of virus particles to separate from host cells when this virus is grown in the absence of exogenously added neuraminidase enzyme (Liu and Air, 1993; Liu *et al.*, 1995; Yang *et al.*, 1997). Internal deletion of the NA gene that abolished sialidase activity was also found to occur on replication of influenza virus in cell lines in which expression of terminal sialic acid residues on the cell surface is drastically reduced (Hughes *et al.*, 2001). This finding suggests that when sialidase activity is not required, it may hinder virus replication by removing receptor molecules that are limited in supply (Hughes *et al.*, 2001). NA protein function was also found to be important for the release of noninfectious influenza virus-like particles from transfected cells, as omission of NA protein expression from these cells resulted in aggregation and cell surface attachment of the particles (Gomez-Puertas *et al.*, 1999, 2000; Mena *et al.*, 1996). When influenza virus with a defective NA gene was passaged in the presence of decreasing amounts of bacterial sialidase, adapted virus was eventually obtained that could replicate in the absence of exogenous sialidase with efficient release of particles and without aggregation (Hughes *et al.*, 2000). This virus still lacked functional sialidase activity. Adaptation resulted from mutations to the HA gene that decreased its affinity for sialic acid-containing receptors (Hughes *et al.*, 2000). This finding suggests that there is a balance between receptor-binding affinity and receptor-destroying activity that must be maintained by the virus for optimal replication, and that shifting this balance to favor receptor binding over receptor destruction causes reattachment and/or self-attachment of progeny virions, which inhibits infection of new cells.

## VI. Perspectives

Much progress has been made in understanding events that occur during the assembly and release of influenza virus particles, but many exciting aspects of this complex process remain to be deciphered. For example, (1) although structural information at the atomic level has provided valuable insight into how M1 protein might organize virus assembly, this structural information is presently limited to the N-terminal half of the M1 protein, and a more complete picture of the assembly process could be obtained if the entire M1 protein structure were to be solved together with viral RNP; (2) new genetic and biochemical strategies have been used to provide evidence in support of the long-suspected notion that influenza virus glycoprotein cytoplasmic tails play a role in virus assembly by making important contacts with internal viral components such as M1 protein. However, the region of M1 protein that participates in these interactions, and the specific nature of these interactions at atomic resolution, must be defined if we are to more fully understand this important step in the assembly process; (3) it is now understood that lipid rafts can serve as platforms for virus assembly by concentrating selected viral proteins into specific sites on the plasma membrane in preparation for virus budding. The mechanism by which the HA and NA glycoproteins select lipid rafts for preferential sorting remains elusive, however; (4) like other enveloped viruses, influenza virus particles appear to be released with the help of cellular machinery that is recruited via viral late budding domains. However, the specific host protein(s) recruited during influenza virus assembly and the way in which these proteins contribute to virus release once they have arrived at virus assembly sites remain to be determined; and (5) progress has been made in the identification of packaging signals that direct incorporation of the segmented influenza virus genome into budding virions. It should now be possible to determine more precisely how these signals function to allow recognition of the RNA segments by the virus assembly apparatus before budding. We look forward to these and other new advances as research in this increasingly dynamic field continues.

## Acknowledgments

Research in the authors' laboratory was supported in part by research grants R37 AI-20201 and R01 AI-23173 from the National Institute of Allergy and Infectious Diseases. A.P.S. was an Associate and R.A.L. is an Investigator with the Howard Hughes Medical Institute.

## References

Ali, A., Avalos, R. T., Ponimaskin, E., and Nayak, D. P. (2000). Influenza virus assembly: Effect of influenza virus glycoproteins on the membrane association of M1 protein. *J. Virol.* **74:**8709–8719.

Area, E., Martin-Benito, J., Gastaminza, P., Torreira, E., Valpuesta, J. M., Carrascosa, J. L., and Ortin, J. (2004). 3D structure of the influenza virus polymerase complex: Localization of subunit domains. *Proc. Natl. Acad. Sci. USA* **101:**308–313.

Arzt, S., Baudin, F., Barge, A., Timmins, P., Burmeister, W. P., and Ruigrok, R. W. (2001). Combined results from solution studies on intact influenza virus M1 protein and from a new crystal form of its N-terminal domain show that M1 is an elongated monomer. *Virology* **279:**439–446.

Arzt, S., Petit, I., Burmeister, W. P., Ruigrok, R. W., and Baudin, F. (2004). Structure of a knockout mutant of influenza virus M1 protein that has altered activities in membrane binding, oligomerisation and binding to NEP (NS2). *Virus Res.* **99:**115–119.

Avalos, R. T., Yu, Z., and Nayak, D. P. (1997). Association of influenza virus NP and M1 proteins with cellular cytoskeletal elements in influenza virus-infected cells. *J. Virol.* **71:**2947–2958.

Barman, S., Adhikary, L., Chakrabarti, A. K., Bernas, C., Kawaoka, Y., and Nayak, D. P. (2004). Role of transmembrane domain and cytoplasmic tail amino acid sequences of influenza a virus neuraminidase in raft association and virus budding. *J. Virol.* **78:**5258–5269.

Barman, S., Adhikary, L., Kawaoka, Y., and Nayak, D. P. (2003). Influenza A virus hemagglutinin containing basolateral localization signal does not alter the apical budding of a recombinant influenza A virus in polarized MDCK cells. *Virology* **305:**138–152.

Barman, S., Ali, A., Hui, E. K., Adhikary, L., and Nayak, D. P. (2001). Transport of viral proteins to the apical membranes and interaction of matrix protein with glycoproteins in the assembly of influenza viruses. *Virus Res.* **77:**61–69.

Barman, S., and Nayak, D. P. (2000). Analysis of the transmembrane domain of influenza virus neuraminidase, a type II transmembrane glycoprotein, for apical sorting and raft association. *J. Virol.* **74:**6538–6545.

Baudin, F., Petit, I., Weissenhorn, W., and Ruigrok, R. W. (2001). *In vitro* dissection of the membrane and RNP binding activities of influenza virus M1 protein. *Virology* **281:**102–108.

Bourmakina, S. V., and Garcia-Sastre, A. (2003). Reverse genetics studies on the filamentous morphology of influenza A virus. *J. Gen. Virol.* **84:**517–527.

Brewer, C. B., and Roth, M. G. (1991). A single amino acid change in the cytoplasmic domain alters the polarized delivery of influenza virus hemagglutinin. *J. Cell Biol.* **114:**413–421.

Brown, D. A., and London, E. (2000). Structure and function of sphingolipid- and cholesterol-rich membrane rafts. *J. Biol. Chem.* **275:**17221–17224.

Brown, E. L., and Lyles, D. S. (2003). A novel method for analysis of membrane microdomains: Vesicular stomatitis virus glycoprotein microdomains change in size during infection, and those outside of budding sites resemble sites of virus budding. *Virology* **310:**343–358.

Bucher, D., Popple, S., Baer, M., Mikhail, A., Gong, Y. F., Whitaker, C., Paoletti, E., and Judd, A. (1989). M protein (M1) of influenza virus: Antigenic analysis and intracellular localization with monoclonal antibodies. *J. Virol.* **63:**3622–3633.

Bucher, D. J., Kharitonenkov, I. G., Zakomirdin, J. A., Grigoriev, V. B., Klimenko, S. M., and Davis, J. F. (1980). Incorporation of influenza virus M-protein into liposomes. *J. Virol.* **36:**586–590.

Bui, M., Whittaker, G., and Helenius, A. (1996). Effect of M1 protein and low pH on nuclear transport of influenza virus ribonucleoproteins. *J. Virol.* **70:**8391–8401.

Bui, M., Wills, E. G., Helenius, A., and Whittaker, G. R. (2000). Role of the influenza virus M1 protein in nuclear export of viral ribonucleoproteins. *J. Virol.* **74:**1781–1786.

Chazal, N., and Gerlier, D. (2003). Virus entry, assembly, budding, and membrane rafts. *Microbiol. Mol. Biol. Rev.* **67:**226–237.

Choppin, P. W., Murphy, J. S., and Tamm, I. (1960). Studies of two kinds of virus particles which comprise influenza A2 virus strains. III. Morphological characteristics: Independence to morphological and functional traits. *J. Exp. Med.* **112:**945–952.

Chu, C. M., Dawson, I. M., and Elford, W. J. (1949). Filamentous forms associated with newly isolated influenza virus. *Lancet* **1:**602–603.

Compans, R. W. (1995). Virus entry and release in polarized epithelial cells. *Curr. Top. Microbiol. Immunol.* **202:**209–219.

Compans, R. W., Content, J., and Duesberg, P. H. (1972). Structure of the ribonucleoprotein of influenza virus. *J. Virol.* **10:**795–800.

Craven, R. C., Harty, R. N., Paragas, J., Palese, P., and Wills, J. W. (1999). Late domain function identified in the vesicular stomatitis virus M protein by use of rhabdovirus-retrovirus chimeras. *J. Virol.* **73:**3359–3365.
Elleman, C. J., and Barclay, W. S. (2004). The M1 matrix protein controls the filamentous phenotype of influenza A virus. *Virology* **321:**144–153.
Elster, C., Larsen, K., Gagnon, J., Ruigrok, R. W., and Baudin, F. (1997). Influenza virus M1 protein binds to RNA through its nuclear localization signal. *J. Gen. Virol.* **78:**1589–1596.
Elton, D., Simpson-Holley, M., Archer, K., Medcalf, L., Hallam, R., McCauley, J., and Digard, P. (2001). Interaction of the influenza virus nucleoprotein with the cellular CRM1-mediated nuclear export pathway. *J. Virol.* **75:**408–419.
Enami, M., and Enami, K. (1996). Influenza virus hemagglutinin and neuraminidase glycoproteins stimulate the membrane association of the matrix protein. *J. Virol.* **70:**6653–6657.
Enami, M., Sharma, G., Benham, C., and Palese, P. (1991). An influenza virus containing nine different RNA segments. *Virology* **185:**291–298.
Freed, E. O. (2002). Viral late domains. *J. Virol.* **76:**4679–4687.
Fujii, Y., Goto, H., Watanabe, T., Yoshida, T., and Kawaoka, Y. (2003). Selective incorporation of influenza virus RNA segments into virions. *Proc. Natl. Acad. Sci. USA* **100:**2002–2007.
Garcia-Sastre, A., Muster, T., Barclay, W. S., Percy, N., and Palese, P. (1994). Use of a mammalian internal ribosomal entry site element for expression of a foreign protein by a transfectant influenza virus. *J. Virol.* **68:**6254–6261.
Garcia-Sastre, A., and Palese, P. (1995). The cytoplasmic tail of the neuraminidase protein of influenza A virus does not play an important role in the packaging of this protein into viral envelopes. *Virus Res.* **37:**37–47.
Garrus, J. E., von Schwedler, U. K., Pornillos, O. W., Morham, S. G., Zavitz, K. H., Wang, H. E., Wettstein, D. A., Stray, K. M., Cote, M., Rich, R. L., Myszka, D. G., and Sundquist, W. I. (2001). Tsg101 and the vacuolar protein sorting pathway are essential for HIV-1 budding. *Cell* **107:**55–65.
Gomez-Puertas, P., Albo, C., Perez-Pastrana, E., Vivo, A., and Portela, A. (2000). Influenza virus matrix protein is the major driving force in virus budding. *J. Virol.* **74:**11538–11547.
Gomez-Puertas, P., Mena, I., Castillo, M., Vivo, A., Perez-Pastrana, E., and Portela, A. (1999). Efficient formation of influenza virus-like particles: Dependence on the expression levels of viral proteins. *J. Gen. Virol.* **80:**1635–1645.
Gottlinger, H. G., Dorfman, T., Sodroski, J. G., and Haseltine, W. A. (1991). Effect of mutations affecting the p6 gag protein on human immunodeficiency virus particle release. *Proc. Natl. Acad. Sci. USA* **88:**3195–3199.
Gregoriades, A. (1980). Interaction of influenza M protein with viral lipid and phosphatidylcholine vesicles. *J. Virol.* **36:**470–479.
Gregoriades, A., and Frangione, B. (1981). Insertion of influenza M protein into the viral lipid bilayer and localization of site of insertion. *J. Virol.* **40:**323–328.
Harris, A., Forouhar, F., Qiu, S., Sha, B., and Luo, M. (2001). The crystal structure of the influenza matrix protein M1 at neutral pH: M1–M1 protein interfaces can rotate in the oligomeric structures of M1. *Virology* **289:**34–44.
Harty, R. N., Brown, M. E., McGettigan, J. P., Wang, G., Jayakar, H. R., Huibregtse, J. M., Whitt, M. A., and Schnell, M. J. (2001). Rhabdoviruses and the cellular ubiquitin–proteasome system: A budding interaction. *J. Virol.* **75:**10623–10629.

Harty, R. N., Brown, M. E., Wang, G., Huibregtse, J., and Hayes, F. P. (2000). A PPxY motif within the VP40 protein of Ebola virus interacts physically and functionally with a ubiquitin ligase: Implications for filovirus budding. *Proc. Natl. Acad. Sci. USA* **97:**13871–13876.

Harty, R. N., Paragas, J., Sudol, M., and Palese, P. (1999). A proline-rich motif within the matrix protein of vesicular stomatitis virus and rabies virus interacts with WW domains of cellular proteins: Implications for viral budding. *J. Virol.* **73:**2921–2929.

Hay, A. J. (1992). The action of adamantanamines against influenza A viruses: Inhibition of the $M_2$ ion channel protein. *Semin. Virol.* **3:**21–30.

Heggeness, M. H., Smith, P. R., Ulmanen, I., Krug, R. M., and Choppin, P. W. (1982). Studies on the helical nucleocapsid of influenza virus. *Virology* **118:**466–470.

Helenius, A. (1992). Unpacking the incoming influenza virus. *Cell* **69:**577–578.

Horimoto, T., Takada, A., Iwatsuki-Horimoto, K., Hatta, M., Goto, H., and Kawaoka, Y. (2003). Generation of influenza A viruses with chimeric (type A/B) hemagglutinins. *J. Virol.* **77:**8031–8038.

Huang, M., Orenstein, J. M., Martin, M. A., and Freed, E. O. (1995). p6Gag is required for particle production from full-length human immunodeficiency virus type 1 molecular clones expressing protease. *J. Virol.* **69:**6810–6818.

Huang, X., Liu, T., Muller, J., Levandowski, R. A., and Ye, Z. (2001). Effect of influenza virus matrix protein and viral RNA on ribonucleoprotein formation and nuclear export. *Virology* **287:**405–416.

Hughes, M. T., Matrosovich, M., Rodgers, M. E., McGregor, M., and Kawaoka, Y. (2000). Influenza A viruses lacking sialidase activity can undergo multiple cycles of replication in cell culture, eggs, or mice. *J. Virol.* **74:**5206–5212.

Hughes, M. T., McGregor, M., Suzuki, T., Suzuki, Y., and Kawaoka, Y. (2001). Adaptation of influenza A viruses to cells expressing low levels of sialic acid leads to loss of neuraminidase activity. *J. Virol.* **75:**3766–3770.

Hughey, P. G., Compans, R. W., Zebedee, S. L., and Lamb, R. A. (1992). Expression of the influenza A virus M2 protein is restricted to apical surfaces of polarized epithelial cells. *J. Virol.* **66:**5542–5552.

Hui, E. K., Barman, S., Yang, T. Y., and Nayak, D. P. (2003). Basic residues of the helix six domain of influenza virus M1 involved in nuclear translocation of M1 can be replaced by PTAP and YPDL late assembly domain motifs. *J. Virol.* **77:**7078–7092.

Hui, E. K., and Nayak, D. P. (2001). Role of ATP in influenza virus budding. *Virology* **290:**329–341.

Jayakar, H. R., Murti, K. G., and Whitt, M. A. (2000). Mutations in the PPPY motif of vesicular stomatitis virus matrix protein reduce virus budding by inhibiting a late step in virion release. *J. Virol.* **74:**9818–9827.

Jin, H., Leser, G. P., and Lamb, R. A. (1994). The influenza virus hemagglutinin cytoplasmic tail is not essential for virus assembly or infectivity. *EMBO J.* **13:**5504–5515.

Jin, H., Leser, G. P., Zhang, J., and Lamb, R. A. (1997). Influenza virus hemagglutinin and neuraminidase cytoplasmic tails control particle shape. *EMBO J.* **16:**1236–1247.

Jones, L. V., Compans, R. W., Davis, A. R., Bos, T. J., and Nayak, D. P. (1985). Surface expression of influenza virus neuraminidase, an amino-terminally anchored viral membrane glycoprotein, in polarized epithelial cells. *Mol. Cell. Biol.* **5:**2181–2189.

Khor, R., McElroy, L. J., and Whittaker, G. R. (2003). The ubiquitin-vacuolar protein sorting system is selectively required during entry of influenza virus into host cells. *Traffic* **4:**857–868.
Kikonyogo, A., Bouamr, F., Vana, M. L., Xiang, Y., Aiyar, A., Carter, C., and Leis, J. (2001). Proteins related to the Nedd4 family of ubiquitin protein ligases interact with the L domain of Rous sarcoma virus and are required for Gag budding from cells. *Proc. Natl. Acad. Sci. USA* **98:**11199–11204.
Kilbourne, E. D., and Murphy, J. S. (1960). Genetic studies of influenza viruses. I. Viral morphology and growth capacity as exchangeable genetic traits: Rapid *in ovo* adaptation of early passage Asian strain isolates by combination with PR8. *J. Exp. Med.* **111:**387–406.
Kretzschmar, E., Bui, M., and Rose, J. K. (1996). Membrane association of influenza virus matrix protein does not require specific hydrophobic domains or the viral glycoproteins. *Virology* **220:**37–45.
Kundu, A., Avalos, R. T., Sanderson, C. M., and Nayak, D. P. (1996). Transmembrane domain of influenza virus neuraminidase, a type II protein, possesses an apical sorting signal in polarized MDCK cells. *J. Virol.* **70:**6508–6515.
Lamb, R. A., and Choppin, P. W. (1983). The gene structure and replication of influenza virus. *Annu. Rev. Biochem.* **52:**467–506.
Lamb, R. A., and Krug, R. M. (2001). *Orthomyxoviridae:* The viruses and their replication. *In* "Field's Virology" (D. Knipe and P. Howley, eds.), 4th Ed., pp. 1487–1532. Lippincott Williams & Wilkins, Philadelphia, PA.
Lamb, R. A., Holsinger, L. J., and Pinto, L. H. (1994). The influenza A virus $M_2$ ion channel protein and its role in the influenza virus life cycle. *In* "Receptor-mediated virus entry into cells" (E. Wimmer, ed.), pp. 303–321. Cold Spring Harbor Laboratory Press, Cold Spring Harbor, NY.
Latham, T., and Galarza, J. M. (2001). Formation of wild-type and chimeric influenza virus-like particles following simultaneous expression of only four structural proteins. *J. Virol.* **75:**6154–6165.
Li, S., Xu, M., and Coelingh, K. (1995). Electroporation of influenza virus ribonucleoprotein complexes for rescue of the nucleoprotein and matrix genes. *Virus Res.* **37:**153–161.
Licata, J. M., Simpson-Holley, M., Wright, N. T., Han, Z., Paragas, J., and Harty, R. N. (2003). Overlapping motifs (PTAP and PPEY) within the Ebola virus VP40 protein function independently as late budding domains: Involvement of host proteins TSG101 and VPS-4. *J. Virol.* **77:**1812–1819.
Lin, S., Naim, H. Y., Rodriguez, A. C., and Roth, M. G. (1998). Mutations in the middle of the transmembrane domain reverse the polarity of transport of the influenza virus hemagglutinin in MDCK epithelial cells. *J. Cell Biol.* **142:**51–57.
Liu, C., and Air, G. M. (1993). Selection and characterization of a neuraminidase-minus mutant of influenza virus and its rescue by cloned neuraminidase genes. *Virology* **194:**403–407.
Liu, C., Eichelberger, M. C., Compans, R. W., and Air, G. M. (1995). Influenza type A virus neuraminidase does not play a role in viral entry, replication, assembly, or budding. *J. Virol.* **69:**1099–1106.
Liu, T., Muller, J., and Ye, Z. (2002). Association of influenza virus matrix protein with ribonucleoproteins may control viral growth and morphology. *Virology* **304:**89–96.
Liu, T., and Ye, Z. (2002). Restriction of viral replication by mutation of the influenza virus matrix protein. *J. Virol.* **76:**13055–13061.

Lohmeyer, J., Talens, L. T., and Klenk, H. D. (1979). Biosynthesis of the influenza virus envelope in abortive infection. *J. Gen. Virol.* **42:**73–88.

Luytjes, W., Krystal, M., Enami, M., Pavin, J. D., and Palese, P. (1989). Amplification, expression, and packaging of foreign gene by influenza virus. *Cell* **59:**1107–1113.

Maisner, A., Klenk, H., and Herrler, G. (1998). Polarized budding of measles virus is not determined by viral surface glycoproteins. *J. Virol.* **72:**5276–5278.

Martin, K., and Helenius, A. (1991). Nuclear transport of influenza virus ribonucleoproteins: The viral matrix protein (M1) promotes export and inhibits import. *Cell* **67:**117–130.

Martin-Benito, J., Area, E., Ortega, J., Llorca, O., Valpuesta, J. M., Carrascosa, J. L., and Ortin, J. (2001). Three-dimensional reconstruction of a recombinant influenza virus ribonucleoprotein particle. *EMBO Rep.* **2:**313–317.

Martin-Serrano, J., Perez-Caballero, D., and Bieniasz, P. D. (2004). Context-dependent effects of L domains and ubiquitination on viral budding. *J. Virol.* **78:**5554–5563.

Martin-Serrano, J., Yarovoy, A., Perez-Caballero, D., and Bieniasz, P. D. (2003). Divergent retroviral late-budding domains recruit vacuolar protein sorting factors by using alternative adaptor proteins. *Proc. Natl. Acad. Sci. USA* **100:**12414–12419.

Martin-Serrano, J., Zang, T., and Bieniasz, P. D. (2001). HIV-1 and Ebola virus encode small peptide motifs that recruit Tsg101 to sites of particle assembly to facilitate egress. *Nat. Med.* **7:**1313–1319.

Mena, I., Vivo, A., Perez, E., and Portela, A. (1996). Rescue of a synthetic chloramphenicol acetyltransferase RNA into influenza virus-like particles obtained from recombinant plasmids. *J. Virol.* **70:**5016–5024.

Mitnaul, L. J., Castrucci, M. R., Murti, K. G., and Kawaoka, Y. (1996). The cytoplasmic tail of influenza A virus neuraminidase (NA) affects NA incorporation into virions, virion morphology, and virulence in mice but is not essential for virus replication. *J. Virol.* **70:**873–879.

Moll, M., Klenk, H. D., Herrler, G., and Maisner, A. (2001). A single amino acid change in the cytoplasmic domains of measles virus glycoproteins H and F alters targeting, endocytosis, and cell fusion in polarized Madin-Darby canine kidney cells. *J. Biol. Chem.* **276:**17887–17894.

Mora, R., Rodriguez-Boulan, E., Palese, P., and Garcia-Sastre, A. (2002). Apical budding of a recombinant influenza A virus expressing a hemagglutinin protein with a basolateral localization signal. *J. Virol.* **76:**3544–3553.

Munro, S. (2003). Lipid rafts: Elusive or illusive? *Cell* **115:**377–388.

Murti, K. G., Brown, P. S., Bean, W. J., Jr., and Webster, R. G. (1992). Composition of the helical internal components of influenza virus as revealed by immunogold labeling/electron microscopy. *Virology* **186:**294–299.

Murti, K. G., Webster, R. G., and Jones, I. M. (1988). Localization of RNA polymerases on influenza viral ribonucleoproteins by immunogold labeling. *Virology* **164:**562–566.

Naim, H. Y., Ehler, E., and Billeter, M. A. (2000). Measles virus matrix protein specifies apical virus release and glycoprotein sorting in epithelial cells. *EMBO J.* **19:**3576–3585.

Naim, H. Y., and Roth, M. G. (1993). Basis for selective incorporation of glycoproteins into the influenza virus envelope. *J. Virol.* **67:**4831–4841.

Nayak, D. P., and Barman, S. (2002). Role of lipid rafts in virus assembly and budding. *Adv. Virus Res.* **58:**1–28.

Nermut, M. V. (1972). Further investigation on the fine structure of influenza virus. *J. Gen. Virol.* **17:**317–331.

Neumann, G., Hughes, M. T., and Kawaoka, Y. (2000a). Influenza A virus NS2 protein mediates vRNP nuclear export through NES-independent interaction with hCRM1. *EMBO J.* **19:**6751–6758.

Neumann, G., Watanabe, T., and Kawaoka, Y. (2000b). Plasmid-driven formation of influenza virus-like particles. *J. Virol.* **74:**547–551.

O'Neill, R. E., Talon, J., and Palese, P. (1998). The influenza virus NEP (NS2 protein) mediates the nuclear export of viral ribonucleoproteins. *EMBO J.* **17:**288–296.

Palese, P., Tobita, K., Ueda, M., and Compans, R. W. (1974). Characterization of temperature sensitive influenza virus mutants defective in neuraminidase. *Virology* **61:**397–410.

Patnaik, A., Chau, V., and Wills, J. W. (2000). Ubiquitin is part of the retrovirus budding machinery. *Proc. Natl. Acad. Sci. USA* **97:**13069–13074.

Pons, M. W., Schulze, I. T., Hirst, G. K., and Hauser, R. (1969). Isolation and characterization of the ribonucleoprotein of influenza virus. *Virology* **39:**250–259.

Pornillos, O., Garrus, J. E., and Sundquist, W. I. (2002). Mechanisms of enveloped RNA virus budding. *Trends Cell Biol.* **12:**569–579.

Pralle, A., Keller, P., Florin, E. L., Simons, K., and Horber, J. K. (2000). Sphingolipid–cholesterol rafts diffuse as small entities in the plasma membrane of mammalian cells. *J. Cell Biol.* **148:**997–1008.

Puffer, B. A., Parent, L. J., Wills, J. W., and Montelaro, R. C. (1997). Equine infectious anemia virus utilizes a YXXL motif within the late assembly domain of the Gag p9 protein. *J. Virol.* **71:**6541–6546.

Rey, O., and Nayak, D. P. (1992). Nuclear retention of M1 protein in a temperature-sensitive mutant of influenza (A/WSN/33) virus does not affect nuclear export of viral ribonucleoproteins. *J. Virol.* **66:**5815–5824.

Roberts, P. C., and Compans, R. W. (1998). Host cell dependence of viral morphology. *Proc. Natl. Acad. Sci. USA* **95:**5746–5751.

Roberts, P. C., Lamb, R. A., and Compans, R. W. (1998). The M1 and M2 proteins of influenza A virus are important determinants in filamentous particle formation. *Virology* **240:**127–137.

Rodriguez-Boulan, E., and Sabatini, D. D. (1978). Asymmetric budding of viruses in epithelial monlayers: A model system for study of epithelial polarity. *Proc. Natl. Acad. Sci. USA* **75:**5071–5075.

Roth, M. G., Compans, R. W., Giusti, L., Davis, A. R., Nayak, D. P., Gething, M. J., and Sambrook, J. (1983). Influenza virus hemagglutinin expression is polarized in cells infected with recombinant SV40 viruses carrying cloned hemagglutinin DNA. *Cell* **33:**435–443.

Ruigrok, R. W., Barge, A., Durrer, P., Brunner, J., Ma, K., and Whittaker, G. R. (2000). Membrane interaction of influenza virus M1 protein. *Virology* **267:**289–298.

Ruigrok, R. W., Calder, L. J., and Wharton, S. A. (1989). Electron microscopy of the influenza virus submembranal structure. *Virology* **173:**311–316.

Sanger, C., Muhlberger, E., Ryabchikova, E., Kolesnikova, L., Klenk, H. D., and Becker, S. (2001). Sorting of Marburg virus surface protein and virus release take place at opposite surfaces of infected polarized epithelial cells. *J. Virol.* **75:**1274–1283.

Scheiffele, P., Rietveld, A., Wilk, T., and Simons, K. (1999). Influenza viruses select ordered lipid domains during budding from the plasma membrane. *J. Biol. Chem.* **274:**2038–2044.

Scheiffele, P., Roth, M. G., and Simons, K. (1997). Interaction of influenza virus haemagglutinin with sphingolipid-cholesterol membrane domains via its transmembrane domain. *EMBO J.* **16:**5501–5508.

Schmitt, A. P., and Lamb, R. A. (2004). Escaping from the cell: Assembly and budding of negative-strand RNA viruses. *Curr. Top. Microbiol. Immunol.* **283:**145–196.

Schubert, U., Ott, D. E., Chertova, E. N., Welker, R., Tessmer, U., Princiotta, M. F., Bennink, J. R., Krausslich, H. G., and Yewdell, J. W. (2000). Proteasome inhibition interferes with Gag polyprotein processing, release, and maturation of HIV-1 and HIV-2. *Proc. Natl. Acad. Sci. USA* **97:**13057–13062.

Schulze, I. T. (1972). The structure of influenza virus. II. A model based on the morphology and composition of subviral particles. *Virology* **47:**181–196.

Sha, B., and Luo, M. (1997). Structure of a bifunctional membrane-RNA binding protein, influenza virus matrix protein M1. *Nat. Struct. Biol.* **4:**239–244.

Shibata, S., Yamamoto-Goshima, F., Maeno, K., Hanaichi, T., Fujita, Y., Nakajima, K., Imai, M., Komatsu, T., and Sugiura, S. (1993). Characterization of a temperature-sensitive influenza B virus mutant defective in neuraminidase. *J. Virol.* **67:**3264–3273.

Shinya, K., Fujii, Y., Ito, H., Ito, T., and Kawaoka, Y. (2004). Characterization of a neuraminidase-deficient influenza a virus as a potential gene delivery vector and a live vaccine. *J. Virol.* **78:**3083–3088.

Simons, K., and Ikonen, E. (1997). Functional rafts in cell membranes. *Nature* **387:**569–572.

Simpson-Holley, M., Ellis, D., Fisher, D., Elton, D., McCauley, J., and Digard, P. (2002). A functional link between the actin cytoskeleton and lipid rafts during budding of filamentous influenza virions. *Virology* **301:**212–225.

Skehel, J. J., and Wiley, D. C. (2000). Receptor binding and membrane fusion in virus entry: The influenza hemagglutinin. *Annu. Rev. Biochem.* **69:**531–569.

Skibbens, J. E., Roth, M. G., and Matlin, K. S. (1989). Differential extractability of influenza virus hemagglutinin during intracellular transport in polarized epithelial cells and nonpolar fibroblasts. *J. Cell Biol.* **108:**821–832.

Smirnov, Y. A., Kuznetsova, M. A., and Kaverin, N. V. (1991). The genetic aspects of influenza virus filamentous particle formation. *Arch. Virol.* **118:**279–284.

Strack, B., Calistri, A., Craig, S., Popova, E., and Gottlinger, H. G. (2003). AIP1/ALIX is a binding partner for HIV-1 p6 and EIAV p9 functioning in virus budding. *Cell* **114:**689–699.

Sugrue, R. J., and Hay, A. J. (1991). Structural characteristics of the M2 protein of influenza A viruses: Evidence that it forms a tetrameric channel. *Virology* **180:**617–624.

Suomalainen, M. (2002). Lipid rafts and assembly of enveloped viruses. *Traffic* **3:**705–709.

Takeda, M., Leser, G. P., Russell, C. J., and Lamb, R. A. (2003). Influenza virus hemagglutinin concentrates in lipid raft microdomains for efficient viral fusion. *Proc. Natl. Acad. Sci. USA* **100:**14610–14617.

Takeda, M., Pekosz, A., Shuck, K., Pinto, L. H., and Lamb, R. A. (2002). Influenza a virus $M_2$ ion channel activity is essential for efficient replication in tissue culture. *J. Virol.* **76:**1391–1399.

Tall, R. D., Alonso, M. A., and Roth, M. G. (2003). Features of influenza HA required for apical sorting differ from those required for association with DRMs or MAL. *Traffic* **4:**838–849.

Tashiro, M., Yamakawa, M., Tobita, K., Seto, J. T., Klenk, H. D., and Rott, R. (1990). Altered budding site of a pantropic mutant of Sendai virus, F1-R, in polarized epithelial cells. *J. Virol.* **64:**4672–4677.

Thomas, D. C., and Roth, M. G. (1994). The basolateral targeting signal in the cytoplasmic domain of glycoprotein G from vesicular stomatitis virus resembles a variety of intracellular targeting motifs related by primary sequence but having diverse targeting activities. *J. Biol. Chem.* **269:**15732–15739.

Timmins, J., Schoehn, G., Ricard-Blum, S., Scianimanico, S., Vernet, T., Ruigrok, R. W., and Weissenhorn, W. (2003). Ebola virus matrix protein VP40 interaction with human cellular factors Tsg101 and Nedd4. *J. Mol. Biol.* **326:**493–502.

Varma, R., and Mayor, S. (1998). GPI-anchored proteins are organized in submicron domains at the cell surface. *Nature* **394:**798–801.

VerPlank, L., Bouamr, F., LaGrassa, T. J., Agresta, B., Kikonyogo, A., Leis, J., and Carter, C. A. (2001). Tsg101, a homologue of ubiquitin-conjugating (E2) enzymes, binds the L domain in HIV type 1 Pr55(Gag). *Proc. Natl. Acad. Sci. USA* **98:**7724–7729.

Vincent, O., Rainbow, L., Tilburn, J., Arst, H. N., Jr., and Penalva, M. A. (2003). YPXL/I is a protein interaction motif recognized by *Aspergillus* PalA and its human homologue, AIP1/Alix. *Mol. Cell. Biol.* **23:**1647–1655.

von Schwedler, U. K., Stuchell, M., Muller, B., Ward, D. M., Chung, H. Y., Morita, E., Wang, H. E., Davis, T., He, G. P., Cimbora, D. M., Scott, A., Krausslich, H. G., Kaplan, J., Morham, S. G., and Sundquist, W. I. (2003). The protein network of HIV budding. *Cell* **114:**701–713.

Wakefield, L., and Brownlee, G. G. (1989). RNA-binding properties of influenza A virus matrix protein M1. *Nucleic Acids Res.* **17:**8569–8580.

Watanabe, K., Handa, H., Mizumoto, K., and Nagata, K. (1996). Mechanism for inhibition of influenza virus RNA polymerase activity by matrix protein. *J. Virol.* **70:**241–247.

Watanabe, T., Watanabe, S., Neumann, G., Kida, H., and Kawaoka, Y. (2002). Immunogenicity and protective efficacy of replication-incompetent influenza virus-like particles. *J. Virol.* **76:**767–773.

Watanabe, T., Watanabe, S., Noda, T., Fujii, Y., and Kawaoka, Y. (2003). Exploitation of nucleic acid packaging signals to generate a novel influenza virus-based vector stably expressing two foreign genes. *J. Virol.* **77:**10575–10583.

Whittaker, G., Bui, M., and Helenius, A. (1996). The role of nuclear import and export in influenza virus infection. *Trends Cell Biol.* **6:**67–71.

Wills, J. W., Cameron, C. E., Wilson, C. B., Xiang, Y., Bennett, R. P., and Leis, J. (1994). An assembly domain of the Rous sarcoma virus Gag protein required late in budding. *J. Virol.* **68:**6605–6618.

Xiang, Y., Cameron, C. E., Wills, J. W., and Leis, J. (1996). Fine mapping and characterization of the Rous sarcoma virus Pr76gag late assembly domain. *J. Virol.* **70:**5695–5700.

Yang, P., Bansal, A., Liu, C., and Air, G. M. (1997). Hemagglutinin specificity and neuraminidase coding capacity of neuraminidase-deficient influenza viruses. *Virology* **229:**155–165.

Ye, Z. P., Baylor, N. W., and Wagner, R. R. (1989). Transcription-inhibition and RNA-binding domains of influenza A virus matrix protein mapped with anti-idiotypic antibodies and synthetic peptides. *J. Virol.* **63:**3586–3594.

Ye, Z., Liu, T., Offringa, D. P., McInnis, J., and Levandowski, R. A. (1999). Association of influenza virus matrix protein with ribonucleoproteins. *J. Virol.* **73:**7467–7473.

Ye, Z. P., Pal, R., Fox, J. W., and Wagner, R. R. (1987). Functional and antigenic domains of the matrix (M1) protein of influenza A virus. *J. Virol.* **61:**239–246.

Ye, Z., Robinson, D., and Wagner, R. R. (1995). Nucleus-targeting domain of the matrix protein (M1) of influenza virus. *J. Virol.* **69:**1964–1970.

Zebedee, S. L., and Lamb, R. A. (1988). Influenza A virus M2 protein: Monoclonal antibody restriction of virus growth and detection of M2 in virions. *J. Virol.* **62:**2762–2772.

Zhang, J., and Lamb, R. A. (1996). Characterization of the membrane association of the influenza virus matrix protein in living cells. *Virology* **225:**255–266.

Zhang, J., Leser, G. P., Pekosz, A., and Lamb, R. A. (2000a). The cytoplasmic tails of the influenza virus spike glycoproteins are required for normal genome packaging. *Virology* **269:**325–334.

Zhang, J., Pekosz, A., and Lamb, R. A. (2000b). Influenza virus assembly and lipid raft microdomains: A role for the cytoplasmic tails of the spike glycoproteins. *J. Virol.* **74:**4634–4644.

Zhirnov, O. P. (1992). Isolation of matrix protein M1 from influenza viruses by acid-dependent extraction with nonionic detergent. *Virology* **186:**324–330.

# INDEX

## D



## E

## F

## G

# I

## O



## P

## S

## T



## U



## V

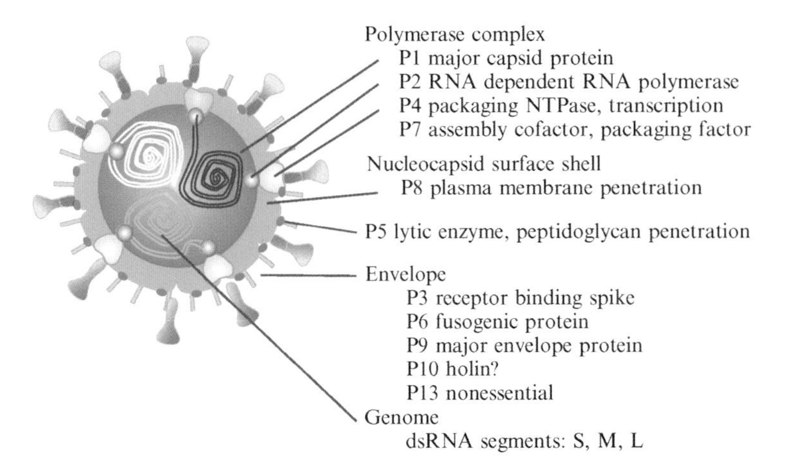


PORANEN *ET AL.*, FIG 1. Architecture of $\phi$6 virion and proposed functions of the proteins.

PORANEN *ET AL.*, FIG 2. Overview of the $\phi$6 assembly pathway. The empty procapsid of $\phi$6 is composed of four proteins (A). P1 forms the structural skeleton of the particle, which is stabilized by P7. The enzymatic components, the packaging NTPase P4 hexamer and the polymerase P2 monomer, are located at or near the fivefold axes. One of the P4 hexamers differs from the others in its physical properties so that P4 is more tightly bound to the procapsid (special vertex, green) than are the others (yellow). The special vertex is active during the packaging of the single-stranded genomic segments (B–E), while transcription is dependent on the P4 hexamers at the other vertices (F). Three structures have been described for the polymerase complex: the empty compressed particle (A), the expanded dodecahedral particle (E), and the mature particle with the rounded appearance (F). In addition, it is suggested that the particle undergoes conformational changes that switch the specificity of the RNA-binding site on the outer surface of the particle (B–D; indicated in green, yellow, and purple outlines). The packaging of the segments is initiated from the 5′ end, which contains a segment-specific *pac* site. The empty particle is preferentially in a conformation that has high-affinity binding site for the s segment (B). The binding or the packaging of the s segment induces a conformational change in the ssRNA-binding site so that the particle preferentially binds the m segment (C), which is subsequently packaged. The m segment then induces a switch in conformation that has high affinity for the l segment (D). The 5′ end of the l segment carries a signal that is needed for the initiation of minus-strand synthesis, and we suggest that the 5′ end of the l segment induces an expansion of the particle (E). After completion of minus-strand synthesis on the l segment the particle can initiate transcription. At this stage the dsRNA density within the expanded

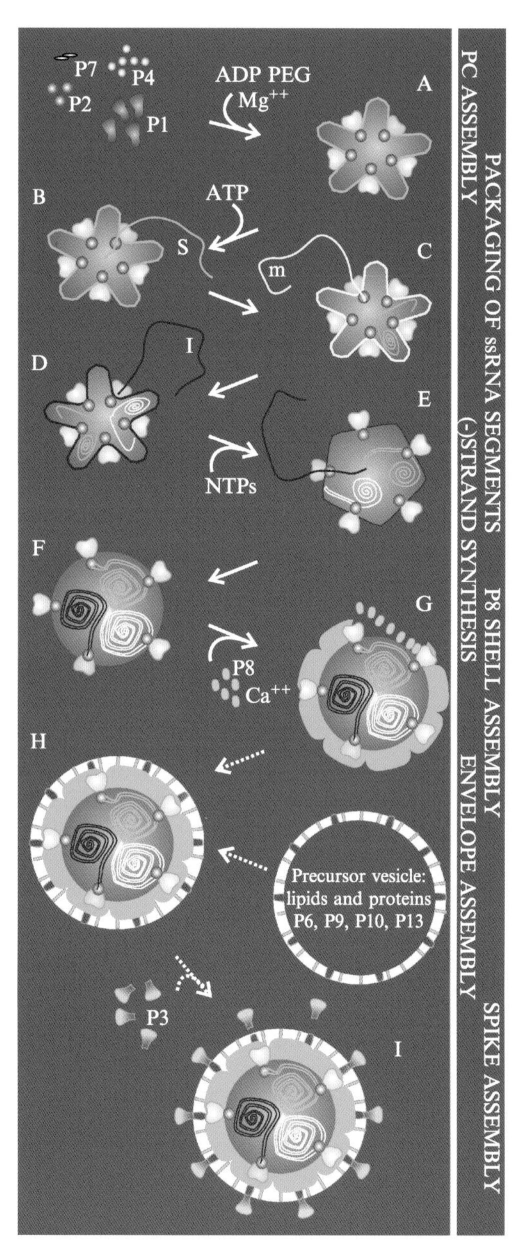


dodecahedral particle reaches its upper limits and we propose that the particle expands to the rounded conformation (F). The nucleocapsid shell (P8) is subsequently assembled around the polymerase complex (G) and the nucleocapsid is enveloped (H). Virions attain infectivity by acquisition of receptor-binding spikes (P3) (I).

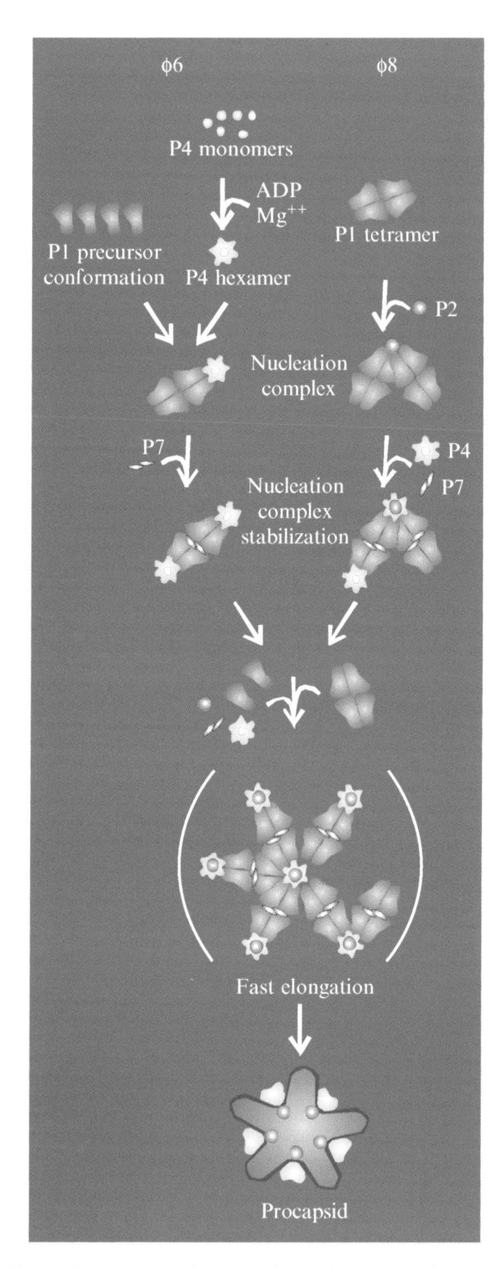


PORANEN *ET AL.*, FIG 3. Summary and comparison of the $\phi6$ and $\phi8$ *in vitro* assembly pathways. Both pathways share a P1 tetramer as an obligatory intermediate and proceed via nucleation-limited polymerization. P1 and P4 association into a nucleation complex initiates $\phi6$ procapsid assembly, whereas the nucleation complex of $\phi8$ contains P1 and P2. In both cases the nucleation complex is then stabilized by P4 and P7 and the procapsid shell is rapidly completed by addition of individual building blocks.

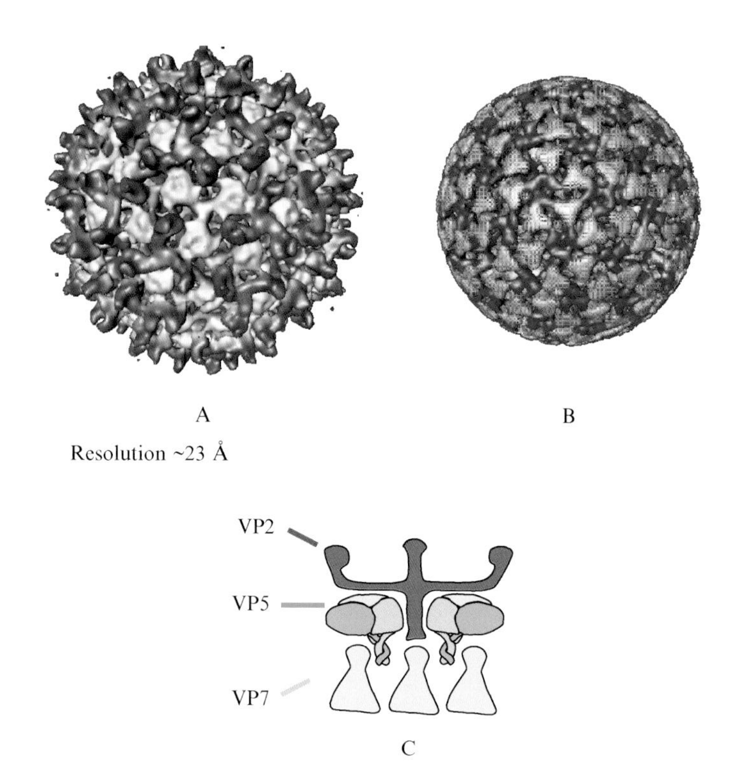


Roy, Fig 2. Surface representations of the three-dimensional cryo-EM structures (22-Å resolution) of BTV. (A) Whole particle showing sail-shaped triskelion propellers (VP2 trimers) in red and globular domains (VP5 trimers) in yellow. (B) A particle with cut-off VP2 density, showing arrangement of globular domains (magenta) close to the surface VP7 layer of the core (purple). (C) Schematic showing a side view of VP2 and VP5 organization in relation to each other and the core surface, depicted from image reconstruction of cryo-EM analysis.

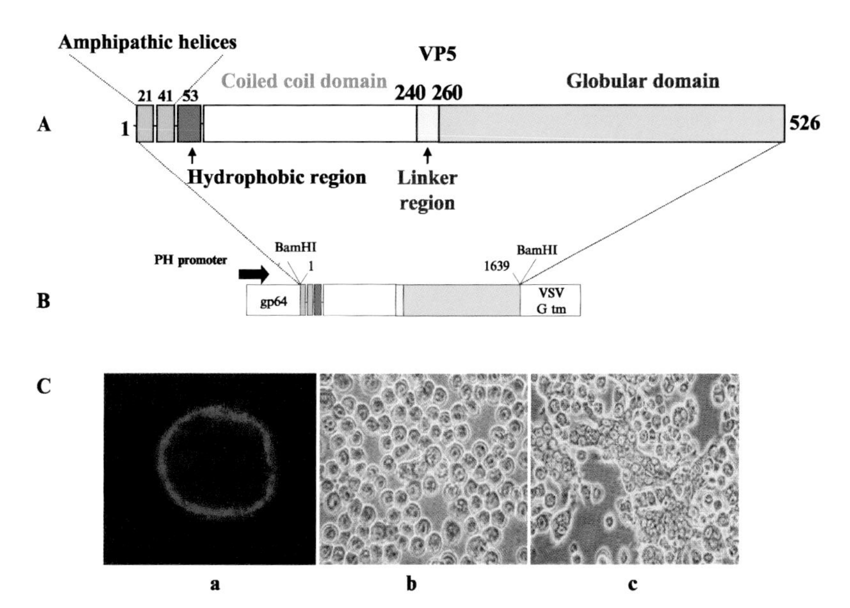


Roy, Fig 3. Arrangement of VP5 domains and fusion activity. (A) Schematic organization of VP5 protein, showing the predicted positions of two amphipathic helices, the coiled-coil domain, and the globular domain that are linked by a hinged region. (B) Schematic of a recombinant baculovirus transfer vector, showing the positions of a signal peptide of the baculovirus gp64 and the C-terminal part of vesicular stomatitis virus glycoprotein (VSV G) flanking the BTV-10 VP5 gene. (C) Confocal microscopy showing the cell surface expression of chimeric VP. (a) Fusion activity of VP5 in *Sf*9 cells after a low-pH shift; (b) note the formation of syncytia (seen by confocal microscopy); (c) absence of fusion activity before the pH shift.

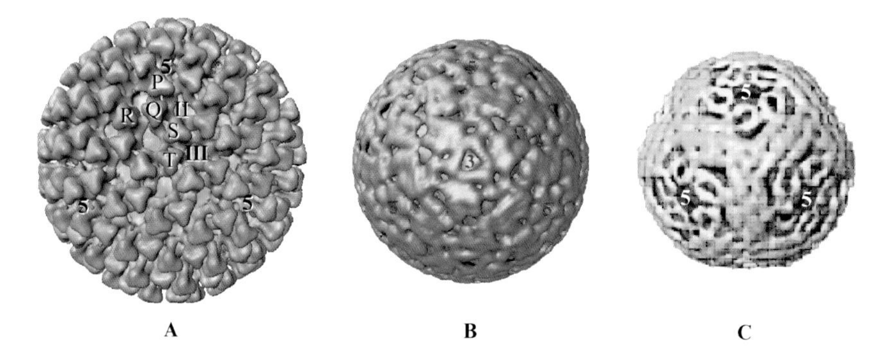


Roy, Fig 4. Surface representations of the three-dimensional cryo-EM structures of BTV-10 core. (A) The whole core (700 Å in diameter) viewed along the icosahedral threefold axis, showing the trimers of VP7 (in blue). The five quasi-equivalent trimers (P, Q, R, S, and T) and the locations of channels II and III are marked. (B) The middle layer of VP3 molecules (210–290 Å) cut away from the core structure. (C) The innermost density of three minor proteins in association with the genome. Ordered strands of genomic RNA are visible at 22-Å resolution.

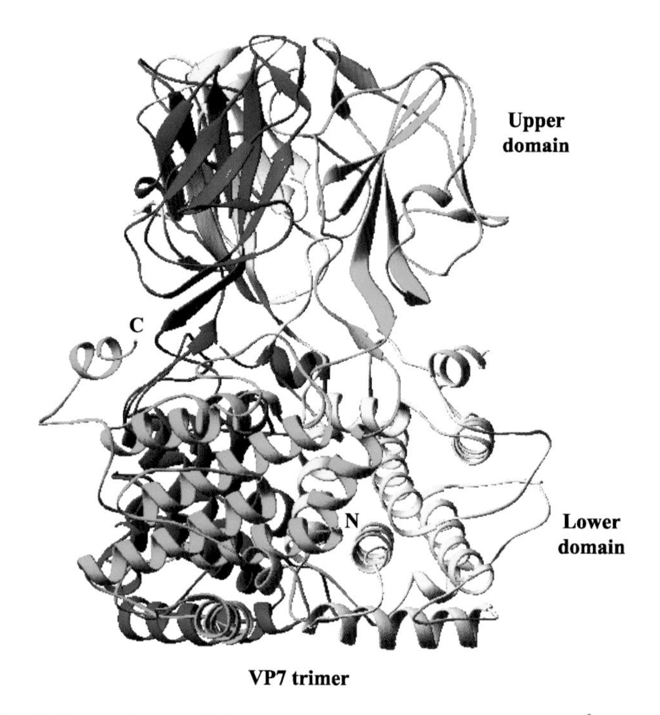


Roy, Fig 5. Trimer image of the VP7 atomic structure solved at 2.8-Å resolution. Two domains of the molecule are indicated. The carboxyl and amino termini are indicated. The view is shown from the side. Note that the flat base of the trimer lies in a horizontal plane in this view.

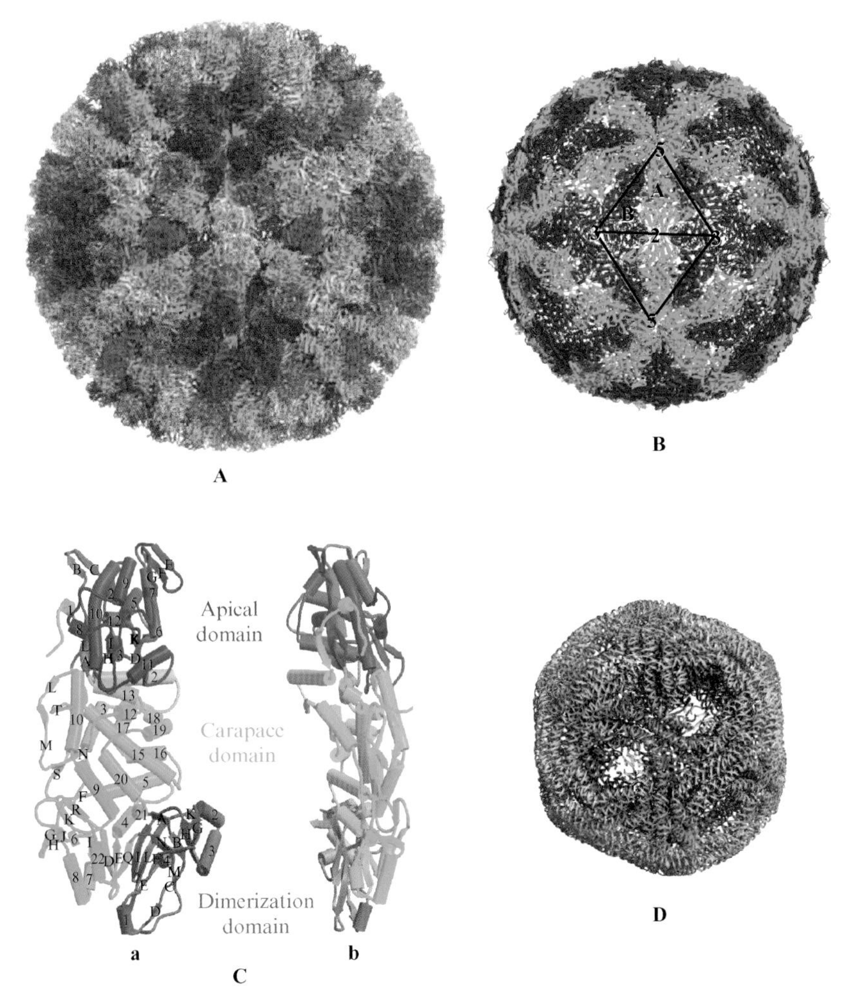


Roy, Fig 6. Structure of the BTV-1 core. (A) Surface of the core structure, showing the arrangement of 260 trimers. Structurally equivalent trimers have been given the same color. (B) Structure of the VP3 layer, showing the arrangement of 120 copies of two conformationally distinct types of VP3 molecules, "A" shown in green and "B" shown in red. Sixty copies of each of the two molecules together form 12 decamers that are linked together, forming a thin protein shell. The icosahedral fivefold, threefold, and symmetry axes are indicated. (C) Structures of two types of VP3 molecule, a (*left*) and b (*right*), as observed in the BTV-1 core. The molecules are similar in overall shape of thin triangular plates, with slight differences in conformation. Note some "curling up" of the b molecule, which facilitates it packing around the icosahedral threefold axes. The details of the fold within the two molecules differ at several places. Note the differences at the top of the molecules (closest to the icosahedral fivefold axes), where a compact structure is seen in molecule a and a more extended structure is seen in molecule b. (D) The ordered genomic dsRNA in the core of BTV. The diagram shows the model that has been built into the four layers of electron density, accounting for approximately 80% of the genome. The empty volume close to the icosahedral fivefold axes is the putative location for transcription complex.

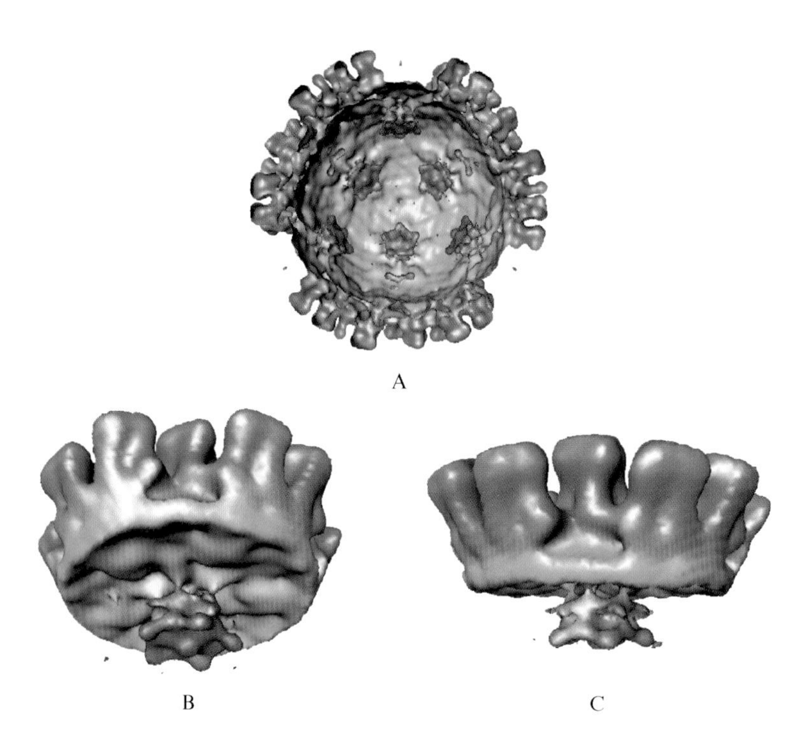


Roy, Fig 7. Organization of internal proteins in core-like particles. (A) Cryo-EM structure of a recombinant core-like particle, showing the inside view of the CLP reconstruction with VP3, VP7, VP1, and VP4. A flower-shaped density feature (red) attached to the inside surface of VP3 (green) at all the fivefold axes is clearly seen. (B and C) Conical cutaway from the reconstruction, providing close-up views of the flower-shaped structure and its interaction with the VP3 layer.

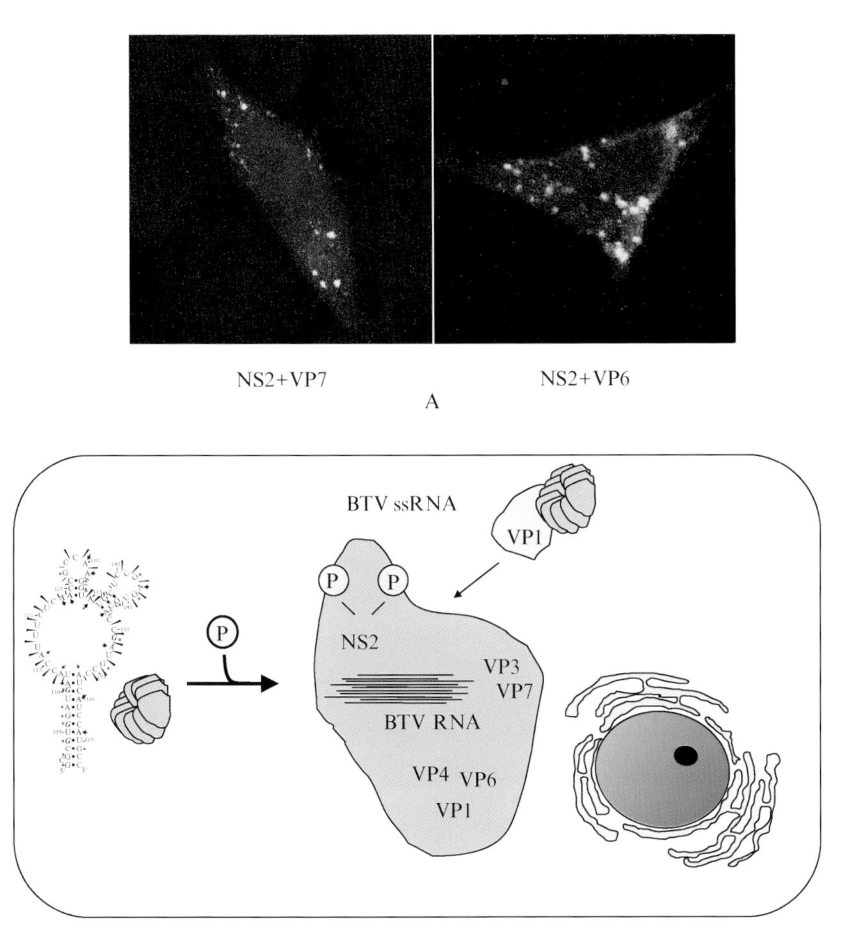


Roy, Fig 8. Interaction of NS2 and viral components. (A) Colocalization of core proteins and NS2 in BTV-infected cells. (B) Schematic showing the interaction of NS2/VIB with BTV proteins and the secondary structure of BTV ssRNA. "P" represents phosphorylation of NS2, which is responsible for the formation of VIBs.

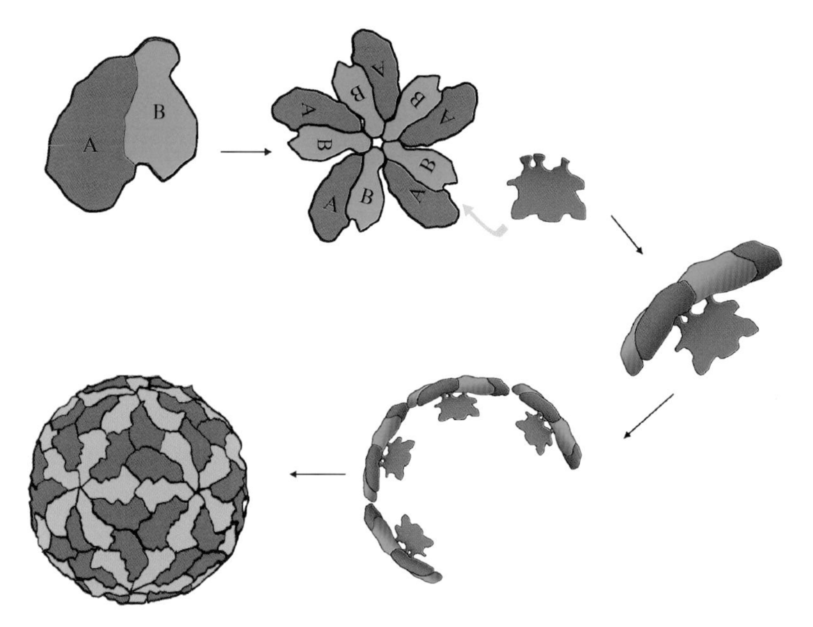


ROY, FIG 10. Schematic showing the growing VP3 decamers and assembly of polymerase complex with the VP3 decamer before formation of the VP3 protein shell.

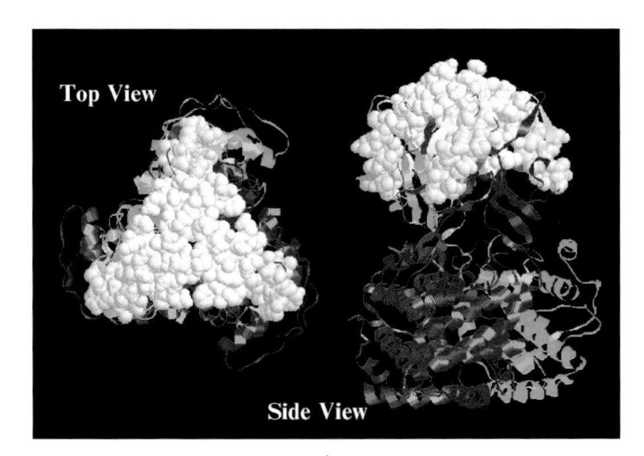


A

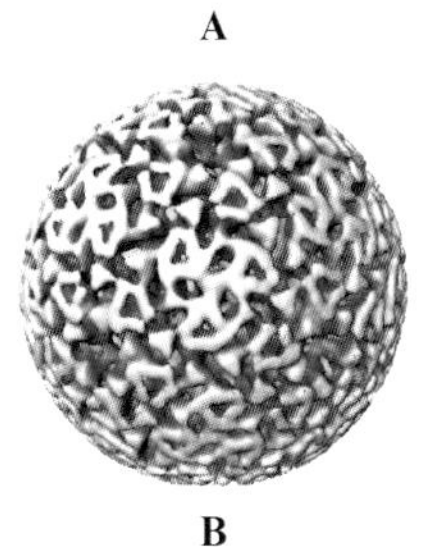

B

ROY, FIG 12. VP7 residues involved in the interactions with outer layer proteins. (A) Top head-on view (*left*) and a side view (*right*). The X-ray structure of the VP7 trimer is shown in ribbon representation. The trimeric subunits are colored red, blue, and green. The locations on VP7 where the triskelion motifs interact are shown in white and where the globular domains interact is shown in yellow. (B) Virion reconstruction, at approximately the same outer radius of the core structure, showing the triangular feature at the type II and type III channels. A set of three fivefold axes is also indicated.

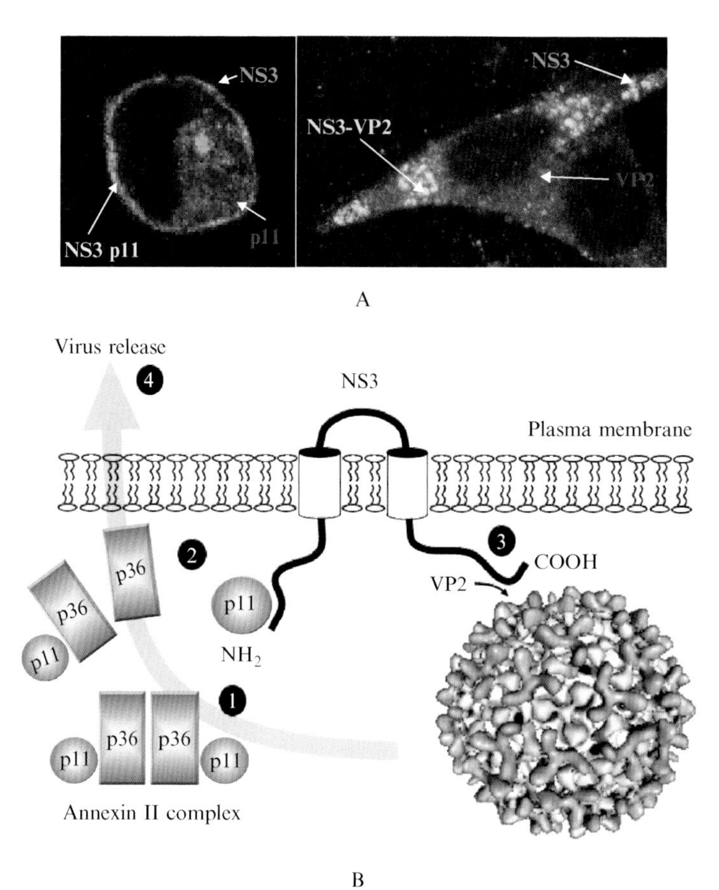


ROY, FIG 13. (A) Confocal microscopy showing (*left*) the merging of p11 and NS3 in BTV-infected HeLa cells and (*right*) the merging of NS3 and VP2 in BTV-infected BHK cells. (B) Schematic model of host protein and NS3 interaction during virus release from the infected cell.

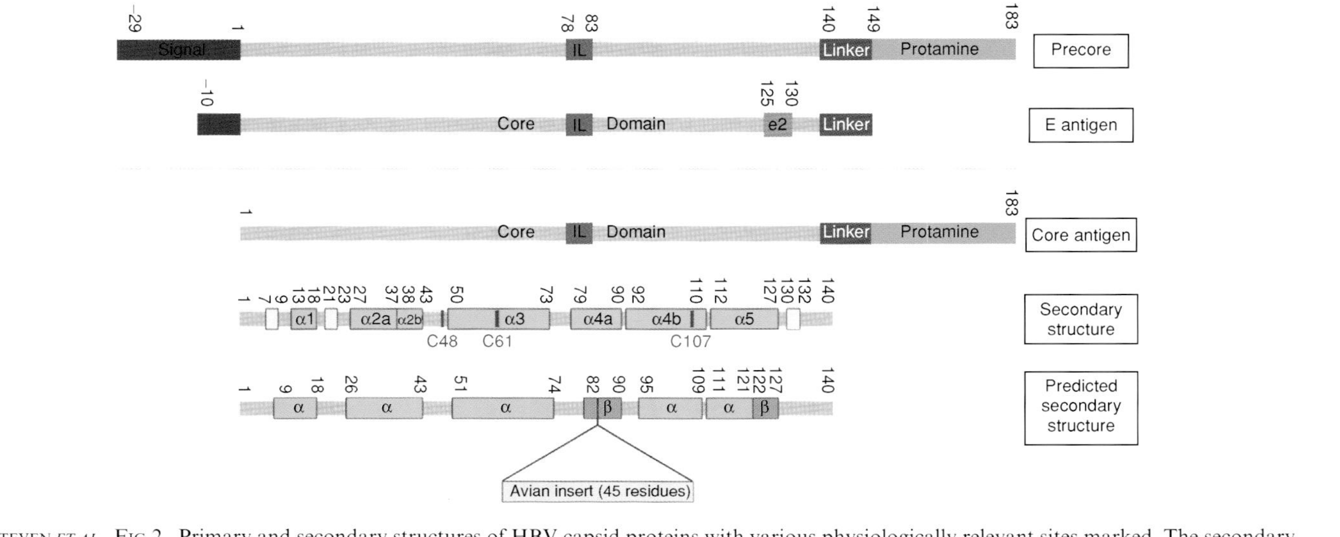


STEVEN *ET AL.*, FIG 2. Primary and secondary structures of HBV capsid proteins with various physiologically relevant sites marked. The secondary structure prediction is from Bringas (1997).

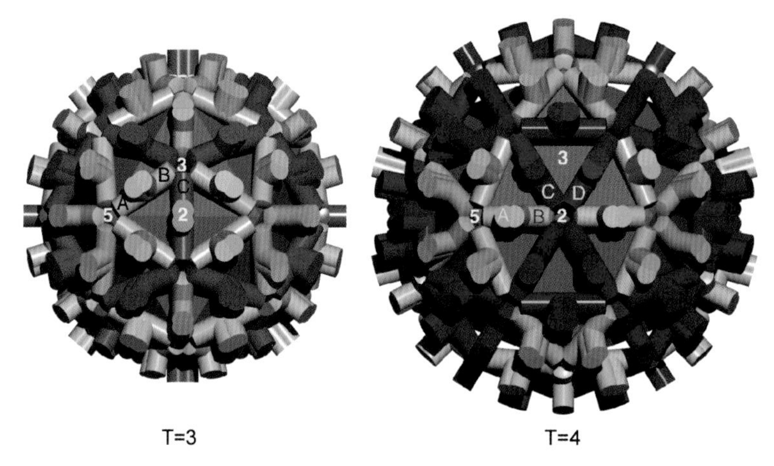


STEVEN *ET AL.*, FIG 4. Models of cAg dimer packing in the $T = 3$ and $T = 4$ capsids. The various quasi-equivalent subunits (A, B, C, etc.) are labeled.

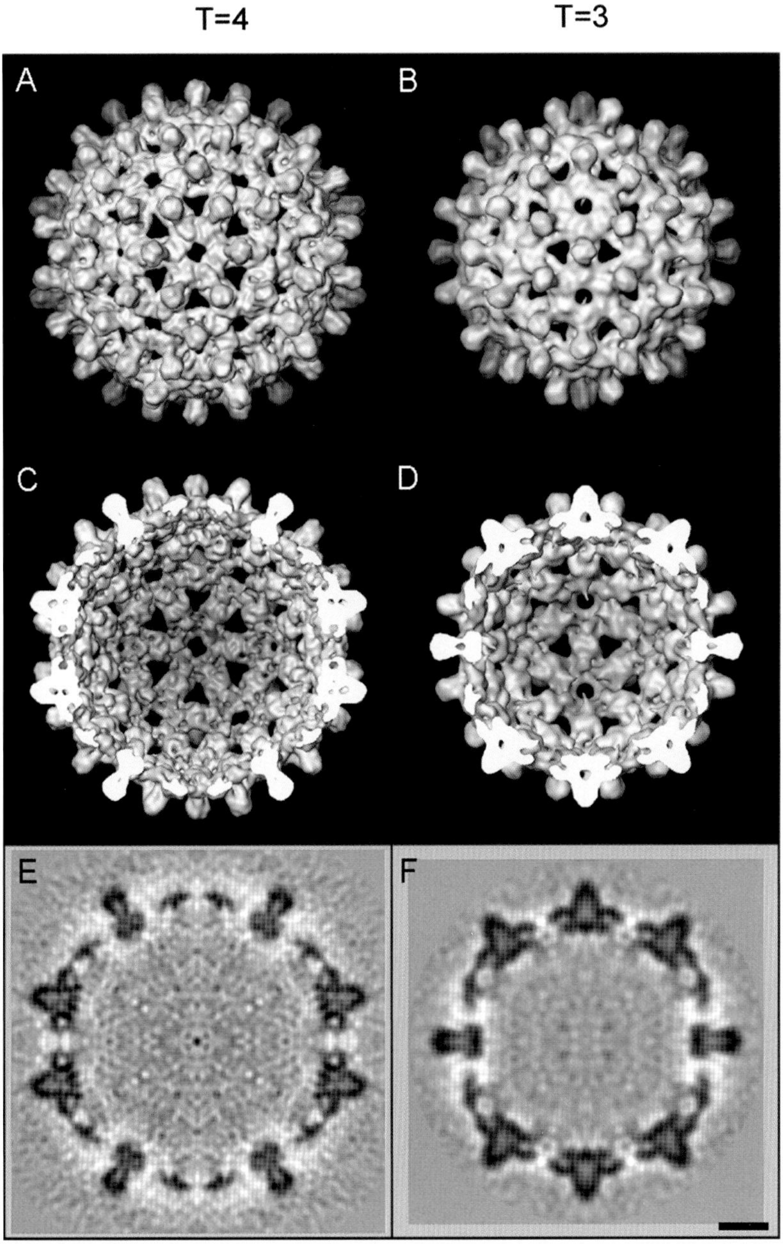

T=4 T=3

A B C D E F

STEVEN *ET AL.*, FIG 5. Cryo-EM reconstructions of the $T = 4$ and $T = 3$ cAg capsids (Cp147). (A, C, and E) $T = 4$ capsid at a resolution of 9.0 Å; (B, D, and F) $T = 3$ capsid at a resolution of 10 Å. From Conway *et al.* (1997) and Watts *et al.* (2002). (A–D) Outside and inside surface renderings. (E and F) Grayscale central sections of capsids viewed along twofold axes. Protein is dark. Scale bar: 50 Å.

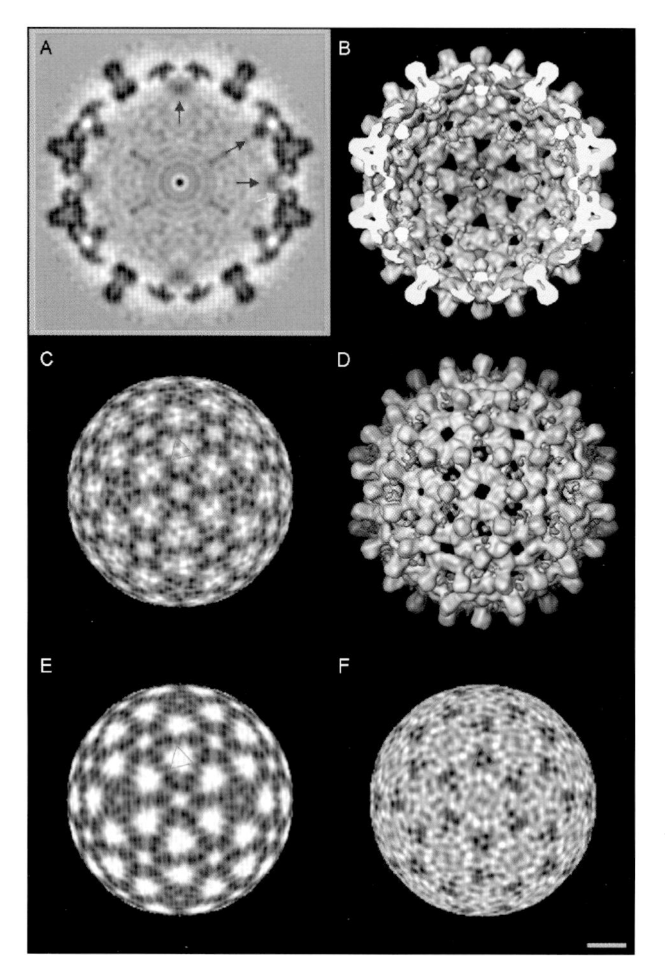


STEVEN *ET AL.*, FIG 6. Localization of the N and C termini of cAg by cryo-EM difference mapping. (A and B) A central section and an interior view of the $T = 4$ capsid for a construct that has a cysteine residue appended to Cp149 and a tetrairidium ($Ir_4$) label attached to Cys-150 (Cheng *et al.*, 1999). The extra density from the $Ir_4$ label is indicated with purple arrows (A) or by purple shading (B). The labels underlie symmetry axes. The blue arrow in (A) marks the site of a C-terminal residue (residue 149). In (C–F), the N terminus is marked by the attachment of an extraneous peptide of eight amino acid residues. This difference density is marked in red in (D). (C, E, and F) A comparison of spherical sections at an appropriate radius to sample the appended peptide. (C) is from the peptide-attached capsid; (E) is from the control (no attachment); and (F) is from the difference map. A triplet of peptide-associated densities is marked with a red triangle in each panel. The N terminus (residue 1) lies at the point where the difference density [red in (D)] meets the capsid surface (blue) in (D). Scale bar: 50 Å.

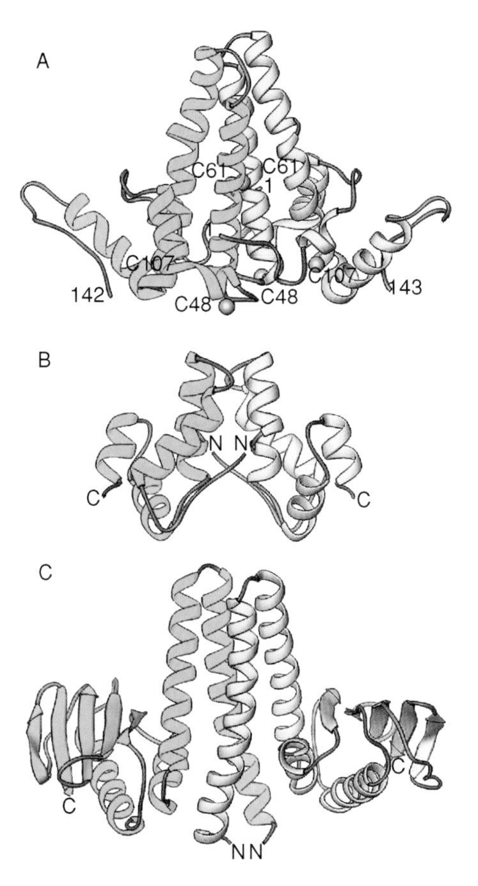


STEVEN *ET AL.*, FIG 7. Comparison of the HBV cAg dimer with proteins that have similar structures. (A) cAg (PDB: 1QGT). Cysteine residues C48, C61, and C107 are indicated as ball-and-stick side chains with enlarged gold sulfur atoms. The two C61 residues form an intermolecular disulfide bond within the cAg dimer. (B) C-terminal domain of the HIV capsid protein, CA (PDB: 1A80); and (C) Spo0B phosphotransferase (PDB: 1IXM). All three proteins dimerize through pairing $\alpha$-helical hairpins to create four-helix bundles. The HIV CA dimer has a short spike, whereas those of cAg and Spo0B are similar in length but of opposite hand. Monomers are green and yellow with the N and C termini labeled or numbered.

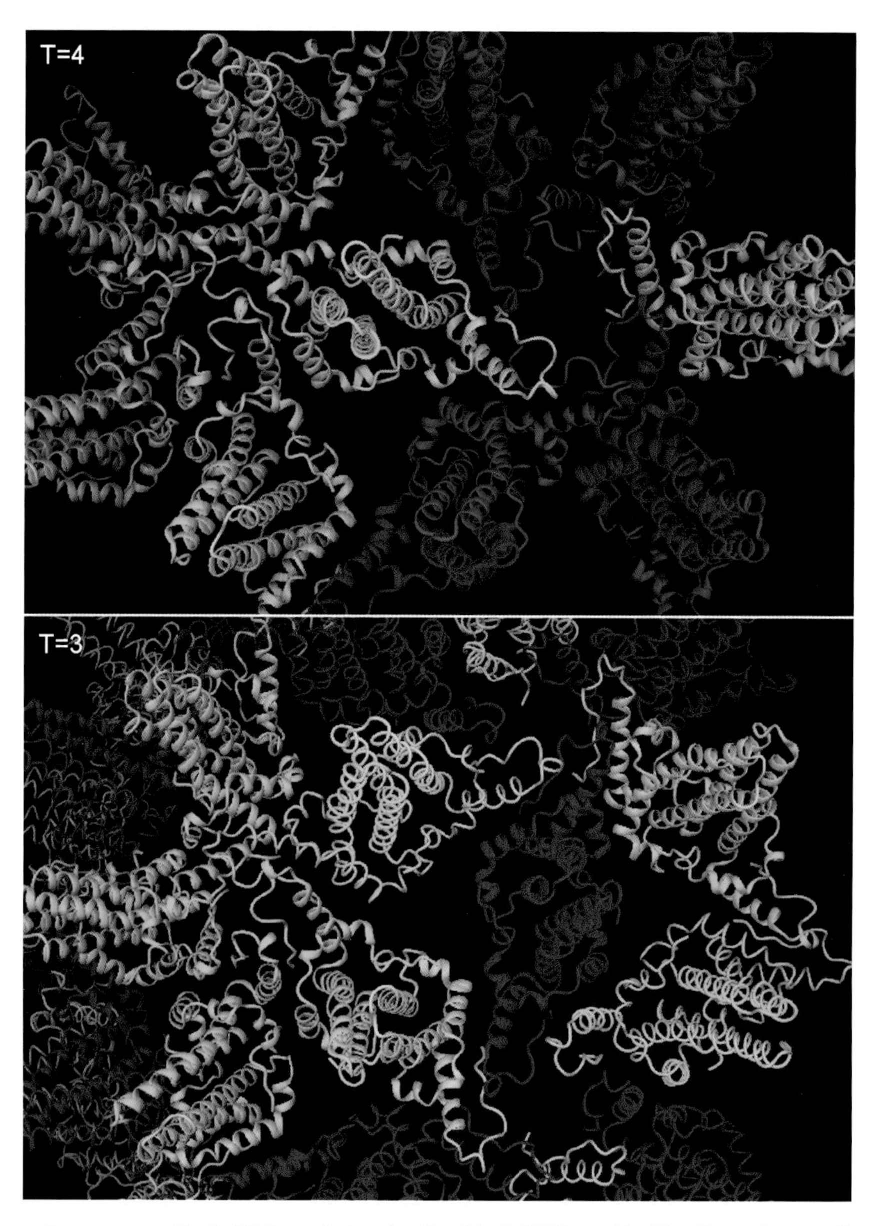


STEVEN *ET AL.*, FIG 8. High-resolution details of both HBV capsids. The dimer, the coordinates of which are from PDB (1QGT; Wynne *et al.*, 1999), was docked into cryo-EM reconstructions of the $T = 4$ and the $T = 3$ capsids (Conway *et al.*, 2003).

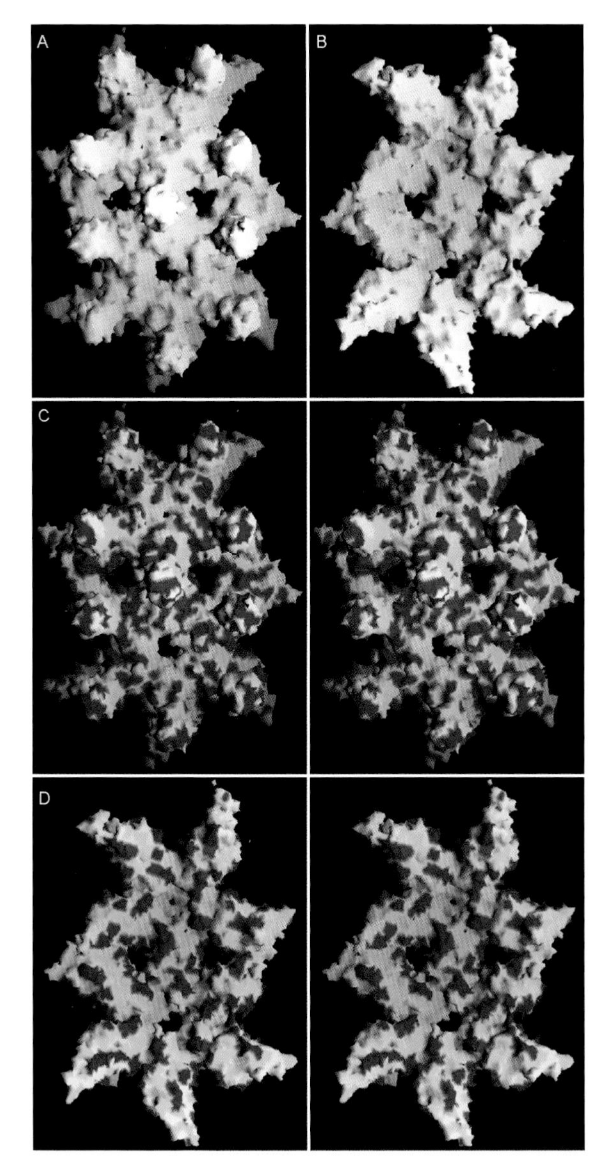


STEVEN *ET AL.*, FIG 9. Distribution of charged residues on the outer and inner surfaces of HBV capsids. (A and B) Outer and inner surfaces, respectively, of a fragment of the $T = 4$ capsid comprising five dimers surrounding a fivefold axis (upper part) plus five more dimers around the adjacent twofold axis (bottom part). (C and D) Stereo pairs of A and B, respectively, color coded for charge. Acidic residues are red, basic residues are blue, polar but uncharged residues are yellow, and hydrophobic residues are green. Calculated with GRASP.

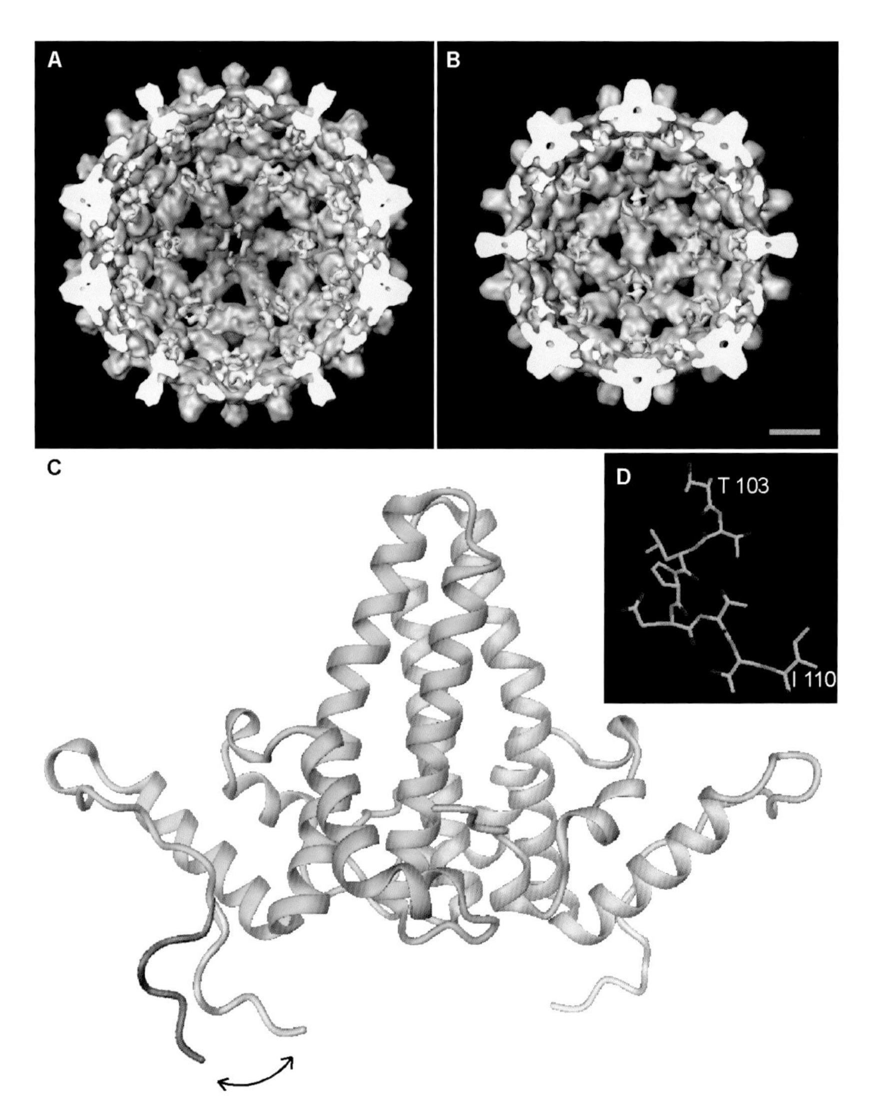


STEVEN *ET AL.*, FIG 11. Location, proposed conformation, and mobility of the cAg linker peptide. (A and B) The linker peptide (yellow) was localized on both the $T = 4$ and the $T = 3$ capsids by cryo-EM difference mapping (Watts *et al.*, 2002). (C) Model of Cp142 with linker peptide attached with hinge about residue 142. (D) Its proposed conformation (extended, non-$\alpha$, non-$\beta$) based on that of a surface peptide of cellobiose dehydrogenase (Hallberg *et al.*, 2000), with which it shares nearly complete sequence identity.

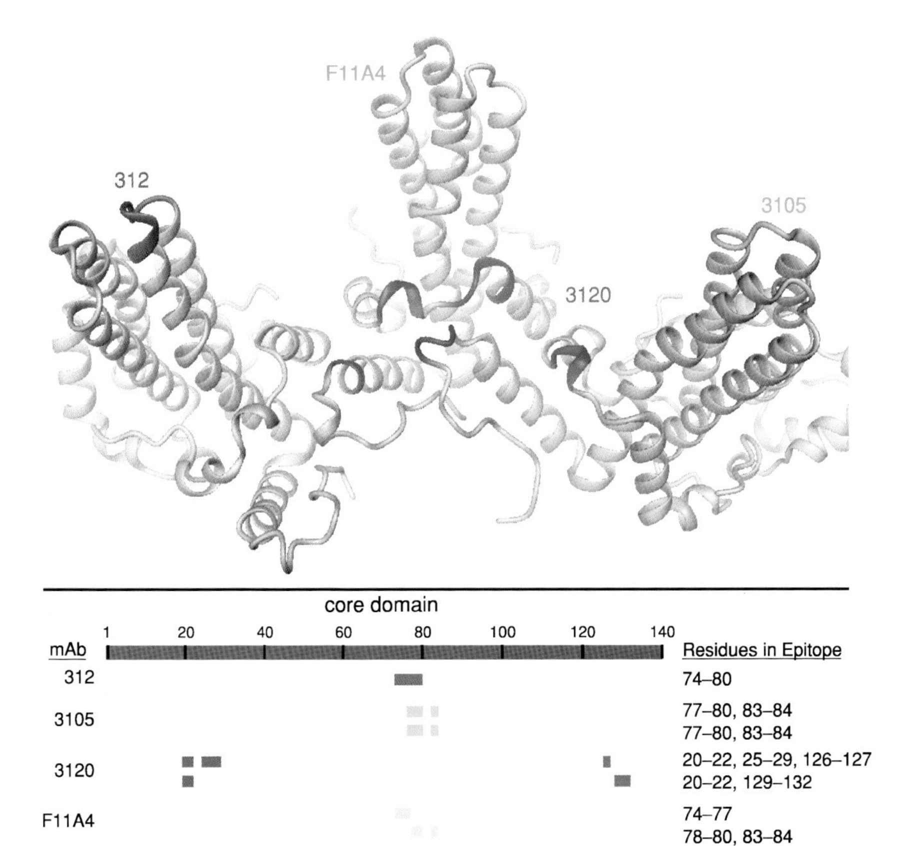


STEVEN *ET AL.*, FIG 14. Mapping the epitopes of four monoclonal anti-cAg IgGs. *Top*: The configurations of peptides involved in each case are marked on a ribbon diagram of three adjacent dimers. The epitope for mAb 3120 involves five peptides from two different subunits on neighboring dimers: on the $T = 4$ capsid, only two of four quasi-equivalent variants of the epitope (C–D and A–A) bind the Fab; on the $T = 3$ capsid, only one of three variants (A–A) binds it (Conway *et al.*, 2003). mAbs 3105 and F11A4 each have contributions from both subunits within a dimer. mAb 312 binds to the immunodominant loop on a single subunit. *Bottom*: Mapping of the positions of these peptides on the core domain sequence. Reproduced from Belnap *et al.* (2003).

STEVEN *ET AL.*, FIG 15. Mapping the epitopes of four monoclonal IgGs on the $T = 4$ HBV capsid. The epitopes are marked as yellow patches on a surface rendering of the capsid surface. The binding aspects of the corresponding four Fabs are shown: orange, mAb 312; magenta, mAb 3120; green, mAb 3105; cyan, mAb F11A4.

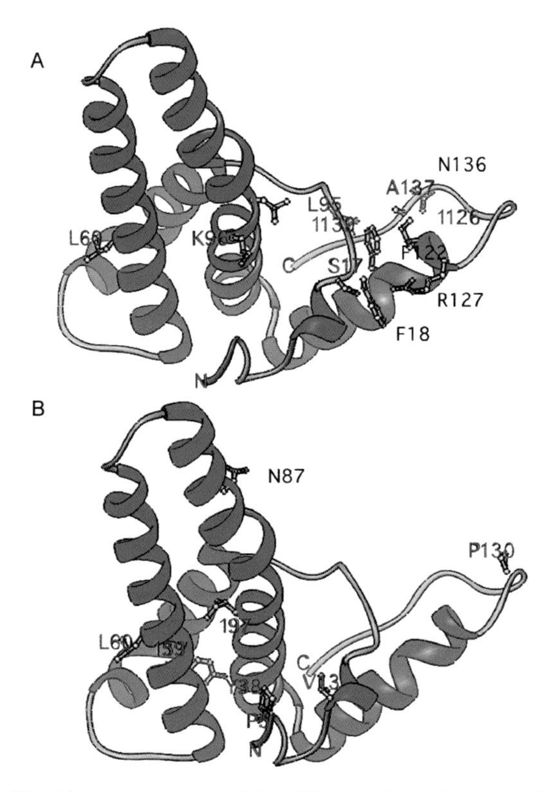


STEVEN *ET AL.*, FIG 16. cAg monomers with residues marked whose mutations disrupt HBV secretion. The mutants were identified by (A) *in vitro* mutagenesis or (B) from *in vivo* natural infections. (A) When mutated to alanine, these residues allow nucleocapsid formation but block envelopment and virion formation. These sites—S17, F18, L60, L95, K96, F122, I126, R127, N136, A137, and I139—form a groove around the base of the spike on the portion of the molecule that faces outward on the assembled capsid. This region has been suggested to interact with sAg during envelopment. (B) The most frequent mutations isolated from chronic carriers of HBV. Residues affected include P5, V13, Y38, I59, L60, N87, I97, and P130. Mutation of I97 to L97 increases the amount of immature genome in secreted particles, whereas low-level secretion of virions is caused by mutations P5 to T5 and L60 to V60. In both cases, the residues affected cluster around the base of the spike, not on it. Residues are represented by side chains, which are green ball-and-stick models. N and C termini along with residues are labeled. Based on work by Bruss and co-workers and Shih and co-workers (see text for references).

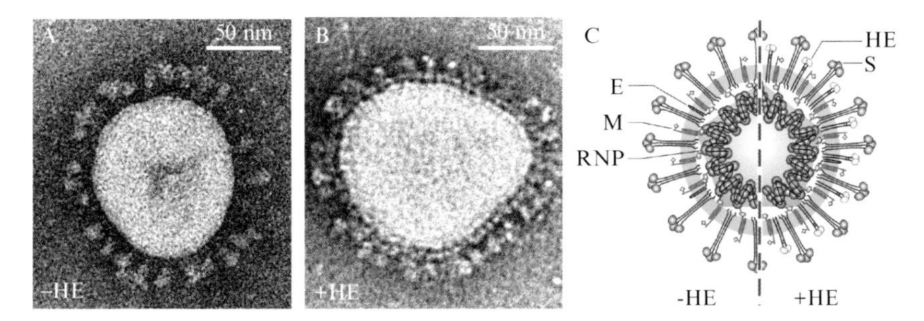


DE HAAN AND ROTTIER, FIG 1. Electron micrographs of mouse hepatitis virus strain A59 (MHV-A59) virions without (A) and with (B) the hemagglutinin-esterase (HE) envelope protein (viruses kindly provided by R. de Groot, Virology Division, Utrecht University, The Netherlands; image courtesy of J. Lepault, VMS-CNRS, Gif-sur-Yvette, France). Large, club-shaped protrusions consisting of spike (S) protein trimers give the viruses their corona solis-like appearance. Viruses containing the HE protein display a second, shorter fringe of surface projections in addition to the spikes. (C) Schematic representation of the coronavirion. The viral RNA is encapsidated by the nucleocapsid (N) protein forming a helical ribonucleoprotein (RNP), which is in turn part of a structure with spherical, probably icosahedral, configuration. The nucleocapsid is surrounded by a lipid bilayer in which the S protein, the membrane glycoprotein (M), and the envelope protein (E) are anchored. In addition, some group 2 coronaviruses contain the HE protein in their lipid envelope as illustrated on the right side of the particle.

→

DE HAAN AND ROTTIER, FIG 2. The coronavirus life cycle. The replication cycle starts with attachment of the virion by its S protein, that is, through the S1 subunit thereof, to the receptors on the host cell. This interaction leads to fusion of the virus envelope with a cellular membrane, for which the S2 subunit is responsible. From the genomic RNA that is released by disassembly of the incoming particle the *pol1a* and *pol1b*, genes are translated, resulting in the production of two large precursors (Pol1a and Pol1ab), the many cleavage products of which collectively constitute the functional replication–transcription complex. Genes located downstream of the *pol1b* gene are expressed from a 3′-coterminal nested set of subgenomic (sg) mRNAs, each of which additionally contains a short 5′ leader sequence derived from the 5′ end of the genome (shown in red). Transcription regulatory sequences (TRSs) located upstream of each gene serve as signals for the transcription of the sgRNAs. The leader sequence is joined at a TRS to all genomic sequence distal to that TRS by discontinuous transcription, most likely during the synthesis of negative-strand sgRNAs. In most cases, only the 5′-most gene of each sgRNA is translated. Multiple copies of the N protein package the genomic RNA into a helical structure in the cytoplasm. The structural proteins S, M, and E are inserted into the membrane of the rough endoplasmic reticulum (RER), from where they are transported to the ER-to-Golgi intermediate compartment (ERGIC) to meet the nucleocapsid and assemble into particles by budding. The M protein plays a central role in this process through interactions with all viral assembly partners. It gives rise to the formation of the basic matrix of the viral envelope generated by homotypic, lateral interactions between M molecules, and it interacts with the envelope proteins E, S, and HE (if present), as well as with the nucleocapsid, thereby directing the assembly of the virion. Virions are transported through the constitutive secretory pathway out of the cell—the glycoproteins on their way being modified in their sugar moieties, whereas the S proteins of some, but not all, coronaviruses are cleaved into two subunits by furin-like enzymes (see text for references).

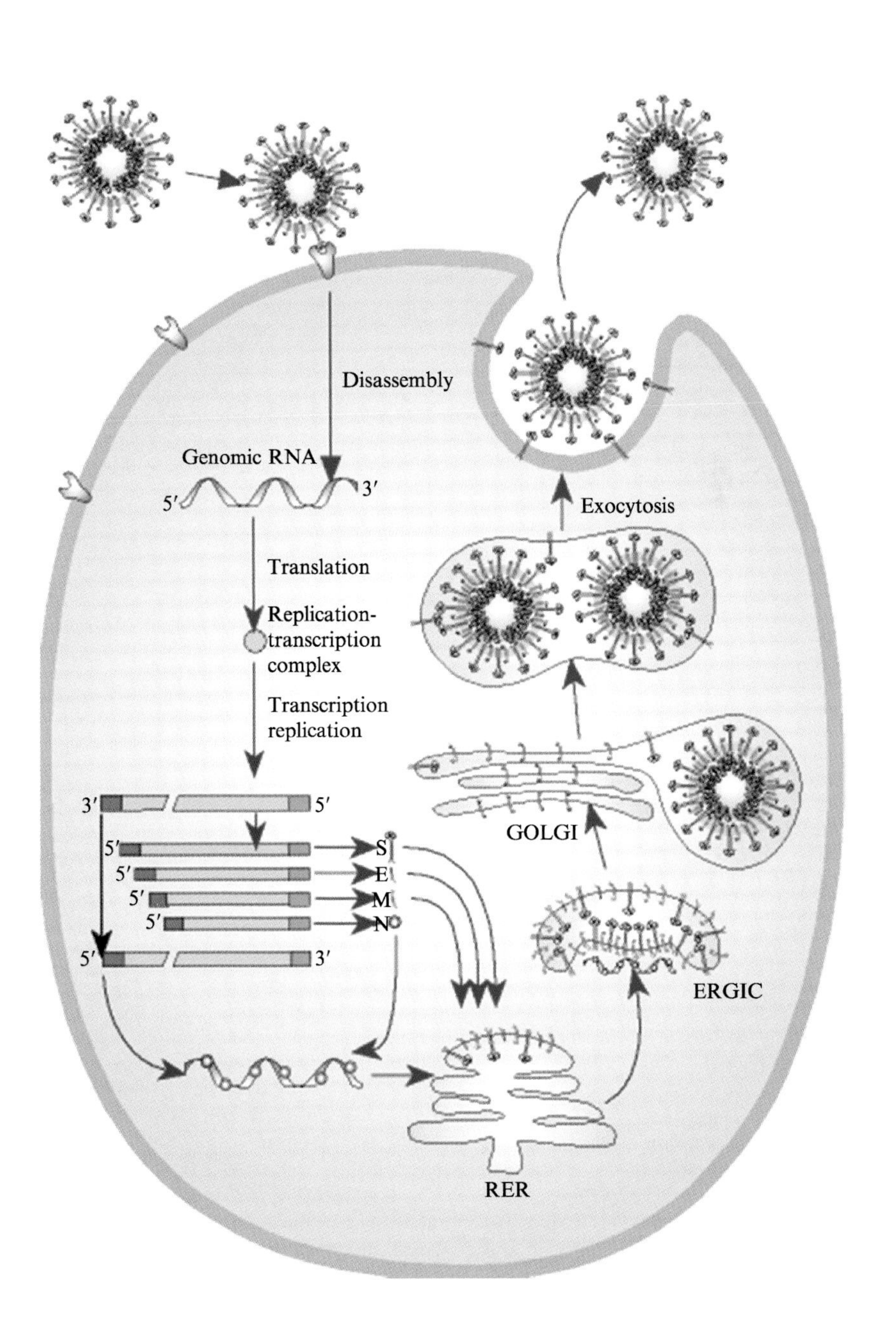

Disassembly
Genomic RNA
5′
3′
Translation
Replication-
transcription
complex
Transcription
replication
3′
5′
5′
5′
5′
5′
S
E
M
N
5′
3′
Exocytosis
GOLGI
ERGIC
RER

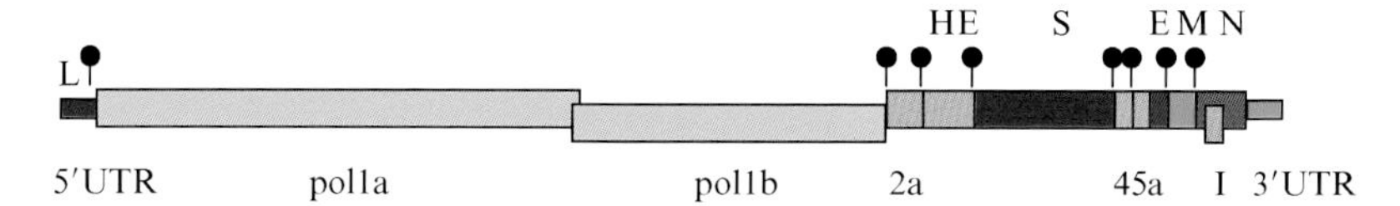


DE HAAN AND ROTTIER, FIG 3. Coronavirus genome organization as illustrated for the group 2 virus MHV. The single-stranded, positive-sense RNA genome contains 5′- and 3′-terminal untranslated regions (UTRs) with a 5′-terminal cap and a 3′-terminal poly(A) tract. The leader sequence (L) in the 5′ UTR is indicated. All coronaviruses have their essential genes in the order 5′-*pol*-S-E-M-N-3′. The *pol1a* and *pol1b* genes comprise approximately two-thirds of the genome. The more downstream *pol1b* gene is translated by translational readthrough, using a ribosomal frameshift mechanism. Transcription regulatory sequences (TRSs) located upstream of each gene, which serve as signals for the transcription of the subgenomic (sg) RNAs, are indicated by circles. The genes encoding the structural proteins HE, S, E, M, and N are specified. Gray boxes indicate the accessory, group-specific genes, in the case of group 2 coronaviruses genes 2a, HE, 4, 5a, and I.

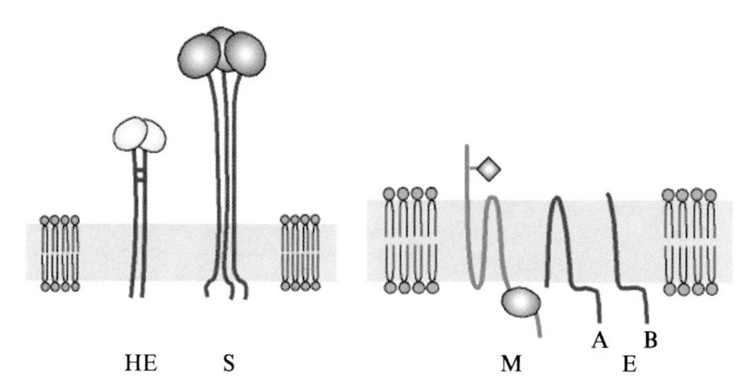


DE HAAN AND ROTTIER, FIG 4. Membrane topology of the coronavirus envelope proteins. The HE and S proteins are both type I membrane proteins with short carboxy-terminal cytoplasmic tails. The HE protein forms disulfide-linked homodimers, whereas the S protein forms non-covalently linked homotrimers. The S1 subunits presumably constitute the globular head, whereas the S2 subunits form the stalk-like region of the spike. The M protein spans the lipid bilayer three times, leaving a small amino-terminal domain in the lumen of intracellular organelles (or on the outside of the virion), whereas the carboxy-terminal half of the protein is located on the cytoplasmic side of the membrane (or inside the virion). In TGEV virions some of the M proteins have their cytoplasmic tail exposed on the outside (not shown). The M protein is glycosylated at its amino terminus (indicated by a diamond). The amphipathic domain of the M protein is represented by an oval. The hydrophilic carboxy terminus of the E protein is exposed on the cytoplasmic side of cellular membranes or on the inside of the virion. The E protein may span the bilayer once (B) or twice (A).

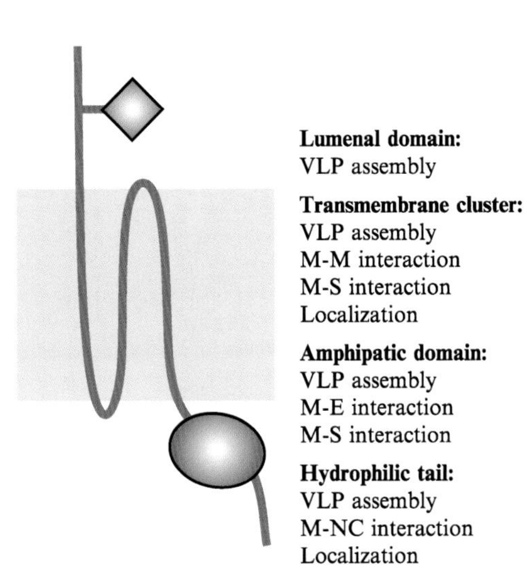


DE HAAN AND ROTTIER, FIG 7. The various domains of the MHV M protein and the processes for which they are important. The amphipathic domain of the M protein is represented by an oval. See text for references.

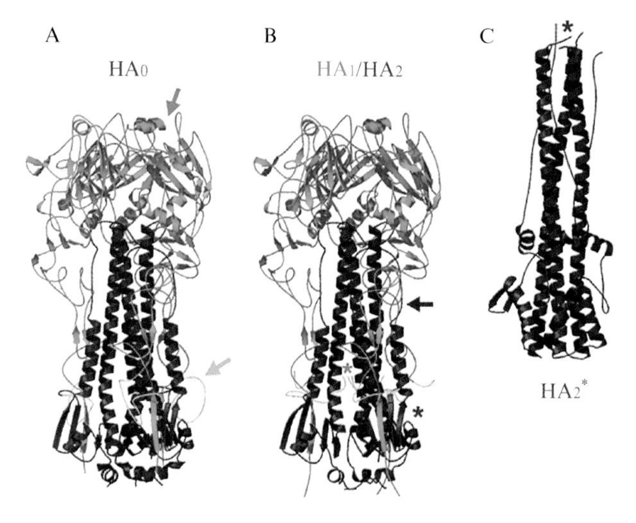


Harrison, Fig 2. The influenza virus hemagglutinin. (A) $HA_0$, before cleavage between $HA_1$ and $HA_2$. The $HA_1$ part of the protein is blue; the $HA_2$ part is purple. The fusion peptide, looped out before cleavage, is yellow. The sialic acid-binding site and the cleavage point are shown by blue and yellow arrows, respectively. (B) The mature HA after cleavage, but before low-pH triggering. The only change from the structure in (A) is insertion of the fusion peptide (the N terminus of $HA_2$, in yellow) into a crevice along the threefold axis (dark yellow asterisk). A purple arrow points to the loop between the shorter, N-proximal helix and the longer, central helix. A purple asterisk indicates the position of residues that will move to the top of the molecule during the low pH-induced transition. (C) $HA_2$ after exposure to low pH. The same structure can be obtained by refolding $HA_2$ expressed in *E. coli.* It is the minimal free-energy state of $HA_2$ unconstrained by covalent association with $HA_1$. The long loop in the prefusion structure (purple arrow in [B]) has now become helical, elevating the N terminus of the protein (the fusion peptide itself is not included in this structure) to the top of the molecule (purple asterisk). A break and reversal of direction in the central $\alpha$-helix of the prefusion trimer likewise projects the C terminus of the protein to the top. The figure is aligned with respect to (B) so this break is roughly at the same height in both panels. In the actual transition, the chain reversal is likely to occur by melting and rezipping of the C-terminal helical segment, as shown in Fig. 3C. For detailed references, see Skehel and Wiley (2000).

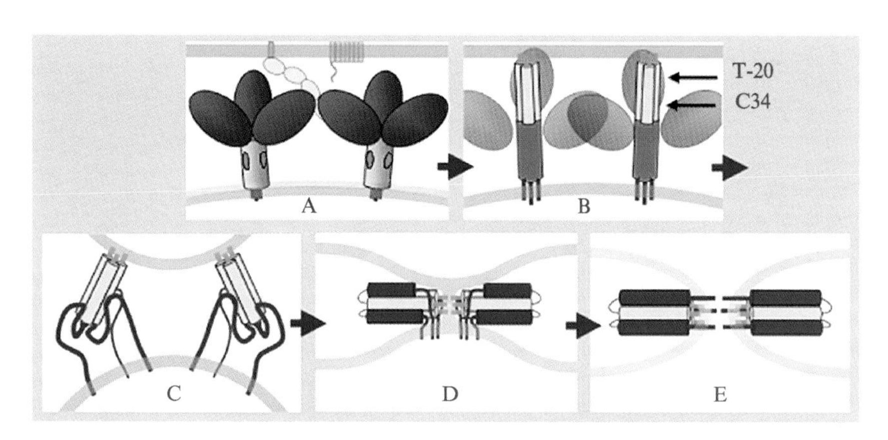


HARRISON, FIG 3. Diagram of membrane fusion mediated by class I viral fusion proteins. (A) Receptor binding (shown here for HIV or SIV Env, where the schematic receptor and coreceptor symbolize CD4 and CXCR4 or CCR5). (B) Dissociation of the receptor-binding domain (in the case of HA, a disulfide bond prevents complete dissociation, but the structural rearrangement requires that $HA_1$ move away from the threefold axis) and projection of the fusion peptide toward the target cell membrane. This state is known as the "prefusion intermediate" (or, sometimes, the "prehairpin intermediate"). The arrows show the positions along the core helical bundle at which, on HIV-1 gp41, the inhibitor peptides T-20 and C34 are expected to bind. (C) Folding back of the C-terminal part of the molecule. This zipping up of a segment of the fusion protein sometimes designated "helical region 2" (HR2); helical region 1, (HR1) forms the inner core of the postfusion trimer) draws the two membranes together as the N-terminal fusion peptide, inserted into the target cell membrane, and the C-terminal transmembrane segment, which is anchored in the viral membrane, are forced by the refolding to approach each other. (D) Hemifusion stalk formation. Provided that they insert only into the outer leaflet of the bilayer, the fusion peptides can migrate into the stalk, as proposed here, and stabilize it. (E) Fusion pore formation. Hemifusion structures flicker transiently into unstable pores, which reseal. The pore can be trapped by the final refolding step, in which the three transmembrane segments snap into place around the inserted fusion peptides. In the case of HA, this step may be driven by formation of a set of interactions that cap the inner core helices.

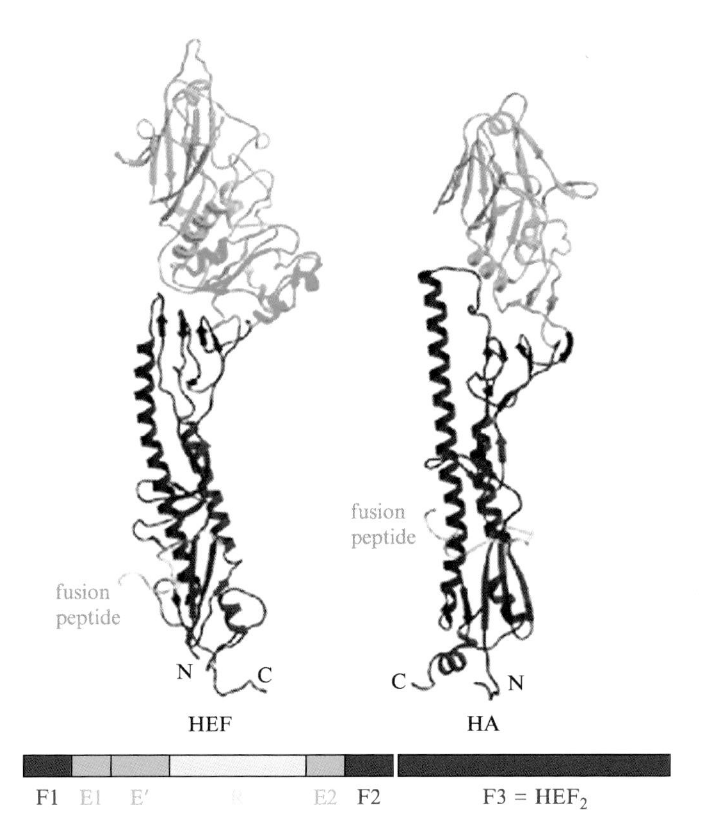


HARRISON, FIG 4. *Top*: The ectodomain of influenza virus C HEF and its structural and apparent evolutionary relationship to influenza A HA. *Bottom*: A key to the color scheme for the HEF polypeptide chain. The N- and C-terminal segments of $HEF_1$ and all of $HEF_2$ (except for the fusion peptide) are in red. The acetylase enzymatic domain (E1 + E″ + E2) is in green; the receptor-binding domain (R) is in blue. Think of the red fragments as an elementary fusion protein and the green and blue fragments as insertions. In HA, most of the acetylase domain has been deleted, except for the E′ fragment, which becomes an adaptor to connect the elementary fusion protein with the receptor-binding domain. Adapted from Rosenthal *et al.* (1998).

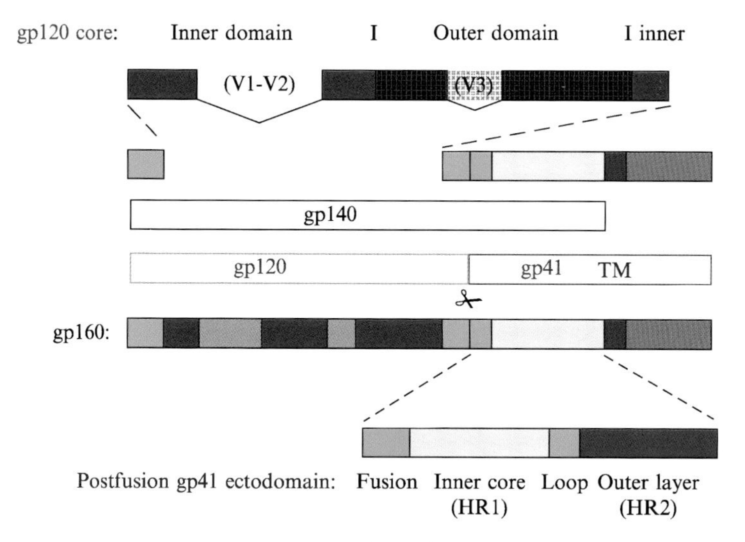


HARRISON, FIG 6. Primary structure of the HIV and SIV envelope proteins. The bar labeled "gp160" represents, schematically, the regions of the envelope precursor. The parts of gp160 corresponding to gp120, gp41, and gp140 (the gp160 ectodomain) are shown as open bars. The scissors symbol shows the furin cleavage point. The gp120 core can be considered a receptor-binding insertion into an elementary fusion protein, as diagrammed in the top two bars. Compare with Fig. 4—the gp120 core is the analog of the R and E′ regions of influenza HA. The principal elements of the gp41 ectodomain are diagrammed in the bottom bar.

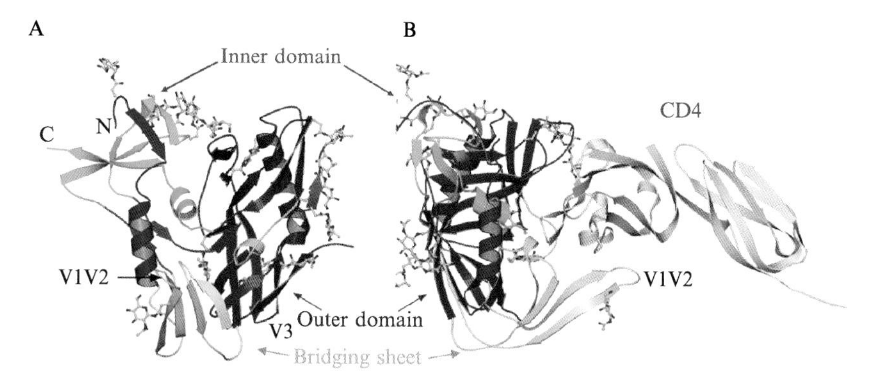


HARRISON, FIG 7. Structure of the gp120 core in the CD4-bound state (Kwong *et al.*, 1998). (A) Ribbon representation, showing the inner and outer domains, linked by a bridging sheet. The locations of the V1–V2 and V3 loops, deleted from the core construct (compare with Fig. 6), are shown. The locations of carbohydrate chains are shown by molecular ball-and-stick representations of those sugars found to be well-ordered in the crystal structure. The view is into the CD4-binding pocket. N and C termini of the core are labeled. (B) Side view of the same structure, with the first two immunoglobulin-like domains of CD4 also shown.

HARRISON, FIG 8. The inner core (HR1) and outer layer (HR2) of the gp41 ectodomain in the postfusion state of the gp41 trimer (Weissenhorn *et al.*, 1997). The structure as determined crystallographically contains six peptides—three inner and three outer. The dashed lines show covalent connectivities that would be present in the intact gp41 trimer. The residue numbers (for gp41, counting from the N terminus of the fusion peptide) for the beginning and end of the inner and outer layer helices are shown for the red subunit.

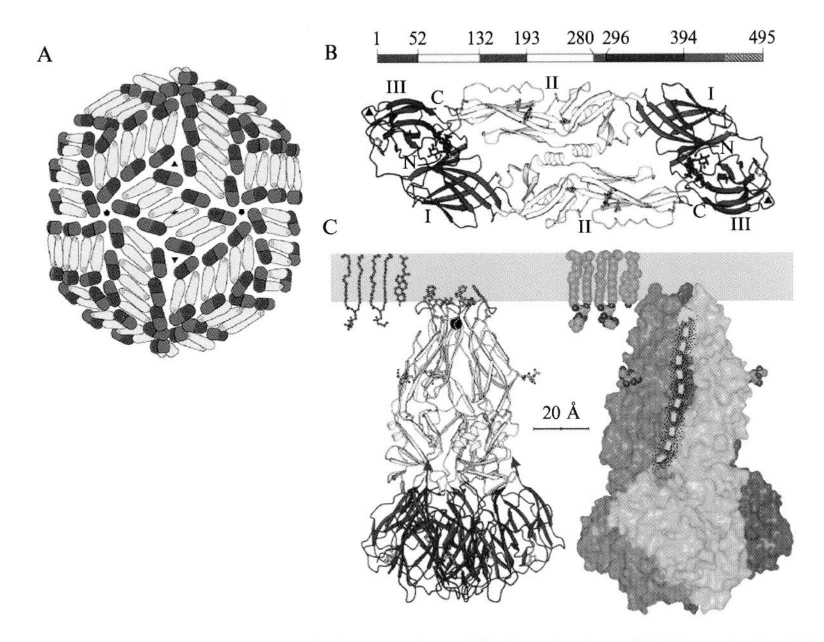


HARRISON, FIG 9. The flavivirus fusion protein E. (A) Organization of E dimers in the virion surface. Each subunit is shown in three colors: domain I in red, domain II in yellow, and domain III in blue—based on the structure described by Kuhn *et al.* (2002). (B) The soluble ectodomain, sE of dengue virus type 2, in the dimeric prefusion conformation found on mature virions (Modis *et al.*, 2003). The domain colors are as in (A). The bar above the ribbon diagram shows the relationship of domains to primary structure. The "stem" segment between residue 394 and the transmembrane anchor is not included in the three-dimensional structure. (C) The sE trimer (Modis *et al.*, 2004). The proteins are shown in relation to a schematic lipid bilayer to illustrate the likely degree of penetration of the fusion loops (*top*) into the membrane. The ribbon diagram (*left*) is colored as in (A) and (B). Arrows at the C terminus of the polypeptide chain suggest its presumed continuing direction. The surface rendering (*right*) includes a dashed arrow to show the proposed course of the stem peptide, which would lead to the transmembrane anchor.

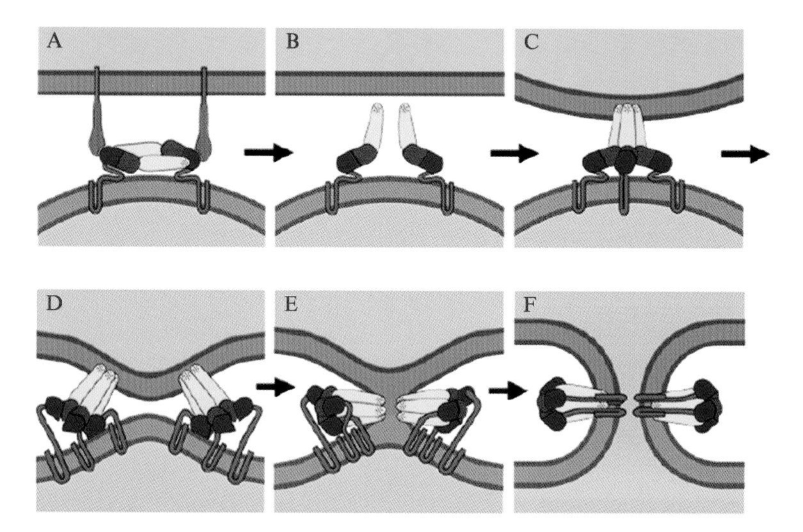


HARRISON, FIG 10. Diagram of membrane fusion by class II viral fusion proteins. (A) Receptor binding through domain III of E (flaviviruses). (B) Lowered pH in an endosome leads to dissociation of the dimer interactions. On release of dimer constraints, monomers can flex outward, presenting their fusion loops to the target cell membrane. (C) Insertion of the fusion loops into the target cell membrane and initial formation of trimer contacts among the projecting domains II. (D) Domain III flips over and the stem zips up along the outside of the trimer. (E) Hemifusion stalk. The diagram shows a proposed role for the inserted fusion loops—stabilization of the hemifusion stalk. (F) Formation of a fusion pore. Completing the zipping up of the stem drives fusion forward, because the cytosolic tails enter the pore and commit it to dilation.

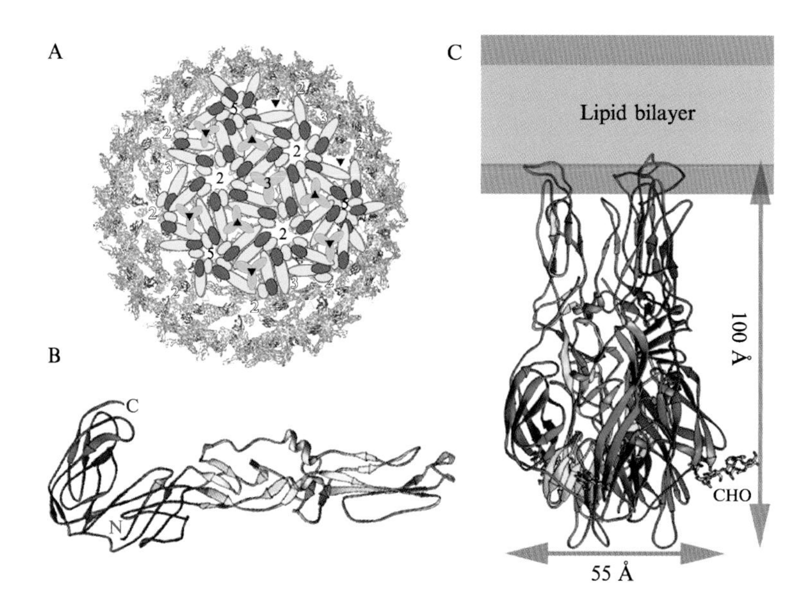


HARRISON, FIG 11. The alphavirus fusion protein E1. (A) Organization of E1 and E2 on the surface of virions. Simplified representations of subunits have been superimposed on a model of the fit of the SFV E1 crystal structure into an image reconstruction of the virion from electron cryomicroscopy (Lescar *et al.*, 2001). E1 is red (domain I), yellow (domain II), and blue (domain III). The E2 trimer (for which only a 9-Å structure is currently known, from electron microscopy) is represented by a green trefoil. It projects outward, capping the fusion loop of E1. Numbers (5, 3, and 2) show the positions of fivefold, threefold, and twofold icosahedral symmetry axes; triangles show the positions of local threefold axes in the T = 4 icosahedral surface lattice. (B) The soluble ectodomain, E1*, of Semliki Forest virus. Domain colors as in (A). (C) The E1* postfusion trimer. Each subunit is a single color. The stem of E1, shorter than the stem of flavivirus E proteins (compare with Fig. 9), would link the C terminus of E1* to the transmembrane anchor. (B) and (C) are adapted from Gibbons *et al.* (2004b).

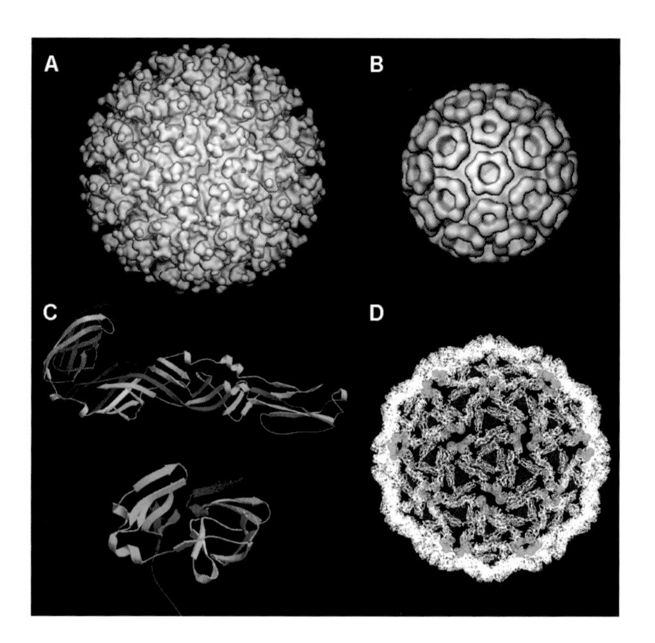


KUHN AND ROSSMANN, FIG 1. (A) Cryo-EM reconstruction of Sindbis virus, showing a surface-shaded view of the virus as viewed down a twofold axis. The glycoproteins E1 and E2 are collectively colored in blue and the bilayer is green. The resolution of the structure is 20 Å. (B) An alphavirus reconstruction in which the surface features and lipid bilayer were subtracted, leaving a surface-shaded view of the nucleocapsid core. (C) Ribbon diagrams showing atomic structures of the Semliki Forest virus E1 protein ectodomain (residues 1–381) and the Sindbis virus capsid protein proteinase domain (residues 106–264). $\beta$-strands are indicated by the arrows, and $\alpha$-helices are indicated by the coiled ribbons. The color gradation indicates the direction of the $C_\alpha$ backbone as it traverses from the amino terminus (blue) to the carboxy terminus (red) of the protein. (D) Fit of the E1 $C_\alpha$ backbone into the cryo-EM density of Sindbis virus, as viewed down the icosahedral threefold axis. E1 is shown in white, and the glycosylation site at Sindbis E2 residue 318 is shown in the magenta density cage; E1 glycosylation residues 139 and 245 are shown in green and red, respectively.

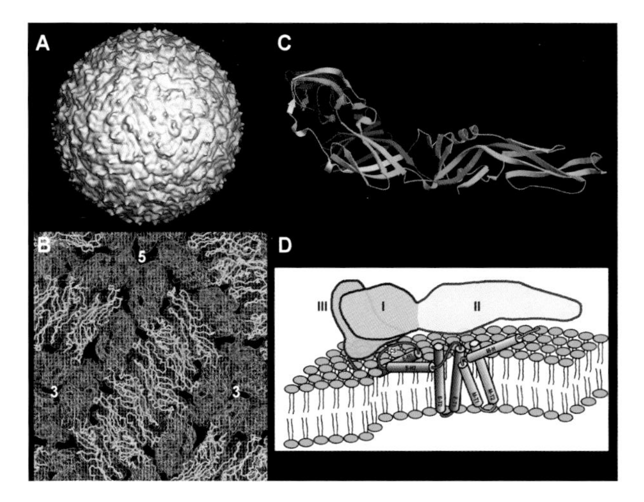


KUHN AND ROSSMANN, FIG 2. (A) Surface-shaded representation of the dengue 2 cryo-EM reconstruction at a resolution of 12 Å, as viewed down a twofold axis. The blue envelope proteins completely obscure the underlying membrane. (B) The fit of E monomers into the dengue structure is shown with domains I, II, and III in red, yellow, and blue, respectively. The limits of one icosahedral asymmetric unit can be defined by connecting the fivefold and threefold axes. (C) Ribbon diagram of the dengue E protein ectodomain (residues 1–395). $\beta$-Strands are indicated by the arrows and $\alpha$-helices by the coiled ribbons. The color gradation indicates the direction of the $C_\alpha$ backbone as it traverses from the amino terminus (blue) to the carboxy terminus (red) of the protein. (D) Diagrammatic drawing of the dengue virus ectodomain and transmembrane domain proteins. The volume occupied by the ectodomain of an E monomer is shown in pink (domain I), yellow (domain II), and lilac (domain III). The stem and anchor helices of E and M are shown in blue and orange, respectively. Helices are identified by the nomenclature described in text. CS represents the linker region between E-H1 and E-H2.

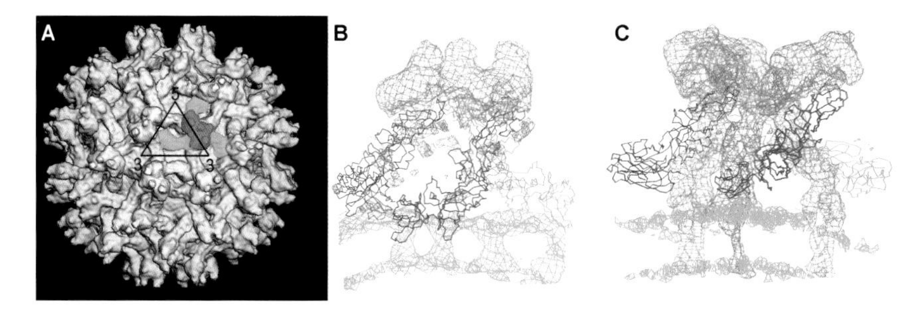


KUHN AND ROSSMANN, FIG 3. (A) Surface-shaded view of an immature dengue virus at a resolution of 16 Å. An icosahedral asymmetric unit is outlined in black. One of the 60 spikes in each particle is identified in color: prM in gray and E in green. (B and C) Comparison of the spike structure of (B) immature dengue particles and (C) mature Sindbis virus. The $C_\alpha$ backbones of the E and E1 glycoprotein trimers (in blue, red, and green) are shown for the dengue and Sindbis particles, respectively. The corresponding densities have been set to zero, leaving the densities (in gray) for the prM and E2 molecules, respectively. The lipid bilayer is shown in green. The slab used for depicting the lipid bilayer is thinner than that used to define the ectodomain trimer.

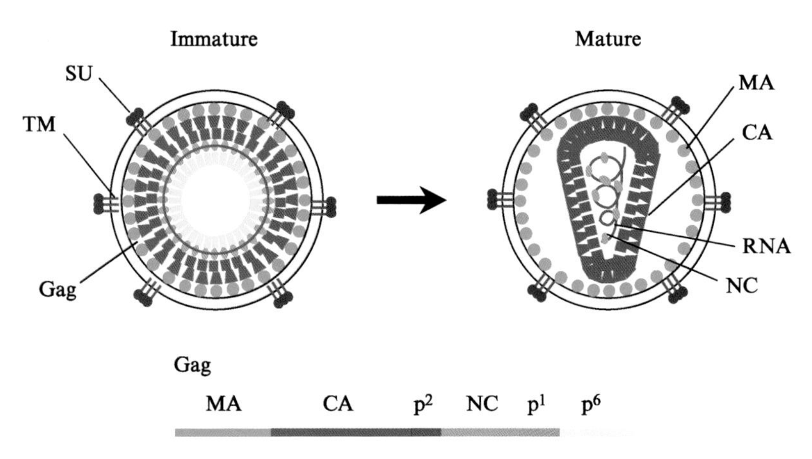


LANMAN AND PREVELIGE, FIG 1. Schematic representation of the maturation stage of the HIV-1 life cycle. The spherical immature virus is composed of the Gag polyprotein. After budding, the virally encoded protease is activated and cleaves Gag at the indicated junctions (*bottom*). The liberated domains are free to rearrange. The MA domain remains membrane-associated while the CA collapses to form the conical core. Contained within the conical core is the diploid viral genome complexed with NC. The locations of p2, p1, and p6 are unknown.

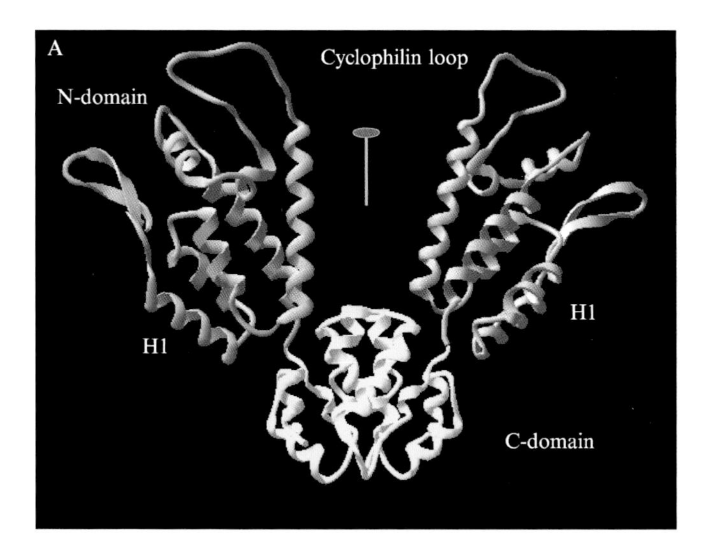


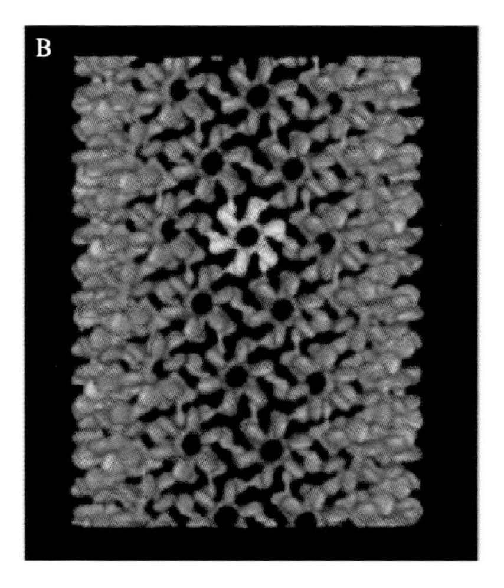


LANMAN AND PREVELIGE, FIG 2. (A) Crystal structure of the N- and C-terminal domains of CA. The crystal structures of the N- and C-terminal domains were solved individually. They are connected by a flexible linker region that is not visualized in the intact protein. The connection is arbitrary and the relative orientation of the two domains is unknown. The C-terminal domain forms a dimer in both the crystal and in solution. (B) Three-dimensional reconstruction of tubes of CA. Three-dimensional reconstructions of tubes of CA polymerized *in vitro* were obtained by cryo-EM. The tubes consist of hexameric protein clusters that are interconnected by dimeric interactions lying slightly below the surface of the tube.

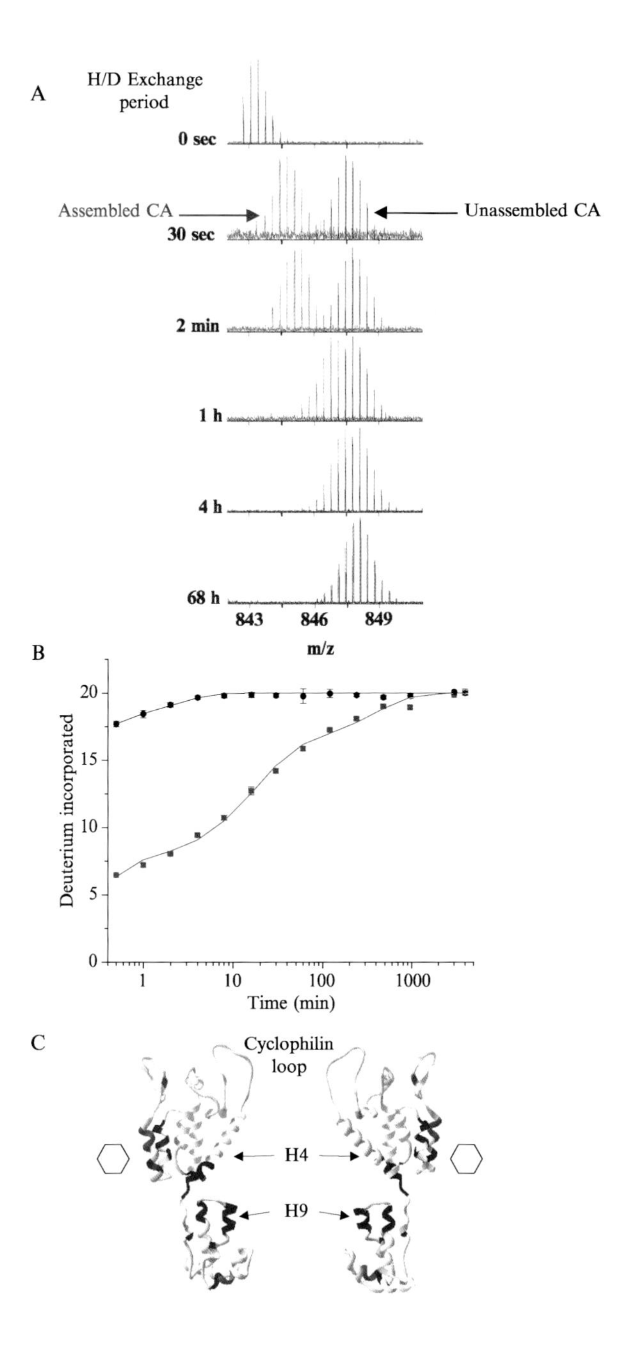

A
H/D Exchange period
0 sec
Assembled CA
Unassembled CA
30 sec
2 min
1 h
4 h
68 h
843
846
849
m/z
B
Deuterium incorporated
20
15
10
5
0
1
10
100
1000
Time (min)
C
Cyclophilin loop
H4
H9

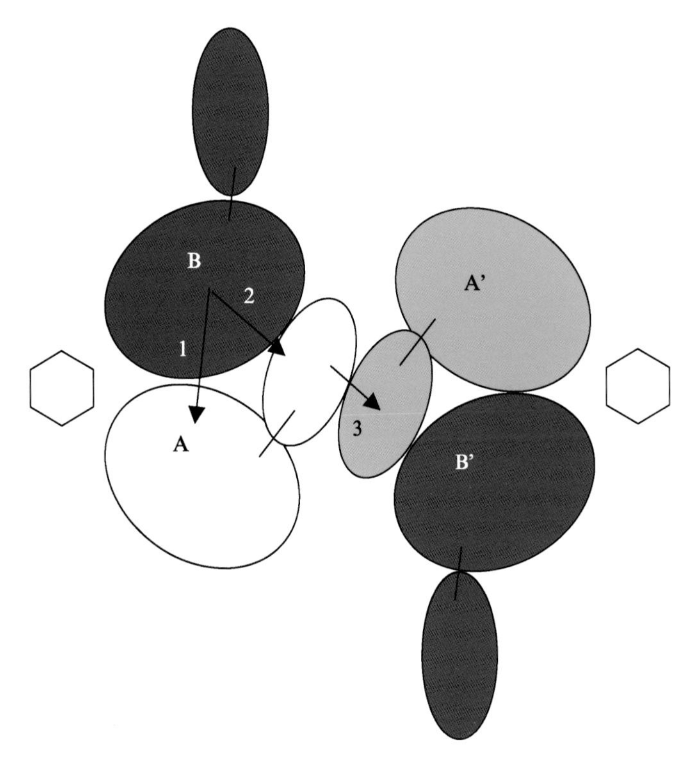


Lanman and Prevelige, Fig 4. Models for the intersubunit interactions in CA tubes. The proposed interactions between CA protein subunits are indicated by arrows and labeled. The homotypic N-terminal domain interaction (1) is mediated by helices I and II. The homotypic C-terminal domain interaction is mediated by helix IX. The heterotypic N-domain:C-domain interaction involves the base of helix IV. Formation of this interaction on dimerization is proposed to be required for continued subunit addition.

Lanman and Prevelige, Fig 3. (A) FT-ICR mass spectra of CA peptide spanning residues 169–189 during exchange. The peptide-spanning CA residues 169–189 lie at the C-terminal domain–dimer interface. As indicated by the time-dependent increase in mass-to-charge ratio, it exchanges rapidly in the free CA (black) and substantially more slowly in the polymerized form (red). (B) Exchange kinetics of peptide 169–189 in the unassembled and assembled form. The centroid of the distributions in (A) were calculated and plotted versus time. The solid line represents a multiexponential fit to the data. (C) The exchange protection in CA as a result of assembly. The degree of exchange protection was calculated from the exponential fits to the data and mapped onto the structure of CA. Darkest blue is most protected on assembly. Red segments are less protected in the assembled form. Yellow represents no change, and white regions were not analyzed. The two representations are rotated by 180°.

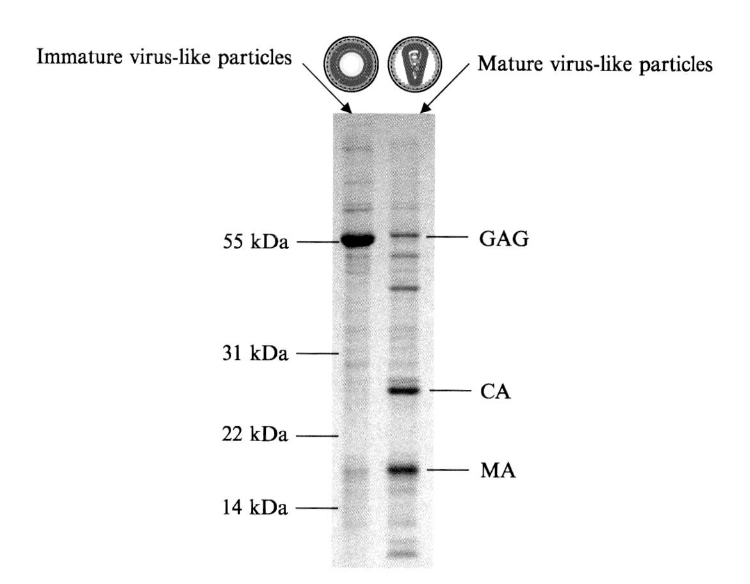


LANMAN AND PREVELIGE, FIG 5. SDS–PAGE analysis of immature and mature virus-like particles. Immature and mature virus-like particles used for hydrogen–deuterium exchange experiments were produced by transient transfection, purified, and analyzed by SDS–PAGE.

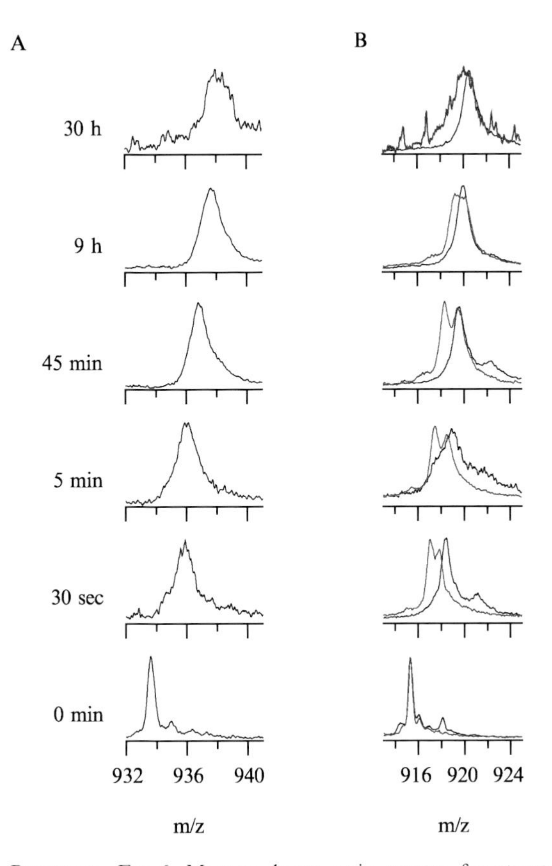


LANMAN AND PREVELIGE, FIG 6. Mass-to-charge ratio spectra for structural proteins in mature HIV capsids during exchange. Mature virus-like particles were exchanged in deuterated phosphate-buffered saline (PBS) and then analyzed by mass spectrometry without further proteolytic digestion. (A) The +16 charge state of MA as a function of exchange time. (B) The +28 charge state of CA in mature virions (red) and free in solution (black) as a function of exchange time.

A)

1
GARASVLSGG ELDKWEKIRL RPGGKKQYKL KHIVWASREL ERFAVNPGLL ETSEGCRQIL GQLQPSLQTG

71
SEELRSLYNT IAVLYCVHQR IDVKDTKEAL DKIEEEQNKS KKKAQQAAAD TGNNSQVSQN YPIVQNLQGQ

141
MVHQAISPRT LNAWVKVVEE KAFSPEVIPM FSALSEGATP QDLNTMLNTV GGHQAAMQML KETINEEAAE

211
WDRLHPVHAG PIAPGQMREP RGSDIAGTTS TLQEQIGWMT HNPPIPVGEI YKRWIILGLN KIVRMYSPTS

281
ILDIRQGPKE PFRDYVDRFY KTLRAEQASQ EVKNWMTETL LVQNANPDCK TILKALGPGA TLEEMMTACQ

351
GVGGPGHKAR VLAEAMSQVT NPATIMIQKG NFRNQRKTVK CFNCGKEGHI AKNCRAPRKK GCWKCGKEGH

421
QMKDCTERQA NFLGKIWPSH KGRPGNFLQS RPEPTAPPEE SFRFGEETTT PSQKQEPIDK ELYPLASLRS

491
LFGSDPSSQ

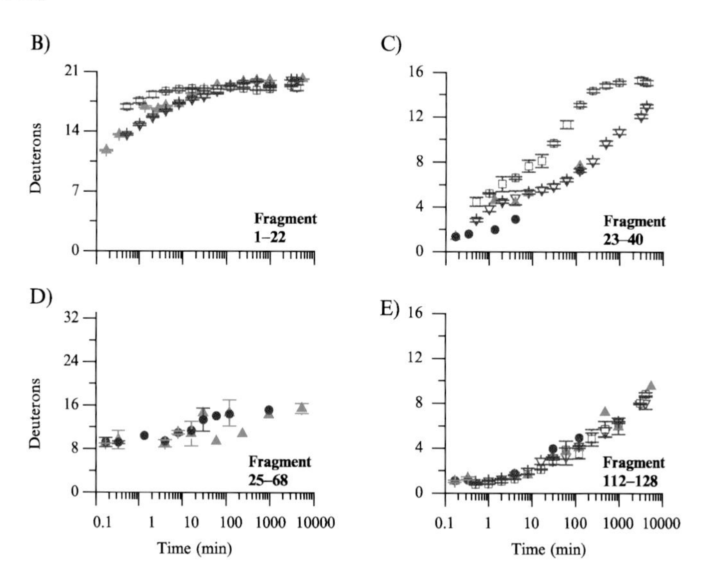


LANMAN AND PREVELIGE, FIG. 7. Peptide fragments produced by digestion of immature and mature virus-like particles. (A) Observed peptide fragments for immature virus-like particles (red dashes), mature virus-like particles (green dots), and both (black solid lines). (B–E) Number-average mass versus hydrogen–deuterium exchange period for immature viral particles (red diamonds), and mature virus-like particles (green triangles) superimposed on the deuterium incorporation profiles for *in vitro*–unassembled (open squares) and *in vitro*–assembled (open triangles) CA.

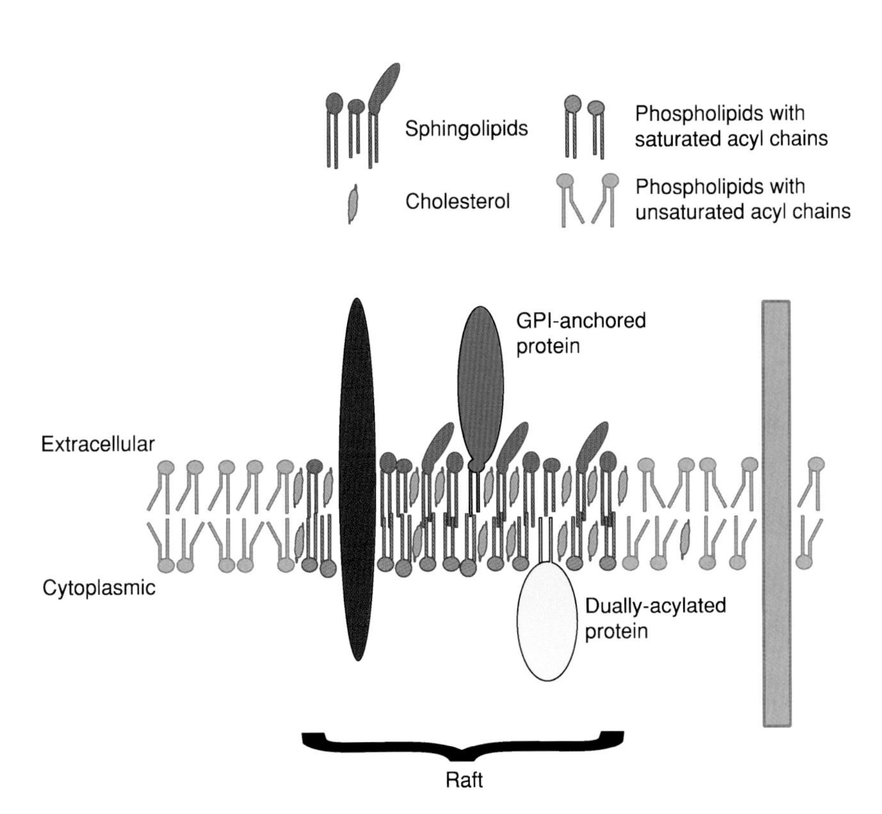


ONO AND FREED, FIG 1. A simplified model of lipid raft structure. Nonraft lipid and protein molecules are shown in gray.

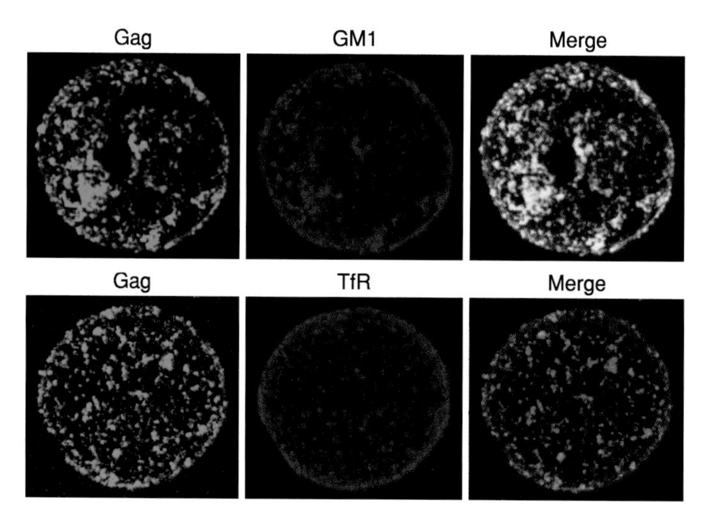


ONO AND FREED, FIG 2. Copatching of HIV-1 Gag proteins with a raft marker but not with a nonraft marker. Jurkat cells expressing HIV-1 Gag were labeled with cholera toxin B subunit (GM1) or antitransferrin receptor (TfR) before fixation. After fixation and permeabilization, Gag proteins were detected with an anti-p17MA antibody (Gag). Images were collected in *z* series with a confocal microscope and projected in 3D volume with Imaris software. Merged images of Gag and GM1 or TfR are also shown (Merge).

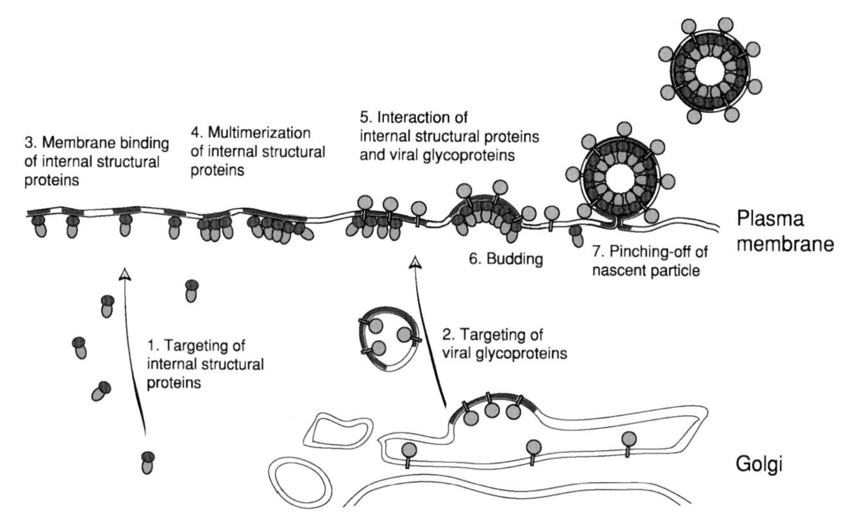


ONO AND FREED, FIG 5. Steps in virus assembly that may be facilitated by rafts. Rafts are shown in red.

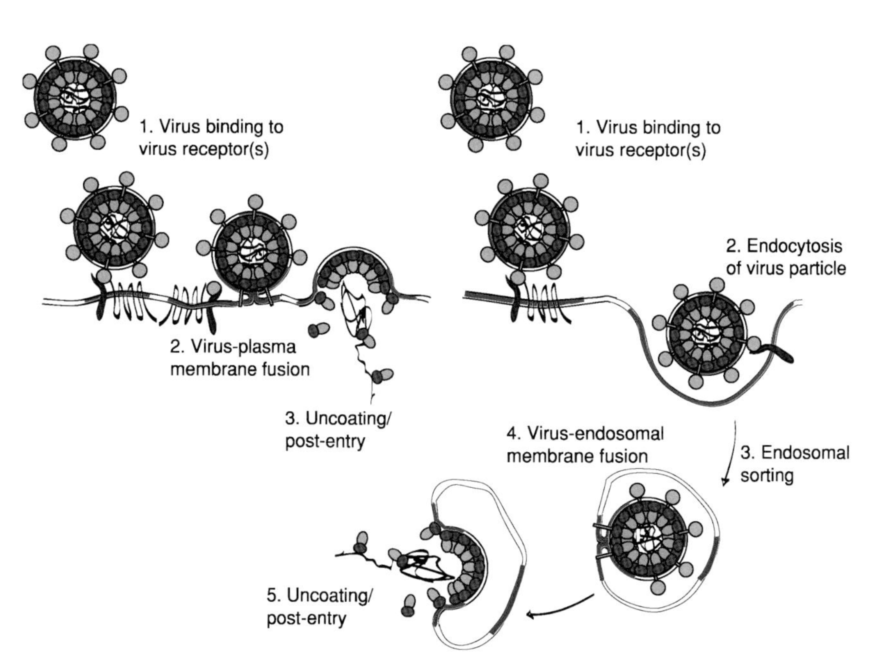


ONO AND FREED, FIG 6. Steps in virus entry that may be facilitated by rafts. Rafts are shown in red. Two modes of virus entry (direct fusion and low pH-triggered fusion) are shown.

A

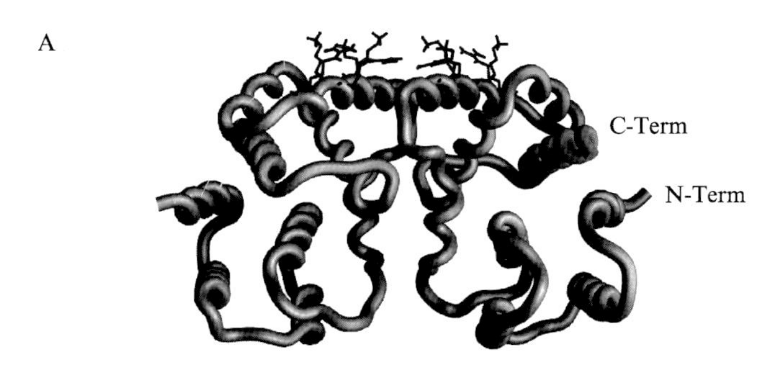


B

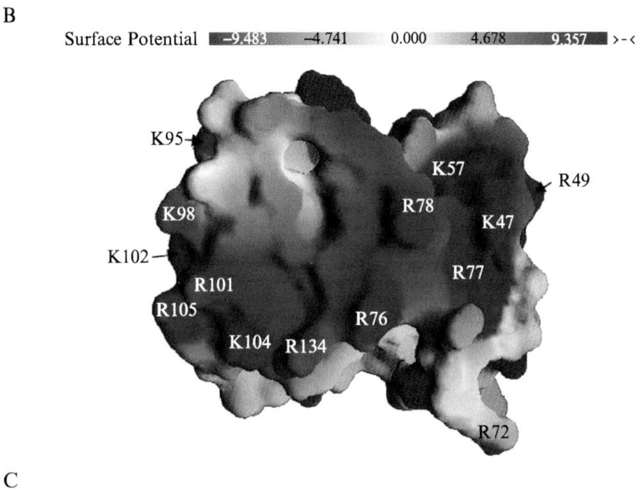


C

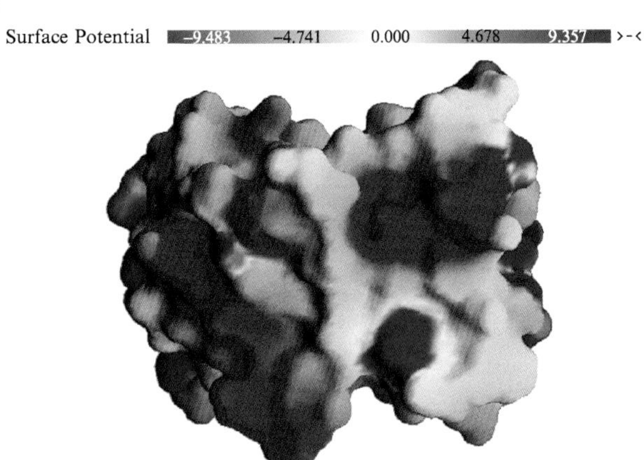


SCHMITT AND LAMB, FIG 3. Atomic structure of the influenza virus M1 protein. (A) Structure of a dimer of M1 residues 1–158. Residues that make up the basic surface of the protein are shown. Taken from Lamb and Krug (2001). (B and C) Surface potential of an M1 monomer. Basic residues are indicated in (A), and acidic residues on the opposite side (180° rotation) are shown in (C). Taken from Arzt *et al.* (2001).